EMULSION POLYMERISATION

Theory and Practice

EMULSION POLYMERISATION

Theory and Practice

D. C. BLACKLEY

Ph.D., B.Sc., F.I.R.I.

Reader in Polymer Science and Technology,
National College of Rubber Technology,
The Polytechnic of North London, London, England

A HALSTED PRESS BOOK

JOHN WILEY & SONS

NEW YORK—TORONTO

PUBLISHED IN THE U.S.A. AND CANADA BY
HALSTED PRESS
A DIVISION OF JOHN WILEY & SONS, INC., NEW YORK

Library of Congress Cataloging in Publication Data

Blackley, D. C.
Emulsion polymerisation.

"A Halsted Press book."
Includes index.
1. Polymers and polymerization. 2. Emulsions.
I. Title.
QD281.P6B447 1975 547′.84 75–16336
ISBN 0–470–07740–9

WITH 134 ILLUSTRATIONS AND 88 TABLES

Printed in Great Britain by Galliard (Printers) Ltd Great Yarmouth

PREFACE

The principal objective of this book is to review critically the published literature on the subject of emulsion polymerisation. Occasional reference is made to results known to the author but as yet unpublished at the time of writing, and also to personal reflections on various aspects of the subject which have occurred to the author as he has pondered the mysteries of emulsion polymerisation over the years. Whilst the last word on the subject is very far from having been written, it is hoped that this book will provide a useful contribution, if only by drawing attention to the more recently published literature.

As the full title indicates, the aim has been to give a reasoned account of some of the theoretical and practical aspects of emulsion polymerisation. Unhappily, these aspects are to a certain extent still unrelated to one another, although the main lines of the relationship can be regarded as having been satisfactorily established. This book is intended for all who have an interest in emulsion polymerisation. It is primarily written for industrial chemists concerned with production, research or development in this field, and for academics who have chosen the seeking of a deeper understanding of this subject as their topic for specialist study. Students, undergraduate and postgraduate, who are required to concern themselves with this subject as a specialisation, should find material here which is relevant to their study, as should scientists and technologists in other fields who wish to acquaint themselves with the present state of knowledge in this area. It is to be hoped that, whether his interest in the subject is mainly theoretical or mainly practical, the serious reader will find in these pages something of interest.

The published literature on emulsion polymerisation is very extensive. Of necessity the present author has had to be selective in at least two respects. Firstly, it has been necessary to restrict the number of items of which note can be taken and upon which comment can be offered. But secondly, it has not been possible to comment upon all those facets of the selected items on which comment would have been desirable. The best that can be hoped for is that, notwithstanding this double selection process, the material actually reviewed is reasonably representative of the whole. It must also be pointed out with regret that it has not been possible to include certain information from industrial sources to which the author has been privileged to have access.

One important fundamental problem which every writer upon a technical subject faces is that of devising a suitable framework within which to expound his subject. The problem is especially acute in the case of emulsion polymerisation, because the subject-matter can be classified in several different ways, each of which is helpful and interesting, but none of which is entirely satisfactory. The author has adopted a method of ordering his material which is explained in the opening chapter of the book. It is hoped that this method of ordering will facilitate easy reference by most readers, whether their interests lie predominantly in the theoretical or in the practical aspects of the subject. The opening chapter also gives some indication of the scope and limitations of the present treatment of the subject within the stated framework.

Although the author's interest in emulsion polymerisation has extended over many years, this book has its immediate origin in a short course of evening lectures given at the National College of Rubber Technology, The Polytechnic of North London. To a certain extent, this book is an expansion of those lectures.

In concluding this preface, the author would like to acknowledge the debt he owes to many friends and colleagues in both industrial and academic circles, who, over the years, have wittingly or unwittingly added to his knowledge and understanding of emulsion polymerisation. Such virtues as this book possesses should probably in large measure accrue to them. The author, however, accepts all responsibility for the statements made here, and, in particular, for any errors of fact or interpretation which may have occurred. It is also a pleasure to acknowledge the unfailing assistance of the library and information services of The Polytechnic of North London, especially in tracing the more obscure original sources; without their help, it would have been very much more difficult to write this book.

D. C. BLACKLEY

CONTENTS

CHAPTER 1

INTRODUCTION

1.1 DEFINITION OF TERMS

Emulsion polymerisation is one of several methods for effecting the process of addition polymerisation. By the latter term is understood a reaction whereby a large number of small molecules become permanently combined together as a large molecule, there being no elimination of any other molecules in the process. The composition of an addition polymer is therefore identical with that of the monomer, except possibly for slight variations associated with the presence of end-groups.

Whilst, in principle, a large number of types of monomer may participate in addition polymerisation reactions, by far the most important group comprises carbon compounds containing one or more multiple carbon–carbon bonds. The overall effect of the polymerisation process is then to open one of the multiple bonds and bring about rearrangement in such a way as to create essentially linear sequences of combined monomer units. In its simplest form, the addition polymerisation of an unsaturated carbon compound can be represented as follows:

$$n\,\mathrm{C}{=}\mathrm{C} \longrightarrow \left[\mathrm{-C-C-}\right]_n$$

Although other methods of effecting addition polymerisation have been applied to many types of compound besides unsaturated carbon compounds, emulsion polymerisation reactions have been almost exclusively applied to the problem of polymerising unsaturated carbon compounds. In what follows, therefore, attention will be confined exclusively to such compounds.

It is well known that the addition polymerisation of unsaturated carbon compounds proceeds by a chain reaction involving three principal steps, namely, the initiation of active centres, the propagation of those active centres with the formation of large molecules, and the ultimate destruction of the active centres with consequent stabilisation of the molecules. The overall reaction may be further complicated by the occurrence of various transfer reactions. It is further well known that propagation can occur through one of three types of active centre: the free radical, the carbonium ion and the carbanion. Polymerisations proceeding by way of these three types of centre are described respectively as free radical, cationic and anionic. It is the case that emulsion polymerisation has been applied almost exclusively to the process of addition polymerisation proceeding by way of free-radical active centres. Although there are indications that a few emulsion polymerisations do proceed by other than a free-radical mechanism, such reactions are at present very much in the minority. Some allusion will be made to them in the final chapter.

With these basic concepts in mind, it is now possible to give a provisional working definition of emulsion polymerisation as *an addition polymerisation process in which the several concurrent propagating centres are isolated from one another*. The form which this isolation takes is usually now regarded as confinement within loci which have originated from the micelles formed by a surface-active substance dissolved in an essentially aqueous medium. Thus, for practical purposes, it is often convenient to regard an emulsion polymerisation as *an addition polymerisation process which proceeds by micellar mechanism, i.e.* in which the micelles, as it were, give rise to individual reaction vessels in which the several simultaneously-occurring propagation reactions can take place independently of one another. However, it must be emphasised that the above working definition of emulsion polymerisation does not require that the separation of individual propagating centres necessarily be due to containment within loci derived from micelles, or that the supporting medium for the reaction be aqueous in nature. As the theoretical aspects of the subject are developed in the course of this book, it remains to be seen to what extent these provisional definitions are acceptable in the light of current controversies concerning the mechanism of emulsion polymerisation.

It should also be noted that the above working definition is not to be regarded as absolutely precise. It is doubtful, for instance, to what extent the isolation of the individual propagating centres is complete in any given reaction. Furthermore, it is possible that some reactions fulfil the criterion for an emulsion polymerisation for some of their course, but not for the whole. It is therefore necessary, in practice, to regard an emulsion polymerisation as one in which most, but not necessarily all, of the polymer is produced by a reaction which satisfies the criterion given by the working definition.

Notwithstanding the above provisos, it remains true that the great majority of industrial emulsion polymerisations are now usually regarded as proceeding by way of a micellar mechanism in which the micelles are formed from surface-active substances dissolved in aqueous media. This being so, the immediate end-product of the reaction belongs to that class of colloid systems known as *latices.* By the term *latex* is meant *a stable dispersion of a polymeric substance in an essentially aqueous medium.*[1] More particularly, the immediate products of the typical emulsion polymerisation reaction are known as *synthetic* latices, to distinguish them, on one hand, from naturally-occurring latices, and, on the other, from latices produced by the dispersion of bulk polymer in an aqueous medium.

1.2 RELATIONSHIP TO OTHER METHODS OF EFFECTING FREE-RADICAL ADDITION POLYMERISATION

It should be clear from the above considerations that emulsion polymerisation is but one of several alternative methods for effecting free-radical polymerisation. It is, in fact, an industrially very important mode of carrying out the reaction; in the United Kingdom alone, approximately 3×10^5 tons of rubbery polymers are currently made by this reaction each year. In order to set emulsion polymerisation into some perspective, it is necessary to consider briefly the characteristics of each of the methods.

Bulk polymerisation is a process in which the monomer is polymerised in bulk form (usually liquid, but occasionally gaseous) with merely the addition of initiating substances and (possibly) the application of heat, ultraviolet light, or high-energy radiation. The method is notable chiefly for the absence of added substances. There is thus no problem of separating the polymer from solvents or from dispersion media. There are no solvents present during the polymerisation reaction to act as chain-transfer agents and thereby reduce the molecular weight of the product. However, the method is beset with many difficulties, the chief of which are that the system becomes intractably viscous as the reaction proceeds, the dissipation of the heat of polymerisation becomes an increasing problem as the viscosity increases, the product is likely to be of variable structure owing to local increases of temperature which may occur, and the polymer which is produced in the latter stages of the reaction is often highly branched or crosslinked.

The maximum permissible rate of polymerisation in a bulk reaction is limited by two factors principally: the problem of heat dissipation, to which reference has already been made in the preceding paragraph, and the polymer molecular weight which it is desired to achieve by means of the reaction. Without going into unnecessary detail, the situation is that, whereas the rate of propagation is kinetically of first order with respect to the concentration of active centres, the rate of termination is usually second order with respect to this concentration because the usual mode of termination is bimolecular interaction between active centres with consequent mutual deactivation of those centres. Thus any attempt to increase the rate of polymerisation by increasing the concentration of active centres (*e.g.* by raising the concentration of an added initiator) is simultaneously accompanied by a larger increase in rate of termination. In the absence of transfer reactions, the degree of polymerisation of the product is determined by the ratio of the rate of chain propagation to the rate of chain termination. Therefore any increase in the rate of propagation is inevitably accompanied by a decrease in the degree of polymerisation of the product. The effect of transfer-to-monomer reactions is to reduce further the attainable molecular weight; in fact, it is readily shown that, for simple free-radical polymerisation reactions from which complications such as branching and crosslinking are absent, the ratio of the rate coefficient for propagation to that for transfer to monomer determines the maximum degree of polymerisation which can be attained by reducing the rate of polymerisation to almost zero.

Solution polymerisation, as the name implies, is polymerisation carried out in the presence of a solvent for the monomer. The solvent may also be a solvent for the polymer as well. If it is, then the end-product of the reaction is a solution of polymer in solvent; if it is not, then the polymer precipitates from the system as it forms, and the result is a slurry of polymer. Under the former circumstance, the polymerisation system will remain fluid over the whole course of the reaction, if sufficient solvent is present. In the latter case, the reaction system is, in part at least, heterogeneous.

The principal reason for using a solvent is to prevent the polymerising system from becoming intractably viscous during the course of the reaction. The problem of heat dissipation is thereby mitigated to some extent. Offsetting these advantages, the process of solution polymerisation is beset with a number of serious disadvantages. Thus it calls for the use of expensive, toxic and often inflammable solvents. Separation of the polymer from the solvent is necessary at the end of the reaction unless, of course, the polymer precipitates during the reaction, thereby forming a slurry from which the polymer can be separated by filtration. (Of course, there will be occasions when separation from the solvent is not necessary because the polymer is required for use in the form of a solution in the solvent which was used as the reaction vehicle.) A further disadvantage is that the molecular weight of the product may be seriously reduced by the reaction of transfer to solvent. For simple reactions uncomplicated by branching and crosslinking, it is found that, for a given concentration of active centres, the reciprocal degree of polymerisation of the product varies linearly with the ratio of solvent to monomer. The slope of the relationship (and therefore the sensitivity of the degree of polymerisation to solvent concentration) depends upon the chemical nature of the solvent. In particular, halogenated hydrocarbon solvents are especially prone to participation in transfer reactions.

Suspension polymerisation is a procedure whereby the monomer is dispersed in a medium (usually aqueous) in which it is insoluble, and the polymerisation allowed to proceed within the individual monomer droplets. Small amounts of protective colloids are usually added to stabilise the dispersion to some extent. Monomer-soluble initiators are used, and the reaction is best regarded as a 'micro-bulk' polymerisation. The kinetics and the mechanism are essentially those for the corresponding bulk polymerisation under the same conditions of temperature and initiator concentration (referred, of course, to the monomer phase only). Droplet size may have a slight effect upon the rate of polymerisation, especially when the droplets are very small.

The process of suspension polymerisation offers several distinct advantages. If properly carried out, the system remains completely fluid over the entire course of the reaction. Heat dissipation is a minor problem only, because the heat can be absorbed and conducted away by the dispersion medium. The resultant polymer is less pure than that obtained by bulk polymerisation, but is not contaminated to the same extent as that produced by emulsion polymerisation. If water is used as the dispersion medium, then problems of expense, toxicity and inflammability are minimised. It is necessary to separate the polymer from the dispersion medium, but this is not usually difficult to accomplish. The main practical problem associated with this process is the prevention of droplet coalescence in the intermediate stages of the reaction when the droplets consist of what is essentially monomer-plasticised polymer, and are therefore very tacky. In the early stages, the droplets are sufficiently fluid to break up continuously as they coalesce; in the latter stages, they are too hard to coalesce; but in the intermediate, semi-solid tacky condition they coalesce readily but re-disperse with difficulty. Careful attention to stirring is necessary if the reaction is to be successfully carried through to completion.

Emulsion polymerisation, as practised industrially, bears a superficial resemblance to suspension polymerisation. The polymerisation is again carried out in a system which comprises monomer and a dispersion medium (usually aqueous) in which the monomer is either virtually insoluble, or else sparingly soluble. A soap-like micelle-generating substance is usually present, and a water-soluble initiator is generally used. Again the system remains perfectly fluid over the entire extent of the reaction. Heat dissipation is no problem, and there is very little tendency for the particles to coalesce as the reaction proceeds, provided the recipe has been correctly formulated. Furthermore, the polymer is obtained in latex form, and this is an obvious advantage if it is desired to apply the polymer as a latex. Apart from these not inconsiderable advantages, the dominant advantage of the emulsion polymerisation process is that, in contrast to the other polymerisation methods, it is possible to obtain polymer of very high molecular weight at relatively fast rates. High product molecular weight is especially necessary when producing synthetic rubbers, for the development of satisfactory elastomer properties requires that the primary polymer molecular weight be high. Furthermore, high conversion rate is, of course, advantageous because it leads to reduced production costs. In fact, the molecular weight of the product obtained by emulsion polymerisation can be too high for practical application. For this reason, it is sometimes necessary to include within the recipe a chain-transfer agent (or 'modifier') in order deliberately to reduce the molecular weight and complexity of the product.

The emulsion polymerisation process has certain disadvantages attending its use, the chief of which is that, if the polymer is required in bulk form, then it has to be separated from the latex. The separation is not always simple. Furthermore, the polymer which is isolated from the resultant latex has associated with it a relatively large proportion of non-polymer constituents, notably insoluble soap derivatives. Whilst for certain polymer applications the presence of such soaps may be an advantage, emulsion polymers are effectively excluded from those applications which require a material of high purity. A further disadvantage of the emulsion polymerisation process is its variability. Whilst this is not such a problem when running recipes for, say, conventional styrene–butadiene copolymerisation to give ultimately grades of copolymer which are suitable for processing as solid rubbers, it becomes increasingly a problem as recipes are developed for special-purpose grades of synthetic latex, and stringent restrictions are imposed upon the levels and types of ancillary recipe ingredients.

A further industrial disadvantage which emulsion polymerisations share with suspension and solution polymerisations, as compared with bulk polymerisations, is that the effective reactor volume available for polymerisation is reduced by the volume of dispersion medium which is present. For typical industrial reactions, the monomer may comprise as little as 40% of the total volume of the reaction system. In such a case, it is therefore necessary to provide reactor volume which is more than twice that required for the monomer itself.

1.3 FURTHER CONSIDERATION OF THE DISTINGUISHING FEATURES OF EMULSION POLYMERISATION PROCESSES

From the remarks of the preceding paragraphs, it will be clear that a typical emulsion polymerisation system will contain at least four basic ingredients: (1) monomer(s); (2) dispersion medium; (3) micelle generator; and (4) initiator. Typical proportions (in parts by weight) for these basic ingredients would be as follows: monomers, 100; dispersion medium, 150; micelle generator, 5; initiator, 0·5. Several other types of substance may also be included, but these four components are essential for the majority of systems of practical interest. Furthermore, since the monomers are sparingly soluble in the dispersion medium, and the solubility limit is usually exceeded many times over, it is clear that there will be present at the beginning of the reaction two separate phases, referred to respectively as the *monomer phase* and the *aqueous phase*. The aqueous phase, containing as it does a micelle-forming solute, can itself be usefully considered as consisting of two phases, the *micelle phase* and the *true aqueous phase*.

With these considerations in mind, it is now possible to set down certain very general but distinctive features of emulsion polymerisation reactions, for which any satisfactory general theory of mechanism must give a satisfactory explanation:

(1) *The rate of conversion of monomer to polymer tends to increase as the volume of the aqueous phase increases if the composition of the aqueous phase remains constant; but the rate of conversion of monomer to polymer does not depend very much upon the quantity of monomer phase.*

It is difficult to cite clear unequivocal support for a relationship of direct proportionality between rate of polymerisation and volume of aqueous phase for conventional free-radical emulsion polymerisations, but there are certainly strong indications of the general truth of the proposition that, other things being equal, an increase in one brings about an increase in the other. However, results obtained by Blackley and Burgar[2] for the aqueous-phase polymerisation of butadiene at 50°C initiated by cobalt(III)

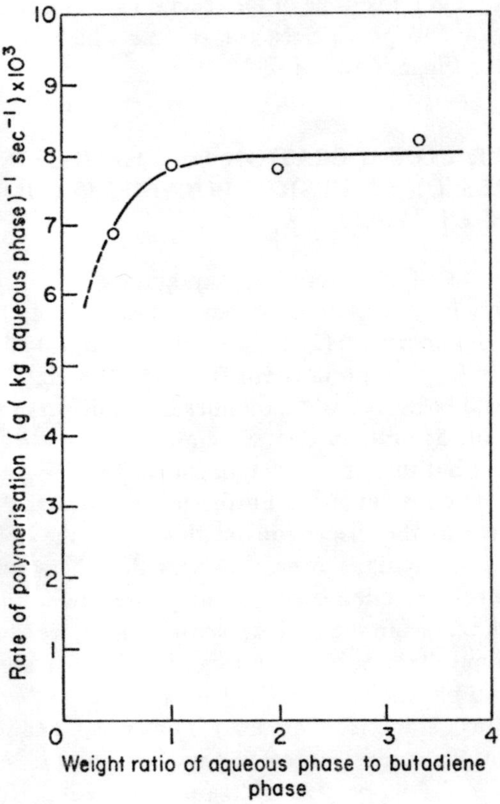

FIG. 1.1. Effect of ratio of aqueous phase to monomer phase for aqueous-phase polymerisation of butadiene at 50°C initiated by cobalt(III) acetylacetonate, according to Blackley and Burgar.[2] (Reaction systems: butadiene 176–300, 2·5% aqueous sodium alkylbenzenesulphonate 2·5, cobalt(III) acetylacetonate 1; sufficient NaOH was added to the aqueous soap solution to raise the pH to 10·4.) Note that, unless otherwise stated, the compositions of reaction systems are given in parts by weight throughout this book. Also, the abbreviation 'pphm' is used for 'parts per hundred parts of monomer'.

acetylacetonate using a sodium alkylbenzenesulphonate as micelle generator, indicate that the rate of conversion of monomer to polymer is directly proportional to the amount of aqueous phase, except at low ratios of aqueous phase to butadiene phase. These results are summarised in Fig. 1.1, in which the ordinate is the rate of polymerisation per unit mass of aqueous phase. The interpretation of these results is complicated by the fact that the initiator partitions between the monomer and aqueous phases, but the general effect of variation of volume of aqueous phase is clear from these results.

Further information on this matter is provided by the results of Blackley and Mustafa[3] for the aqueous-phase polymerisation of butadiene at 50°C in the presence of a sodium alkylbenzenesulphonate as micelle generator and rhodium(III) chloride as catalyst. Again, this is not a conventional emulsion polymerisation system, but it is nevertheless one of considerable theoretical interest, and one to which further allusion will be made in a later chapter. The point to which attention is drawn here is that the rate of conversion of butadiene to polybutadiene is strictly proportional to the weight of aqueous phase over a range of monomer/aqueous phase weight ratio from about 1/4 to about 2. The exactness of the proportionality is evident from the summary of results given in Fig. 1.2.

Concerning the effect of quantity of monomer phase upon the absolute rate of conversion of monomer to polymer, it is again difficult to cite precise results for conventional free-radical emulsion polymerisations. General experience seems to indicate that the absolute rate of conversion is very much less dependent upon the quantity of monomer phase than in the case of, say, a bulk polymerisation, where a relationship of direct proportionality obtains if the composition of the monomer phase is constant. In some instances, the absolute conversion rate in emulsion polymerisation reactions seems to be almost independent of the quantity of monomer. In others, the variation is less than that which would be expected on a basis of direct proportionality. As an example of the latter, the results of Bakker[4] on the emulsion polymerisation of styrene may be cited. He found that, for polymerisations using potassium myristate as the micelle generator and potassium persulphate as the initiator, the time required for the polymerisation of 50% of the monomer was proportional to the square root of the initial amount of monomer. For an absolute rate directly proportional to the amount of monomer, the time for 50% conversion should be independent of initial amount of monomer, whereas for an absolute rate independent of amount of monomer, the time for 50% conversion should be directly proportional to the initial amount

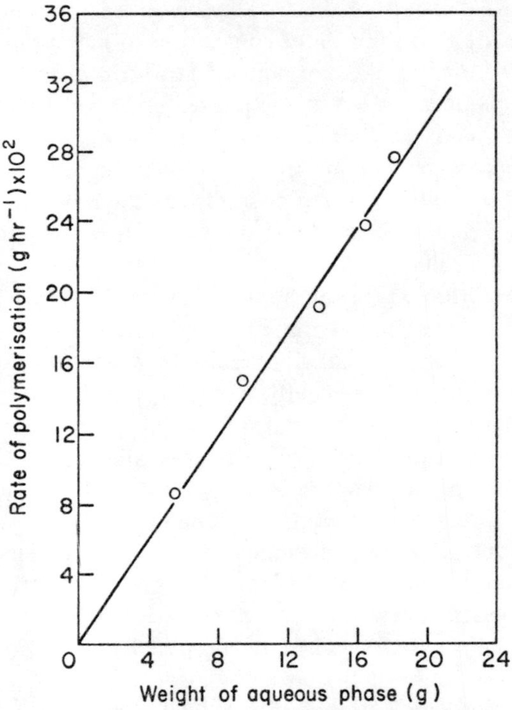

FIG. 1.2. Effect of amount of aqueous phase upon rate of aqueous-phase polymerisation of butadiene at 50°C initiated by rhodium(III) chloride, according to Blackley and Mustafa.[3] (The reaction systems comprised varying amounts of butadiene and an aqueous phase of fixed composition such that the total volume of the reaction system was approximately constant at 20 ml. The composition of the aqueous phase was as follows: water 100, sodium dodecylbenzenesulphonate 1, rhodium(III) chloride 1.)

of monomer. As an example of the former behaviour, the following quotation from Bovey *et al.*[5] (based on unpublished observations by Bovey and Kolthoff) may be cited: 'When a series of emulsion polymerizations of styrene is carried out with equal volumes of the aqueous phase (containing potassium persulfate initiator and fatty acid soap as the emulsifier), but with the monomer phase varied in volume over a wide (32-fold) range, it is found that the initial rate of polymerization, calculated as grams of polymer formed per milliliter of aqueous phase per unit time, is entirely independent of the monomer volume.'

(2) *The rate of conversion of monomer to polymer depends upon the concentration of micelle generator in the aqueous phase.*

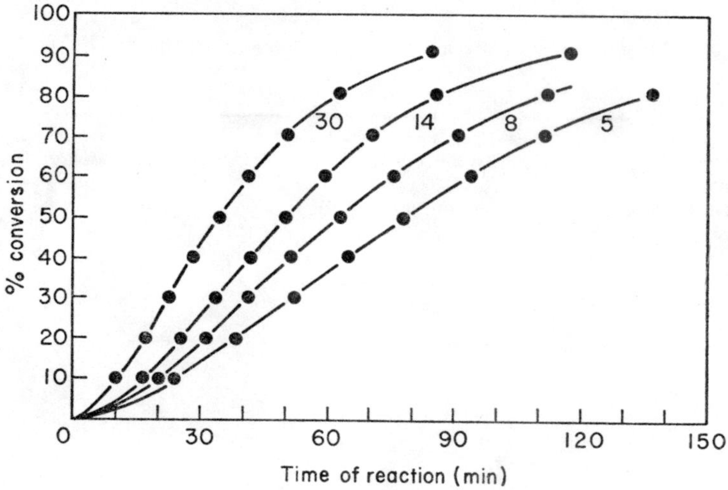

FIG. 1.3. Effect of amount of soap upon conversion–time relationship for persulphate-initiated emulsion polymerisation of styrene at 40°C, according to Bakker.[4] (Reaction systems: styrene 100, water 800, $K_2S_2O_8$ 0·6; the soap was potassium myristate, and the levels in pphm are given against the respective curves.)

In the early days of the study of emulsion polymerisation, it was tempting to think of the micelle-generating substance (which was, and still is, usually a soap) as merely stabilising, initially, the dispersion of monomer in the aqueous phase, and then, subsequently, the dispersion of polymer particles in the aqueous phase. Undoubtedly it does perform these two functions, but its effect upon the reaction is more profound than either of these functions would suggest. This is indicated, for instance, by the results of Bakker[4] for the emulsion polymerisation of styrene at 40°C in the presence of varying concentrations of soap in the aqueous phase. The conversion–time curves for various levels of soap are shown in Fig. 1.3. Clearly the rate of polymerisation increases markedly as the level of soap increases. This dependence of conversion rate upon the level of soap is further emphasised if, as has been done in Fig. 1.4, the time required for, say, 50% conversion of the monomer is plotted against soap level. The relationship between the two is seen to be approximately hyperbolic in this instance.

(3) *The rate of conversion of monomer to polymer is not much affected by the mere addition of water to the aqueous phase, provided not too much water is added.*

TABLE 1.1

Effect of addition of water upon emulsion polymerisation of styrene at 40°C (according to Bakker)[4]
(amount of styrene in reaction system: 50 g)

	Potassium persulphate: 0·3 g *Potassium myristate:* 3 g		*Potassium persulphate:* 0·3 g *Potassium myristate:* 4 g	
Volume of aqueous phase (ml)	200	400	200	400
Time (min) required for 50% conversion	72·5	69	63·5	64·5

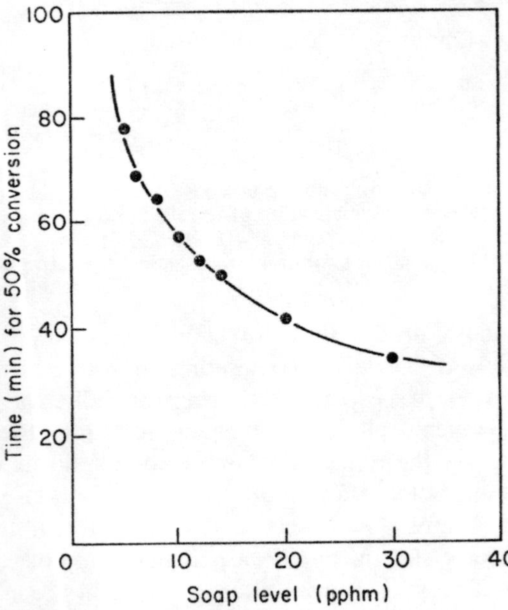

FIG. 1.4. Effect of amount of soap upon time required for 50% conversion of styrene at 40°C using potassium myristate as soap and potassium persulphate as initiator, plotted from data given by Bakker.[4] (The compositions of the reaction systems were as given in the caption to Fig. 1.3.)

The truth of this proposition is illustrated for the polymerisation of styrene at 40°C by the results given in Table 1.1, again from the report by Bakker.[4] These results show that, provided the total amount of soap and initiator remains fixed, increasing the volume of the aqueous phase by merely adding more water has little effect upon the time required for 50% conversion.

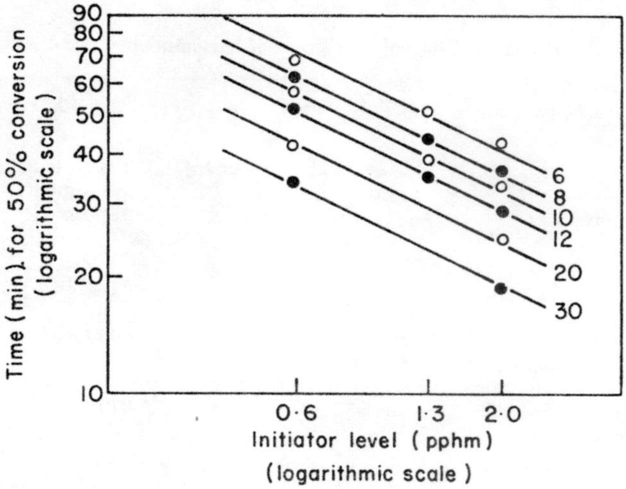

FIG. 1.5. Effect of amount of persulphate initiator upon time required for 50% conversion of styrene at 40°C at various levels of soap, according to Bakker.[4] (Reaction systems: styrene 100, water 800, potassium myristate—as indicated by numbers against curves, $K_2S_2O_8$—variable.)

(4) *The rate of conversion of monomer to polymer depends upon the concentration of the initiator, even though this is water-soluble and in many instances confined exclusively to the aqueous phase.*

The results of Bakker[4] given in Fig. 1.5 show clearly the joint effect of both initiator concentration and soap level upon the rate of emulsion polymerisation of styrene at 40°C. In fact, Bakker found that, using potassium persulphate as initiator and either potassium myristate or potassium stearate as soap, the time required for 50% conversion was inversely proportional to the square roots of both initiator and soap levels.

(5) *The molecular weight of the polymer obtained is considerably higher than that obtained from a bulk polymerisation proceeding at the same rate at the same temperature.*

Reference has already been made to this fact when considering the relationship of emulsion polymerisation to the other methods which may be used for effecting free-radical addition polymerisation. Some results for the polymerisation of styrene in both bulk and emulsion, attributed to Ewart,[6] are given in Table 1.2. The principal point which emerges from these results is that, although the reaction in emulsion proceeds nearly twenty times faster than that in bulk, the degree of polymerisation of the

TABLE 1.2

Comparison of bulk and emulsion polymerisation of styrene at 60°C (according to Ewart[6])

	Bulk	*Emulsion*
Initiator	0·0428 mole l^{-1} dibenzoyl peroxide	0·010 mole l^{-1} potassium persulphate
Micelle generator	—	0·027 mole l^{-1}
Rate	3·28% per hour	60% per hour
Free radical concentration	$2{\cdot}7 \times 10^{16}\ l^{-1}$	—
Concentration of polymer particles	—	$1{\cdot}35 \times 10^{18}\ l^{-1}$
Degree of polymerisation	800	37 000
Average effective spherical diameter per free radical	4 100 Å	—
Average diameter of latex particles	—	1 070 Å

emulsion polymer is nearly five times greater than that of the bulk polymer. As Burnett[7] has pointed out, the degrees of polymerisation obtained from an emulsion reaction are not merely greater than would be expected on the basis of normal bulk-polymerisation kinetics; they can also be much greater than the limit which, theoretically, should be set by the ratio of the rate coefficient for propagation to that for the reaction of transfer to monomer. In the case of styrene polymerising at 60°C, this should be of the order of 10^3 (corresponding to a molecular weight of the order of 10^5), and polymer of this limiting degree of polymerisation should only be obtained if the polymerisation occurs very slowly. As Table 1.2 shows, the degree of polymerisation of the emulsion polystyrene is nearly forty times this limiting value, and this for a polymer formed at a reaction rate of 60% per hour. Burnett further points out that it is most unlikely that the ratio of propagation to transfer coefficients is much affected by change in reaction medium, and that the high degree of polymerisation is probably explained by the prevalence of transfer-to-polymer reactions due to the relatively low ratio of monomer to polymer at the actual reaction locus.

(6) *The particles of the latex produced by emulsion polymerisation are generally at least one order of magnitude smaller than the droplet size of the original dispersed monomer.*

The particle size of the latex produced by emulsion polymerisation depends very much upon the details of the recipe used. The lower limit of average particle diameter is in the region of 500 Å. It is difficult to set an

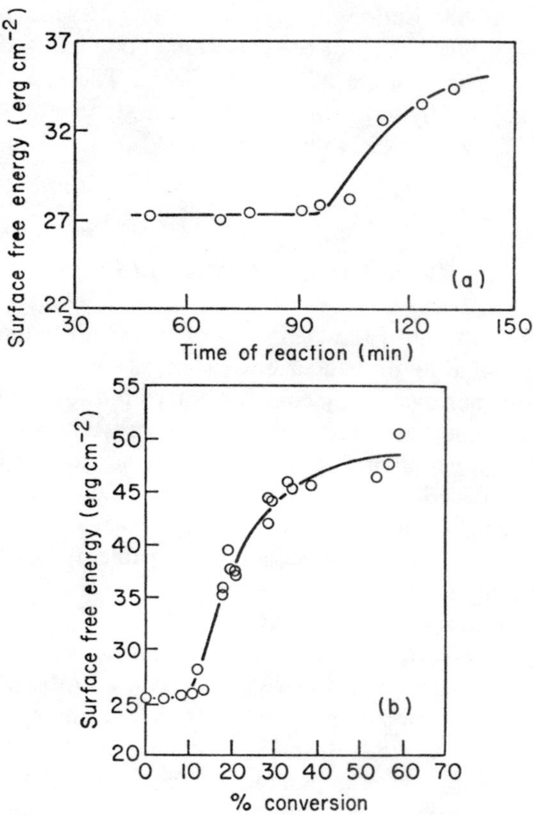

FIG. 1.6. Effect of conversion upon surface free energy of aqueous phase–air interface during emulsion polymerisation. Fig. 1.6(a), from Bakker,[4] refers to persulphate-initiated polymerisation of styrene at 40°C, using potassium myristate as soap. Fig. 1.6(b), from Harkins,[8] refers to persulphate-initiated copolymerisation of styrene and isoprene at 50°C using sodium oleate as soap.

upper limit, but certainly it is not easy to make (by emulsion polymerisation) a latex having a mean particle size of, say, 5000 Å. On the other hand, the sizes of droplets in a well-stirred monomer dispersion are typically of the order of 10^4–10^5 Å.

(7) *The surface free energy of the aqueous phase undergoes an increase at a certain stage in the reaction.*

This change, which is often manifested practically by a reduction in the tendency of the reacting system to foam, is illustrated in Fig. 1.6(a) for the emulsion polymerisation of styrene in the presence of potassium myristate, and in Fig. 1.6(b) for the emulsion copolymerisation of styrene and isoprene in the presence of sodium oleate. The rise in surface free energy is not necessarily large (in the case of the styrene polymerisation to which Fig. 1.6(a) refers, it is about 8 ergs cm^{-2}), but it is quite typical of emulsion polymerisation reactions.

1.4 NOTE ON THE NATURE OF MICELLES

It has been pointed out at the beginning of this chapter that the manner in which the isolation of concurrent propagation reactions is usually considered to be achieved is by confinement within loci which have been derived from the micelles which surface-active substances form when they are dissolved in aqueous media. The primary evidence for the formation of micelles is provided by the observation that many of the properties of very dilute solutions of surface-active substances vary in an unusual manner with concentration. This is especially noticeable for the property of surface free energy at the air–water interface, the variation of which with concentration is illustrated diagrammatically in Fig. 1.7. When the surface-active substance is first added to the solution, the surface free energy falls approximately linearly with concentration. However, a situation is soon reached where the surface free energy undergoes no further drop, thus becoming almost independent of concentration.

The conventional molecular explanation for this effect takes account of the fact that the molecules of surface-active substances usually have two distinct regions: one, a polar region which has an affinity for the aqueous medium, and another, a non-polar region, which has no such affinity. The adjectives 'hydrophilic' and 'hydrophobic' are often used to distinguish the two regions, respectively. Such molecules are conventionally represented by a small rod with a 'blob' at one end, the rod representing the hydrophobic portion, and the 'blob' the hydrophilic portion. Many of the unusual properties of such molecules arise from the tendency of the hydrophobic portion to be expelled from the aqueous phase whilst at the same time the polar portion is retained by the aqueous phase. It is inappropriate to discuss in detail whether the reduction in free energy which accompanies the expulsion of the hydrophobic portion is due to a loss of enthalpy or a gain of entropy. A loss of enthalpy might, for instance, be brought about by the action of electrical forces between water molecules squeezing out the hydrophobic part of the molecule, whereas a gain in

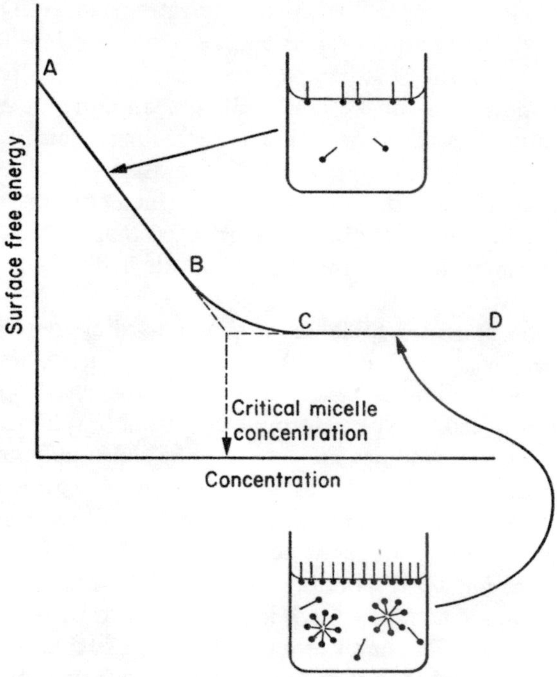

FIG. 1.7. Variation of surface free energy with concentration for dilute aqueous solution of a surface-active substance, together with illustration of molecular interpretation. (—● denotes a molecule of surface-active substance.)

entropy might occur because expulsion of the hydrophobic part may mean the loss of a 'template' around which water molecules may be ordered.

Whatever may be the primary motivation for the expulsion of the hydrophobic portion of the molecule from the aqueous phase, a consequence of this dual character within a single molecule is a strong tendency for the molecule to be adsorbed at interfaces between water and air or other media. The hydrophobic part is orientated away from the water, and the polar part is situated within the water. The descending portion of the surface free energy–concentration curve (AB of Fig. 1.7) is interpreted as being due to the gradual filling of the air–water interface with adsorbed molecules, the air–water interface being in effect gradually replaced by an air–oil interface of lower surface free energy. However, when sufficient solute has been added to saturate the surface, no further reduction in surface free energy can occur. This is the interpretation of the flat portion of the surface free energy–concentration curve (CD of Fig. 1.7). For the system to stabilise itself after the water–air interface is saturated, a new mechanism must be operative whereby the hydrophobic part of the solute molecules can be removed from an aqueous environment. Micelle formation has been postulated as such a mechanism, and, indeed, this phenomenon satisfactorily accounts for most of the observed effects. The micelle is to be regarded as a structure which comprises many individual solute molecules, these being so arranged that all the hydrophobic parts are in the interior of the micelle and all the hydrophilic parts are on the exterior surface.

There has been considerable discussion regarding the shape of the micelles formed by surface-active substances. The general view taken now is that, in dilute solution at least, they are approximately spherical, and that errors introduced solely on account of assuming spherical shape will be small. More precise studies indicate that the shape may not be exactly spherical. Thus, for instance, Tartar[9] has obtained evidence that the micelle formed by sodium dodecyl sulphate is an oblate spheroid, the ratio of major to minor axis for which is 1·13.

The concentration corresponding to the point of intersection of the two lines AB and CD of Fig. 1.7 is commonly known as the *critical micelle concentration*. The implication of this name is that no micelles are present below this concentration, and that micelle formation commences only when this concentration of surface-active substance is present. This view almost certainly represents an oversimplification; the formation of micelles is probably a type of reversible reaction, and therefore in principle the presence of a few micelles is to be expected long before the critical micelle concentration is attained. For practical purposes, however, the micelle concentration can be regarded as the concentration below which no appreciable amount of the solute is present as micelles, and above which most of the solute in excess of that required to attain the critical micelle concentration is present in the form of micelles. All the available evidence suggests that, although the transition from the one situation to the other is not absolutely sharp, it nevertheless does occur over a very narrow range of concentration. At concentrations below the critical value, it is therefore sufficient to take account of only dissolved 'monomolecular' or 'monomeric' soap and soap adsorbed at the air–water interface, whereas above the critical micelle concentration the solute must be regarded as being present in three forms, namely, adsorbed soap, micellar soap and dissolved 'monomeric' soap.

One further matter needs to be noted here in connection with micelles, and that is their ability to imbibe small quantities of non-polar substances. This phenomenon is known as *solubilisation*. Its consequence is that many non-polar liquids which are almost insoluble in water appear to be rather more soluble in aqueous solutions of surface-active substances. What in fact happens is that the solute passes into the hydrophobic interior of the micelles, the driving force for this process being essentially the gain of entropy which accompanies the mixing of two non-polar liquids. The phenomenon of solubilisation will be considered in more detail in a later chapter. It is sufficient to remark at this point that the phenomenon is usually regarded as being of crucial importance for emulsion polymerisation processes, since it is among solubilised monomer molecules that the principal polymerisation reactions of the typical emulsion polymerisation are commonly believed to occur in the early stages of the process.

1.5 EVIDENCE FOR THE MICELLAR MECHANISM FOR EMULSION POLYMERISATION

Before proceeding to consider in more detail the various aspects of emulsion polymerisation, it is desirable briefly to review the evidence which is available in support of the view that the micelles and particles derived from micelles are the principal locus for the propagation step of an emulsion polymerisation. Not unnaturally, this was a subject of concern in several of the earliest studies on the subject, and although such studies will be considered in more detail in the two succeeding chapters, it is appropriate to state here the outline of the case. Reference to the final chapter of this book will show that the issue cannot yet be regarded as completely settled.

In brief, the position is that four distinct loci can readily be envisaged for the propagation step in the early stages of the reaction: (1) the monomer droplets; (2) the monomer droplet–aqueous phase interface; (3) the true aqueous phase; and (4) the micelle phase. By a process of elimination, the conclusion is reached that only the micelle phase as the principal locus is consistent with the observed features of the reaction. Furthermore, this view is found to be consistent with several other qualitative features of the reaction, and can be used as the basis for theories which, to a limited extent, can account for certain quantitative aspects of the reaction.

Of the four possibilities, the obvious locus would be the monomer droplets. However, this is ruled out by several of the observed features of the reaction, reference to some of which has already been made. The following are the more important:

(1) The emulsified droplets of monomer are very much larger than are the particles of the latex which is subsequently produced.

(2) The initial rate of polymerisation is independent of the volume of monomer phase, provided the volume and composition of the aqueous phase is kept constant.

(3) The rate of polymerisation increases as the volume of the aqueous phase is increased, if the composition of the aqueous phase is kept constant.

(4) The rate of polymerisation depends very much upon the amount of micelle-generating substance initially present in the system, but, after a certain stage, the addition of neither further micelle generator nor further initiator causes any significant increase in the rate of polymerisation.

(5) Typical emulsion polymerisation systems require the presence of water-soluble initiators, and not of monomer-soluble initiators.

(6) The kinetics of emulsion polymerisation reactions, including the relationship between degree of polymerisation and rate of polymerisation, differ markedly from those of bulk polymerisation. In no sense can an emulsion polymerisation be regarded as merely a 'micro-bulk' polymerisation.

Considerations such as these clearly preclude the emulsified monomer droplets as the principal locus of reaction. They also effectively preclude the interface between the monomer and the aqueous phase. For, on this view, the reaction would commence in the interface, and then the growing chains would penetrate the monomer droplets. Whilst this view would accommodate the need for water-soluble initiators, it clearly cannot account for the other five characteristics listed above.

The true aqueous phase is discounted as a principal locus on two main grounds:

(i) Emulsion polymerisation can proceed quite smoothly and rapidly with many monomers, even though they are virtually insoluble in water.

(ii) In the case of the emulsion copolymerisation of two monomers, it is the *less* water-soluble monomer which becomes incorporated in the polymer to a *greater* extent than in a bulk polymerisation employing the same overall monomer feed composition. An early example of this interesting and significant observation is provided

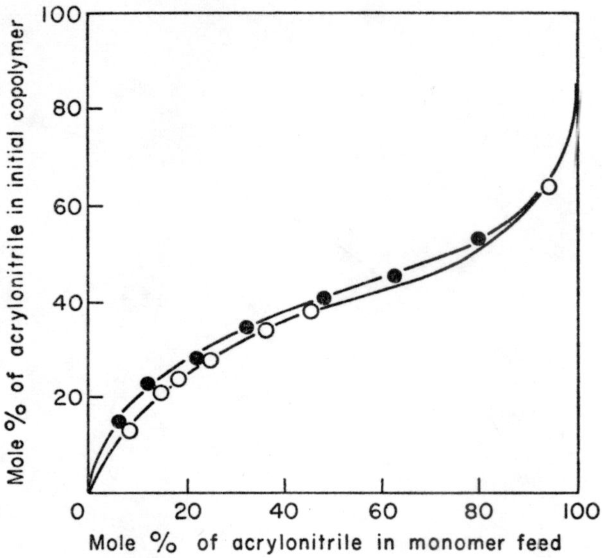

FIG. 1.8. Copolymer–monomer feed composition curves for initial copolymers obtained from styrene and acrylonitrile by copolymerisation at 75°C in both bulk and emulsion, according to Fordyce and Chapin.[10] Significance of points: ● copolymers prepared by bulk polymerisation using 0·05% of dibenzoyl peroxide as initiator; ○ copolymers prepared by emulsion polymerisation using 0·2% of potassium persulphate as initiator.

by the results of Fordyce and Chapin,[10] summarised in Fig. 1.8. Various mixtures of styrene and acrylonitrile were polymerised at 75°C both in bulk, using dibenzoyl peroxide as initiator, and in emulsion, using potassium persulphate as initiator. The results show that, for a given overall ratio of acrylonitrile to styrene in the monomer feed, the copolymer formed initially contains slightly, but significantly, less acrylonitrile if the copolymer is formed in emulsion than if it is formed in bulk. Acrylonitrile is, of course, very much more soluble in water than is styrene. The differences in copolymer composition are, however, admittedly rather slight, although they are remarkably consistent. Further early evidence on this point comes from the results of Fordyce and Ham[11] on the pair styrene–itaconic acid. The latter is an unsaturated dicarboxylic acid ($HO_2C.C(:CH_2).CH_2CO_2H$) which is relatively soluble in water but has a very low solubility in styrene. Here, then, are a pair of monomers, one of which is present almost entirely in the aqueous

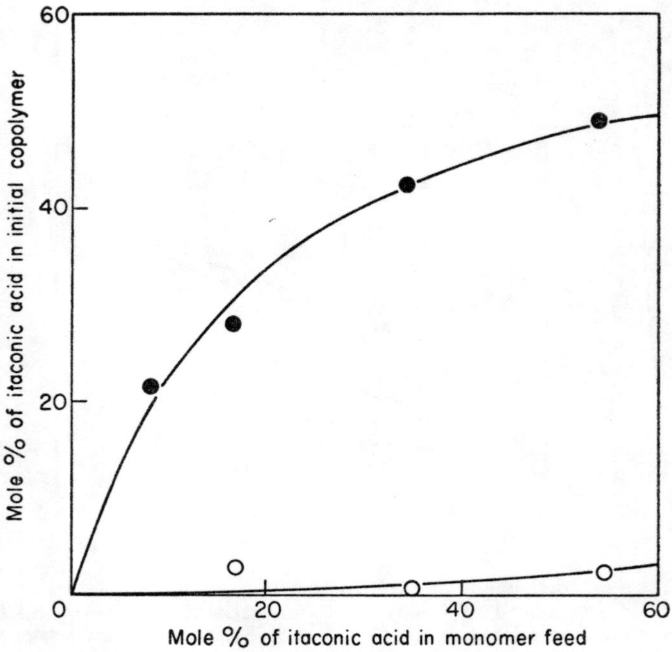

FIG. 1.9. Copolymer–monomer feed composition curves for initial copolymers obtained from styrene and itaconic acid by copolymerisation at 70°C in both solution and emulsion, according to Fordyce and Ham.[11] Significance of points: ● copolymers prepared by solution polymerisation in dioxan using 0·1% of dibenzoyl peroxide as initiator; ○ copolymers prepared by emulsion polymerisation without added initiator.

phase, and the other of which is almost entirely absent from that phase. The results shown in Fig. 1.9 for copolymerisation at 70°C in both emulsion and solution indicate quite clearly that the emulsion copolymer contains almost none of the water-soluble monomer. That this is not apparently due to any inability to copolymerise with styrene is revealed by the curve for the copolymer produced by solution polymerisation. Clearly, if the aqueous phase were the principal locus of the propagation reaction, then it would be expected that the *more* water-soluble of two monomers would become preferentially incorporated in the copolymer during the course of the reaction.

Although these observations effectively preclude the aqueous phase as a principal locus, it must be pointed out that there are some emulsion polymerisation reactions, notably those employing the more water-soluble monomers, for which the initial reaction almost certainly does occur in the aqueous phase. An example of considerable industrial importance is that of vinyl acetate polymerisation.

It will be apparent from the above considerations that the case for the micellar mechanism is essentially negative in nature. Historically, the micellar mechanism has been accepted largely because all the alternatives which could be readily envisaged appeared to be unsatisfactory, and because in general terms the micellar mechanism seemed to give a reasonable explanation of the observed phenomena. Direct evidence for the micellar mechanism is, however, hard to find, and this fact has not gone unnoticed in recent years as other possible plausible alternatives, not included in the above list, have been given serious consideration.

It will also be apparent that, in reality, there are two quite separate issues to be decided in formulating an outline reaction mechanism for an emulsion polymerisation reaction. The first concerns the question of whether the principal polymerisation locus is the monomer droplets, or whether it is some other species of particle. It is generally accepted that, in most cases, it is necessary to postulate some other species of particle; this is the general conclusion to which much of the above evidence points. The second issue concerns the origin of that other species of particle. In the older literature, the assumption is often implicitly made that these two issues are virtually one, the view being taken that the most obvious origin of the particles is the soap micelles. In recent years, however, the origin of the particles has been increasingly questioned, although the necessity for postulating their existence has not. As will be seen in the final chapter of this book, the present tendency is to regard the soap micelles as having no direct relevance for the process of locus nucleation.

1.6 ORDERING OF SUBJECT-MATTER IN THE PRESENT BOOK

In the preface, attention has been drawn to the fact that the subject-matter of emulsion polymerisation can be presented in several essentially distinct ways, each of which is helpful and interesting, but none of which is entirely satisfactory. The method adopted here is to order much of the basic experimental information under the heading of the particular type of recipe ingredient to which it chiefly relates. There are thus chapters on initiation systems, surface-active substances, modifiers, monomers, and so on. These chapters contain much of what may be regarded as the 'hard core' of basic information on the effect of recipe ingredients upon the reaction. These chapters are preceded by chapters dealing with historical and theoretical aspects of the subject, and are succeeded by chapters which deal with certain recent developments which it was considered inappropriate to include within earlier chapters. It was originally intended to include a chapter on procedural aspects, with special emphasis upon industrial processes for carrying out emulsion polymerisation reactions. However, the large amount of material which was available for consideration under the other headings has reluctantly necessitated the omission of such a chapter.

There is a serious disadvantage in ordering the 'hard core' of basic information under the heading of the particular recipe ingredient to which it principally refers. This is that it is difficult to avoid overlap and duplication when one comes to consider the monomer component of the recipe. For it is clearly impossible to carry out emulsion polymerisation experiments on initiators, micelle generators or modifiers without having a particular monomer present. Thus it is that much of the information which might otherwise have been considered in the chapter on monomers will, in fact, be found elsewhere.

It is also appropriate to mention here what has been deliberately omitted from the present treatment. The point of view adopted has been to regard emulsion polymerisation as a generalised mode of reaction by means of which free-radical addition polymerisation can be accomplished. Since emulsion polymerisation is itself a large subject now, it has been necessary to omit detailed consideration of matters such as the kinetics of bulk free-radical polymerisation or the characterisation of the latex produced by emulsion polymerisation and of the polymer which it contains. In fact, the reader is assumed to have a knowledge of the basic facts of both free-radical polymerisation kinetics and polymer characterisation. No attempt has been made to deal systematically with the subject of emulsion copolymerisation. Free-radical copolymerisation is itself a large subject, and to have given any worthwhile treatment of its accomplishment by emulsion reaction would have meant enlarging the book quite unreasonably. Not only would the more conventional aspects of copolymerisation have had to be considered, but also the very important reaction of graft copolymerisation in emulsion systems.

It is also appropriate to note that the literature which has been reviewed here is almost exclusively that which is available in the English language, either in the original or in translation.

REFERENCES

1. D. C. Blackley, *High Polymer Latices*, Applied Science Publishers, London, 1966, p. 1.
2. D. C. Blackley and W. F. H. Burgar, unpublished results; W. F. H. Burgar, Ph.D. Thesis, C.N.A.A., 1973.
3. D. C. Blackley and I. Mustafa, unpublished results; I. Mustafa, Ph.D. Thesis, C.N.A.A., 1972.
4. J. Bakker, *Philips Research Report*, 1952, **7**, 344.
5. F. A. Bovey, I. M. Kolthoff, A. I. Medalia and E. J. Meehan, *Emulsion Polymerization*, Interscience, New York, 1955, p. 166.
6. R. H. Ewart, quoted by H. Mark and A. V. Tobolsky, *Physical Chemistry of High Polymeric Systems*, Interscience, New York, 1950, p. 451.
7. G. M. Burnett, *Mechanism of Polymer Reactions*, Interscience, New York, 1954, pp. 320–1.
8. W. D. Harkins, *J. Am. chem. Soc.*, 1947, **69**, 1428.
9. H. V. Tartar, *J. phys. Chem., Ithaca*, 1955, **59**, 1195.
10. R. G. Fordyce and E. C. Chapin, *J. Am. chem. Soc.*, 1947, **69**, 581.
11. R. G. Fordyce and G. E. Ham, *J. Am. chem. Soc.*, 1947, **69**, 695.

CHAPTER 2

HISTORICAL ASPECTS

2.1 INTRODUCTION: EARLY ATTEMPTS AT HETERO-PHASE ADDITION POLYMERISATION

As Hohenstein and Mark have pointed out in their extensive review of the polymerisation of olefins and diolefins in suspension and emulsion,[1] the earliest observations on the polymerisation of unsaturated compounds were made as early as 1838, but referred almost entirely to polymerisation in the bulk liquid phase. What was observed was a gradual transition from liquid monomer to a viscous or solid polymer, the transition taking place under the influence of heat, of light or of a catalyst. There seems to be general agreement that the earliest literature references to the polymerisation of monomers in the form of an aqueous emulsion are those which originated from Farbenfabriken Bayer in the years 1909 to 1915.[2–5]

Hohenstein and Mark state that there were two main reasons for the desire to carry out diene polymerisation in the presence of water. In the first place, it was common practice at the time to effect polymerisation in the bulk phase using metallic sodium as catalyst. This led to a very heterogeneous product, and also presented the rather difficult problem of completely removing the alkali metal from the final polymer. Indeed, it is interesting to note that Whitby and Katz,[6] writing in 1933, state that 'As applied before and during the war, sodium as a catalyst led to polymerization products which were lacking in plasticity, elasticity and strength and were, moreover, difficult to vulcanize. The alkali metal reacted to a certain extent with the diene, and the polymerisation process proceeded in an irregular or turbulent way and was difficult to control.'

But, in the second place, it was observed that all naturally-occurring rubbers are found in the form of latices. It was therefore concluded that polymerisation in the plant must take place under the relatively mild conditions of an aqueous system, without the application of elevated temperatures and high pressures, and certainly without the use of such catalysts as metallic sodium or alkali metal alkyls. Hohenstein and Mark consider that, of these two reasons, the second was the more important. However, in retrospect it must be pointed out that the formation of naturally-occurring diene polymers does not proceed by way of the addition polymerisation of the corresponding monomeric diene, but by a condensation polymerisation of a phosphate derivative of the diene. (For a description of the generally-accepted biochemical pathway leading to the formation of naturally-occurring polyisoprenes, and of the evidence for that pathway, the reader is referred to the summary which has been given by the present author.[7]) It therefore appears that a technically-useful process was in fact arrived at by a false logic. As will be seen below, the aim of reproducing the physiological conditions occurring in the plant is mentioned in some of the earlier patent disclosures. The position therefore seems to be that the early workers were, unknown to themselves, striving to carry out a free-radical addition polymerisation under conditions similar to those under which a naturally-occurring condensation reaction occurs.

The purpose of this present chapter is to consider in some detail a few of the earliest processes for heterogeneous polymerisations involving aqueous media, and to take note of some of the information which became available up to the period just after the end of World War II when much of the data accumulated in the course of the American synthetic rubber programme became freely available through the medium of the scientific journals.

2.2 FURTHER CONSIDERATION OF EARLIEST PROCESSES

As typical of the earliest processes for the polymerisation of unsaturated monomers in heterogeneous aqueous phase, we consider that of Gottlob[5] published in 1915 under the title 'Caoutchouc substance and process of making same'. In the introduction, it is pointed out that, whereas rubbery substances can be produced by polymerising butadiene or its homologues either in bulk or in solution, with or without the addition of catalysts, a superior product is obtained if the hydrocarbons are emulsified or intimately mixed during the process of polymerisation with aqueous colloidal solutions of substances such as egg albumin, starch, gelatin, etc. It is claimed that 'During the process the emulsification or continuous stirring causes a more uniform intermixture and distribution of the colloidal solutions, and an intermixture and distribution of the rubber during its formation with such solutions and with the unchanged hydrocarbons, *thus giving a novel resulting product having peculiar latex-like properties, comparable with the natural latex of rubber-producing plants*' (the italics are not in the original). The products yielded by the process are further stated to be 'contained' in a similar condition to that of natural rubber in the latex, and also that they can be separated from the resultant latices in a manner similar to that by which natural rubber is separated from its latex. The chief technical advantage arising from the process is said to consist in 'the improved properties of the caoutchouc substances thus obtained.... They are very tenacious, elastic and very little sticky and they can be worked more easily and more satisfactory in their further application. They are normally free from resin but contain the organic substance, *e.g.* starch, albumin, etc., intimately incorporated therewith, and also contain small amounts of the monomeric butadiene hydrocarbon and of the dimeric hydrocarbon derived from the latter.'

By way of illustration, a process is described in which a solution of 7 parts of egg albumin, starch or gelatin, in 500 parts of water is heated with 300 parts of isoprene in an autoclave for some weeks at 60°C, the mixture being continuously stirred. A latex-like liquid is said to be obtained, from which the rubber component gradually separates while the egg albumin coagulates. Alkali may be present during the polymerisation process, but in this case the rubber component can be separated from the reaction product only after acidification. No details of the properties of the polymeric product are given.

From this example, it is apparent that the polymerisation occurred very slowly indeed. Hohenstein and Mark[1] carried out a number of polymerisations following closely the examples of this and of another early patent, and substantially verified the claims of the patents. They observed a very slow, partial conversion of the monomer (isoprene, dimethylbutadiene) into a polymer latex. The total amount of polymer formed varied between 40% and 80%; the duration of the reaction was as much as six weeks in certain cases. They state that 'The results, in general very erratic and almost irreproducible, create the impression that the reaction under such conditions could be considered a *suspension polymerisation* catalysed by the oxygen of the air, which was never specifically excluded in any of the examples. In order to check this conclusion we repeated a few experiments of this type with deaerated monomer and deaerated water under nitrogen and found that under these conditions only *extremely slow* polymerization can be observed. In some instances conversion was not achieved at all' (the italics are in the original). Hohenstein and Mark, writing in 1946, therefore reached the conclusion that the early practice, as disclosed in patents of the type under consideration, relates not to what we now know as emulsion polymerisation, but rather to a process of suspension polymerisation, in which the colloidal additives acted as protective colloids for the dispersion of the monomer in water, and in which initiation of the reaction occurred by the action either of atmospheric oxygen or of products formed by interaction with atmospheric oxygen.

Other points which emerge from study of the early literature on 'emulsion' polymerisation include the following:

(1) A patent to Klatte and Rollett in 1914[8] was significant in that it mentioned the use of added oxygen-liberating compounds as polymerisation activators.
(2) In 1912, Kondakov[9] described the use of a mixture of two different monomers, so that, as Duck[10] points out, all the essential elements of emulsion polymerisation were known (if not exploited) just prior to the commencement of World War I.
(3) Ostromislensky[11,12] in 1915–16 reported the results of polymerisation experiments using vinyl halides as monomers, and discussed the advantages of the presence of an inert diluent. Hohenstein and Mark[1] consider that, as there is no mention of the use of soap or of other micelle-forming substances, the observations refer to 'uncatalysed' or photocatalysed polymerisation in solution or suspension.

2.3 EXAMPLES OF FIRST TRUE EMULSION POLYMERISATION PROCESSES

Disclosures in the patent literature of processes which would now be regarded as true emulsion polymerisations first began to appear in about 1929. In that year, Dinsmore,[13] working for The Goodyear Tire & Rubber Company, was granted a patent entitled 'Synthetic rubber and method of making it'. Attention is drawn to the fact that synthetic rubber, when manufactured by the then usual processes, was unsatisfactory for many purposes, since it lacked toughness and elasticity. 'It is presumed', says Dinsmore, 'that these defects are due to the fact that the polymerized hydrocarbon, like the overly milled natural rubber, does not have a definite globular structure.' It is not clear what exactly these words are intended to convey; presumably the reference is to certain views prevailing at that time, concerning the supposed structure of the natural rubber latex particle, and also concerning the structure of material like the smoked sheet obtained from natural rubber latex by coagulation. Dinsmore then claims to have made the discovery that 'by treating the basic unsaturated hydrocarbon from which the rubber is to be produced, with an emulsifying agent so as to break up the hydrocarbon into definite globules, a product much superior to any synthetic rubber heretofore produced is obtained'.

Dinsmore takes pains to point out that the validity of his patent is in no way affected by the truth of the theory he puts forward to explain the effectiveness of his process. His view is that the polymer produced by his method has the form of small globules which are surrounded by outer shells of the protective colloid, thereby maintaining a permanent globular structure. The protective colloid is also considered to have a catalytic influence on the polymerisation. It is therefore argued that the outer layers of the globule become tough whereas the inner regions, being less accessible to the catalytic influence of the protective colloid, remain in a more or less liquid state. From this cause is said to derive the elasticity and toughness of the finished product.

As thus described, the process would appear to be one of suspension polymerisation, especially if the alleged catalytic influence of the protective colloid is ignored. The reason for regarding it as an emulsion polymerisation is that the two examples given call for the addition of a soap, presumably added as an emulsifying agent in the first instance, as well as of a proteinaceous protective colloid. Thus the second of the examples describes the following procedure for obtaining from dimethylbutadiene a product which is said to have similar properties to natural rubber: 'Take 30 cc. of the hydrocarbon and emulsify it in 30 cc. of water containing 10 drops of oleic acid neutralized with potassium hydroxide. To this solution add 15 cc. of two percent solution of egg albumin. Seal the mixture in a tube and maintain it at a temperature of about 50° to 70°C. for a period of six months. The material may then be removed, coagulated with alcohol, and dried in a vacuum. The product obtained is tough, elastic and may be cured by means of sulphur and a suitable accelerator of vulcanization in a manner similar to that employed to cure the natural product.' It should be noted that neither example requires the addition of an initiating substance, and in consequence polymerisation rates are extremely slow. Initiation probably occurs by way of adventitious peroxides produced by interaction between atmospheric oxygen and either the monomer or the fatty-acid soap.

Similar processes are described in a United States patent granted to

Luther and Heuck[14] of I.G. Farbenindustrie A.G. in 1932. This patent claims the conversion of butadiene and its homologues into 'products of latex character' by emulsifying the monomer in aqueous media with emulsifying agents of soap-like character, and allowing polymerisation to occur. No substances which would now be recognised as initiators are deliberately added, although the curious statement is made that 'the polymerization may be effected not only by heat treatment but also by the exposure to light of short wavelength, *and also by the action of special polymerizing agents such as acrylic esters, condensation products of acrolein, alkylamines, arylamines and the like* or by simultaneous employment of several methods of polymerization' (the italics are not in the original). 'It is usually preferable to conduct the polymerization in the presence of buffer systems with a fixed hydrogen-ion concentration, because the polymerization is thereby accelerated, and premature coagulation of the emulsion is prevented. A suitable hydrogen-ion concentration for this operation is between pH = 4 and 8·5 such as is obtained, for example, by the addition of acetic acid and sodium acetate or of secondary and tertiary sodium phosphate.' The examples include as emulsifiers not only fatty-acid soaps but also the sodium salt of an *iso*butylated naphthalene-sulphonic acid, and Turkey-red oil.

A notable advance was recorded in a second United States patent granted to Luther and Heuck,[15] also in 1932 (actually slightly earlier than the patent referred to above, although the filing date was more than one year later). This patent claims the simultaneous use of emulsifiers and polymerisation initiators. Essentially, the process consists in the use of an oxidising agent, such as hydrogen peroxide, as a polymerisation catalyst and then the subsequent addition of a reducing agent or other substance to destroy excess oxidising agent. As examples of reducing agents, the following are cited: sodium bisulphite, sodium hydrosulphite, sodium formaldehyde sulphoxylate, stannous chloride, hydroxylamine, hydrazine (and its salts) and hydrogen in the presence of catalysts. As examples of other substances are given phenols, naphthols, sols of metals and metal oxides, peroxidases and amines such as aniline. Two other benefits are said to accrue from the process. Firstly, the polymer is to some extent lightened in colour, and, secondly, it is stabilised against oxidation by residual traces of, for example, hydrogen peroxide. The latter advantage arises because of the peroxide-destroying effect of the subsequent additive and would be expected to be especially marked in cases where the additive was a substance, such as phenol or an amine, which is known to have a strong antioxidant effect for hydrocarbon polymers.

By way of illustration of the process, the following example is given, all parts being by volume: 18 parts of oleic acid are dissolved in 200 parts of isoprene, and the solution emulsified by shaking with 350 parts of an aqueous solution containing 0·5% of ammonia and 0·5% glue. The resultant emulsion is said to be stable and to have a pH of 8·9 (this may mean pH 8–9). Polymerisation is effected by adding 5 parts of 30% hydrogen peroxide and heating for two days at 50°C, when a latex-like product is obtained. (The sharp reduction in polymerisation time as compared with earlier processes should be noted.) 30 parts of a 10% aqueous solution of sodium hydrosulphite are then added to destroy any hydrogen peroxide which is still present. The resultant latex, it is claimed, can be coagulated in the usual manner to give a white product of good stability. As expected, the use of phenol or aniline to destroy the hydrogen peroxide results in a somewhat darker final product.

In the years following the patents of Dinsmore and of Luther and Heuck, there were a large number of further patents containing a multitude of disclosures and claims, some of which claims conflict with each other. Writing in 1933, Whitby and Katz[6] gave the following summary of substances mentioned in patents as useful for effecting the emulsion polymerisation of butadiene and its homologues:

(1) *Emulsifying agents*: Sodium *iso*propyl-β-naphthalenesulphonate, sodium butyl-α-naphthalenesulphonate, tetrahydronaphthalene-β-sulphonic acid, oleyl-β-diethylaminoethylamine, sodium oleate, sodium linoleate, sodium stearate, sulphite cellulose waste liquor, sulphonated mineral oils, organic sulphonic acids, bile salts and derivatives, and magnesium oleate.

(2) *Protective colloids*: Carrageen moss extract, gum arabic, linseed meal extract, blood serum, gelatin, glue, casein, egg albumin, cow's milk, starch and dextrin.

(3) *Polymerisation catalysts*: Hydrogen peroxide, either alone or in the form of a loose compound with urea, sodium perborate, potassium percarbonate, persulphates, organic and metallic peroxides, perbenzoates, perchlorates, ozonides, and the salts of multivalent heavy metals such as cobalt, lead and manganese. The use of acid anhydrides, such as acetic and benzoic anhydrides, for increasing the rate of polymerisation of dienes catalysed by inorganic peroxides is noted, as is the claim that the presence of about 5% of organic halogen compounds, such as carbon tetrachloride, hexachloroethane, chloral hydrate and trichloroacetic acid, has an advantageous

effect upon the emulsion polymerisation process by increasing the rate of polymerisation, by enhancing the yield, and also by improving the plasticity and adhesive properties of the product.

(4) *Electrolytes and plasticisers*: The use of suitable electrolytes in conjunction with added colloids is claimed as advantageous in improving the plasticity, milling properties and, in many cases, the strength of the product. Suitable electrolytes are said to be calcium caseinate, sodium sulphate, sodium bisulphite, potassium iodide, acetic acid, hydrochloric acid, phosphoric acid and malonic acid. Other substances claimed to improve the plasticity of the rubber produced include polymerised or oxidised castor, linseed and other oils, and unsaturated ketones such as methylmethylene ethyl ketone.

(5) *Substances which modify the surface tension*: Amyl, heptyl and benzyl alcohols, and various ketones. These additives, which modify the properties of air–water interfaces, are said to have advantageous effects upon diene emulsion polymerisation systems, especially in the presence of added colloids.

Among the several examples given by Whitby and Katz illustrating the use of some of the substances listed above, is the following formulation:

Butadiene	200 parts (by weight?)
Milk	90
Glue	4
Methylcellulose	4
Sodium butyl-α-naphthalenesulphonate	2
Sodium *iso*propyl-β-naphthalenesulphonate	1
Castor oil	4
Urea–hydrogen peroxide adduct	1

The ingredients are emulsified together and heated for 7 days at a temperature between 50°C and 55°C. The product, obtained in quantitative yield, is said to be a white polymer, stable in air. The effect of the urea–hydrogen peroxide initiator is apparent from the fact that, without it, the yield under similar conditions was found to be less than 20%.

2.4 EARLY PAPERS IN SCIENTIFIC JOURNALS

Hohenstein and Mark[1] note that, in contrast to the large number of patents for emulsion polymerisation processes which appeared during the period 1930–40, very few papers on the subject were published in the scientific journals. They draw attention to papers by Dogadkin and his co-workers,[16,17] which appeared in 1936 and were concerned with the polymerisation of butadiene in the presence of soap, peroxides and other catalytic substances at various temperatures. The kinetics of the reaction were reported upon.

The hypothesis that the principal locus of the emulsion polymerisation reaction is within the aqueous phase and not the monomer phase seems first to have been formulated by Fikentscher in 1938, both at a meeting of the Verein Deutscher Chemiker and subsequently in a published paper.[18] A paper on the polymerisation of butadiene vapour on the surface of water containing a small amount of hydrogen peroxide was published by Gee, Davies and Melville[19] in 1939. These papers will be considered in detail in the following chapter. A brief description and intepretation of emulsion polymerisation appeared in 1939 in a chapter written by Trommsdorff.[20]

Mention must also be made of an extensive section on emulsion polymerisation which appeared in a book by Talalay and Magat,[21] published in 1945. This book is concerned with what could be learnt from the Russian literature regarding the production of synthetic rubber, up to the time of publication. Although relatively little could be said of the mechanism of the emulsion polymerisation reaction, this contribution is of the greatest interest in that it describes the practice of emulsion polymerisation as it appears to have been carried out in Russia, thereby giving access to much information which is probably not available elsewhere in the English language. Some information is also given concerning the kinetics of emulsion polymerisation. Graphs are given showing the effect of increasing levels of both hydrogen peroxide and diazoaminobenzene on the rate of emulsion polymerisation of butadiene. In both cases, an order of reaction of 0·5 was observed, although, when increasing amounts of diazoaminobenzene were used as initiator, the rate was observed first to increase (in accordance with an order of reaction of 0·5), and then to pass through a maximum and fall again.

Although several other papers appeared in the literature during the period 1940–46, some of which will be discussed in detail in the following chapter, it is convenient to conclude this summary of historical aspects by a further reference to the paper of Hohenstein and Mark[1] which appeared in 1946. After summarising the known facts up to that date, they present very clearly the distinguishing features of the three types of free-radical addition polymerisation in which the monomer is diluted. These are the processes which we now know respectively as solution polymerisation, suspension polymerisation and emulsion polymerisation. It is clearly recognised that, in the latter process, monomer can be present as monomer droplets, as a molecular solution in the water and also as a solubilisate within the micelles which exist in a soap solution of sufficient concentration. Furthermore, it is recognised that the amount of polymer formed in each of these three phases will depend primarily upon the way in which monomer and initiator are distributed between these phases, and also possibly to some extent upon the accessibility and reactivity of the monomer in these three phases. The typical emulsion polymerisation is clearly stated to be one in which polymer formation occurs predominantly in the micellar phase. The impossibility of making hard-and-fast distinctions is noted, for the so-called emulsion polymerisation of a monomer, such as vinyl acetate, which is appreciably soluble in water, is recognised as being in fact, a superposition of solution, suspension and emulsion polymerisations.

REFERENCES

1. W. P. Hohenstein and H. Mark, *J. Polym. Sci.*, 1946, **1**, 127, 549.
2. F. Hofmann and K. Delbrück/Farbenfabriken Bayer A.G., German Pat. 250,690(1909).
3. F. Hofmann and K. Delbrück/Farbenfabriken Bayer A.G., German Pat. 254,672(1912).
4. F. Hofmann and K. Delbrück/Farbenfabriken Bayer A.G., German Pat. 255,129(1912).
5. K. Gottlob/Farbenfabriken Vorm. Friedr. Bayer & Co., U.S. Pat. 1,149,577(1915).
6. G. S. Whitby and M. Katz, *Ind. Engng Chem.*, 1933, **25**, 1204, 1338.
7. D. C. Blackley, *High Polymer Latices*, Applied Science Publishers, London, 1966, Ch. IV.
8. F. Klatte and A. Rollett, German Pat. 281,688(1914).
9. I. Kondakov, *Rev. Gén. Chim.*, 1912, **15**, 408.
10. E. W. Duck, 'Emulsion Polymerisation', in *Encyclopedia of Polymer Science and Technology*, John Wiley, New York, 1966, Vol. 5, pp. 801–59.
11. I. I. Ostromislensky, *J. Russ. Phys. Chem. Soc.*, 1915, **47**, 1472, 1928, 1947.
12. I. I. Ostromislensky, *J. Russ. Phys. Chem. Soc.*, 1916, **48**, 1071.
13. R. P. Dinsmore/The Goodyear Tire & Rubber Co., U.S. Pat. 1,732,795(1929).
14. M. Luther and C. Heuck/I.G. Farbenindustrie A.G., U.S. Pat. 1,864,078(1932).
15. M. Luther and C. Heuck/I.G. Farbenindustrie A.G., U.S. Pat. 1,860,681(1932).
16. V. Balandina, K. Berezan, A. Dobromyslova, B. Dogadkin and M. Lapuk, *Bull. Acad. Sci. U.S.S.R.*, 1936, Ser. 7, 397, 423.
17. K. Berezan, A. Dobromyslova and B. Dogadkin, *Bull. Acad. Sci. U.S.S.R.*, Ser. 7, 409.
18. H. Fikentscher, *Angew. Chem.*, 1938, **51**, 433.
19. G. Gee, C. B. Davies and H. W. Melville, *Trans. Faraday Soc.*, 1939, **35**, 1298.
20. E. Trommsdorff, in *Chemie und Technologie der Kunststoffe* (ed. R. Houwink), Akademische Verlagsgesellschaft mbH, Leipzig, 1939, pp. 304ff.
21. A. Talalay and M. Magat, *Synthetic Rubber from Alcohol: A Survey Based on the Russian Literature*, Interscience, New York, 1945, pp. 123–44.

CHAPTER 3

EARLY THEORIES OF EMULSION POLYMERISATION

3.1 SCOPE AND PLAN OF CHAPTER

The purpose of this chapter is to review those contributions to understanding of the mechanism of emulsion polymerisation which appeared in the literature prior to and contemporaneously with the publication of the Smith–Ewart theory. This theory was first described in a paper published in 1948.[1] Although various previous attempts to explain the quantitative features of emulsion polymerisation kinetics had appeared in the literature before this, the Smith–Ewart theory represents the first attempt to deal with the problem thoroughly and rigorously, and using a model which, in its main features, accords well with the generally-held present-day views on the mechanism. This theory has played such an important part in providing a framework for much of the subsequent discussion of emulsion polymerisation mechanism and kinetics that it has been deemed desirable to devote a separate chapter (Chapter 4) to an exposition of it and of the more obvious refinements of it which have been proposed from time to time.

The plan which has been adopted for this present chapter is to consider first some of the earliest contributions to the understanding of mechanism and kinetics, and then to proceed to a discussion of the Harkins theory. This latter theory can fairly be regarded as the qualitative precursor of the quantitative Smith–Ewart theory. The Harkins theory, which is largely descriptive in character, established a sound model for the mechanism of the reaction; Smith and Ewart then used this model as the basis for their calculations concerning the rate at which emulsion polymerisation reactions occur, and for their predictions concerning the manner in which the various operative factors affect that rate. This chapter also includes a discussion of an early, but now largely discredited, quantitative theory for emulsion polymerisation, namely, the so-called 'phenomenological' theory of Montroll. It concludes with a discussion of two other early attempts to seek a quantitative understanding of the process, namely, those associated with the names of Corrin and of Haward.

3.2 SOME INITIAL CONTRIBUTIONS TO UNDERSTANDING OF THE MECHANISM OF EMULSION POLYMERISATION

3.2.1 Fikentscher's views on the principal locus of reaction

It was pointed out in the preceding chapter that Fikentscher is usually regarded as the first person to realise that the principal locus of an emulsion polymerisation reaction is not in the monomer droplets but in the aqueous phase. As far as is known, the only extant written record of his views is contained in a short paper published in 1938[2] with the title 'Emulsions-polymerisation und technische Auswertung'. After pointing out that the existence of natural rubber latex provided the main impetus for the development of emulsion polymerisation processes for diolefins, and that this method of effecting polymerisation was then subsequently extended to vinly monomers, Fikentscher proceeds to summarise his views on the mechanism of emulsion polymerisation in a short paragraph, a rough translation of which is as follows: 'On the basis of experimental evidence, we incline to the opinion that in emulsion polymerisations it is not, as is frequently assumed, the part of the vinyl compound as droplets, but it is the part dissolved in the water of the emulsion which is polymerised, and this is continually replenished from the emulsified liquid droplets, and therefore throughout the whole polymerisation remains constant as long as such droplets are present.' There is thus clear recognition that the polymerisation takes place in the aqueous phase, and that the emulsified droplets merely serve as reservoirs of monomer, making monomer available to the aqueous phase as required. There is, however, no discussion of the precise location of the polymerisation reaction within the aqueous phase.

One piece of experimental evidence which seems to have influenced Fikentscher in his thinking was the observation that most vinyl monomers are soluble to some extent in water, and that, under the influence of heat and possibly of added 'accelerators' (*Beschleuniger*), such solutions become cloudy due to the polymerisation of the monomer. Presumably a general resemblance between such dispersions and the more concentrated latices obtained by emulsion polymerisation was construed as evidence for the aqueous phase as the locus of the latter reaction. It should be noted, however, that this particular piece of evidence seems to relate to systems to which no micelle-forming substances had been added.

Hohenstein and Mark,[3] writing in 1946, make the following interesting personal comment regarding the views on the mechanism of emulsion polymerisation prevailing in Germany and in the U.S.A. in the period between 1930 and 1940: 'One of the authors (H.M.) had an opportunity to discuss the problem of emulsion polymerization in the period between 1935 and 1938 with Drs. H. Fikentscher, H. Hopff, and E. Valko in Ludwigshafen am Rhein. At that time they offered several arguments in favor of polymerization taking place preponderantly in the aqueous phase. Valko even considered it as highly probable that the monomer, solubilized in the micelles of the soap solution, was most favorably exposed to the action of a water-soluble catalyst and, therefore, might be considered as the principal site of the reaction. At a seminar on high polymers in Kansas City in September, 1945, Dr. C. F. Fryling told us that he had, at the same time, independently arrived at very similar conclusions on the basis of his own observations. It appears, therefore, that some of the more recent developments were anticipated to a certain extent in the unpublished work between 1930 and 1940.'

Notwithstanding that the main outlines of the modern theory of the mechanism of emulsion polymerisation were probably enunciated well before 1940, Mark and Raff[4] stated in a standard reference work published in 1941 that 'it is not yet known exactly how emulsion polymerizations proceed'. They noted the general assumption that it is the emulsified monomer droplets which polymerise from the interfacial boundary towards the interior, but that Fikentscher dissented from this view.

3.2.2 Gee, Davies and Melville's study of the polymerisation of butadiene at water–air interfaces[5]

This work, a published account of which appeared in 1939, was concerned with the polymerisation of butadiene vapour at the surface of aqueous solutions of hydrogen peroxide. The course of the polymerisation was followed by observing the decrease in the surface free energy of the solution. On prolonged contact of butadiene with hydrogen peroxide solution, a slight cloudiness was observed in the solution, indicating that polymerisation can occur in the solution. However, the rate of this reaction was considered to be negligible compared to the total rate of polymerisation of butadiene under these conditions. The gas phase was discounted as an important locus of polymerisation on the grounds that evidence from studies of photochemical polymerisation in the gas phase indicated that, under the conditions of temperature and pressure used in these experiments, the kinetic chain length of the reaction would be so small that only negligible polymerisation would occur in the gas phase. It was therefore supposed that what was being studied was a polymerisation occurring at the interface between the aqueous solution and the gas phase.

The effects upon the reaction rate of variables such as temperature, butadiene pressure and hydrogen peroxide concentration were studied. Some typical results are summarised in Fig. 3.1. The conclusions reached were that the reaction showed no induction period, that the lifetime of the propagating centre was small compared with the overall time for the reaction, and that the rate of polymerisation was proportional to the first power of the hydrogen peroxide concentration and to the first power of the butadiene pressure. The kinetics of the reaction were shown to be consistent with the view that initially the reaction involves interaction between one hydrogen peroxide molecule and one butadiene molecule. The product of this interaction (which Gee, Davies and Melville describe as a 'complex') then initiates a polymerisation chain reaction, termination of which is by interaction with monomer. It is an obvious requirement that the outcome of the terminative interaction must differ from that of the propagative interaction.

It is arguable to what extent this paper represents a contribution to early understanding of the mechanism of emulsion polymerisation. Its relevance is certainly questionable in that, as there was no soap or other surface-active substance added to the systems studied, there can have been no micelles present, initially at least. The view of the present writer is that this contribution is relevant in that it shows clearly that the free-radical polymerisation of monomers which are almost completely insoluble in water can take place within an aqueous environment, even though it is only within the surface layers. Furthermore, the reaction occurs notwithstanding that the initiating substance is contained exclusively within the aqueous phase. These observations were in themselves quite important at the time they were published, considering that the prevailing view among polymer chemists at that time was that polymerisation in an emulsion system proceeds within the monomer droplets.

3.2.3 Experiments of Fryling and Harrington on the solubilisation of monomers in soap solutions[6]

This study had its origin in the observation that the addition of small quantities of acrylonitrile to an aqueous soap solution lowers the pH from

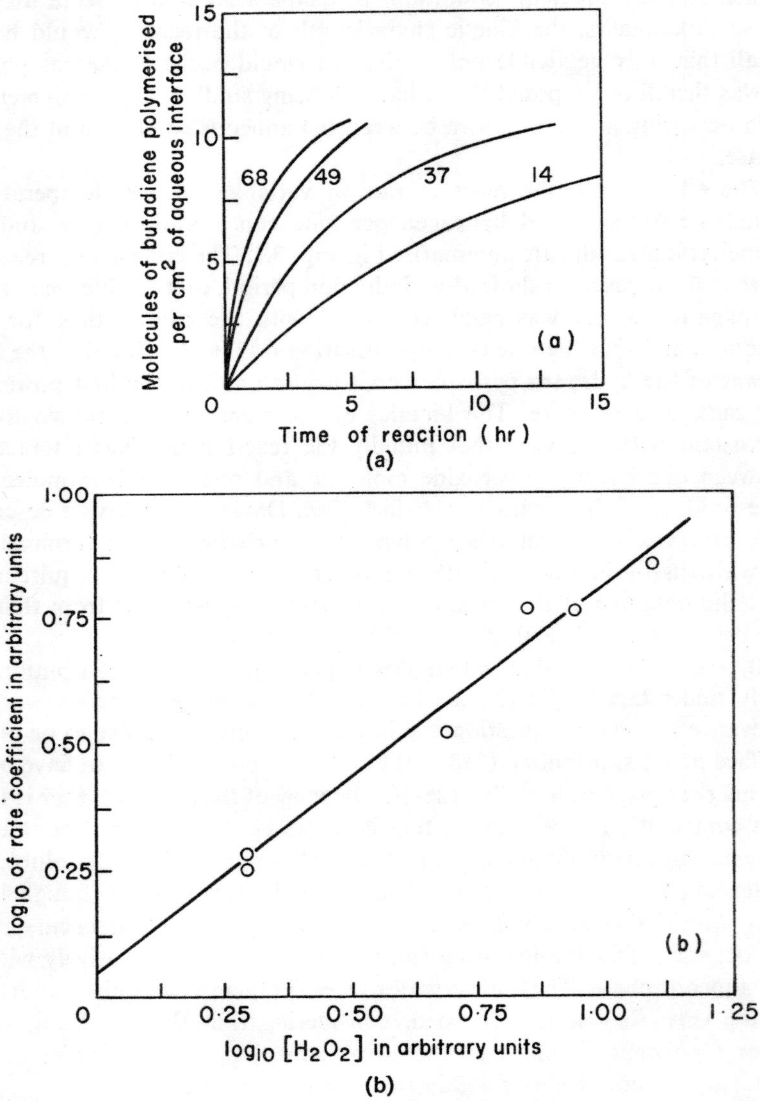

FIG. 3.1. Effect of certain variables upon polymerisation of butadiene at a water–air interface initiated by hydrogen peroxide, according to Gee, Davies and Melville:[5] (a) Effect of butadiene pressure upon conversion–time curve for reaction at 60°C. Numbers against curves indicate butadiene pressures in psi. (b) Effect of hydrogen peroxide concentration upon rate coefficient for reaction at 75°C, butadiene pressure being 25 psi.

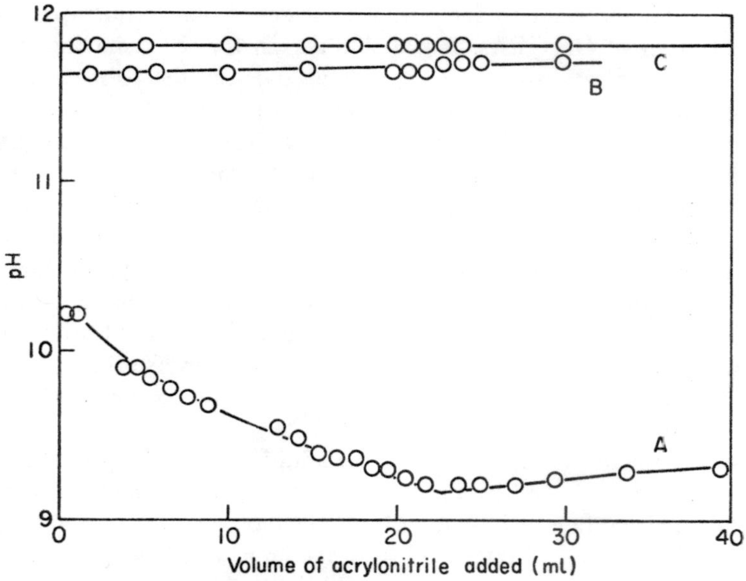

FIG. 3.2. Effect of addition of acrylonitrile upon the pH of aqueous sodium myristate solution at 28°C, according to Fryling and Harrington.[6] The initial volume of soap solution was 200 ml in each case. The soap concentrations and extents of neutralisation were as follows: A: 2% soap, 100% neutralisation; B: 1% soap, 125% neutralisation; C: 1% soap, 185% neutralisation.

9·5 to 8·5. Further experiments showed that the addition of acrylonitrile to a soap solution is first accompanied by a decrease in pH, followed by an increase in pH when larger quantities of monomer are added. Typical results are shown in Fig. 3.2. That these effects are associated with the disturbance of ionic equilibria involving the soap, notably hydrolysis, is evident from the observation that soap solutions containing excess alkali undergo almost negligible pH change on addition of acrylonitrile (see Fig. 3.2).

Fryling and Harrington interpreted the initial pH decrease as due to solubilisation of monomer within the soap micelles. The suggestion which they appear to make is that the process of solubilisation into the micelles is accompanied by a reduction in the concentration of fatty-acid anions available for participation in the hydrolysis reaction. If this latter reaction is represented as

$$RCO_2^- + H_2O \rightleftharpoons RCO_2H + OH^-$$

then it is clear that the removal of RCO_2^- ions will cause the equilibrium to shift from the right to the left, with a consequent reduction in the concentration of OH^- ions. A possible reason for the reduction in the effective concentration of RCO_2^- ions is given as the formation of a complex with monomer within the micelles. It is also noted in this paper that Kolthoff had observed similar pH decreases in sodium myristate solutions which were sufficiently dilute not to contain any appreciable number of micelles. The explanation for the effect given by Kolthoff was, it is said, that the addition of the organic liquid alters the general nature of the environment in which the soap is dissolved. In particular, the solubility of the free fatty acid, RCO_2H, in the aqueous phase would be increased, and so therefore would be its effective concentration. This would also have, as a consequence, a tendency to reverse the hydrolysis and so reduce the pH. On balance, this seems to be the more likely explanation for the observed effect.

The increase of pH accompanying the addition of further quantities of monomer was attributed by Fryling and Harrington to the tendency for the undissociated fatty acid to dissolve in the monomer phase. This would result in the removal of fatty acid from the aqueous phase, that is, from the phase in which is occurring the equilibrium which determines the pH. The above equilibrium would then shift from left to right, with a consequent increase in pH.

This paper is of considerable interest in indicating an awareness of the importance of soap micelles in accounting for certain aspects, at least, of the behaviour of emulsion polymerisation systems. It is in marked contrast to the view which regards the soap micelles as being mere reservoirs of monomeric soap molecules which can be made available if required as a result of changes taking place elsewhere in the system.

However, the paper is also of interest in recognising clearly the fact that the polymerisation process starts in the aqueous phase. The following experimental evidence is offered by Fryling and Harrington in support of this view:

'(1) With an undispersed acrylonitrile–water–potassium persulfate system, it can be visually demonstrated that polymerization takes place exclusively in the aqueous phase. Acrylonitrile, purified for polymerization, is added to a 6% solution of potassium persulfate at room temperature. A portion of the acrylonitrile dissolves in the aqueous phase while the greater portion forms a well-defined, clear upper phase. After standing quietly for a short time, the aqueous phase becomes cloudy and eventually opaque, due to formation of polyacrylonitrile. No evidence of polymer is detectable in the upper phase; as the result of an interesting phenomenon, it can be observed that no polymer is formed at the interface. The aqueous phase remains clear and transparent for about 5 mm. below the interface, as though acrylonitrile, diffusing into the aqueous phase to replenish that removed by polymerization, has had insufficient time to undergo polymerization during its diffusion through the upper 5 mm. of the aqueous phase. However, if the partially polymerized system is stirred, the polyacrylonitrile is preferentially wetted by acrylonitrile and passes immediately into the upper phase. Consequently if such a system were submitted to continuous agitation, its appearance might lead to the erroneous conclusion that polymerization had taken place either in the acrylonitrile phase or at the interface.

(2) "Emulsion" polymerization of synthetic rubber will take place even though the mixture of monomers is separated from the other ingredients of the recipe by the vapour phase. The monomers evaporate, pass into the aqueous phase, and copolymerize without any discernible evidence of a monomer phase in contact with the aqueous solution. Under these conditions the initiation of polymerization is a homogeneous process, and no part of it (*i.e.* the initiation) can be attributed to interfacial reactions.

(3) Water-soluble initiators, such as potassium persulfate, hydrogen peroxide, and sodium perborate, are effective initiators in emulsion polymerization of synthetic rubber.

(4) In many recipes stable emulsification of the organic phase is not obtained until after polymerization starts.

(5) The emulsion polymerization velocity of many recipes at low soap concentrations is almost linearly proportional to the concentration of the soap. This catalytic effect appears to favor the theory of initiation of polymerization in the aqueous phase.

(6) The utility of various substances as polymerization emulsifiers follows approximately the same order as their solubilizing action, as measured for propylene by McBain and Soldate. It is not intended to imply that a numerical correlation exists between solubilization and emulsion polymerization velocity; rather, the more useful polymerization emulsifiers are found among those that exhibit considerable solubilization.'

This lengthy extract from Fryling and Harrington's paper has been reproduced here to show that the outlines of an acceptable theory of the mechanism of emulsion polymerisation existed in the literature in early 1944. The main features of emulsion polymerisations distinguishing them from other methods for effecting free-radical addition polymerisation are clearly enunciated; the importance of the aqueous phase as a locus of reaction, and not merely as an inert vehicle, is recognised; so also is that of the micelle-forming tendency of the surface-active substances present in the reaction system. The monomer–aqueous phase interface as a principal locus for the reaction is dismissed on the grounds of inadequate evidence: 'The prevalent idea that emulsion polymerization occurs at the interface seems to be founded solely on the observation that polymerization is faster in emulsion systems than in homogeneous systems, and that orientation in the interface provides a mechanism to explain this increased rate.'

3.2.4 Contributions from Hohenstein and his co-workers[7–10]

In the first of these papers,[7] Hohenstein, Vingiello and Mark present results to show that suspension or 'pearl' polymerisations can be considered as essentially bulk polymerisations with very effective water cooling, the monomer being broken up into small globules. They are thus to be distinguished from typical emulsion polymerisations. Hydrocarbon-soluble initiators, such as dibenzoyl peroxide, are far more effective than are water-soluble compounds such as hydrogen peroxide or sodium persulphate. The overall rate of monomer consumption was reported as being approximately proportional to the square root of the concentration of hydrocarbon-soluble initiator, this concentration being expressed in terms of the total volume of oil phase. The molecular weight of polystyrene obtained by polymerisation at 85°C using 1% of initiator was reported as 85 000, this being very much lower than that of the polymer which is obtained from a typical emulsion polymerisation. It was noted, however, that the rate of monomer consumption in a suspension polymerisation is significantly greater than that in a bulk polymerisation carried out under equivalent conditions.

In the second paper,[8] Hohenstein, Siggia and Mark describe experiments on the polymerisation of styrene in water and in aqueous soap solutions, and attempt to interpret the results in terms of a mechanism and location for the various elementary steps of the overall reaction. The monomer chosen for the investigation was styrene, the precautionary note being added that, because its behaviour is much simpler than that of systems generally used in industrial processes, care should be exercised in drawing conclusions regarding one special case from observations on another. Visual observation of unstirred soap-free systems provides evidence that in these systems polymerisation occurs in the aqueous phase. Styrene left in contact with aqueous potassium persulphate solution for several days was observed to cause the aqueous phase to become cloudy, with small white flakes of polymer settling down to the bottom of the aqueous phase after a long time. This observation in itself did not exclude the styrene–water interface as a locus of reaction, but if monomers of higher water solubility were used (such as methacrylic acid esters, vinyl acetate or acrylonitrile), then the cloudiness in the aqueous phase appeared very much more rapidly although the area of the interface was the same as for the styrene system. This was taken as an indication that at least the formation of active nuclei took place in the aqueous phase. Whether under the conditions of this experiment the nuclei grew in the aqueous phase or whether they diffused to the interface and then grew there at the expense of the bulk liquid monomer could not be decided from these observations. However, the observation that separation of the monomer phase from the aqueous phase, allowing contact only through the vapour phase, led to the development of cloudiness in the aqueous phase in just the same way as before, was taken as strong evidence that propagation as well as initiation takes place in the aqueous phase. Furthermore, the rate at which cloudiness appeared was again roughly proportional to the solubility of the monomer in water. This was taken to indicate that the rate of development of cloudiness was a measure of the rate of polymer formation, and that the rate of evaporation and diffusion of monomer into the aqueous phase were sufficient to make monomer available as fast as it was required.

Further evidence for the correctness of the above view was found in the observation that the addition of small amounts of inhibitor to the monomer phase in the second experiment has no noticeable effect upon the course of the polymerisation, whereas the addition of the same amount of inhibitor to the aqueous phase causes the appearance of the cloudiness to be considerably delayed. Furthermore, the length of the induction period brought about in this way is roughly proportional to the amount of inhibitor added, as is also the case when inhibitor is added to monomer undergoing polymerisation in bulk or in solution. Increase of temperature at constant inhibitor concentration was observed to reduce the induction period, the variation being in accordance with the normal Arrhenius relationship, as illustrated in Fig. 3.3. An activation energy of 25 kcal mole^{-1} was calculated for styrene polymerisation from the observed

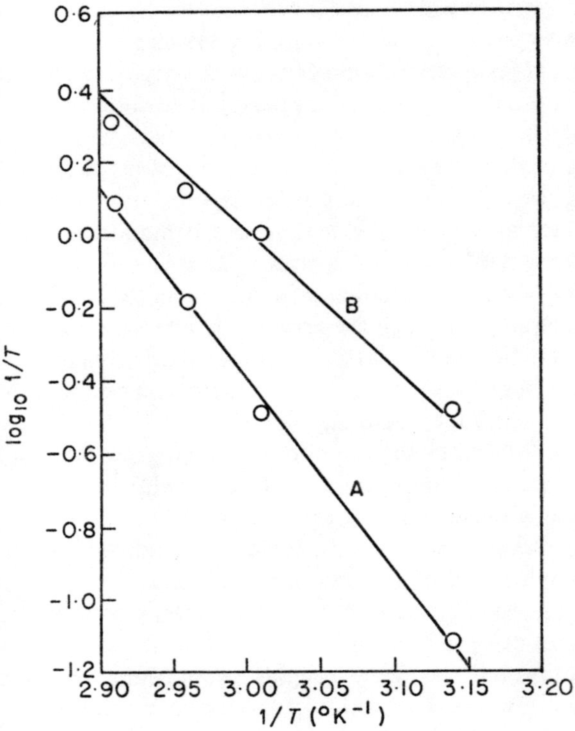

FIG. 3.3. Arrhenius-type plot showing effect of temperature upon length of induction period for persulphate-initiated polymerisation of styrene in aqueous solution in presence of a fixed concentration of inhibitor (hydroquinone), according to Hohenstein, Siggia and Mark.[8] Curve A refers to polymerisation in aqueous solution free of soap; curve B refers to polymerisation in 2% aqueous ammonium oleate solution.

variation with temperature. But this calculation was made on the assumption that all other factors influencing the rate of nucleus formation, such as solubility of the monomer in water, were independent of temperature. The activation energy was therefore probably underestimated to some extent. But in spite of this, it agrees reasonably well with reported values for the polymerisation of styrene in bulk, in solution and in suspension.

Addition of a soap to the aqueous phase in the second type of experiment was found to have two effects. Firstly, polymerisation occurred more rapidly than before, but, secondly, the temperature variation of the inhibition period was such as to indicate (see Fig. 3.3) that the activation energy was significantly less in the presence of soap than in its absence. A result of 17 kcal $mole^{-1}$ was reported. The corresponding collision constant prefactors were $1{\cdot}2 \times 10^{16}$ in the absence of soap and $1{\cdot}8 \times 10^{11}$ in the presence of soap. The drop in activation energy is attributed in part at least to 'orientation and polarization of the monomer within the soap micelle'. It is not clear, however, exactly what is meant by this expression, nor why such orientation and polarisation should lead to reduced energy of activation.

The second paper also describes experiments on the emulsion polymerisation of styrene under conditions of agitation. The typical conversion–time curve is as shown in Fig. 3.4, in which the amount of polymer formed as a percentage of the monomer initially present is plotted as a function of the time of reaction. A number of separate regions were distinguished. Initially (the stage OA of Fig. 3.4), the formation of polymer is very slow; it then increases gradually (the stage AB), reaching a comparatively long, linear stage (BC) within which the rate of polymer formation is virtually of zero order with respect to monomer concentration; eventually the rate falls as monomer becomes exhausted (the stage CD).

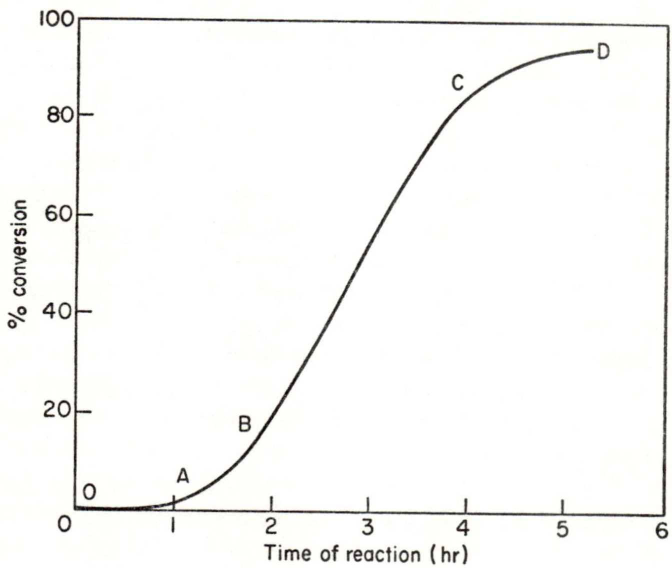

FIG. 3.4. Typical conversion–time curve for emulsion polymerisation of styrene, according to Hohenstein, Siggia and Mark.[8] See text for significance of letters appended to curve. (No details are given of either reaction system or reaction conditions, except that the reaction was carried out with agitation.)

Hohenstein, Siggia and Mark observed that they were never able to eliminate entirely the initial period of inhibition and retardation, although it could be reduced considerably by carefully purifying the monomer, by excluding oxygen, and by carrying out the reaction under purified nitrogen or *in vacuo*. The presence of this initial phase of the reaction was attributed to the presence of residual impurities. The strong upward turn of the curve of Fig. 3.4 between A and B was attributed to the disappearance from the system of the inhibiting substances, similar behaviour being observed in bulk polymerisations to which inhibitor had been deliberately added. These workers also reported that the polymer formed during the stage AB had a lower degree of polymerisation than that produced later in the reaction; this was attributed to premature chain termination, or possibly chain transfer, by residual inhibitor or possibly inhibitor reaction products. The linear stage of the reaction, B to C, was regarded as indicating the establishment of a steady state, during which active centres are formed, propagate, and then terminate by mutual interaction, all other chain-terminating substances having by this time been eliminated from the system. The final stage from C to D is characterised by a progressive slowing-up of the reaction, due mainly to monomer depletion, but also partly to a reduction in the amount of initiator remaining. These interpretations are summarised in Table 3.1 which is taken with only minor modification from the paper under review.

The third paper in this group[9] is by Siggia, Hohenstein and Mark. It describes the results of further experiments on the emulsion polymerisation of styrene under conditions of agitation. The details of the reaction systems used in this study are given in Table 3.2. Conversion–time curves are shown in Fig. 3.5 for emulsions of different average droplet sizes. In each case, a definite period of retardation was observed, which increased with the initial droplet size of the emulsion. Various explanations were offered for this effect. Thus if the inhibitor partitions mainly in the monomer phase, it can be argued that it will take more time to diffuse out of a large droplet than out of a small one; hence the critical inhibitor concentration, below which the reaction proceeds rapidly, is reached sooner in an emulsion of small droplet size. Alternatively, if the inhibitor partitions mainly in the aqueous phase, then it can be argued that it could be more rapidly destroyed in systems having large specific interfacial area (*i.e.* small droplet size). Or again, it could be that emulsions containing a higher soap:monomer ratio are capable of faster nucleus formation, and therefore consume the inhibitor more rapidly.

Results for the average and instantaneous polymer molecular weights

TABLE 3.1

Interpretation of various stages of an emulsion polymerisation reaction (according to Hohenstein, Siggia and Mark[8])

Characterisation of period	*Rate during this period*	*Reason for the observed rate*	*Degree of polymerisation of polymer formed during this period*	*Reason for the degree of polymerisation*	*Particles of emulsion*	*Reason for behaviour of particles*
Inhibition 0–A†	Zero	All active nuclei are destroyed by the inhibitor before they can start to grow into a chain	Zero	Propagation reaction has no chance to set in and produce chains	Maintain their initial size, typically 4·0 μm	No polymerisation takes place
Acceleration A–B	Rapidly increasing	Inhibitor disappears and gives nuclei the chance to grow into chains	Below average	Reaction products of inhibitor and activated monomer act as chain breakers	Start slowly to decrease in size	Polymerisation of smaller particles and in soap micelles changes the ratio soap to monomer in favour of soap and shifts the distribution curve to smaller average particle sizes
Steady-state reaction B–C	Appreciable and almost constant	Steady state is reached; all inhibitor is consumed	Around and above average	No chain breakers are left, and there is still enough monomer for rapid propagation	Decrease steadily in size	
Deceleration C–D	Rapidly decreasing	Monomer concentration decreases; decomposition products of the catalyst act as inhibitors	Below average	Lack of monomer and the presence of chain breakers (decomposition products of the catalyst) prevent the formation of long chains	Reach a limiting value of diameter, typically 0·4 μm	The reaction decelerates; there are no large monomer particles left which can break up into smaller ones

† For the significance of these letters in relation to the conversion-time curve, refer to Fig. 3.4.

TABLE 3.2

Details of reaction systems used by Siggia, Hohenstein and Mark[9] in their further investigations of the emulsion polymerisation of styrene under conditions of agitation

(All emulsions contained: 180 ml, *i.e.* 10 moles, of water
19 g, *i.e.* 0·18 mole, of styrene
2·8 g, *i.e.* 0·034 mole, of sodium perborate.
Molar ratio water:monomer, 55·5.
Molar ratio monomer:initiator, 5·3.)

	Amounts of ammonium oleate			
	g	*Moles*	*Molar ratio monomer : soap*	*Molar ratio soap : monomer*
(1)	0·9	$3{\cdot}0 \times 10^{-3}$	60	$1{\cdot}67 \times 10^{-2}$
(2)	1·0	$3{\cdot}3 \times 10^{-3}$	54·5	$1{\cdot}83 \times 10^{-2}$
(3)	1·1	$3{\cdot}6 \times 10^{-3}$	50	2.00×10^{-2}
(4)	1·2	$4{\cdot}0 \times 10^{-3}$	45	$2{\cdot}22 \times 10^{-2}$
(5)	1·3	$4{\cdot}3 \times 10^{-3}$	42	$2{\cdot}38 \times 10^{-2}$
(6)	1·4	$4{\cdot}6 \times 10^{-3}$	39	$2{\cdot}56 \times 10^{-2}$
(7)	1·5	$5{\cdot}0 \times 10^{-3}$	36	$2{\cdot}78 \times 10^{-3}$

formed in a styrene emulsion polymerisation are summarised in Fig. 3.6. These results are for viscosity-average molecular weights determined from intrinsic viscosity measurements. Comparison with the corresponding conversion–time curves shows that the molecular weight of the polymer formed during the period immediately following the retardation stage is comparatively low. This is explained by the occurrence of premature chain termination or of chain transfer due to the presence of residual inhibitor or of its reaction products. As the reaction reaches the linear stage, so the molecular weight of the polymer formed increases. Lower-molecular-weight polymer is again formed in the later stages of the reaction. Two reasons offered for this are as follows:

(1) As more monomer is consumed, its concentration decreases and falls to rather low values towards the end of the reaction. This affects the ratio of the rate of propagation to the rate of termination in such a way as to lower the average degree of polymerisation. This explanation is far from convincing, if only because it fails to take account of the very high ratios of polymer to monomer which must be present in the final stages of the reaction.

(2) The decomposition products of the initiator act as chain-terminating or chain-transfer agents, and therefore tend to reduce the degree of polymerisation of the product. If, as seems likely, towards the

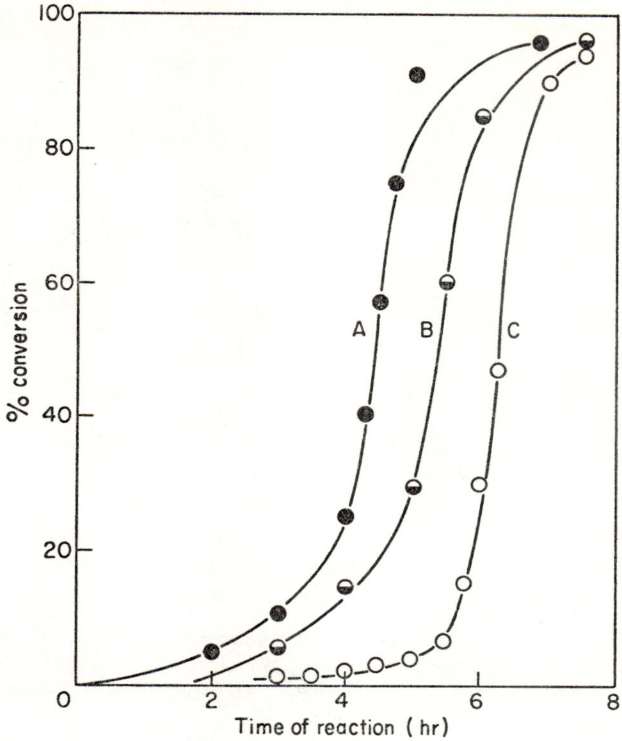

FIG. 3.5. Effect of initial emulsion droplet size upon course of emulsion polymerisation of styrene initiated by sodium perborate in presence of ammonium oleate, according to Siggia, Hohenstein and Mark.[9] The curves refer to emulsions of different average droplet size as follows: A, 0·4 μm; B, 1·6 μm; C, 3·6 μm. (No details are given of either reaction systems or reaction conditions, except the nature of the soap and initiator, and the fact that the reactions were carried out with agitation.)

end of the reaction there were more of such substances present than in the earlier stages, then this could account for the observation that a polymer of lower average molecular weight is formed in the later stages.

Whatever may be the true explanation of these observations, as Siggia, Hohenstein and Mark point out, these results demonstrate the extreme heterogeneity of the polymer molecular sizes produced if an emulsion polymerisation is taken to virtual completion. For not only is a spread of sizes produced at each stage in the reaction, but upon this heterogeneity

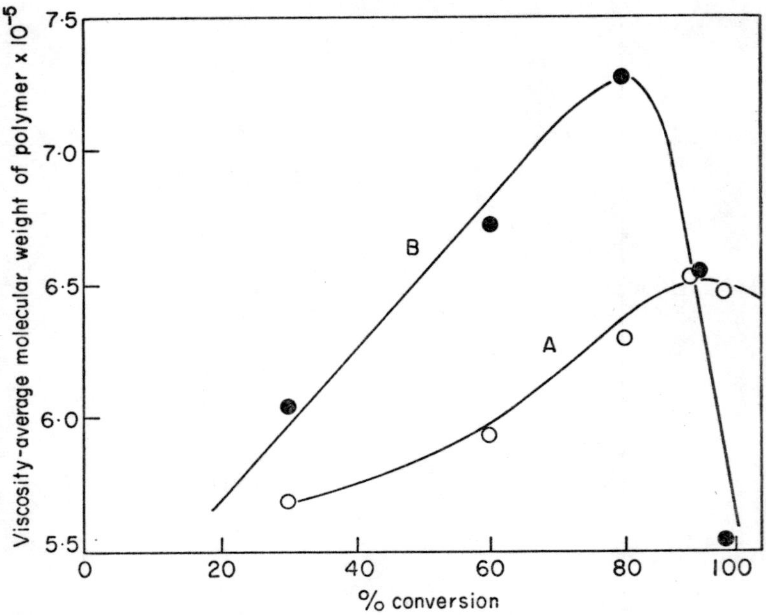

FIG. 3.6. Effect of conversion upon (A) average molecular weight of polymer produced, and (B) instantaneous molecular weight of polymer produced, for sodium perborate-initiated emulsion polymerisation of styrene, according to Siggia, Hohenstein and Mark.[9]

is then superimposed the non-uniformity of the polymer formed at different periods of the reaction.

Among the other interesting results given in this paper are data for the effect of soap level upon the average droplet size in the initial emulsion. These results, which are reproduced in Fig. 3.7, show that the droplet size falls very rapidly as the soap level is increased from 0·4% up to 0·8% in the aqueous phase. Furthermore, calculations based on these studies indicate that, in the systems used, only a small portion of the soap present is initially adsorbed on the monomer droplets, and that this proportion falls as the amount of soap falls because the droplet size increases sharply. Further allusion will be made to these results subsequently.

Most of the observations and results noted above are also contained in the second of two lengthy papers published by Hohenstein and Mark in 1946.[10] This latter paper is also of interest in that it contains references to redox systems for the initiation of emulsion polymerisations, to the influence of oxygen upon polymerisation rate, to the promoting effect

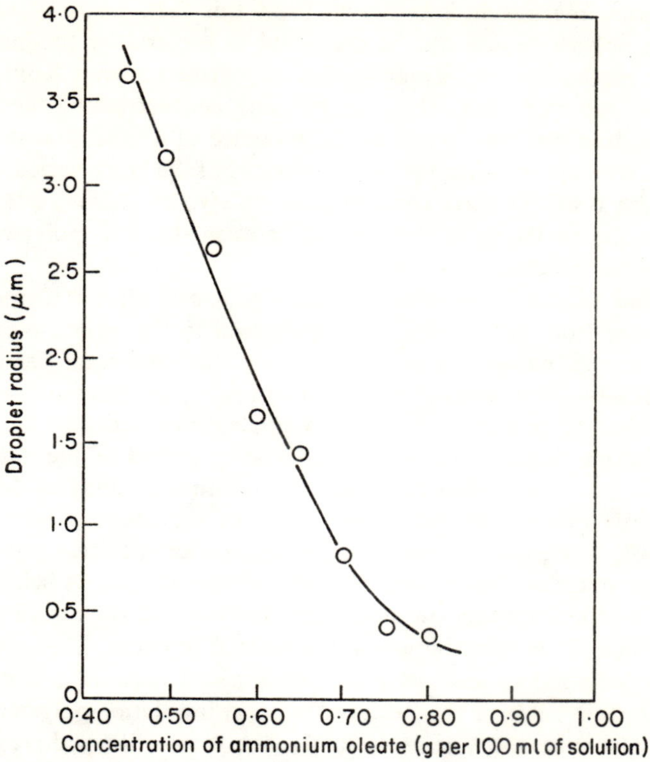

FIG. 3.7. Effect of soap level upon initial droplet size in a styrene emulsion polymerisation system, according to Siggia, Hohenstein and Mark.[9]

displayed by certain heavy-metal ions, and to the effects of modifiers. Further reference will be made to these matters in subsequent chapters under appropriate headings.

3.2.5 Contributions from Kolthoff and Dale[11] and from Price and Adams[12]

These papers, both of which appeared in 1945, are concerned with the mechanism and kinetics of the emulsion polymerisation of styrene. Kolthoff and Dale[11] showed that the rate of emulsion polymerisation of styrene at 30°C initiated by potassium persulphate in the absence of oxygen is approximately proportional to the square root of the initiator concentration. The obvious resemblance to the situation pertaining in bulk and solution polymerisations was noted. It is not clear, however, what exactly were the conclusions drawn from this fact, unless it was that emulsion polymerisation can be presumed to proceed by fundamentally the same mechanism as bulk and solution polymerisations. Kolthoff and Dale also reported that, at given initiator concentration, the rate of polymerisation was unaffected by the presence of a chain-transfer agent such as a mercaptan, when the concentrations of the latter varied over the range from 0 to 2·0 parts per 100 parts of styrene. Again, attention is drawn to the similarity in this respect between the bulk polymerisation and emulsion polymerisation of styrene.

Price and Adams[12] confirmed that rate of polymerisation at temperatures in the range from 40° to 60°C is proportional to the square root of the initiator concentration. Again, the initiator was potassium persulphate. Polymerisations were carried out under nitrogen; the reaction was found to be completely inhibited by air. It was also observed that the rate of polymerisation under these conditions is independent of the amount of styrene in the system, and that the concentration of initiator falls only very slightly throughout the whole course of the reaction from 0% to about 100% conversion. The zero-order dependence of emulsion polymerisation rate upon monomer concentration was interpreted as being due to limitation of rate by the low solubility of styrene in the aqueous phase within which the polymerisation was presumed to occur. This interpretation was preferred to any alternative view which required a mechanism essentially different from that believed to occur in solution polymerisation, where, of course, the rate is certainly not zero-order with respect to monomer concentration. The observation that, as with solution polymerisations, the rate is proportional to the square root of the initiator concentration was again taken as evidence that the polymerisation proceeds by essentially the same mechanism in emulsion as in solution.

Price and Adams also confirmed that the addition of dodecyl mercaptan to the polymerising system has little effect upon the rate of reaction, but does reduce both the molecular weight of the polymer formed and the induction period before the reaction starts.

It is of interest to note the view which Price and Adams take of the possible role of the soap: 'The purpose of the emulsifying agent might therefore be three-fold. It may serve to subdivide the monomer so that an extremely large surface area would be exposed to water to maintain equilibrium solubility. Secondly, it may considerably increase the equilibrium solubility of the monomer. Thirdly, it may act as a colloid-stabilizer, largely preventing the coalescence of the original monomer droplets with the polymer particles thus maintaining a supply of substantially pure

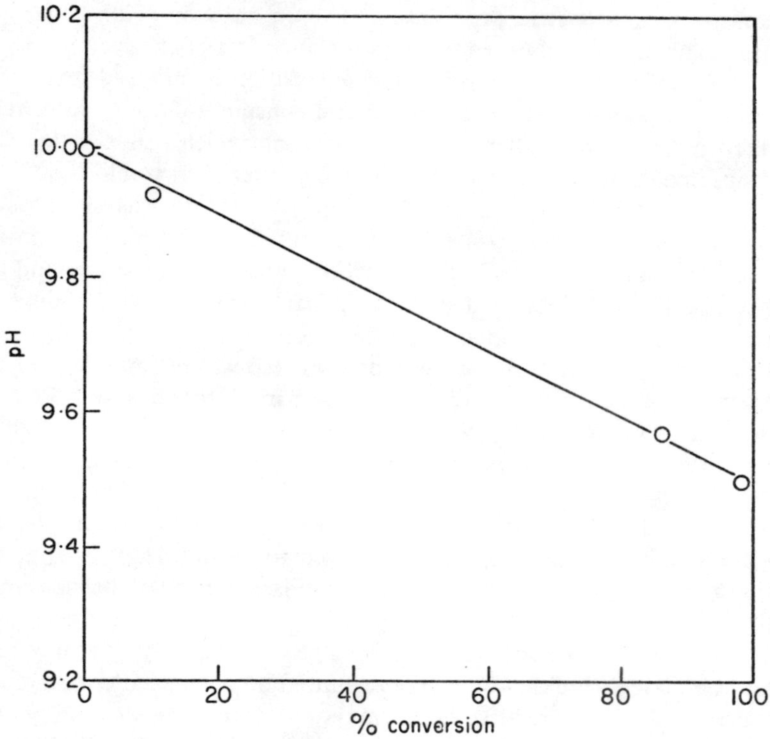

FIG. 3.8. Variation of pH with conversion during emulsion polymerisation of styrene initiated by potassium persulphate in presence of a potassium fatty-acid soap, according to Price and Adams.[12] (Exact details are not given for either the reaction system or the reaction conditions.)

styrene throughout a major portion of the polymerization.' All direct reference to the micelles or to their possible function as loci of reaction is absent, although an indirect reference is implied by the surmise that the equilibrium solubility of the monomer might be increased by the presence of a soap; this is clearly an allusion to the phenomenon of solubilisation.

Price and Adams also noted that the pH of the initial soap solution falls slightly during the course of the reaction. Their results are shown in Fig. 3.8. No discussion of this observation is given.

3.2.6 Work of Frilette and Hohenstein on the polymerisation of styrene in soap solutions[13]

In this investigation, the course of the polymerisation of styrene in a saturated solution of styrene in 2% potassium oleate was followed by means of a colorimetric procedure based on reaction with potassium permanganate. The published curves for the course of the reaction show that an initial period of inhibition and retardation is followed by a stage in which the polymerisation rate is fast and constant (although it must be pointed out that the scatter of the points upon which some of the conversion-time curves are based is very much greater than would have been desirable). Rather curiously, the rate of polymerisation was observed to fall rapidly at about 50% conversion to almost zero. No explanation for this feature is given. Although the induction period was decreased and the polymerisation rate increased by raising the temperature, the tailing-off in rate was always observed at about 50% conversion.

The energy of activation for the initiation step was estimated from the temperature dependence of the induction period to be about 17 kcal mole^{-1}. This was compared with a value of about 18 kcal mole^{-1} which had been reported earlier by Vinograd for the emulsion polymerisation of styrene.

Frilette and Hohenstein expressed the view that in emulsion systems, which contain monomer in both solubilised form and droplet form, the initiation reaction may take place in the solubilised portion of the monomer, and also possibly at the interface between droplet and aqueous phase. They comment that '...one is tempted to assume that the initiation may take place in both loci, and that both types of initiation may occur concurrently or successively. Their relative extent may well depend on the ability of a specific monomer to be solubilized in a certain system.' They further add the precaution that 'It may be desirable to reconsider the concept of a mechanism generally valid for all emulsion polymerization reactions, since it may prove necessary to distinguish between specific mechanisms pertaining to individual monomers and systems.'

3.2.7 Other contributions

Kolthoff and Dale,[14] continuing their studies of the emulsion polymerisation of styrene, published a paper in 1947 on the factors which affect the length of the induction period which is usually observed. They concluded that the induction period is caused by the presence of oxygen, and that it is (a) inversely proportional to the amount of potassium persulphate used as initiator, (b) proportional to the amount of oxygen initially present, provided that the oxygen pressure is maintained constant at its initial value, and (c) almost independent of the amount of soap in the charge. From experiments carried out at different initial oxygen pressures, it was further observed that, for a given amount of oxygen, the induction period is somewhat dependent upon the initial oxygen pressure, *i.e.* that the rate of oxygen disappearance is not quite independent of the oxygen partial pressure in the system. From the fact that the length of the oxygen induction period is nearly independent of the amount of soap, it was inferred that the principal locus of initiation is in the true water phase, and not in the soap micelles.

Further results presented by Bovey and Kolthoff[15] showed that the reaction of oxygen during the induction period is essentially zero-order with respect to the oxygen pressure, nearly independent of the soap concentration, and directly proportional to the initiator concentration. The oxygen consumed during the induction period was largely accounted for by two reactions: (a) by the formation of a polymeric styrene peroxide, and (b) by a minor reaction in which benzaldehyde and formaldehyde are produced. A copolymer was isolated having a composition corresponding to a 50:50 copolymer of oxygen and styrene. The main oxygen-consuming reaction was believed to be, in effect, a styrene–oxygen copolymerisation reaction. The rate at which this reaction occurred was about one-thousandth of the normal rate of emulsion polymerisation of styrene.

A paper with the title 'The role of emulsifying agents in polymerization', published by Staudinger[16] in 1948, is of some interest in the present connection for two reasons. In the first place, it enunciates clearly the major problems which have to be resolved in seeking a general understanding of the mechanism of emulsion polymerisation. The most probable locus for the initiation step is considered to be the true aqueous phase. Whilst some propagation could occur within the true aqueous phase, this is rejected as a principal locus because of the very low concentration of monomer which is present in many cases. The primary locus of propagation is considered to be the micelles, which, as the reaction proceeds, give way to what must now be described as polymer particles.

But the second point of interest is that Staudinger draws attention to a possibility to which occasional allusion will be found elsewhere in the early literature. Because the monomer which is consumed in the polymerisation reaction can only be replenished by diffusion in from outside the polymerisation locus, conditions of high polymer concentration at the polymerisation locus will prevail throughout the whole course of the reaction. The propagation reaction therefore proceeds in a homogeneous environment of high viscosity. Conditions are thus appropriate for the operation of the Trommsdorff 'gel' effect, *i.e.* suppression of bimolecular termination of growing radicals relative to their propagation, because termination requires the interaction of two macromolecular species whereas propagation requires the interaction of only one macromolecular species and one monomeric species. The consequence is an enhancement of both the degree of polymerisation of the product and of the rate at which it is formed; and, as has been observed in Chapter 1, the conjunction of these two characteristics is a distinguishing feature of emulsion polymerisation.

Staudinger expresses the view that the combination of high rate of polymerisation and high degree of polymerisation is mainly due to the operation of this 'gel' effect, and further states that the effect can only be operative in those cases, such as styrene and methyl methacrylate, where the polymer is soluble in, or swellable by, the monomer. On the other hand, instances are known of an acceleration of rate of polymerisation in bulk and solution systems, in which the polymer is insoluble in the reaction medium. This acceleration is associated with the separation of the polymer as an insoluble phase, and is usually explained in terms of reduced termination rate relative to propagation rate due to the partial immobilisation of macromolecular free radicals within the insoluble phase.

Whether or not the operation of the Trommsdorff 'gel' effect is the principal reason for the combination of high rate and degree of polymerisation which is so characteristic of polymerisations carried out in emulsion systems, it will become clear in the course of this book that this effect has frequently been invoked to account qualitatively for various features of emulsion polymerisation reactions.

3.3 THE HARKINS THEORY[17–20]

The most important qualitative theory of emulsion polymerisation which has appeared in the literature is undoubtedly that of Harkins. A brief outline of this theory is contained in two short notes[17,18] which appeared in 1945 and 1946, respectively. The theory is developed in much greater detail in two longer papers[19,20] which appeared in 1947 and 1950, respectively. The essential features of the theory are as follows:

(1) *The principal function of the monomer droplets is to act as a reservoir of monomer*, from which monomer molecules can diffuse into the aqueous phase, and from this into either soap micelles or latex particles containing monomer and polymer.

(2) *The principal loci of polymerisation are two in number.* The principal locus for the initiation of polymer particle nuclei is regarded as the extremely small amounts of solubilised monomer contained in the

micelles formed from the surface-active substance in the system. The principal locus for the formation of polymer is regarded as the polymer particles themselves. A polymer particle nucleus is envisaged as being formed by polymerisation of the monomer contained within a micelle. The polymer particle then tends to absorb monomer from the surrounding aqueous phase, because monomer has been consumed within the reaction loci by the polymerisation process. Further polymerisation occurs as monomer is absorbed, and, throughout this time, the polymer–monomer particle grows in size and soon exceeds the size of the micellar 'core' from which it was derived.

(3) *A small amount of particle initiation can occur within the true aqueous phase.* This locus is regarded as being responsible for the formation of almost all the polymer particle nuclei when no soap is present, but becomes progressively less significant as the amount of micellar soap in the system increases. The amount of polymerisation occurring in the monomer droplets is regarded as being a very minor proportion of the whole.

(4) *Growth of a polymer–monomer particle leads to an increase in its surface area.* It therefore tends to adsorb monomolecular soap from the true aqueous phase, and this in turn tends to lead to the dissociation of micelles containing monomer in which polymerisation has not yet started. Thus 'micellar' soap tends continuously to be changed into 'adsorbed' soap (*i.e.* soap which is adsorbed at the interface between a polymer–monomer particle and the aqueous phase) as the polymerisation proceeds. Although Harkins writes as though a sharp distinction can be made between, on the one hand, micelles in which polymerisation may be starting and, on the other, monomer-swollen polymer particles upon whose surfaces soap is adsorbed, it is clear that no such sharp distinction can really be made. In fact, the theory seems to picture a continuous transition from the micelle, containing as yet uninitiated monomer, right up to the final latex particle which, if the reaction has been taken to completion, contains no unreacted monomer. The minute polymer–monomer nuclei which form in the early stages of the reaction are probably best regarded as 'latent' latex particles.

(6) *Continual adsorption of micellar soap on to growing polymer–monomer particles eventually leads to the disappearance of micellar soap as such.* This stage is reached relatively early in the reaction (*e.g.* between 10% and 20% conversion).

(7) *Continual imbibition of monomer into growing polymer–monomer particles eventually leads to the disappearance of the monomer droplets as a separate phase.* This happens after the micellar soap has disappeared, and in consequence the system now contains only monomer-swollen polymer particles.

(8) Although not explicitly stated (Harkins' interest being chiefly in the role of the surface-active substances in emulsion polymerisation systems), it is assumed that *the primary initiation of the active polymerisation centres occurs within the aqueous phase.* The most reasonable view is that the centres are generated within the true aqueous phase, and then pass into the micelles and growing polymer–monomer particles by a process such as, for instance, random diffusion.

The main features of the Harkins theory are illustrated diagrammatically in Fig. 3.9. The separate parts of this figure relate to the various stages of the reaction system from before initiation up to complete conversion. At (a) is shown the system before initiation has occurred. The micelle-forming substance is present mainly as micelles; a small amount is adsorbed on the monomer droplets, and a very small amount is molecularly dissolved. Some micelles contain solubilised monomer; a few may be free of monomer. The monomer droplets are very large relative to the micelles. The surface free energy at the aqueous phase–air interface is low due to the adsorption of surface-active molecules. The monomer droplets may tend to coalesce if the system is not agitated. At (b) is shown the system in the initial stages of the reaction. Free radicals are generated in the aqueous phase and diffuse into micelles. In consequence, some polymer is now present within some of the micelles. Otherwise the system is as in (a). At stage (c), all the monomer-swollen micelles have disappeared, the amount of soap adsorbed on the monomer droplets has fallen, and so too has the concentration of molecularly-dissolved soap. The latent latex particles are growing in size at the expense of monomer droplets, which have become fewer in number and smaller in size. The surface free energy at the air interface has increased, the number of reaction loci is constant, and so too, therefore, is the rate of polymerisation. Monomer is continually diffusing through the aqueous phase from droplets to loci. At stage (d), the monomer droplets have disappeared, and the system comprises a dispersion of polymer particles containing some unreacted monomer. The rate of polymerisation gradually falls off due to monomer depletion at the reaction loci. Stage (e) represents the situation at the conclusion of the reaction. Polymerisation is complete. The system now comprises a dispersion of small polymer particles stabilised with the molecules of the original micelle-forming substance. The initiator free radicals are presumed to have been destroyed by the addition of a suitable inhibitor, and the initiating substance to have been completely decomposed. It should be noted that the diagrams given in Fig. 3.9 differ from those originally given by Harkins, in that the micelles are shown as spherical in shape whereas Harkins' diagrams show them as laminar. It is perhaps for this reason that Harkins made a sharper distinction than would probably be made by most today between, on the one hand, the micelle containing solubilised monomer and, on the other, the latent latex particle.

It is now necessary to review briefly the evidence upon which Harkins based his theory. It will be taken for granted that the solubilisation of low-molecular-weight hydrophobic molecules by aqueous solutions of surface-active substances is an established phenomenon, and that therefore some of the monomer in a typical emulsion polymerisation system is inevitably solubilised if there are micelles within the aqueous phase. A description and discussion of some of the facts pertaining to solubilisation will be deferred until Chapter 7, where the matter of surface-active substances in emulsion polymerisation reactions will be considered in greater detail.

That *negligible quantities of polymer form in the actual droplets of emulsified monomer* was established by experiments on isoprene or styrene polymerising in soap solution containing initiator. For example, polymerisation mixtures comprising 10 ml of isoprene, 15 ml of an aqueous solution containing 1% potassium laurate and 1% potassium persulphate, and 0·01 ml of lauryl mercaptan were placed in sealed tubes and shaken in a thermostat at 46°C for various lengths of time. The tubes were then centrifuged and the aqueous layer frozen in liquid nitrogen. The monomer layer was then decanted and the polymer content determined by evaporating off the monomer. The published results show that, in one case for instance, although 40% of the monomer had been converted to polymer, the ratio of polymer to monomer in the monomer phase was only 0·01.

That *the principal locus of the polymerisation is within the aqueous phase* is perhaps indicated most clearly by the results of measurements of the total volume of the apparent aqueous phase as a function of extent of polymerisation (by 'apparent aqueous phase' is meant that part of the system which excludes the bulk monomer phase). The results shown in Fig. 3.10 were obtained for the copolymerisation of styrene and isoprene.

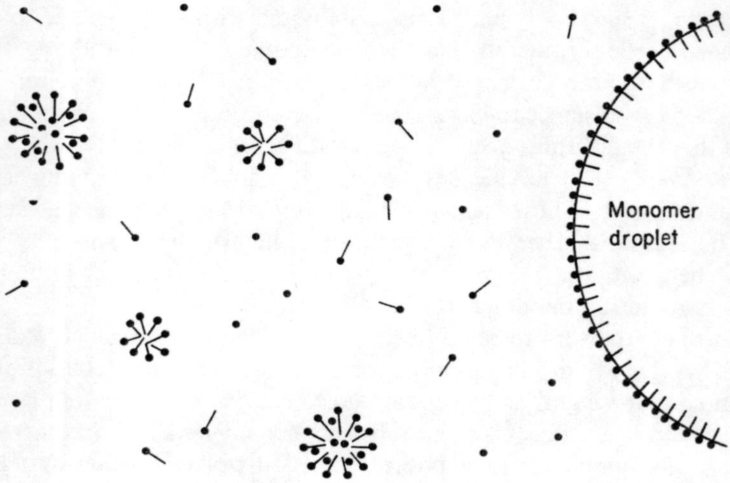

(a)

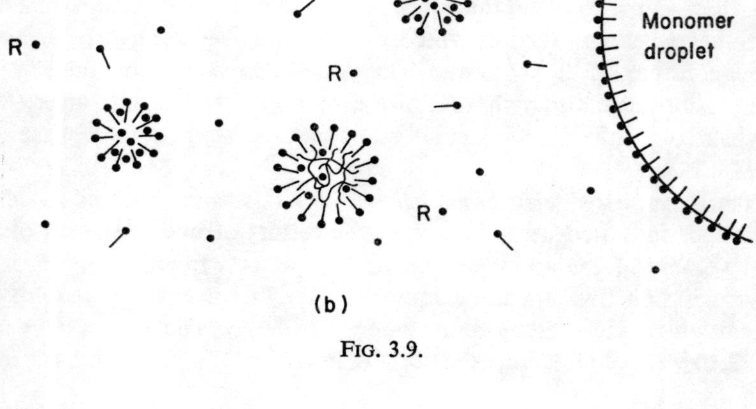

(b)

FIG. 3.9.

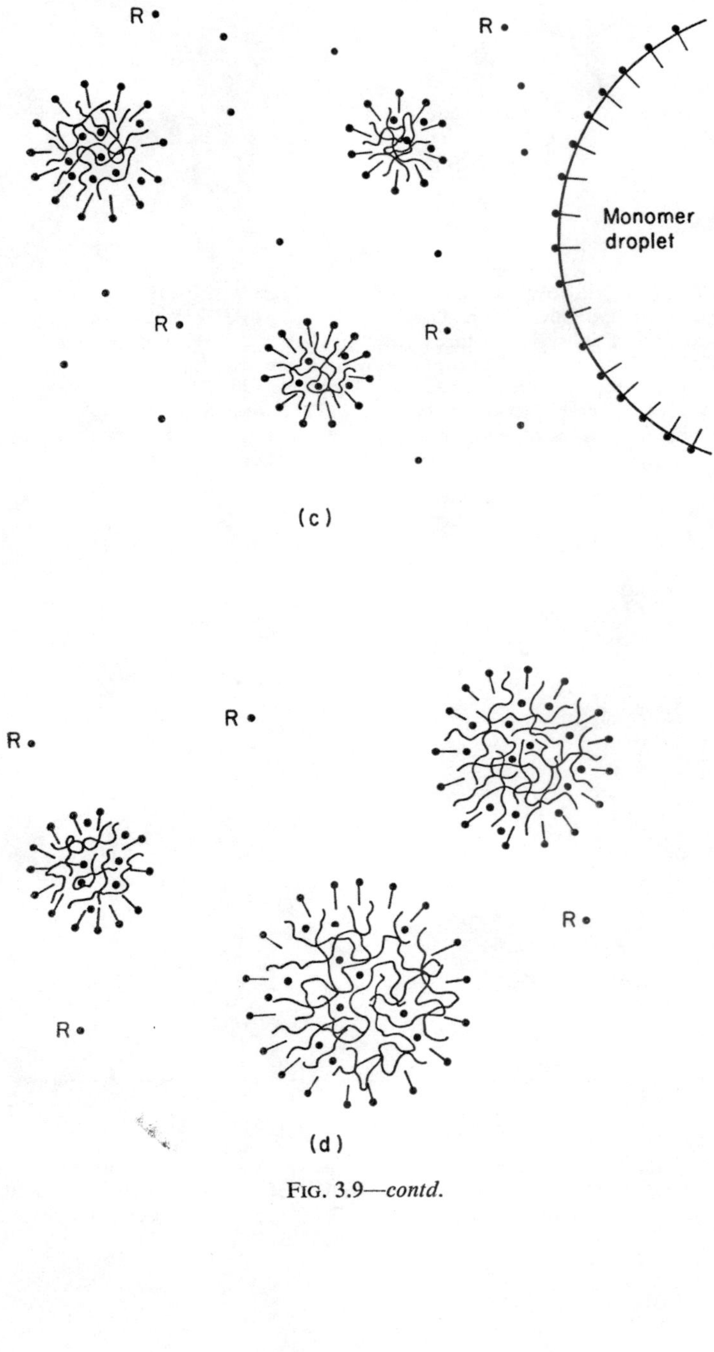

FIG. 3.9—*contd.*

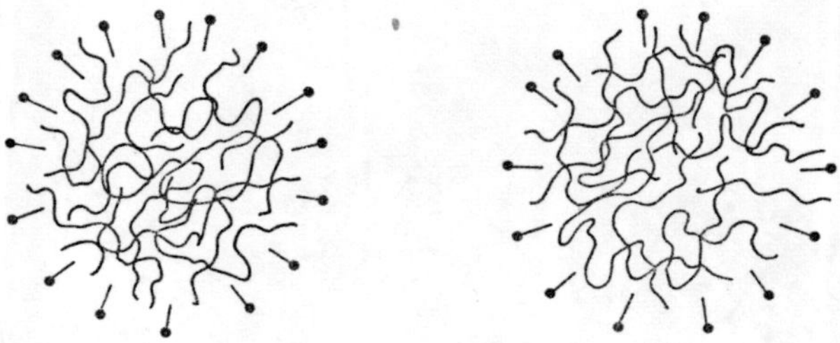

(e)

FIG. 3.9—*contd.* Illustrating the main features of the Harkins theory for the mechanism of emulsion polymerisation. The diagrams refer to different stages of the reaction as follows: (a) initial system before initiation; (b) initial stage of polymerisation, when micellar soap and monomer droplets are both present; (c) after micellar soap has disappeared, but monomer droplets are still present; (d) after monomer droplets have disappeared; (e) at completion of reaction. (—● denotes a molecule of surface-active substance; ● denotes a monomer molecule; ⁓ denotes a polymer molecule; R· denotes a free radical.)

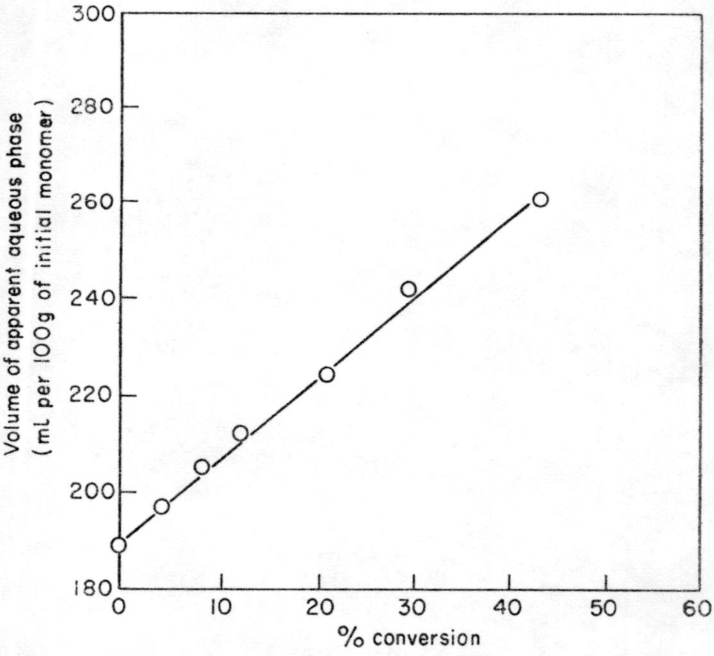

FIG. 3.10. Increase of volume of apparent aqueous phase with conversion during persulphate-initiated emulsion copolymerisation of styrene and isoprene, according to Harkins.[19] (Reaction system: styrene 75, isoprene 25, water 180, potassium myristate 3, $K_2S_2O_8$?)

The fact that the total volume of the apparent aqueous phase increases linearly with extent of conversion over the range for which results were given (0–50% conversion) is most obviously interpreted in terms of a reaction which is taking place within the aqueous phase, and which is drawing upon the monomer droplet phase for the replenishment of depleted reactants. Furthermore, agreement with the quantitative prediction of the theory seems to be reasonable. Thus the conversion of 40 g of monomer into polymer is accompanied by an increase in total apparent aqueous phase volume of about 90 ml. If it is assumed that the volume of the 40 g of polymer is about 45 ml, then this would mean that contained in the apparent aqueous phase is also approximately 45 ml of unreacted monomer. According to the Harkins theory, this is imbibed within the polymer particles.

The theory clearly requires that *an appreciable proportion of the soap content of the initial monomer emulsion should be in the form of micelles.* This raises again a question to which passing allusion has been made earlier, namely, in such an emulsion, what proportion of the total amount of surface-active substances present is adsorbed at the monomer–aqueous phase interfaces, and what proportion is contained within the aqueous phase itself? The results of Siggia, Hohenstein and Mark,[9] given in Table 3.3, are immediately relevant to this question. They show clearly that the proportion of soap adsorbed at the interfaces is always rather small, that the amounts of both free and adsorbed soap increase as the total amount of soap in the emulsion increases, and that the ratio of adsorbed soap to free soap increases as the total amount of soap increases. But a further question which must be answered is whether these results provide evidence that the aqueous phase of the emulsion contains sufficient soap for micelles to form, *i.e.* is the critical micelle concentration exceeded? Siggia, Hohenstein and Mark state that all their emulsions contained 180 ml of water. Their results therefore indicate that the concentrations of free soap, based on the volume of aqueous phase, ranged from about $1{\cdot}6 \times 10^{-2}$ to about $2{\cdot}2 \times 10^{-2}$ mole l^{-1}. The soap used was ammonium oleate, for which a critical micelle concentration of the order of 10^{-3} mole l^{-1} would be expected. Thus, even in the emulsions investigated by Siggia, Hohenstein and Mark, which contained rather less soap than conventional emulsion polymerisation systems (their emulsions contained from 0·4 to 0·8 g of ammonium oleate per 100 ml of aqueous phase), the critical micelle concentration was exceeded by a factor of more than 10. Thus the aqueous phase must have contained an abundant supply of micelles.

TABLE 3.3

Effect of increasing soap concentration in a styrene emulsion upon the droplet size produced, and upon the distribution of soap between the phases present (according to Siggia, Hohenstein and Mark[9])

Moles of soap	*Droplet radius* (μm)	*Number of droplets*	*Total surface area of droplets* (cm^2)	*Moles of adsorbed soap*	*Moles of free soap*
$3{\cdot}0 \times 10^{-3}$	3·6	$1{\cdot}1 \times 10^{11}$	$1{\cdot}5 \times 10^{5}$	$1{\cdot}0 \times 10^{-4}$	$2{\cdot}9 \times 10^{-3}$
$3{\cdot}3 \times 10^{-3}$	3·2	$1{\cdot}6 \times 10^{11}$	$1{\cdot}9 \times 10^{5}$	$1{\cdot}3 \times 10^{-4}$	$3{\cdot}2 \times 10^{-3}$
3.6×10^{-3}	2·6	$2{\cdot}8 \times 10^{11}$	$2{\cdot}3 \times 10^{5}$	$1{\cdot}5 \times 10^{-4}$	$3{\cdot}4 \times 10^{-3}$
4.0×10^{-3}	1·6	1.2×10^{12}	$3{\cdot}6 \times 10^{5}$	$2{\cdot}4 \times 10^{-4}$	$3{\cdot}8 \times 10^{-3}$
$4{\cdot}3 \times 10^{-3}$	1·4	$1{\cdot}8 \times 10^{12}$	$4{\cdot}3 \times 10^{5}$	$3{\cdot}0 \times 10^{-4}$	$4{\cdot}0 \times 10^{-3}$
$4{\cdot}6 \times 10^{-3}$	0·8	$1{\cdot}0 \times 10^{13}$	$7{\cdot}7 \times 10^{5}$	$5{\cdot}1 \times 10^{-4}$	$4{\cdot}1 \times 10^{-3}$
$5{\cdot}0 \times 10^{-3}$	0·4	$8{\cdot}0 \times 10^{13}$	$1{\cdot}5 \times 10^{6}$	$1{\cdot}0 \times 10^{-3}$	$4{\cdot}0 \times 10^{-3}$

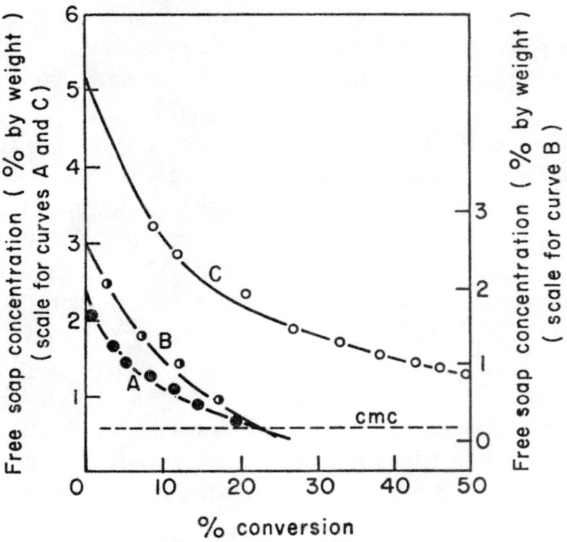

FIG. 3.11. Disappearance of free soap during the course of the persulphate-initiated emulsion polymerisation of styrene, according to Harkins.[19] Curves A and B are for reaction systems containing respectively 5 pphm of potassium laurate and potassium myristate; curve C is for a reaction system containing 10 pphm of potassium laurate. The broken line represents the critical micelle concentration (cmc) for both soaps, the soap concentration scales having been adjusted to allow for the difference in cmc between potassium laurate and potassium myristate.

Evidence for the *disappearance of micellar soap at a comparatively early* stage of the reaction comes from the observation that the surface free energy of the aqueous phase–air interface rises sharply at a point between about 10% and 20% conversion. Results illustrating this effect have already been given in Fig. 1.6 (p. 15). This observation lends support to the Harkins theory because its obvious interpretation is that the increasing area of interface presented by the growing polymer–monomer particles adsorbs increasing numbers of soap molecules, thereby leading to partial desorption from other interfaces in the system. These latter include the air–aqueous phase interface; desorption from here will be accompanied by a rise in surface free energy.

Direct evidence for the disappearance of micellar soap is provided by the results of Harkins,[19] which are reproduced in Fig. 3.11. It is seen that in systems containing about 5 pphm† of soap, the concentration of free

† Throughout this book, the abbreviation 'pphm' is used with the meaning 'parts per hundred parts of monomer', it being understood that all parts are by weight unless otherwise stated.

soap fell below the critical micelle concentration at about 23% conversion, whether the soap was potassium laurate or potassium myristate. It must therefore be presumed that micelles had disappeared from the aqueous phase by about this conversion. For systems containing 10 pphm of potassium laurate, the free soap concentration did not fall to the critical level even at the highest conversions investigated; it must therefore be presumed that micelles were present throughout. However, it is interesting to observe that increasing the amount of soap in the system caused the rate of consumption of free soap to be increased, even though the critical level was not attained. Thus at 5 pphm of potassium laurate, about 2·2% of free soap on the whole system had disappeared by 23% conversion, whereas at 10 pphm of the same soap the figure rose to about 3.2%. It should be noted that the results shown in Fig. 3.11 were obtained by a method which depends upon the disappearance of the fluorescence of certain dyes when micelles cease to be present. Harkin comments that 'The dye method used is not very accurate in the presence of latex but gives the general shape of the curves.'

According to the Harkins theory, *monomer droplets as a separate phase should disappear at some stage in the reaction.* Fig. 3.12 shows the results

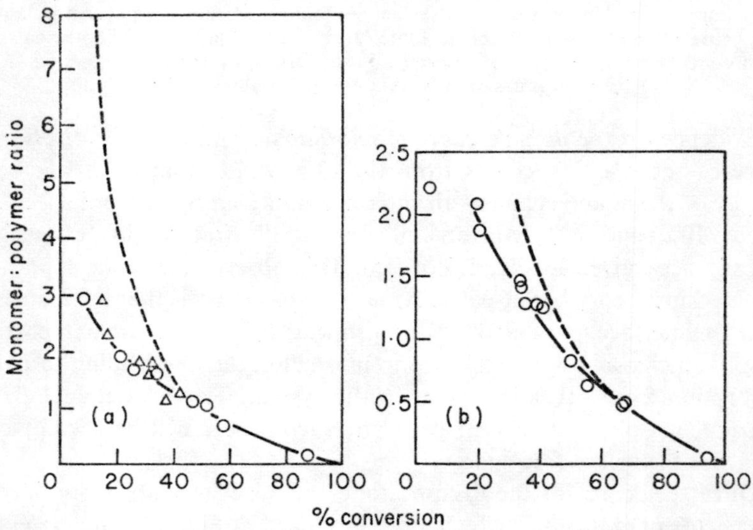

FIG. 3.12. Effect of conversion upon monomer:polymer ratio in reaction locus for emulsion polymerisation of styrene, according to Harkins.[19] Fig. 3.12(a) gives results obtained by a spectrophotometric procedure, whilst those in Fig. 3.12(b) were obtained by a gravimetric method. In both cases, the broken line gives the expected ratio if all the monomer were imbibed within the polymer particles.

of two attempts to investigate this matter for the case of styrene emulsion polymerisation. In each case, the monomer droplet phase was eliminated from the reacting system by centrifuging. Then, in one case, the monomer: polymer ratio was estimated by a spectrophotometric method, and in the other by a gravimetric method. In each case, a broken line is superimposed on the experimental results, indicating the overall monomer:polymer ratio as a function of conversion. Whilst the two sets of results do not correspond exactly, nevertheless they present the same general picture of (a) an initial stage of the reaction in which the monomer:polymer ratio at the polymerisation loci is very much less than that pertaining in the system as a whole, (b) an intermediate stage in which the overall ratio is falling much more rapidly than that in the polymerisation loci, and, finally, (c) a stage from 50–60% conversion where the ratio in the polymerisation loci is indistinguishable from that in the system as a whole. The final stage is interpreted as one from which monomer droplets as a separate phase are absent; the conversion corresponding to their disappearance is clearly of the order of 50%. The intermediate stage is interpreted as one in which monomer is diffusing into the loci faster than it is being polymerised, because the more polymer which is produced the more becomes available for imbibing unreacted monomer. Further experimental evidence concerning the variation of the monomer: polymer ratio during the emulsion polymerisation of styrene is given in a paper by Herzfeld *et al.*[30]

The next matter to which attention has to be turned is that of the *diffusion of monomer molecules from the emulsified droplets through the aqueous phase and into the polymerisation loci.* This process is of crucial importance because, as will be seen from results to be discussed when the phenomenon of solubilisation is considered in more detail (see Chapter 7), the monomer-containing micelles in the initial polymerisation system contain far fewer monomer molecules than do the eventual polymer molecules which are produced. There must therefore be replenishment of monomer which has reacted, if the micellar mechanism is to be credible. Harkins has reported results for the diffusion of ethyl benzene into a 12% solution of potassium laurate at 40°C. The method involved the use of a microscope to measure the variation with time of the diameter of small drops of the oil immersed in soap solution. The results are given in Fig. 3.13, from which it appears that the rate of decrease of diameter increases as the diameter of the droplet falls. Harkins also reported that increase in temperature favours an increase in rate of diffusion, the rate being about twice as high at 40°C as at 25°C. Harkins has also presented

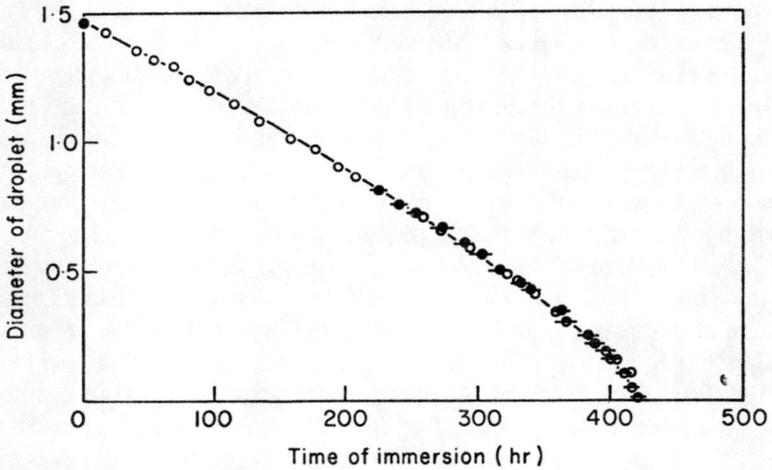

FIG. 3.13. Variation of oil-droplet diameter with time of immersion for diffusion of ethyl benzene from droplets into a 12% aqueous potassium laurate solution at 40°C, according to Harkins.[20]

evidence to show that the capture of monomer molecules by micelles may be a slow process, in the sense that even though the micelle is unsaturated by monomer, the probability of it taking up a molecule which approaches it is rather low.

One further matter to which consideration must be given in this discussion of the Harkins theory is his prediction that *the steady rate of polymerisation should be 'nearly' proportional to the initial concentration of soap.* His argument is that, if the critical micelle concentration is low relative to the initial soap concentration in the aqueous phase, then the number of micelles present in unit volume of solution should be proportional to the concentration of soap. Although not all the micelles will subsequently become polymer particles, it might be expected that a fixed proportion would, regardless of the initial soap concentration. If this were the case, then the number of polymerisation loci would be proportional to the soap concentration, and so therefore would be the steady rate of polymerisation. Factors which would perturb this relationship of proportionality could include (a) the critical micelle concentration becoming non-negligible as the level of soap is lowered, and (b) changes in the viscosity of the aqueous phase as the soap concentration changes, leading to changes in the rate of diffusion of monomer from droplets to reaction loci. The validity of this argument will be discussed in more detail when the Smith–Ewart theory is considered in the next chapter. It is sufficient to point out here that the weakness of the argument lies in the assumption that the number of eventual reaction loci is directly proportional to the number of micelles initially present in the aqueous phase.

The final matter to which reference must be made concerns the explanation which the Harkins theory gives for the combination of high rate of polymerisation and high degree of polymerisation in the product. Reduced to its simplest terms, it is that if, at any one time, a given reaction locus contains very few propagating centres, then the opportunities for termination relative to propagation are greatly reduced. The conventional kinetics of free-radical polymerisation show that, under these circumstances, the rate of polymerisation will be enhanced, and so too will the degree of polymerisation. If, as there is reason to believe, the majority of loci contain either one or no propagating centre, then an extreme situation will arise in which each individual propagation reaction is being carried out in its own reaction vessel in isolation from all the others which are proceeding concurrently. The majority of technical emulsion polymerisations probably proceed in this way, at least for part of the reaction. (Harkins himself says[19] that in order to extend his theory to take account of the free-radical nature of the polymerisation mechanism, it would be necessary to know the average number of propagating centres present in a single locus at any one instant. His own view is that, if this number is the same for all loci and droplets, regardless of the amount of monomer they contain, then it is probably small, being often less than unity.)

In effect, carrying out the polymerisation in emulsion provides a method whereby the overall concentration of propagating centres (and therefore the reaction rate) can be increased without incurring the penalty of short kinetic chain length. However, it has already been noted in Chapter 1 that degrees of polymerisation can be far higher than the limiting values predicted from the ratio of the rate coefficient for propagation to that for the reaction of transfer to monomer. It is most unlikely that this ratio would differ for an emulsion reaction from that for a bulk or solution reaction; furthermore, the elementary kinetic analysis shows that this limiting value should be unaffected by changes in monomer concentration, for the limit derives from the balance of two competing reactions, both of which are kinetically of the first order in monomer concentration. As pointed out in Chapter 1, the most likely explanation for the very high molecular weights given by emulsion polymerisation reactions is to be found in the high ratio of polymer to monomer existing at the propagation loci, leading to transfer-to-polymer reactions with consequent branching and possibly crosslinking. The Trommsdorff 'gel' effect may also contribute to the combination of high rate and high molecular weight, although the extent to which it is effective is open to dispute. This effect can only be operative if more than one active centre are propagating simultaneously at a given locus. As has been pointed out above, it is probable that this may not be the case in a typical emulsion polymerisation for some at least of its course.

3.4 QUALITATIVE EXTENSIONS OF THE HARKINS THEORY

The type of emulsion polymerisation for which the Harkins theory purports to give a mechanistic picture is sometimes referred to as an 'ideal' emulsion polymerisation system. The monomers are essentially insoluble in water, and the polymer is miscible with the monomer in all proportions. It so happens that the emulsion polymerisations of styrene and of butadiene fulfil both these conditions, as also does the emulsion copolymerisation of the two at all monomer–feed ratios. Since the intense interest in emulsion polymerisation which was aroused during World War II was concerned mainly with these two monomers, it is hardly surprising that this type of system has tended to be regarded as the ideal from which all others deviate to a lesser or greater extent.

It is now desirable to take note of the modifications to the Harkins theory which become necessary when one or other or both of the above simplifying conditions are not fulfilled. The following remarks are largely based upon the excellent summary which has been given by Duck.[21]

If a monomer is appreciably soluble in water, then the true aqueous phase can become an important locus for the propagation reaction, at least in the early stages of the reaction. In effect, the polymerisation commences as a solution polymerisation in water. The tendency of the surface-active substances in the system to form micelles becomes correspondingly less important, and their effect as colloid stabilisers becomes more important. This follows because the oligomers formed in the initial stages of the polymerisation will tend to precipitate out of the aqueous solution and form potential polymerisation loci if the precipitated oligomers can be sufficiently stabilised against massive coagulation.

Other complications occur if the polymer is insoluble in its own monomer, as is the case of, for instance, acrylonitrile, vinyl chloride and vinylidene chloride. The reaction commences in the way described by the Harkins theory, but polymer soon begins to precipitate as a separate phase within the latent latex particles. The polymerisation will now probably proceed upon the surface of the insoluble polymer on which the monomer is adsorbed. Some polymerisation will also occur within the body of the particles if the monomer is partially miscible with the polymer. The micelles will disappear in the usual way as the particles grow, but the monomer cannot be taken up by the particles in great quantity. Because the specific interfacial area within the system increases rapidly as polymerisation proceeds, and because more unreacted monomer remains as droplets than in the ideal case, the monomer droplets tend to become unstable and to coalesce to form larger particles. The reduction in monomer surface area is then accompanied by a drop in rate of diffusion of monomer into the aqueous phase, and so the polymerisation rate falls. The period of particle formation may therefore be followed by a period of retardation. As the particles continue to grow, albeit more slowly than in an equivalent ideal case, collisions with monomer droplets become more frequent, and this may overcome to some extent the monomer deficiencies caused by reduced rate of diffusion. If this is the case, the rate of polymerisation may increase steadily. If these conditions prevail, agitation can have a great effect upon the rate of polymerisation.

3.5 STUDIES OF BAXENDALE AND CO-WORKERS ON THE EMULSION POLYMERISATION OF METHYL METHACRYLATE[22–26]

It is now necessary to refer to an early study of the polymerisation of methyl methacrylate both in aqueous solution and in emulsion, which was carried out by Baxendale and his co-workers. The results of this study were published in 1946 at about the same time as the two brief initial outlines of the Harkins theory appeared in the literature. These results provide further experimental support for the essential correctness of that theory.

Initiation of polymerisation was achieved using the redox couple iron(II)–hydrogen peroxide. In the case of reactions occurring in aqueous solution, it was found that the addition of a surface-active substance was accompanied by a considerable increase in both the rate and the extent of the polymerisation. Results for the addition of cetyltrimethylammonium bromide are given in Fig. 3.14. This observation was explained by the suppression of the mutual termination reaction through discouragement of the precipitation of growing polymer molecules from the aqueous

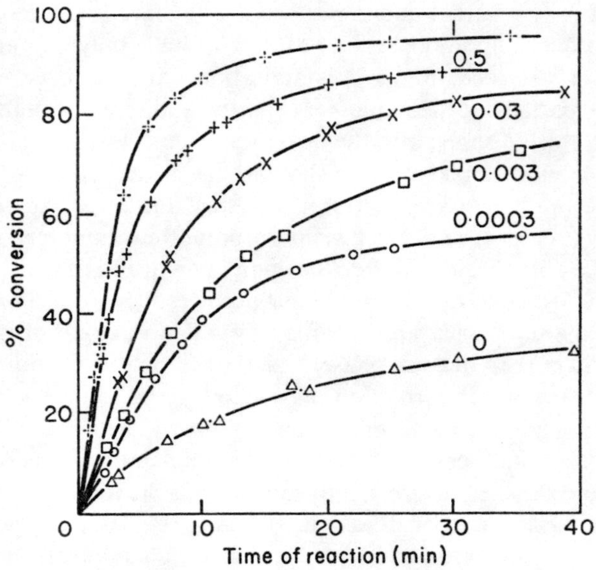

FIG. 3.14. Effect of addition of cetyltrimethylammonium bromide upon conversion–time curve for polymerisation of methyl metacrylate in aqueous solution at 25°C initiated by hydrogen peroxide–iron(II) redox couple, according to Baxendale, Evans and Kilham.[23] (The initial concentration of methyl methacrylate was 0.1 mole l^{-1}. The numbers appended to the curves give the concentrations of cetyltrimethylammonium bromide in %.)

solution. It was argued that, as the polymer molecules grow, they will tend to become insoluble. This will therefore give added impetus to the termination reaction, since it will tend to bring together polymer chains which are still active, and so encourage them to deactivate one another by mutual interaction. The function of the added surface-active substance was therefore seen to be essentially that of a protective colloid, and its effect as being to reduce the apparent value for k_t, the rate of coefficient for the mutual termination reaction. Apart from the complicating influence of added surface-active substance, the reaction mechanism proposed for the polymerisation is quite conventional. In accordance with the conclusions drawn from the proposed kinetic analysis of this mechanism, the ratio $k_p/\sqrt{(k_1 k_t)}$ appeared to be constant over the greater part of the reaction at any given concentration of surface-active substance (k_p is the rate of coefficient for the propagation step, and k_1 the rate coefficient for the reaction between iron(II) and hydrogen peroxide). However, the value of this ratio was observed to increase markedly with the addition

TABLE 3.4

Effect of cetyltrimethylammonium bromide upon ratio $k_p/\sqrt{(k_1k_t)}$ for polymerisation of methyl methacrylate in 0·1 mole l^{-1} aqueous solution at 25°C (according to Baxendale, Evans and Kilham[23])

% Cetyltrimethylammonium bromide	$k_p/\sqrt{(k_1k_t)}$
0	0·153
0·0003	0·41
0·003	0·53
0·03	0·83
0·5	0·98
1·0	1·72
3·0	1·22

of cetyltrimethylammonium bromide (see Table 3.4), and this increase was interpreted as due to a reduction in the effective value of k_t.

Emulsion polymerisations were carried out using cetyltrimethylammonium bromide as the micelle generator. Reactions were carried out over a wide range of monomer-aqueous phase volume ratio. In no case was there any evidence of precipitation of the polymer; the final mixtures appeared to be quite homogeneous. The effect which varying the amount of monomer had upon the conversion–time curve is illustrated in Figure 3.15. It was considered that, under the conditions of agitation and emulsifier concentration used in these experiments, there was approximate proportionality between the total volume of non-aqueous phase and the total interfacial area created between the two phases. It therefore appears from the results of Fig. 3.15 that variation in the amount of non-aqueous phase, and thus of interface, produced a family of curves which branch off from a single stem, *i.e.* the initial rate of polymerisation was independent of the amount of non-aqueous phase and of interfacial area. Furthermore, analysis of the results on the basis of a reasonable reaction mechanism showed that the ratio $k_p/\sqrt{(k_1k_t)}$ (see above for the definitions of the symbols) was not constant throughout the course of a given reaction, but increased in a manner consistent with a large increase in viscosity occurring at the polymerisation loci.

The conclusion reached regarding the mechanism of the emulsion polymerisation was that initiation occurs in the aqueous solution, and that the growing polymer chains readily absorb monomer from the monomer droplet phase. The principal locus for the propagation step was regarded as being a 'new phase of polymer–monomer solution',

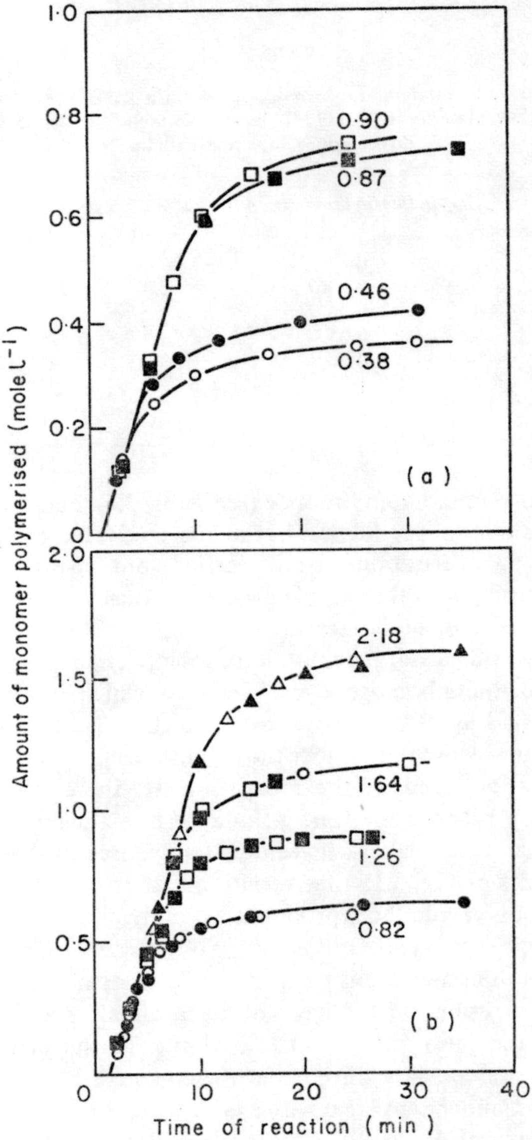

FIG. 3.15. Effect of variation of amount of monomer upon conversion–time curve for emulsion polymerisation of methyl methacrylate at 25°C initiated by hydrogen peroxide-iron(II) redox couple in the presence of cetyltrimethylammonium bromide, according to Baxendale, Evans and Kilham.[26] (The numbers appended to the curves gives the monomer concentrations in mole l^{-1}. The curves of Fig. 3.15(a) are for reaction systems which contained 1% of cetyltrimethylammonium bromide; those of Fig. 3.15(b) are for systems for which the concentration was 3%. The initial concentrations of H_2O_2 and $FeSO_4$ were both $9{\cdot}63 \times 10^{-5}$ mole l^{-1} for (a) and $9{\cdot}48 \times 10^{-5}$ mole l^{-1} for (b).)

which is elsewhere described as comprising 'highly swollen micelles with dead polymer molecules and monomer'. Monomer from the droplets was regarded as being readily available to these propagating centres, so that propagation is not impeded, but the monomer droplet phase rapidly disappears. Termination is suppressed because of the high viscosity at the propagation locus, and therefore both the rate of polymerisation and the molecular weight of the product are enhanced (*i.e.* the Trommsdorff 'gel' effect is operative). The high molecular weight of the polymer produced leads to an even greater viscosity at the reaction locus, implying that the termination reaction will be further suppressed relative to propagation. Baxendale, Evans and Kilham[26] have provided evidence to show that the intrinsic viscosity of the polymethyl methacrylate produced by emulsion polymerisation rises as the reaction proceeds, and that it is always much higher than that of polymer obtained by polymerisation in aqueous solutions. Their results on this point are shown in Fig. 3.16.

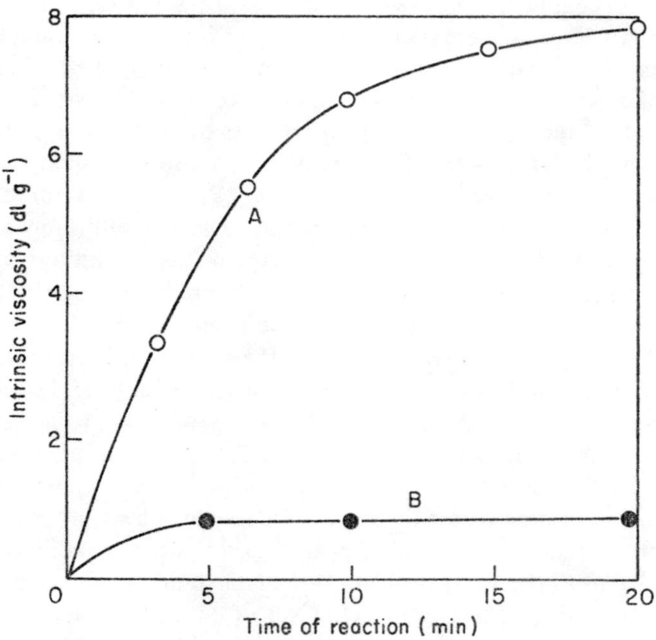

FIG. 3.16. Effect of reaction time upon intrinsic viscosity of polymethyl methacrylates produced by emulsion polymerisation (A) and by aqueous solution polymerisation (B), according to Baxendale, Evans and Kilham.[26] (Details of reaction systems are not given.)

3.6 THE MONTROLL 'PHENOMENOLOGICAL' THEORY[27]

This theory, which appeared in 1945, is now only of academic interest because the mechanistic model upon which it is based is now generally regarded as defective. It is, however, of interest in being one of the earliest attempts to give a quantitative account of the phenomenon of emulsion polymerisation. The concern is exclusively with the polymerisation of styrene.

The theory is based upon a model which regards the principal locus of the reaction as the interface between monomer droplets and aqueous phase. The argument is as follows: 'Since the catalysts used in styrene polymerization are generally inorganic salts insoluble in styrene, it is highly improbable that the reaction could occur inside the styrene globules. If the reaction zone were in the aqueous phase, the rate should be independent of the total interfacial surface, which does not seem to be the case.' (The allusion here is to the fact that the reaction rate increases as the soap level is increased, and that the increased amount of soap results in the formation of a monomer emulsion of smaller droplet size.) 'When the reaction zone is at the water–styrene interface, the change in volume of a styrene globule is proportional to the surface area of the globule, or $dV/dt = -4\pi kr^2$, where k is the proportionality constant. Since the sphere volume is given by $V = 4\pi r^3/3$, this leads to $dr/dt = -k$, and the radius diminishes at a constant rate. This law of disappearance of monomer was observed (microscopically) in soap solutions of many organic liquids, including styrene, by Vinograd, Fong and Sawyer.' Harkins[19,20] has drawn attention to these results reported by Vinograd, Fong and Sawyer, and says that, using their method, he found that the rate of decrease of droplet radius is *not* constant with time, but increases progressively. Harkins' results have already been noted in Fig. 3.13.

Attention is confined to monomer emulsions whose initial distribution of droplet size is such that the distribution of radii has a single maximum. The state of agitation within the system is assumed to be constant, and to be sufficient to maintain a unimodal distribution of droplet radii at all times.

For Montroll, the basic problem, the understanding of which provides the key to understanding the mechanism of emulsion polymerisation, is the apparent inhibition or retardation period which the styrene emulsion polymerisation displays, together with its apparent dependence upon the droplet size of the initial monomer emulsion. The results of Siggia, Hohenstein and Mark,[9] given in Fig. 3.5, illustrate this effect. This initial retardation is assumed to be due to the presence of an inhibitor which is distributed between the monomer and the aqueous phase. The inhibitor dissolved in the aqueous phase is progressively destroyed by interaction with the initiator or its decomposition products, and inhibitor then diffuses from the monomer droplets into the aqueous phase in an endeavour to restore the partition equilibrium. It is then assumed that there is a critical concentration of inhibitor in the monomer droplets, below which the monomer polymerises at a normal rate. It is shown by Montroll that, theoretically, diffusion from a large droplet is slower than from a small droplet. Therefore the small droplets in the emulsion start to polymerise before the larger ones, and emulsions of small average droplet radius should polymerise faster than those of larger radius, provided both have the same initial inhibitor concentration. This is in accordance with what is actually observed.

The theory also accounts for the gradual acceleration of the reaction to a steady state during its initial stage. For, since the initial emulsion has a distribution of droplet sizes, it is clear that the number in which the inhibitor concentration has dropped below the critical value increases progressively with time, and therefore so also does the amount of monomer which becomes available for polymerisation. Furthermore, since there are initially relatively few very small droplets, the extent of polymerisation changes very slowly at first. As time goes on, it is the larger droplets which become available for polymerisation. Thus not only does more monomer become available, but it becomes available at an increasing rate. It is also suggested that, as the small droplets polymerise, so they give up their soap, thus increasing the emulsifier concentration and reducing the sizes of the larger droplets, causing polymerisation to commence in them sooner than it would otherwise. The summation of these effects gives a reaction rate which gradually increases until all the droplets are reactive, after which time there can be no further increase in rate.

The mathematical development of this theory is complex. It would be inappropriate to give here other than the barest outline, since the theory is now considered to be founded upon an untenable reaction mechanism. The reader who desires further details is referred to the original paper.

Montroll's first step is to apply classical diffusion theory to a spherical droplet suspended in a medium in which it is insoluble, and containing a solute which partitions between droplet and suspension medium. Furthermore, the substance is supposed to be consumed from the medium instantaneously. It is then shown that the time, t^*, required for the solute concentration in a droplet of radius R to fall to a critical value c^* is approximately given by

$$t^* = \frac{R^2}{\pi^2 D} \log_e \left(\frac{6c_0}{\pi c^*}\right)$$

where c_0 is the initial concentration in the droplet, and D is the diffusion constant for the solute in the medium. (It should be noted that the equation for t^* given in Montroll's original paper appears to be in error.) This equation implies that, if c^* is the critical inhibitor concentration below which polymerisation can occur, then after time t has elapsed all droplets whose radii are less than

$$R \simeq \sqrt{\left[\frac{\pi^2 D t}{\log_e (6c_0/\pi c^*)}\right]}$$

will be polymerising.

The next matter considered is the effect of soap concentration upon the distribution of droplet sizes. On the basis of experimental results, the mean radius is taken as varying linearly with soap concentration, and the standard deviation as independent of soap concentration. Various functions describing the distribution of droplet radii are assumed, namely, a Gaussian distribution and two rectangular distributions which approximate to it. On the basis of experimental evidence, it is considered that the actual distributions are unimodal and that therefore so must be the model distributions used for the purposes of calculation.

The results of calculations are presented for a variety of conditions (droplet size distribution, average droplet size, inhibitor present or not, etc.), and remarkable similarity with experimental observation is claimed for the predicted progress of polymerisation when the droplets contain inhibitor. Figure 3.17 shows curves predicted for two types of initial droplet radius distribution and three average droplet sizes. Besides values for mean droplet radius and standard deviation for the distribution of radius, it is also necessary to assume a value for a parameter which is a measure of the time required for inhibitor to diffuse out of a monomer droplet of unit radius. Thus, in effect, the equations predicting conversion as a function of time contain one adjustable parameter.

The two types of initial droplet radius distribution assumed for the curves of Fig. 3.17 were a single rectangular distribution of the same mean and standard deviation as that of the Gaussian distribution to which it is taken as an approximation, and three adjacent rectangular distributions. The full lines in Fig. 3.17 refer to the single rectangular distribution, and the broken lines to the threefold rectangular distribution. It is seen that,

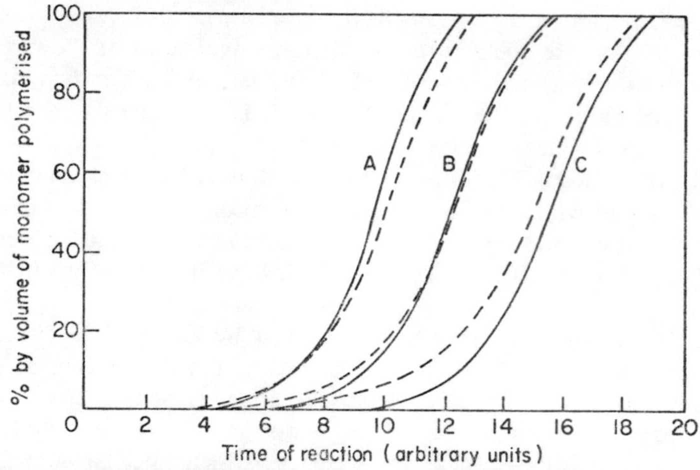

FIG. 3.17. Conversion–time curves for an emulsion polymerisation reaction as predicted by the Montroll theory.[27] The full curves are for a simple rectangular distribution of initial droplet sizes, whilst the broken curves are for an initial distribution comprising three adjacent rectangular distributions. The three pairs of curves A, B and C refer respectively to initial mean droplet radii of 3 μm, 3·5 μm and 4 μm.

except at very short reaction times, the predicted conversion–time curve is little affected by the assumption made regarding the initial distribution of droplet sizes. But clearly, the mean droplet size has a profound effect upon the course of the reaction, and, furthermore, the nature of this effect accords very well with that which is apparently observed in practice. Montroll states that the standard deviation of the distribution of droplet sizes also has a greater effect upon the conversion–time curve than does the nature of the distribution. He also presents curves which show the effect of varying the parameter which characterises the time required for inhibitor to diffuse out of a droplet of unit radius. As would be expected, the choice of value for this parameter has a considerable effect upon the placing of the conversion–time curve along the time axis, although not upon its general shape. A larger value for this parameter displaces the curve in the direction of longer times.

3.7 OTHER EARLY QUANTITATIVE THEORIES

It is one thing to have a general understanding of the mechanism of a complex reaction such as emulsion polymerisation; it is quite another matter to be able to predict, for instance, the actual rate of reaction and its dependence upon the various formulation variables from a knowledge of, say, (a) the rate coefficients for the individual steps of the monomer polymerisation reaction as revealed by studies of bulk polymerisation, and (b) the nature of the aqueous phase. It is this aspect of the subject which is the primary concern of the Smith–Ewart theory,[1] to which attention is turned in the following chapter. It is desirable to conclude this present chapter with reference to two theories for the kinetics of emulsion polymerisation which appeared in the literature at about the same time as did the Smith–Ewart theory, but which have been largely superseded by it.

The first of these theories was published by Corrin[28] in 1947. It is concerned only with that part of the reaction which occurs after free micellar soap has disappeared from the aqueous phase. Corrin makes the comment that 'sufficient data have not as yet been accumulated to allow a theoretical treatment for yields at which micellar soap is still present.' Two stages of the reaction are considered: (a) that in which a separate monomer phase still exists although micellar soap does not, and (b) that in which a separate monomer phase is no longer present. The second is dealt with first, on the basis that monomer will polymerise in the separate reaction loci according to a law which holds for certain bulk polymerisations initiated by the decomposition of added initiators. This law is $-\mathrm{d}c/\mathrm{d}t = \mathrm{k}c^{3/2}$, where c is the concentration of monomer at time t, and k is a constant. (This equation can be justified theoretically if both initiation and propagation are first-order with respect to monomer; the joint dependence of $\mathrm{d}c/\mathrm{d}t$ upon c is then of the form of the product of $c^{1/2}$ and c.) Integration of this equation between time t_0, when the separate monomer phase has just disappeared, and a subsequent time t gives

$$\frac{1}{\sqrt{(c)}} - \frac{1}{\sqrt{(c_0)}} = \frac{\mathrm{k}}{2}(t - t_0)$$

where c_0 is the monomer concentration in the reaction loci at time t_0, and $c(<c_0)$ is the value at time t. If c is expressed as grams of monomer dissolved per gram of polymer, then, when the percentage conversion is Y, we must have that $c = (100 - Y)/Y$, since no separate monomer phase is present. Similarly, $c_0 = (100 - Y_0)/Y_0$, where Y_0 is the conversion at which the monomer droplet phase first disappeared. Substitution of these values in the above equation gives, with straightforward rearrangement,

$$t = t_0 + \frac{2}{\mathrm{k}}\left\{\sqrt{\left(\frac{Y}{100 - Y}\right)} - \sqrt{\left(\frac{Y_0}{100 - Y_0}\right)}\right\}$$

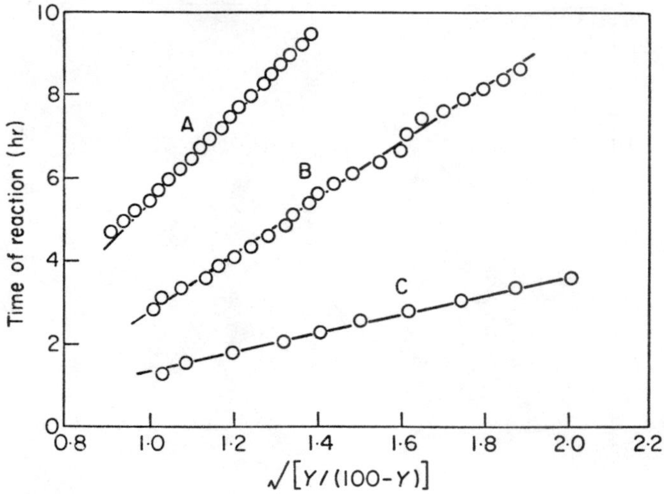

FIG. 3.18. Relationship between t and $\sqrt{[Y/(100 - Y)]}$ for persulphate-initiated emulsion polymerisation of styrene at three different temperatures, according to Corrin.[28] (Reaction system comprised styrene polymerising in 2·8% sodium oleate solution in presence of potassium persulphate and 0·5 pphm of dodecyl mercaptan. Curves A, B and C refer to polymerisations at 35°C, 42·5°C and 50°C respectively.)

This equation shows that, after the monomer droplets have disappeared, a graph of t against $\sqrt{[Y/100 - Y)]}$ should be linear. Furthermore, if linear, k can be estimated from the slope. That such a plot can be linear for the emulsion polymerisation of styrene is shown in Fig. 3.18, which is taken from Corrin's paper.

For the stage of the reaction between the disappearance of the soap micelles and the disappearance of the monomer droplets, Corrin states that the rate of conversion of monomer to polymer (by which he presumably means $\mathrm{d}Y/\mathrm{d}t$) is equal to $\mathrm{k}(m/Y)^{3/2}Y$, if the law $-\mathrm{d}c/\mathrm{d}t = \mathrm{k}c^{3/2}$ is obeyed at the reaction loci. Here m is the number of grams of monomer dissolved in the Y grams of polymer which have so far formed from 100 grams of monomer present initially. Thus if, as is indeed observed throughout this stage of the reaction, $\mathrm{d}Y/\mathrm{d}t$ is constant, this implies that $(m/Y)^{3/2}Y$ is also constant. But m/Y is the monomer:polymer ratio, P, for the reaction system excluding the monomer droplets. Thus if $\mathrm{d}Y/\mathrm{d}t$ is constant, the relationship between $P^{3/2}$ and $1/Y$ should have the form of a straight line of slope $(\mathrm{d}Y/\mathrm{d}t)/\mathrm{k}$ which passes through the origin. Corrin tested this prediction, taking the value of k as that estimated from the

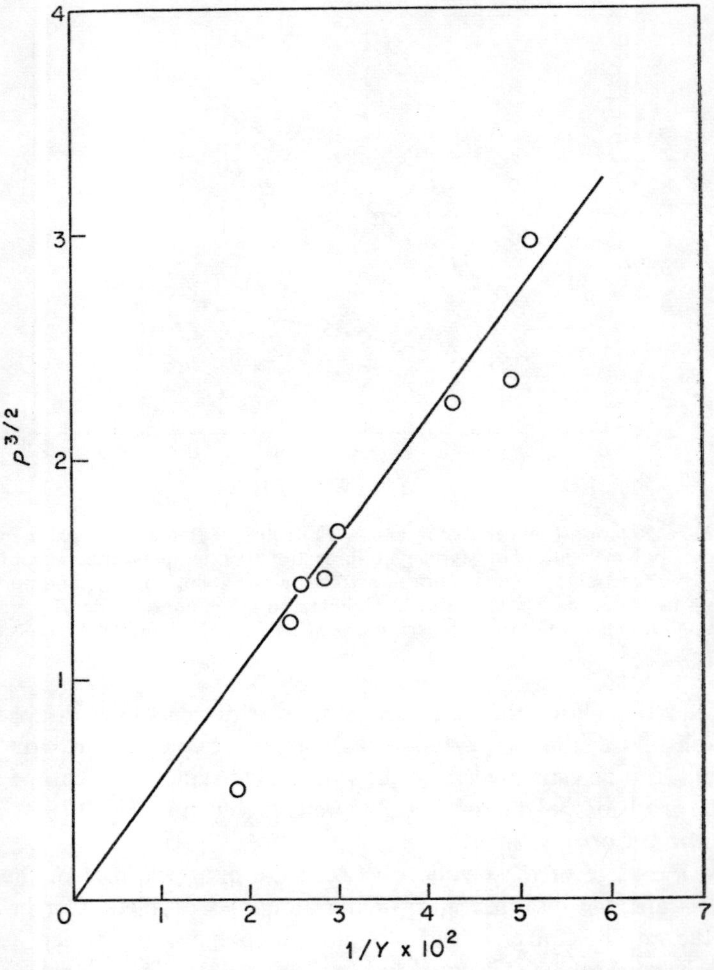

FIG. 3.19. Observed and predicted variation of monomer:polymer ratio (P) with % conversion (Y) for persulphate-initiated emulsion polymerisation of styrene at 42·5°C during stage between disappearance of micelles and disappearance of monomer droplets, according to Corrin.[28] (See text for further details.)

results for the later stage of the reaction when the monomer droplets had disappeared. Corrin's plot is reproduced in Fig. 3.19. The agreement between prediction and observation is satisfactory, considering that the prediction line was based upon data obtained quite independently of those represented by the points.

What is not, however, entirely clear is exactly why Corrin supposed that $\mathrm{d}Y/\mathrm{d}t$ is equal to $k(m/Y)^{3/2}Y$ in the region where monomer droplets are present. Corrin gives no details of his argument. That $\mathrm{d}Y/\mathrm{d}t$ is in fact equal to $k(m/Y)^{3/2}Y$ can be demonstrated by the following argument. $\mathrm{d}c/\mathrm{d}t$ is clearly equal to $-k(m/Y)^{3/2}$ if, as has been supposed above, c is expressed as grams of monomer dissolved in one gram of polymer. (In the case where monomer is present as a separate phase, it must, of course, be understood that c refers to the reaction loci only.) Thus $k(m/Y)^{3/2}$ is equal to the rate of conversion of monomer to polymer, expressed as grams of monomer polymerised in unit time per gram of polymer which has already formed. Consider now a reaction system which initially contained 100 grams of monomer. When the percentage conversion is Y ($< Y_0$), such a system will contain Y grams of polymer, and thus the total rate of conversion of monomer to polymer at this stage is $k(m/Y)^{3/2}Y$ grams per unit time. Since the system contained initially 100 grams of monomer, it follows that $k(m/Y)^{3/2}Y$ is the rate of change with time of the percent conversion at the instant in question, *i.e.* $\mathrm{d}Y/\mathrm{d}t$. It is also of interest to note that $c = (100 - M - Y)/Y$ when $Y < Y_0$, where M is the percentage of the initial monomer which is present as droplets. Thus $cY = 100 - M - Y$. Differentiation with respect to time, followed by rearrangement and substitution of $c = m/Y$ and $\mathrm{d}c/\mathrm{d}t = -k(m/Y)^{3/2}$, gives

$$\frac{\mathrm{d}Y}{\mathrm{d}t} = \frac{kY(m/Y)^{3/2}}{1 + m/Y} - \frac{1}{1 + m/Y}\frac{\mathrm{d}M}{\mathrm{d}t}$$

This expression does not appear to be reducible to that given by Corrin, even if one notes that always $Y + m + M = 100$.

The second of the theories to which reference must be made is that of Haward,[29] which was published in 1949. His paper discusses the effect which compartmentalisation can have upon the kinetics of a simple free-radical polymerisation. Haward first endeavours to estimate the diameter of what he calls the 'critical drop' for a bulk free-radical polymerisation reaction. By the term 'critical drop' he means that spherical volume of monomer which contains on average one propagating centre. Using published values of rate coefficients for various common monomers, he concludes that these diameters are generally within the range 0·1–1 μm, *i.e.* the 'critical drop' is of 'colloidal' dimensions. It is therefore to be expected that compartmentalisation into individual reaction loci of this size, as is believed to occur in an emulsion polymerisation reaction, would have an effect upon the kinetics of the reaction. Haward predicts that, if the active centres are generated *singly* within a system of small drops, then the rate of polymerisation and the degree of polymerisation of the product should increase, because of the suppression of the mutual termination reaction. Furthermore, with decreasing droplet size, the reaction rate should become proportional to the number of drops into which the system is divided. But if the active centres are generated *in pairs* within isolated drops, then the rate of reaction and degree of polymerisation should be reduced.

Haward considers several cases in some detail, of which the one which is of greatest interest for the present subject is that where the droplets are completely isolated and the active centres are formed singly. It does not matter whether the centres are initiated inside the droplets or whether they reach the droplets by diffusion in from an external continuous phase. In either case, it is assumed that the rate at which active centres become present in a droplet is proportional to the volume of the droplet. All droplets are assumed to have the same diameter. It is required to consider the number of such droplets which contain i active centres; such a droplet will be referred to as a type-i droplet.

A droplet will gain a centre to give a droplet containing $i + 1$ centres if one new centre is acquired; it will lose two centres to give a droplet with $i - 2$ centres if a single mutual termination reaction occurs within the droplet. The number, n_i, of droplets containing exactly i centres is determined by the equilibrium between four processes:

(1) creation from type $i - 1$ droplets by acquisition of one centre;
(2) creation from type $i + 2$ droplets by mutual termination;
(3) loss to type $i + 1$ droplets by acquisition of one centre;
(4) loss to type $i - 2$ droplets by mutual termination.

The rates at which type-i droplets are formed and lost by processes (1) and (3) are written as kN_Avn_{i-1} and kN_Avn_i, respectively, where k is a rate coefficient characterising either initiation within a droplet or diffusion in of an active centre from without, v is the volume of a single droplet, and N_A is the Avogadro number. The rate at which type-i droplets are formed by process (2) is written as

$$\frac{k_t}{2}n_{i+2}\cdot\frac{i+1}{N_Av}\cdot\frac{i+2}{N_Av}\cdot N_Av, \quad i.e. \text{ as } \frac{k_t}{2}n_{i+2}\frac{(i+1)(i+2)}{N_Av}$$

where k_t is the rate coefficient for mutual termination. (The factor $\frac{1}{2}$ is introduced to allow for the fact that two active centres disappear in each termination step.) A similar expression is written for the rate at which type-i droplets are lost by process (4). It is then assumed that a steady state exists for the number of droplets containing i active centres. This implies that

$$kN_Avn_{i-1} + \frac{k_t}{2} n_{i+2} \frac{(i+1)(i+2)}{N_Av} = kN_Avn_i + \frac{k_t}{2} n_i \frac{(i-1)i}{N_Av}$$

This is a recurrence relationship which is not readily amenable to exact treatment. However, application to the case $i = 0$ (for which droplets processes (1) and (4) cannot occur) gives

$$\frac{k_t}{2} n_2 \frac{1 \cdot 2}{N_Av} = kN_Avn_0$$

i.e.

$$n_2 = \frac{k}{k_t} N_A{}^2v^2n_0$$

Similarly, without approximation, it is readily found by putting $i = 1$ and $i = 2$ respectively that

$$n_3 = \frac{k}{3k_t} N_A{}^2v^2(n_1 - n_0)$$

and

$$n_4 = \frac{k}{6k_t} N_A{}^2v^2 \left\{\frac{k}{k_t} N_A{}^2v^2n_0 - (n_1 - n_0)\right\}$$

From these equations, it follows that n_2, n_3, $n_4, \ldots$ become negligible relative to n_1 and n_0 as v becomes very small. The same conclusion can be reached for higher values of i.

Furthermore, since n_3 cannot be negative, it follows that $n_1 \geqslant n_0$. And, since n_4 cannot be negative, it follows that

$$n_1 - n_0 \leqslant \frac{k}{k_t} N_A{}^2v^2n_0$$

Thus $n_1 \to n_0$ as $v \to 0$. Consequently, since n_3, $n_4, \ldots$ are negligible as $v \to 0$, it follows that for compartmentalisation of the reacting system into very small volumes, and acquisition of active centres singly, the situation is that, *at any given instant, approximately one-half of the droplets will contain one active centre and the other half will contain none.* As Haward points out, this conclusion can also be arrived at by a different argument. If the droplet size is small, the chance of a single droplet retaining two active centres for any appreciable length of time is small, because mutual termination would occur rapidly in view of the proximity of the radicals. Hence the history of any particular droplet will consist of its changing from $i = 0$ to $i = 1$ and then back again to $i = 0$. If the acquisition of individual active centres is a random process with respect to time, then, in the long run, the length of time during which the droplet contains one radical must equal the length of time during which it contains none. Thus, for a large number of droplets at any one instant, on average one-half will contain one centre and the other half none.

Under the conditions envisaged in the preceding paragraph, the number of active centres in the system is equal to n_1. The rate of polymerisation in a single droplet containing one active centre, expressed in mole l^{-1} sec^{-1}, is equal to $k_p[M][M\cdot]$, where k_p is the rate of coefficient for propagation in l mole^{-1} sec^{-1}, $[M]$ is the bulk monomer concentration in mole l^{-1}, and $[M\cdot]$ is the concentration of active centres in mole l^{-1}. This latter concentration is equal to $1/N_A v$, since there is only one centre present, v being expressed in litres. The total rate of polymerisation within the system as a whole, expressed as moles of monomer converted per litre of droplets per second, is therefore given by

$$-\frac{d[M]}{dt} = \frac{k_p[M]}{2N_A v} = \frac{Nk_p[M]}{2N_A}$$

where N is the number of droplets into which 1 litre of monomer has been subdivided (*i.e.* $Nv = 1$). The factor 2 enters because, on average, for every droplet in which reaction is occurring, there is another in which no reaction is occurring, *i.e.* on average, at any given time, reaction is occurring in only one-half of the total monomer volume. This result shows clearly that *under the supposed conditions of very small droplet volume and acquisition of active centres singly, the rate of reaction within 1 litre of monomer increases in direct proportion to the number of droplets into which the system has been subdivided*, or, what is the same thing, in inverse proportion to the droplet volume. This is really a consequence of the rather obvious fact that if, on average, one-half of the droplets contain one active centre and the other none, then the effective concentration of active centres in a given volume of monomer must be directly proportional to the number of droplets into which that volume is subdivided.

A similar relationship obtains if the rate of polymerisation within the system as a whole is expressed in units such as moles per second or grams per minute. For the rate in moles per second is merely the product of $-d[M]/dt$ and the total volume of monomer in the system, and that volume is supposed to remain constant as the degree of subdivision of the system is varied. Again, a rate in, say, grams per minute is a simple constant multiple of the rate in moles per second.

It is also necessary to consider the average degree of polymerisation of the polymer which is formed. Consider a single droplet of volume v (expressed in litres) which can contain either one radical or none. The rate of acquisition of new active centres is kN_Av per second. In the same time, the number of monomer molecules converted into polymer is $k_p[M][M\cdot]N_Av/2$ (the factor 2 occurring because the droplet is active for only half the time), *i.e.* $k_p[M]/2$. The average number of monomer units per polymer molecule is the ratio of this latter number to one-half the number of active centres acquired (one-half because only one-half of the acquired centres initiate a new polymer chain—the other half terminate an existing chain). The number-average degree of polymerisation of the product should therefore be given by

$$\bar{p}_n = \frac{k_p[M]/2}{kN_Av/2} = \frac{k_p[M]}{kN_Av} = \frac{Nk_p[M]}{kN_A}$$

This result indicates that, like the rate of polymerisation, *the degree of polymerisation should increase in direct proportion to the number of droplets into which the system has been subdivided.*

It should be emphasised that the Haward theory is concerned solely with the effects of compartmentalisation upon the kinetics of a simple bulk free-radical polymerisation. Whilst it is useful in predicting some of the effects which would be expected to accompany straightforward compartmentalisation, its applicability to emulsion polymerisation systems is limited on at least two counts. In the first place, it gives no indication as to how to calculate the degree of compartmentalisation in an emulsion polymerisation system, and, secondly, it makes no allowance for the possible effects, direct or indirect, of compartmentalisation upon the values of the rate coefficients for the individual steps of a free-radical polymerisation.

REFERENCES

1. W. V. Smith and R. H. Ewart, *J. chem. Phys.*, 1948, **16**, 592.
2. H. Fikentscher, *Angew Chem.*, 1938, **51**, 433
3. W. P. Hohenstein and H. Mark, *J. Polym. Sci.*, 1946, **1**, 127.
4. H. Mark and R. Raff, *High Polymeric Reactions*, Interscience, New York, 1941, p. 270.
5. G. Gee, C. B. Davies and H. W. Melville, *Trans. Faraday Soc.*, 1939, **35**, 1298.
6. C. F. Fryling and E. W. Harrington, *Ind. Engng Chem.*, 1944, **36**, 114.
7. W. P. Hohenstein, F. Vingiello and H. Mark, *India Rubb. Wld*, 1944, **110**, 291.
8. W. P. Hohenstein, S. Siggia and H. Mark, *India Rubb. Wld*, 1944, **111**, 173.

9. S. Siggia, W. P. Hohenstein and H. Mark, *India Rubb. Wld*, 1945, **111**, 436.
10. W. P. Hohenstein and H. Mark, *J. Polym. Sci.*, 1946, **1**, 549.
11. I. M. Kolthoff and W. J. Dale, *J. Am. chem. Soc.*, 1945, **67**, 1672.
12. C. C. Price and C. E. Adams, *J. Am. chem. Soc.*, 1945, **67**, 1674.
13. V. J. Frilette and W. P. Hohenstein, *J. Polym. Sci.*, 1948, **3**, 22.
14. I. M. Kolthoff and W. J. Dale, *J. Am. chem. Soc.*, 1947, **69**, 441.
15. F. A. Bovey and I. M. Kolthoff, *J. Am. chem. Soc.*, 1947, **69**, 2143.
16. J. J. P. Staudinger, *Chem. & Ind.*, 1948, 563.
17. W. D. Harkins, *J. chem. Phys.*, 1945, **13**, 381.
18. W. D. Harkins, *J. chem. Phys.*, 1946, **14**, 47.
19. W. D. Harkins, *J. Am. chem. Soc.*, 1947, **69**, 1428.
20. W. D. Harkins, *J. Polym. Sci.*, 1950, **5**, 217.
21. E. W. Duck, 'Emulsion polymerisation', in *Encyclopedia of Polymer Science and Technology*, John Wiley, New York, 1966, Vol. 5, pp. 801–59.
22. J. H. Baxendale, M. G. Evans and G. S. Park, *Trans. Faraday Soc.*, 1946, **42**, 155.
23. J. H. Baxendale, M. G. Evans and J. K. Kilham, *Trans. Faraday Soc.*, 1946, **42**, 668.
24. J. H. Baxendale, S. Bywater and M. G. Evans, *Trans. Faraday Soc.*, 1946, **42**, 675.
25. J. H. Baxendale, S. Bywater and M. G. Evans, *J. Polym. Sci.*, 1946, **1**, 237.
26. J. H. Baxendale, M. G. Evans and J. K. Kilham, *J. Polym. Sci.*, 1946, **1**, 466.
27. E. W. Montroll, *J. chem. Phys.*, 1945, **13**, 337.
28. M. L. Corrin, *J. Polym. Sci.*, 1947, **2**, 257.
29. R. N. Haward, *J. Polym. Sci.*, 1949, **4**, 273.
30. S. H. Herzfeld, A. Roginsky, M. L. Corrin and W. D. Harkins, *J. Polym. Sci.*, 1950, **5**, 207.

CHAPTER 4

THE SMITH–EWART THEORY

4.1 INTRODUCTION

This chapter is concerned with what has come to be regarded as the canonical quantitative theory of emulsion polymerisation. The Smith–Ewart theory was first published in 1948,[1] and since then has been the subject of continuing discussion and refinement. The theory bases its model upon the Harkins mechanism of emulsion polymerisation, and then endeavours to predict the rate of reaction and its dependence upon the concentrations of certain of the components of the reaction system.

This chapter will deal exclusively with theoretical predictions, with no allusion to attempts to confirm the theory experimentally. The purpose is twofold: to expound the theory and its developments, emphasising clearly the assumptions and conclusions; and thereby to provide a framework within which to discuss, in subsequent chapters, some of the effects which have been observed when the levels of certain of the recipe ingredients are varied.

The plan of this chapter is as follows: Consideration will first be given to the simplest case considered by Smith and Ewart (their so-called 'Case 2'). This is the one which has received widest attention in the literature, because the model upon which it is based is believed to accord most closely with the actual situation in many systems and because its predictions are reasonably amenable to experimental verification. Next will be given an exposition of the more general Smith–Ewart theory. The chapter will then conclude with a discussion of some of the more recent developments and refinements of the theory, such as those associated with the names of Stockmayer, O'Toole, Napper and Parts, Parts, Moore and Watterson, and Ugelstad, Mørk and Aasen. The recent theory developed by Gardon is regarded as being sufficiently extensive and interesting to deserve a separate chapter for its consideration.

4.2 THE SMITH–EWART 'CASE 2' THEORY

The model for the Smith–Ewart theory is essentially that given by the Harkins theory. The initiating substance is partitioned exclusively in the aqueous phase, and thus the initiating radicals are considered to be generated in the aqueous phase. The monomer is virtually insoluble in the aqueous phase, but is solubilised within the micelles which are formed from the micelle-generating substance added to the system. Initiation occurs when a micelle which contains monomer acquires a free radical from the aqueous phase. After the acquisition of a radical, a micelle is regarded as having been lost from the system, and a particle gained. After polymerisation has started, an additional phase is therefore present, consisting of very small polymer particles which are swollen with monomer and upon which are adsorbed molecules of the micelle-generating substance. These particles serve as the principal loci of polymerisation. As polymerisation proceeds, so they grow and adsorb more of the surface-active molecules, until eventually micelles as a separate phase disappear and the number of reaction loci is now constant.

The theory is concerned with predicting the steady rate of polymerisation after the number of loci has reached its maximum. This rate, expressed in terms of mass of monomer converted to polymer in unit time, is seen as the product of two terms: (a) the rate of polymerisation at a single locus, and (b) the number of loci. The theory has therefore to concern itself with the factors which influence each in turn of these two quantities.

The general theory allows for the possibility that the activity of a free radical in a polymer particle can be transferred back to the aqueous phase *e.g.* by the diffusion of a propagating species out of the particle, or by a transfer reaction at the interface. For the so-called 'Case 2', two simplifying assumptions are made:

(1) that there is no mechanism whereby the transfer of activity out of a polymerisation locus can occur; and
(2) that the time which elapses between the entry of a second radical and termination by mutual interaction with the first radical present is very short relative to the average time which elapses between the entry of successive radicals.

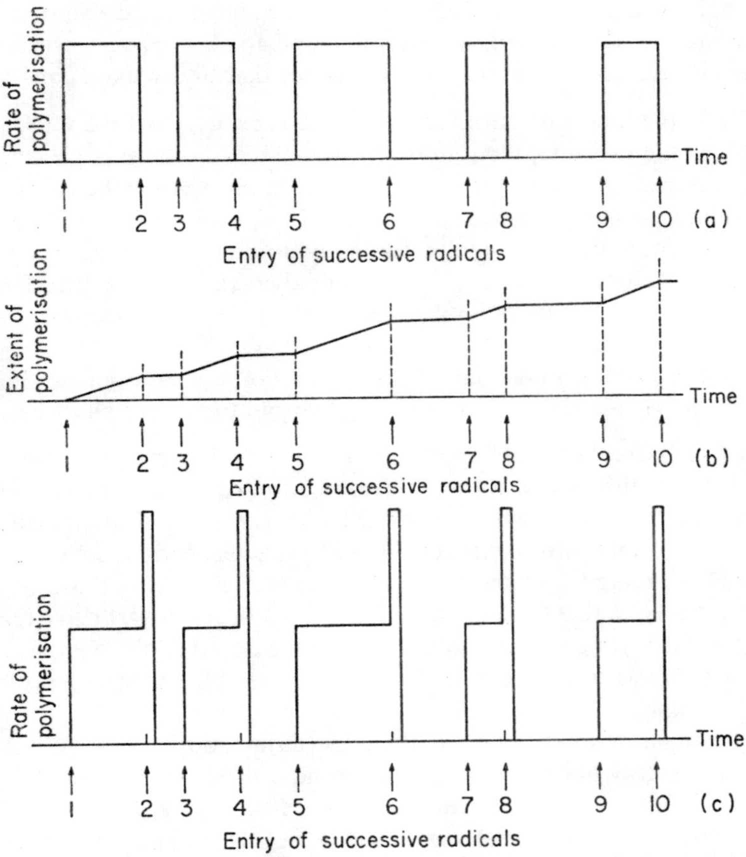

FIG. 4.1. Variation with time of (a) rate of polymerisation, and (b) extent of polymerisation, at a single reaction locus, according to 'Case 2' of the Smith–Ewart theory. The arrows against the time scale refer to the times of entry of successive radicals into the locus. At (c) is shown rate of polymerisation at a single locus as a function of time if mutual termination of radicals is rapid but not instantaneous.

The situation occurring at any single polymerisation locus can therefore be represented diagrammatically as in Fig. 4.1(a) and (b). These show, respectively, (a) rate of polymerisation, and (b) amount of monomer polymerised as functions of time. Radicals enter the locus at a steady constant average rate, but the entry is random and so the time interval which elapses between each successive entry also varies in random fashion. No radical is ever lost from a locus other than by mutual interaction with the next radical entering. At the entries of radicals numbers 1, 3, 5,. . ., the polymerisation rate immediately rises to a constant, steady rate, which continues until the entries of radicals numbers 2, 4, 6,..., respectively, when it falls abruptly to zero. It is obvious that under these conditions:

(1) in the long run, each locus is active for about half the time, since, in the long run, the sum of the time intervals elapsing between the entries of successive odd-numbered and even-numbered radicals must equal the sum of the time intervals elapsing between the entries of successive even-numbered and odd-numbered radicals;
(2) in the long run, the average rate of polymerisation at a given locus is one-half the rate of polymerisation occurring in the locus when it is active; and
(3) in a system containing a large number of loci, at any given instant almost exactly one-half will be active and the other half quiescent.

It will be recalled (see Chapter 3) that Haward[2] reached the same conclusions for the case of a bulk polymerisation compartmentalised into a very large number of unconnected cells in which mutual termination is very rapid. It will also be recalled that Haward justified his conclusion by an analytical argument and by considerations analogous to those given in (1) above. Note that the assumption which Smith and Ewart make regarding the exclusive generation of radicals in the aqueous phase implies that the acquisition of active centres by the loci (in the Haward sense) probably occurs singly.

It should also be noted that, if termination does not occur instantaneously upon the entry of a second radical, then the variation of rate of polymerisation with time at a given locus will not be as shown in Fig. 4.1(a), but rather as shown in Fig. 4.1(c). Upon entry of the second radical, the polymerisation rate will rise immediately to twice its former value, and remain at that level until termination occurs, when the rate will fall sharply to zero. The situation depicted diagrammatically in Fig. 4.1(c) is that which would pertain if the average time for termination to occur were short relative to the average interval between the entry of successive radicals, but not so short that termination could be considered to occur instantaneously. Of course, if the average time for termination to occur is long relative to the average interval between the entry of successive radicals, then a third radical can enter before the first two have terminated, then possibly a fourth, and so on. If this is the case (and it probably is during the later stages of an emulsion polymerisation), then the average value of i can far exceed the Smith–Ewart 'Case 2' value of 0·5.

It is now necessary to consider the relationship between the overall rate of conversion of monomer to polymer and the number of loci. Smith and Ewart simply state that, under the conditions of 'Case 2', the rate of polymerisation per unit volume of aqueous phase is given by

$$\frac{\mathrm{d}M}{\mathrm{d}t} = \tfrac{1}{2}k_p[M]N \tag{4.1}$$

N is the number of loci per unit volume of aqueous phase, $[M]$ must clearly be the molal concentration of the monomer *at the polymerisation locus*, and k_p is the rate coefficient for the propagation step, presumably in litres per mole per second. What is not clear is the exact meaning of $\mathrm{d}M/\mathrm{d}t$. (The omission of any square brackets enclosing M should be noted.) The argument given in Chapter 3 (p. 88) shows that, for a single locus,

$$-\frac{\mathrm{d}[M]}{\mathrm{d}t} = \frac{k_p[M]}{2N_A v} \tag{4.2}$$

where N_A is the Avogadro number and v is the volume of the locus. This expression clearly gives the rate in moles of monomer converted per second per litre of dispersed monomer. Thus $k_p[M]/2v$ must be the rate in number of molecules of monomer converted per second per litre of dispersed monomer. Suppose now that the volume of monomer dispersed in unit volume of aqueous phase is V litres. Then clearly $V = Nv$, so that the rate of polymerisation in number of molecules per second per unit volume of aqueous phase is equal to $k_p[M]V/2v$, *i.e.* to $k_p[M]N/2$. This is the expression given by Smith and Ewart. It is therefore apparent that, if the above meanings are assigned to k_p and to $[M]$, then *the Smith–Ewart expression for* $\mathrm{d}M/\mathrm{d}t$ *gives the rate of conversion of monomer to polymer expressed as number of monomer molecules converted per second in unit volume of aqueous phase.* This rate is certainly proportional to the rate expressed in conventional units, such as per cent per hour, if the overall composition of the reaction system remains unchanged throughout the course of the reaction. For if the volume of aqueous phase remains fixed and no incremental additions of monomer are made, then the rate in per cent per unit time is in effect the product of $\mathrm{d}M/\mathrm{d}t$ and that volume of aqueous phase which is associated with 100 parts by weight of initial monomer, a suitable adjustment being made for the difference in units of monomer mass in the two cases.

It is perhaps also worth pointing out that, under the supposed conditions for 'Case 2', the rate of polymerisation per single locus is quite independent of the composition of the system, so long as it is such that reaction loci are present and radicals are generated. In particular, the rate of polymerisation is quite independent of the rate at which radicals become available to the loci. If the availability of radicals is reduced, then the intervals between successive entries are longer, but the locus is still active for one-half of the time and so the total growth time available is the same, as is the average number of growing radicals present at any one time. A similar argument applies if the availability of radicals is increased. Thus it is clear that the effects upon rate of recipe variables such as initiator level and soap level must be sought in their effects upon the total number of loci.

Smith and Ewart also point out that, for 'Case 2', a very simple relationship can be written down for the average time which elapses between the entry of successive radicals, and therefore for the lifetime of the growing polymer chains. For if ρ is the total number of radicals which enter N loci in unit time, then ρ/N is the number of radicals which, on average, enter each single locus in unit time. The average time which elapses between each successive entry must therefore be N/ρ, and this must also be equal to the average time for which each growing chain polymerises. (The factor 2 which appears in Smith and Ewart's formula seems to be an error.) The number-average degree of polymerisation of the product should be proportional to the average lifetime, if complicating influences are absent. The above argument shows clearly that, for a given rate of entry of radicals, the average degree of polymerisation is directly proportional to the number of loci. Smith and Ewart do not, however, draw attention to the fact that this simple model also predicts an exponential form for the distribution of number of polymer molecules with respect to degree of polymerisation. This follows because, if the entry of successive radicals into a given locus is truly random, then the number entering in a given time interval will follow a Poisson distribution, and the intervals between successive entries will follow an exponential distribution. In fact, if the mean rate of entry of radicals into a single locus is equal to λ, then the parameter of the exponential distribution is also λ. Were it possible to find an emulsion polymerisation system such that the chain length of the product is determined solely by the time which elapses between the entry of successive radicals, then investigation of the nature of the molecular weight distribution of the product could form the basis of a very sensitive test of some of the ideas which underlie the Smith–Ewart theory.

It is now necessary to consider the Smith–Ewart procedure for calculating the number of polymerisation loci which are present during that stage of the reaction following the disappearance of micellar soap. The following additional assumptions are made concerning the micelle-generating substances in the system:

(1) Initially, almost all the soap is present as micelles, *i.e.* only negligible amounts are adsorbed at monomer–droplet interfaces or are monomolecularly dissolved in the aqueous phase.
(2) At all subsequent stages, the soap is distributed entirely between the micelles (if still present) and the growing polymer particles. Thus if S is the total mass of soap contained in the system per unit volume of aqueous phase, it is regarded as consisting of an amount, S_m, present in micellar form plus an amount, S_p, adsorbed on the surfaces of the particles, *i.e.* $S = S_m + S_p$ at all times.
(3) The interfacial area, ϖ, occupied by unit mass of soap is the same whether the soap is in a micelle or adsorbed on a particle, so long as micelles are present.

The situation is, then, that initially $S_p = 0$, and as locus nucleation proceeds, so S_p increases at the expense of S_m. The process of nucleation of new loci terminates when $S_m = 0$, *i.e.* when the micelles have disappeared. Polymerisation then proceeds at a steady rate, which is determined by the total number of particles which have been formed. Throughout the period up until $S_m = 0$, the following relationships will hold:

$$\frac{A}{S} = \frac{A_m}{S_m} = \frac{A_p}{S_p} = \varpi \tag{4.3}$$

where A, A_m and A_p are, respectively, the total interfacial area of micelles plus particles, the interfacial area of micelles, and the interfacial area of particles.

Other assumptions which are made in calculating the maximum number of polymerisation loci are as follows:

(a) Throughout the stage when new loci are being nucleated, the ratio of monomer to polymer in a particle remains constant. This assumption is examined in more detail below. It is generally regarded as valid as long as a separate monomer droplet phase is present throughout the stage of locus formation; this is usually the case unless excessive amounts of soap are present.
(b) The rate of polymerisation within a single particle is constant throughout this stage. This assumption is reasonable given that the monomer:polymer ratio is constant.

Taken together, these two assumptions lead to the conclusion that the rate of increase of volume of a particle, *i.e.* dv/dt, is a constant independent of its size. This constant is denoted by μ.

Further comment is necessary regarding assumption (a) of the preceding paragraph, for it poses a number of questions. Thus whilst it is clearly plausible to assume a restricted swelling of the latent latex particles by the residual monomer if the polymer which forms is crosslinked, it is not immediately obvious why, if the polymer is not crosslinked, the unreacted monomer should not be rapidly imbibed in unlimited quantity by the polymer particles which form initially. That the composition of the monomer-swollen particles is essentially constant until the monomer droplet phase disappears is suggested by the results of Meehan[3] for the variation of butadiene vapour pressure with conversion during the emulsion polymerisation of butadiene. His conclusion was that the amount of butadiene dissolved in the polymer particles depends only upon the amount of polymer available and not upon the size of the particles. These observations raise the further question as to whether the limitation upon the imbibition of monomer is due to the non-attainment of equilibrium, or whether it truly represents an equilibrium condition. Further observations reported by Meehan[3] suggest the latter. Premature short-stopping of the reaction during the stage of constant butadiene vapour pressure did not result in any fall in butadiene vapour pressure over a period of at least twenty-four hours; some fall would have been expected had the butadiene been continuously imbibed into the polybutadiene particles.

The probable explanation for the limited equilibrium swelling of the latent latex particles of an uncrosslinked polymer was given by Morton, Kaizerman and Altier.[4] They proposed that the restraining force which counterbalances the swelling tendency arises from the increase in total interfacial free energy which accompanies the swelling of the particles. Thus, for hydrocarbon polymers at least, the position of equilibrium is determined largely by the balance between the increase in entropy accompanying the mixing of monomer and polymer, and the free energy increase accompanying the swelling of the particles. The enthalpy changes accompanying mixing have a relatively minor effect.

For the details of the calculation of the equilibrium swelling, the reader is referred to the original paper, or to the summary given by Blackley.[5] The conclusion is reached that at equilibrium the larger particles should contain proportionately more monomer than the smaller ones, and that a low interfacial free energy per unit area between particles and aqueous phase will favour imbibition. (The former prediction has its origin in a fact which can very simply be verified for geometrically similar particles of different sizes. This is that if they imbibe an increment of liquid, then the ratio of the incremental increase in surface area to the incremental increase in volume is inversely proportional to the linear size of the particle.) Both these conclusions have been confirmed experimentally. Thus Morton, Kaizerman and Altier found that saturation of the particle surface with a layer of soap permitted polystyrene particles to imbibe more than twice as much styrene as would have been possible without the added soap. Confirmation of the prediction that large particles tend to imbibe proportionately more monomer than small ones has been reported by Allen[6] for the absorption of methyl methacrylate into the particles of natural rubber latex.

Morton, Kaizerman and Altier point out that, in the absence of added soap, the interfacial free energy between particles and aqueous phase will increase as polymerisation proceeds and the particle grows. This effect helps to counteract the tendency for the particles to imbibe a higher concentration of monomer as they grow in size. The balancing-out of these effects is given as the reason why, during any given polymerisation, the monomer:polymer ratio within the particles changes very little as long as a separate monomer droplet phase is present.

Clearly, the arguments of the preceding paragraphs apply not only to the stage of particle nucleation, but also to all subsequent stages of the reaction where monomer is present as a separate phase. In particular, the implication is that the term $[M]$ of eqn (4.1) can be regarded as constant over the whole of the reaction for which monomer droplets are present, *i.e.* not only during the stage of particle nucleation under consideration here, but also during the post-nucleation stage up until the monomer droplets as a separate phase disappear.

Regarding the attainment of equilibrium between monomer droplets and particles, Morton, Kaizerman and Altier provide evidence for believing that equilibrium is attained rapidly. Their conclusion is that the rate of polymerisation is not affected by the rate of diffusion of monomer into the particles, since the particles are, for practical purposes, saturated with monomer as long as a separate monomer droplet phase is present.

It is difficult to make an exact computation of N. Smith and Ewart solve the problem for two idealised situations, one which should give more particles and the other which should give fewer particles than in the actual situation. The calculation for *too many* particles is made on the assumption that, *so long as micelles are present, they capture all the radicals that are available for capture.* This assumption gives too many particles, since some of the radicals will be captured by existing polymer particles. The calculation for *too few* particles assumes that *a given interfacial area always has the same effectiveness in capturing free radicals regardless of the size (and therefore curvature) of the particle upon which it is situated.* This basis gives too few particles because it is known from classical diffusion theory that the flux of diffusing matter across unit area of interface is inversely proportional to the radius of curvature at the interface.

(i) *Calculation for too many loci.* Using the stated assumption, N will be given by ρt^*, where t^* is the time which elapses between the start of the reaction and the disappearance of micellar soap, *i.e.* the time when $A_p = A = \varpi S$ and $A_m = 0$. Now the volume, $v(\tau, t)$, at time t of a particle which was first initiated at time τ must be

$$v(\tau, t) = v_0 + \mu(t - \tau) \tag{4.4}$$

where v_0 is the initial volume of the micelle. v_0 is assumed to be negligible, so that $v(\tau, t) = \mu(t - \tau)$. What is required is the surface area, $a(\tau, t)$, at time t of a particle which was first initiated at time τ. This is very simply related to v, since $a = 4\pi r^2$ and $v = 4\pi r^3/3$, where r is the radius of the particle (which is assumed to be spherical). Elimination of r between these two gives $a = \sqrt[3]{(36\pi v^2)}$. Thus

$$a(\tau, t) = \sqrt[3]{[36\pi\mu^2(t - \tau)^2]} \tag{4.5}$$

This can be written

$$a(\tau, t) = \theta(t - \tau)^{2/3} \tag{4.6}$$

where $\theta = \sqrt[3]{(36\pi\mu^2)}$, and is constant if μ is constant. It is now necessary to calculate the aggregate surface area of particle interface, within unit volume of aqueous phase, which is generated up to a certain time t, which is less than t^*. To do this, consider the contribution to the aggregate surface area made by particles which were first initiated in unit volume of aqueous phase in the interval between time τ and time $\tau + \delta\tau$. They will number $\rho\delta\tau$ in all, and each will have a surface area approximately equal to $a(\tau, t)$ at time t. Their total contribution to the aggregate interfacial area will therefore be $\rho a(\tau, t)\delta\tau$. The aggregate interfacial area must therefore be the sum of all such contributions taken over the interval from $\tau = 0$ to $\tau = t$. Thus if $A_p(t)$ denotes the aggregate interfacial area of particles contained within unit volume of aqueous phase at time t, we shall have that

$$A_p(t) = \int_0^t \rho a(\tau, t)\,\mathrm{d}\tau = \int_0^t \rho\theta(t - \tau)^{2/3}\,\mathrm{d}\tau \tag{4.7}$$

Evaluation of the integral readily gives

$$A_p(t) = (3/5)\rho\theta t^{5/3} \tag{4.8}$$

It is then argued that the time t^* at which micellar soap disappears must be such that $A_p(t^*) = A = \varpi S$. Thus

$$(3/5)\rho\theta t^{*5/3} = \varpi S \tag{4.9}$$

so that

$$t^* = \left(\frac{5\varpi S}{3\rho\theta}\right)^{3/5} \tag{4.10}$$

and

$$N = \rho t^* = \rho^{2/5}\left(\frac{5\varpi S}{3\theta}\right)^{3/5} \tag{4.11}$$

Inserting the value for θ gives

$$N = (5/3)^{3/5}(36\pi)^{-1/5}\mu^{-2/5}\varpi^{3/5}\rho^{2/5}S^{3/5} \tag{4.12}$$

The approximate value of $(5/3)^{3/5}(36\pi)^{-1/5}$ is 0·528, which is normally rounded up to 0·53. Thus the result is obtained that

$$N \simeq 0{\cdot}53\mu^{-2/5}\varpi^{3/5}\rho^{2/5}S^{3/5} \tag{4.13}$$

It should be noted that, in taking $\rho\theta(t - \tau)^{2/3}\delta\tau$ as the total contribution to the aggregate surface area at time t from those particles which were initiated in the interval between time τ and $\tau + \delta\tau$, Smith and Ewart are assuming that the particle grows steadily throughout the interval $t - \tau$. This seems to imply that the particle contains one active centre throughout the whole of the interval, and that none of the radicals made available in the interval $\delta\tau$ (or in any other interval during the stage of locus formation) is wasted in a termination reaction. This apparently conflicts with the conclusion that a radical is present, on average, for only one-half of the time. Whilst presumably allowance for termination within loci already formed could be made by taking suitably-modified values of ρ and μ (and hence of θ), the real resolution of the apparent conflict is to be found in the observation that the calculation under present consideration is concerned with the stage of locus formation, whereas the prediction of one-half a radical per locus on average refers to the steady state when no more new loci are being formed. It will appear from subsequent chapters that in fact the number of micelles present initially (and indeed almost up until the end of locus formation) exceeds by several orders of magnitude the number of loci eventually formed. Thus, throughout the greater part of the stage of locus nucleation, it can reasonably be assumed that most of the radicals which enter micelles and particles actually enter micelles rather than particles. Hence termination reactions will be rare. These considerations do, however, raise a further difficulty which, as far as the present writer is aware, has never been properly resolved. It appears that, at the close of the locus-forming stage, there should be a transition from a situation of approximately one radical per locus to a situation of, on average, approximately one-half a radical per locus. Other things being equal, the overall rate of polymerisation should pass through a maximum at the close of the initial stage, falling to a steady rate of approximately one-half of what it was at the end of the initial stage. As far as the present writer is aware, such rate maxima have never been observed, nor any adequate explanation given for their general absence.

(ii) *Calculation for too few loci.* From the assumption for this case, it follows that the rate of creation of new reaction loci at any instant is given by

$$\frac{\mathrm{d}N}{\mathrm{d}t} = \rho \frac{A_m}{A} \tag{4.14}$$

since only those radicals captured by the interface which is located on micelles are capable of creating new loci. Use of the relationships $A_m + A_p = A = \varpi S$ gives

$$\frac{\mathrm{d}N}{\mathrm{d}t} = \rho - \frac{\rho A_p}{\varpi S} \tag{4.15}$$

Now the number of new particles which are formed in the interval of time between τ and $\tau + \delta\tau$ may be expressed as $(\mathrm{d}N/\mathrm{d}\tau)\delta\tau$, where $\mathrm{d}N/\mathrm{d}\tau$ is the rate of creation at time τ. But, as for the previous case, the surface area at a subsequent time, t, of a particle which was created at time τ is given by eqn (4.6). The contribution of the particles formed in the interval $\delta\tau$ to the aggregate interfacial area of particles at time τ is therefore equal to

$$\theta(t - \tau)^{2/3} \frac{\mathrm{d}N}{\mathrm{d}\tau} \delta\tau \tag{4.16}$$

The aggregate interfacial area of all particles present in the system at times $t \leqslant t^*$ is therefore

$$A_p(t) = \int_0^t \theta(t - \tau)^{2/3} \frac{\mathrm{d}N}{\mathrm{d}\tau} \mathrm{d}\tau \tag{4.17}$$

Substituting this expression for $A_p(t)$ in the equation for $\mathrm{d}N/\mathrm{d}t$ gives

$$\frac{\mathrm{d}N}{\mathrm{d}t} = \rho - \frac{\rho\theta}{\varpi S}\int_0^t (t-\tau)^{2/3}\frac{\mathrm{d}N}{\mathrm{d}\tau}\,\mathrm{d}\tau \tag{4.18}$$

Solution of this equation for N as a function of t is not straightforward. The equation itself is of the type known as a Volterra equation of the second kind. It is of the form

$$\phi(t) = f(t) + \lambda\int_0^t H(\tau, t)\phi(\tau)\,\mathrm{d}\tau \tag{4.19}$$

In this case, $\phi(t) = \mathrm{d}N/\mathrm{d}t$, $\phi(\tau) = \mathrm{d}N/\mathrm{d}\tau$, $H(\tau, t) = (t-\tau)^{2/3}$, $f(t) = \rho$, and $\lambda = -\rho\theta/\varpi S$. The standard procedure for solving is to write

$$\phi(t) = \sum_{s=0}^{\infty}\lambda^s\phi_s(t), \qquad \phi(\tau) = \sum_{s=0}^{\infty}\lambda^s\phi_s(\tau) \tag{4.20}$$

and then to evaluate the functions $\phi_0(t), \phi_1(t), \phi_2(t), \ldots$, as follows: Substitute eqns (4.20) in eqn (4.19) to give

$$\sum_{s=0}^{\infty}\lambda^s\phi_s(t) = f(t) + \lambda\int_0^t\left\{H(\tau, t)\sum_{s=0}^{\infty}\lambda^s\phi_s(\tau)\right\}\mathrm{d}\tau \tag{4.21}$$

i.e.

$$\begin{aligned}\phi_0(t) + \lambda\phi_1(t) + \lambda^2\phi_2(t) + \cdots &= f(t) + \lambda\int_0^t H(\tau, t)\phi_0(\tau)\,\mathrm{d}t\\ &\quad + \lambda^2\int_0^t H(\tau, t)\phi_1(\tau)\,\mathrm{d}\tau + \cdots\end{aligned} \tag{4.22}$$

Equating coefficients of the various powers of λ gives

$$\left.\begin{aligned}\phi_0(t) &= f(t)\\ \phi_1(t) &= \int_0^t H(\tau, t)\phi_0(\tau)\,\mathrm{d}\tau\\ \phi_2(t) &= \int_0^t H(\tau, t)\phi_1(\tau)\,\mathrm{d}\tau\\ &\vdots\\ \phi_s(t) &= \int_0^t H(\tau, t)\phi_{s-1}(\tau)\,\mathrm{d}\tau\end{aligned}\right\} \tag{4.23}$$

But, in this case, $f(t) = \rho =$ constant, and $H(\tau, t) = (t - \tau)^{2/3}$. Thus

$$\left.\begin{aligned} \phi_0(t) &= \rho \\ \phi_1(t) &= \rho \int_0^t (t-\tau)^{2/3}\,\mathrm{d}\tau = (3/5)\rho t^{5/3} \\ \phi_2(t) &= \int_0^t (t-\tau)^{2/3}\cdot(3/5)\rho\tau^{5/3}\cdot\mathrm{d}\tau = \rho t^{10/3}\frac{\{\Gamma(5/3)\}^2}{\Gamma(13/3)} \\ &\vdots \qquad\qquad \vdots \qquad\qquad \vdots \\ \phi_s(t) &= \int_0^t (t-\tau)^{2/3}\phi_{s-1}(t)\,\mathrm{d}t = \rho t^{5s/3}\frac{\{\Gamma(5/3)\}^s}{\Gamma(1+5s/3)} \end{aligned}\right\} \tag{4.24}$$

where $\Gamma(s)$ is the gamma function of s. [That the expression for $\phi_s(t)$ is as shown can be proved by induction. For, if the result is correct for $\phi_{s-1}(t)$, then

$$\phi_s(t) = \int_0^t (t-\tau)^{2/3}\,\rho\tau^{5(s-1)/3}\,\frac{\{\Gamma(5/3)\}^{s-1}}{\Gamma(1+5\overline{s-1}/3)} = \rho t^{5s/3}\frac{\{\Gamma(5/3)\}^s}{\Gamma(1+5s/3)}$$

i.e. it is also true for $\phi_s(t)$. But it is true for $s = 0, 1, 2$; therefore it is true for $s = 3, 4, \ldots$. Those wishing to verify the above integrals who are not familiar with gamma and beta functions may care to note that

$$\int_0^1 u^{r-1}(1-u)^{s-1}\,\mathrm{d}u = \frac{\Gamma(r)\Gamma(s)}{\Gamma(r+s)}$$

and that $\Gamma(s) = (s-1)\Gamma(s-1)$.] Thus the solution of eqn (4.18) is

$$\frac{\mathrm{d}N}{\mathrm{d}t} = \sum_{s=0}^{\infty} \lambda^s \rho t^{5s/3}\frac{\{\Gamma(5/3)\}^s}{\Gamma(1+5s/3)} \tag{4.25}$$

where $\lambda = -\rho\theta/\varpi S$. Substitution for λ, and simple rearrangement, gives

$$\frac{\mathrm{d}N}{\mathrm{d}t} = \rho\sum_{s=0}^{\infty}\frac{x^s}{\Gamma(1+5s/3)} \tag{4.26}$$

where $x = -\Gamma(5/3)\rho\theta t^{5/3}/\varpi S, = -\Gamma(5/3)\rho\sqrt[3]{(36\pi\mu^2)}t^{5/3}/\varpi S$. Smith and Ewart then assume that equating $\mathrm{d}N/\mathrm{d}t$ to zero gives the point at which all the micellar soap disappears and after which no new loci are formed. They state that, by using a method of successive approximation, it can be shown that the value of x for which dN/dt is zero is $-2{\cdot}13$. From the equation by which x is defined, it follows that

$$t^* = (2{\cdot}13)^{3/5}\{\Gamma(5/3)\}^{-1/5}(36\pi)^{-1/5}\mu^{-2/5}\varpi^{3/5}\rho^{-3/5}S^{3/5} \tag{4.27}$$

To obtain N, it is noted that

$$N = \int_0^t \frac{dN}{dt}\,dt \tag{4.28}$$

and that, if the value of dN/dt is substituted from eqn (4.26), then the integration resolves itself into the integration of a sum of terms, each one of which is of the form $\psi(s)t^{5s/3}$, and therefore is simple in principle. In fact, Smith and Ewart found that only the first few terms are of importance in determining the magnitude of the result. By substituting for the upper limit of the integration from eqn (4.27), they reported the final result as

$$N = 0{\cdot}370\mu^{-2/5}\varpi^{3/5}\rho^{2/5}S^{3/5} \tag{4.29}$$

One rather surprising feature of these two calculations is the small difference between the results. Both are of the form

$$N = \chi\mu^{-2/5}\varpi^{3/5}\rho^{2/5}S^{3/5} \tag{4.30}$$

The calculation for too many loci gives $\chi = 0{\cdot}53$, whilst that for too few gives $\chi = 0{\cdot}37$. Smith and Ewart therefore gives their final answer in the form of eqn (4.30), stating that, in the real situation, χ is a constant whose value must lie between 0·37 and 0·53.

The final result for the rate of polymerisation in molecules of monomer per unit time per unit volume of aqueous phase during the stage when locus nucleation is complete, but a separate monomer droplet phase still remains, is therefore

$$\frac{dM}{dt} = \frac{\chi}{2}k_p[M]\mu^{-2/5}\varpi^{3/5}\rho^{2/5}S^{3/5} \tag{4.31}$$

where k_p is the rate constant for propagation, $[M]$ is the concentration of monomer at the polymerisation locus, μ is the rate of growth of the particles during the stage of locus nucleation, ϖ is the adsorption area of unit mass of soap, ρ is the rate at which radicals enter the micelles and loci contained in unit volume of aqueous phase during locus nucleation, S is the total mass of soap present initially in unit volume of aqueous phase, and $0{\cdot}37 < \chi < 0{\cdot}53$.

Insofar that k_p and (M) are constant during the stage of the reaction between the disappearance of the micelles and the disappearance of the monomer droplets, the rate of polymerisation will appear to be of zero order with respect to the monomer concentration in the system as a whole. The question of possible variations of μ and ρ during this stage of the reaction is not relevant since, as the above analysis shows, it is their values during the particle-nucleation stage which are important. ϖ is, of course, a constant, as too is S for a given system. In fact, eqn (4.31) predicts the steady-stage rate of polymerisation from an interesting combination of factors, some of which relate to that stage itself, and some to the preceding stage of particle nucleation.

ρ will be proportional to (and indeed may be equal to) the rate of generation of free radicals in unit volume of the aqueous phase during the stage of particle nucleation. It will be constant if the particle-nucleation stage is short compared to the half-life of the initiator. If the decomposition of the initiator is first-order with respect to its own concentration during this stage of the reaction, then ρ will be proportional to the initial concentration of initiator.

For a system for which all these assumptions are valid, the Smith–Ewart theory leads to the conclusions that:

(1) the steady rate of polymerisation should be of order 2/5 with respect to the initial initiator concentration;
(2) the steady rate of polymerisation should be of order 3/5 with respect to the concentration of the micelle-generating substance; and
(3) the steady rate of polymerisation per unit mass of soap should be inversely proportional to the 2/5th power of the concentration of micelle-forming substance; thus a given mass of soap should be more efficient at promoting polymerisation in a dilute solution than in a concentrated solution.

It is hardly necessary to point out that these conclusions refer to concentrations expressed in terms of the initial aqueous phase only, and not in terms of the whole system. A further conclusion to which the theory leads, and which is independent of the above assumptions, is that there will be no polymerisation if there are no micelles present in the system. This conclusion is, of course, implicit in the Harkins theory.

A comment is necessary regarding the order of reaction with respect to initiator concentration. The Smith–Ewart prediction is 2/5, and this is very close to, indeed experimentally indistinguishable from, the order of 1/2 which is common in bulk and solution polymerisations. The latter order of reaction has its origin in the fact that the termination of growing chains occurs by the mutual interaction of two radicals. The Smith–Ewart exponent of 2/5 has nothing whatsoever to do with the termination mechanism. Essentially, it arises from the geometrical properties of geometrically-similar particles, namely, the relationship between surface area and volume. The order of reaction with respect to soap concentration arises from the same cause, but of course this presents no apparent analogy with bulk or solution polymerisation.

It is appropriate at this point to make further reference to Harkins' prediction that the rate of polymerisation should be 'nearly' proportional to the initial concentration of soap. Reference was made to this prediction in Chapter 3, where it was pointed out that its basis was the twofold assumption that (a) the number of micelles initially present in unit volume of aqueous phase is approximately proportional to the soap concentration, and that (b) the number of eventual polymerisation loci formed is a fixed proportion of the number of micelles initially present. Assumption (a) is probably justified, but (b) is not. The Smith–Ewart calculations show that N is, in fact, proportional to $S^{3/5}$ and not to S, if the non-micellar soap in the system is a negligible fraction of the whole. It is a further consequence that the number of loci formed from a given mass of initial soap is not independent of S, but is in fact proportional to $S^{-2/5}$, and therefore falls as S increases. It may therefore be enquired what is the physical reason for the discrepancy between the reasonable prediction of Harkins and the result of the Smith–Ewart calculations.

The full explanation depends to some extent upon which of the two Smith–Ewart models is used for the calculation of N. But it seems to the present writer that the origin of the discrepancy lies in the fact that, if ρ remains constant, then the larger is S, the smaller is the proportion of micelles which can be activated initially. Thus, at any given time during the period of locus nucleation, the ratio of active particles to the sum of particles plus residual micelles must fall as S increases. In itself, this would lead to a situation in which the number of loci eventually produced would be independent of S. That this is not likely to be the actual case becomes clear if one recalls that a given radical is presented with the possibility of being captured by (i) a micelle, (ii) an active particle, and (iii) an inactive particle. As S increases, it will be expected that the number ratio (i):{(ii) + (iii)} will increase, and so a given radical is more likely to be used in creating a new particle than in re-activating an existing quiescent particle or in de-activating an existing active particle. The situation is further complicated by the fact that the more new loci are formed, the more rapidly will the micellar soap be consumed by adsorption upon growing particles. Thus the true relationship between N and S is likely to be somewhere between $N \propto S$ and $N \propto S^0$; The Smith–Ewart theory gives the relationship as $N \propto S^{3/5}$.

Some comment regarding the Montroll theory (see Chapter 3, Section 3.6) is also desirable. It will be recalled that Montroll's view was that increasing amounts of soap caused enhancement of the rate of polymerisation because the droplet size of the initial monomer emulsion was reduced. It is no doubt true that, for given conditions of agitation, the droplet size of the monomer emulsion does fall with increasing additions of soap, but this would now be regarded as an ancillary effect which is virtually unconnected with the increase in reaction rate which accompanies an increase in soap level.

4.3 THE MORE GENERAL SMITH–EWART THEORY

Although Smith and Ewart regard their 'Case 2' as the one of major interest, they also consider two other situations. One, which they refer to as 'Case 1', is where the average number of growing chains per reaction locus is very small relative to unity; the other, 'Case 3', is where the number of radicals per locus is large relative to unity.

Smith and Ewart commence their treatment by writing down a recurrence relationship for the numbers of radicals which may be present in a single locus. The possible processes of which they take account are as follows:

(1) The transfer of radicals from the aqueous phase into the loci. The rate of this transfer into a single locus is written as ρ/N, where, as above, ρ is the total rate of entry of radicals into all loci, and N is the number of actual or potential loci which are available at the particular time under consideration.

(2) The transfer of radicals from reaction loci back into the aqueous phase. The rate of transfer out of a single locus is written in the form $k_0 a(i/v)$, where i is the number of radicals in the locus, v is the volume and a is the surface area of the locus, and k_0 is the rate coefficient for the process; i/v is therefore the concentration of free radicals at the locus.

(3) Loss of free radicals by mutual termination. The rate of loss of radicals from a single locus by this process is written as $2k_t i[(i-1)/v]$, where k_t is the rate coefficient for mutual termination. This expression can be justified in several ways. Smith and Ewart regard

$(i-1)/v$ as the concentration of free radicals with which any of the i free radicals in a locus can react. (If large numbers of radicals are present, as in a bulk polymerisation reaction, then the rate of termination can be written with sufficient accuracy as the product of a rate coefficient and the square of the radical concentration. But when small numbers may be present, it is necessary to take account of the fact that each of the i radicals present has only $i-1$ partners with which it may interact.)

If these three are the only events which affect the numbers of radicals in the various reaction loci, and if $n_0, n_1, n_2, \ldots, n_i, \ldots$ represent the numbers of loci containing, respectively, 0, 1, 2, ..., i, ... radicals, then the rate at which loci containing n_i radicals are produced is given by

$$n_{i-1}\frac{\rho}{N} + n_{i+1}k_0 a\,\frac{i+1}{v} + n_{i+2}k_t\,\frac{(i+2)(i+1)}{v} \tag{4.32}$$

The rate at which loci containing n_i radicals are lost is similarly given by

$$n_i\frac{\rho}{N} + n_i k_0 a\,\frac{i}{v} + n_i k_t\,\frac{i(i-1)}{v} \tag{4.33}$$

If it is assumed that the number of different types of loci are stationary, then the above two expressions may be equated to give a general recurrence relationship which $n_0, n_1, n_2, \ldots$ must satisfy. This relationship will be referred to hereafter as the 'Smith–Ewart recurrence equation'.

It is interesting to compare this recurrence relationship with that of Haward[2] which was considered in Chapter 3 (Section 3.6). The chief point of difference is that Haward does not take into account the possibility that active centres may be lost from the reaction loci by diffusion out. For this reason, his recurrence relationship contains only four terms as against the six in the Smith–Ewart relationship. Furthermore, the rate of formation of loci containing i centres from those containing $i-1$ is written by Haward in the form $kN_A v n_{i-1}$. The analogous term in the Smith–Ewart theory is written less controversially as $n_{i-1}\rho/N$. However, the expressions used for the rate at which loci containing i centres disappear by mutual termination are virtually identical, if adjustment is made for possible differences in the units used.

It is now necessary to consider the use which Smith and Ewart make of their recurrence equation. They make no attempt to obtain a general solution, which they consider would be difficult. Instead they consider three limiting cases.

4.3.1 Case 1: Number of free radicals per polymer particle is small compared with unity

This situation will occur if the rate at which radicals are transferred out of loci by diffusion is great compared with the rate at which they are gained from the aqueous phase, *i.e.* if $\rho/N \ll k_0a/v$. Then, of the multiplicity of recurrence relationships, the only one of significance is

$$n_1 k_0 \frac{a}{v} = n_0 \frac{\rho}{N} \tag{4.34}$$

this being the steady-state condition for loci containing only one radical. But since $n_0 \cong N$, it follows that

$$n_1 \cong \frac{\rho v}{k_0 a} \tag{4.35}$$

and thus the rate at which polymerisation will occur within the N loci as a whole is $k_p[M]\rho v/k_0a$. If the N loci are contained within unit volume of aqueous phase, then this will be the rate of polymerisation per unit volume of aqueous phase. Furthermore, n_1 can also be written as $c_P V$, where V is the total volume of dispersed polymer in unit volume of aqueous phase and c_P is the concentration of radicals, expressed as number per unit volume of dispersed polymer. This gives $k_p[M]c_P V$ as an alternative expression for the rate of polymerisation.

It is next necessary to consider the locus of the termination reaction. In principle, termination can take place either in the aqueous phase or in the polymer particles. If termination occurs mainly in the aqueous phase, and a steady-state concentration, c_W, exists in the aqueous phase, then

$$R = 2k_t c_W^2 \tag{4.36}$$

where R is the rate of generation of radicals per unit volume in the aqueous phase. (Of course, the possibility exists that the rate coefficient for termination in the aqueous phase differs from that in the particles, but Smith and Ewart do not make this distinction.) This gives $c_W = \sqrt{(R/2k_t)}$. Thus if $\alpha, = c_P/c_W,$ denotes the partition coefficient for radicals between particles and aqueous phase, then $c_P = \alpha\sqrt{(R/2k_t)}$ and hence the rate of polymerisation per unit volume of aqueous phase can be written

$$\frac{\mathrm{d}M}{\mathrm{d}t} = k_p[M]V\alpha\sqrt{\left(\frac{R}{2k_t}\right)} \tag{4.37}$$

The average lifetime, T, of a radical in polymer particles is, under the present assumptions, equal to the number of free radicals in polymer particles divided by the rate at which they are terminated in the aqueous phase. Therefore

$$T = \frac{c_P V}{2k_t c_W{}^2} = \frac{\alpha V}{2k_t c_W} = \frac{\alpha V}{\sqrt{(2k_t R)}} \tag{4.38}$$

the last step following by substitution from eqn (4.36). Thus, if the supposed conditions prevail, neither rate of polymerisation nor lifetime of radicals depends upon the size of the polymer particles, but only upon their total volume.

Alternatively, if termination takes place in the polymer particles, and if these particles are so small that termination takes place very rapidly every time two free radicals are in the same particle, then the rate of termination is simply twice the rate at which free radicals enter those polymer particles which already contain a free radical. Thus the rate of loss of radicals by this mechanism is $2n_1\rho/N$, *i.e.* $2\rho^2 v/k_0 Na$ (substituting for n_1 from eqn (4.35)). If a steady-state concentration of radicals exists, then this expression can be equated to R, giving

$$\rho = \sqrt{\left(\frac{Rk_0 Na}{2v}\right)} \tag{4.39}$$

and hence

$$\frac{dM}{dt} = \frac{k_p[M]\rho v}{k_0 a} = k_p[M]\sqrt{\left(\frac{RNv}{2k_0 a}\right)} = k_p[M]\sqrt{\left(\frac{RV}{2k_0 a}\right)} \tag{4.40}$$

The average lifetime of the growing radicals is approximately n_1 divided by the rate of termination. Thus

$$T = \frac{n_1}{2\rho^2 v/k_0 Na} = \frac{N}{2\rho} = \sqrt{\left(\frac{V}{2k_0 aR}\right)} \tag{4.41}$$

the last step following by substituting ρ from eqn (4.39) and noting that $Nv = V$.

To obtain a criterion by which to decide which of these two termination mechanisms actually applies, Smith and Ewart state that it is necessary to make an assumption concerning the relationship between c_W and ρ/N. The simplest assumption to make is that the rate of entrance of free radicals into a particle is that given by the rate of diffusion of radicals from an infinite medium in which the concentration is c_W into a particle of radius r having zero concentration of the radicals. This assumption gives

$$\rho/N = 4\pi \mathrm{D} r c_W \tag{4.42}$$

where D is the diffusion constant of the radicals through the aqueous phase. Then the rate of termination of radicals in the polymer particles, which according to the above considerations can be written as $(2Nv/k_0a)(\rho/N)^2$, will be equal to

$$\frac{2V}{k_0a}(4\pi \mathrm{D}rc_{\mathrm{W}})^2 \tag{4.43}$$

If this rate is small compared with the rate of termination in the aqueous phase, then it will be the first of the termination mechanisms which will preponderate. The condition for this is

$$2k_tc_{\mathrm{W}}^2 \gg \frac{2V}{k_0a}(4\pi \mathrm{D}rc_{\mathrm{W}})^2 \tag{4.44}$$

i.e.

$$k_t \gg \frac{V}{k_0a}(4\pi \mathrm{D}r)^2 \tag{4.45}$$

i.e.

$$k_t \gg 4\pi \mathrm{D}^2V/k_0 \tag{4.46}$$

the last step following because $a = 4\pi r^2$.

4.3.2 Case 2: Number of free radicals per polymer particle is approximately equal to 0.5

It has already been argued on general grounds that this will be the case if (a) there is no readily-available mechanism by which the activity of a growing radical can be transferred out of the polymerisation locus back into the aqueous phase, and (b) mutual termination of a propagating centre occurs almost instantaneously when a second radical is acquired by the locus. Smith and Ewart show that this conclusion also follows more rigorously by application of their recurrence equation, if the appropriate mathematical embodiments of conditions (a) and (b) are written into the recurrence equation. Condition (a) obviously implies that $k_0 = 0$. Condition (b) can be restated in the form $\rho/N \ll (k_ti(i-1)/v)_{i=2}$, *i.e.* as $\rho/N \ll k_t/v$. Thus, if $\beta = k_tN/v\rho$, this condition becomes $\beta \gg 1$.

Substituting $k_0 = 0$ and $\beta = k_tN/v\rho$ in the Smith–Ewart recurrence equation gives

$$n_{i-1} + \beta n_{i+2}(i+2)(i+1) = n_i + \beta n_i i(i-1) \tag{4.47}$$

Smith and Ewart show that a satisfactory approximate solution to this recurrence equation when $\beta \gg 1$ is

$$\frac{n_{i-1}}{n_i} = 1 + \beta i(i-1) \tag{4.48}$$

But substituting $i = 0$ in eqn (4.47) gives $n_2 = n_0/2\beta$, and substituting $i = 1$ gives $n_0 + 6\beta n_3 = n_1$. Also, eqn (4.48) gives

$$n_2 = (1 + 6\beta)n_3 \tag{4.49}$$

Thus

$$n_3 = \frac{n_2}{1 + 6\beta} = \frac{n_0}{2\beta(1 + 6\beta)} \tag{4.50}$$

and therefore

$$n_1 = n_0 + \frac{3n_0}{1 + 6\beta} = n_0\frac{4 + 6\beta}{1 + 6\beta} \tag{4.51}$$

In this way, approximate values can be obtained for as many of the n_i as are desired. It should be noted, however, that $\beta \gg 1$ gives immediately that $n_1 = n_0$, and $n_2 \cong 0$, $n_3 \cong 0$, etc. Thus application of the recurrence equation confirms that, under the supposed conditions, approximately equal numbers of loci will contain one and no propagating radical, and that almost no loci will contain more than one propagating radical.

So far as the rate of polymerisation is concerned, the important quantity is the total number of propagating free radicals present in unit volume of polymer particles. This number is clearly equal to $\sum in_i$, the total number of loci being given by $N = \sum n_i$. Smith and Ewart state that, using these and the above relationships, the total number of propagating radicals can be expressed as the series

$$\frac{N}{2}(1 + \beta^{-1} - \tfrac{1}{3}\beta^{-2} + \cdots) \tag{4.52}$$

If $\beta \gg 1$, this clearly approximates to $N/2$.

4.3.3 Case 3: Number of free radicals per polymer particle is large compared with unity

This situation will occur if radicals enter the loci much more rapidly than they are destroyed by mutual termination, *i.e.* if $\rho/N \gg k_t/v$. Smith and Ewart assume that the actual system can be approximated by one in which all the particles contain the same number, n, of free radicals. This is reasonable if n is sufficiently large. Making this assumption, it is possible to write down the steady-state condition directly as

$$\frac{\rho}{N} = 2k_t\frac{n^2}{v} \tag{4.53}$$

i.e.

$$n = \sqrt{\left(\frac{\rho v}{2k_t N}\right)} \tag{4.54}$$

The overall rate of polymerisation per unit volume of aqueous phase is therefore

$$\frac{dM}{dt} = k_p[M]Nn = k_p[M]\sqrt{\left(\frac{\rho N v}{2k_t}\right)} = k_p[M]\sqrt{\left(\frac{\rho V}{2k_t}\right)} \quad (4.55)$$

whilst the average lifetime of a polymerising radical is the ratio of the number of radicals present in a locus to the rate at which they are lost by mutual termination, *i.e.*

$$T = \frac{n}{2k_t n^2/v} = \sqrt{\left(\frac{Nv}{2k_t\rho}\right)} = \sqrt{\left(\frac{V}{2k_t\rho}\right)} \quad (4.56)$$

Smith and Ewart point out that, if the exposed interfacial surface area is large enough, so that almost every free radical produced in the aqueous phase enters the particles (*i.e.* $\rho = R$), then there will be no advantage in having a large number of very small particles, since the rate and lifetime depend only upon the total volume of the polymer particles, and not upon the number of particles into which that volume is subdivided. The system is then effectively a suspension polymerisation in which the free radicals are supplied from the aqueous phase. But if, of course, an appreciable fraction of the free radicals is terminated in the aqueous phase, then the system is inherently inefficient in terms of utilisation of radicals.

4.4 DEVELOPMENTS OF THE SMITH–EWART THEORY

4.4.1 Stockmayer–O'Toole treatment[7,8]

Stockmayer[7] has given a general solution to the Smith–Ewart recurrence equation involving Bessel functions. It will be recalled that the recurrence relationship is obtained by equating expressions (4.32) and (4.33). If the substitutions $\epsilon = v\rho/k_tN$ and $m = k_0a/k_t$ are made, it is found that the recurrence equation becomes

$$\epsilon n_{i-1} + mn_{i+1}(i+1) + n_{i+2}(i+1)(i+2) = n_i\{i(i-1) + mi + \epsilon\} \quad (4.57)$$

(Note that ϵ is the reciprocal of the parameter β introduced by Smith and Ewart.)

Stockmayer's procedure is to seek a function, $f(\xi)$, of a parameter ξ such that its expansion as a power-series in ξ has n_i as the coefficient of ξ^i, *i.e.* such that

$$f(\xi) = \sum_{i=0}^{\infty} n_i \xi^i \quad (4.58)$$

Differentiation with respect to ξ gives

$$
\begin{aligned}
f'(\xi) &= \sum_{i=0}^{\infty} i n_i \xi^{i-1} \\
f''(\xi) &= \sum_{i=0}^{\infty} i(i-1) n_i \xi^{i-2}
\end{aligned}
\tag{4.59}
$$

Stockmayer then argues that $f(\xi)$ must be such as to satisfy the second-order ordinary differential equation

$$(1 + \xi)f''(\xi) + mf'(\xi) - \epsilon f(\xi) = 0 \tag{4.60}$$

That this is so can be shown by substituting for $f(\xi)$, $f'(\xi)$ and $f''(\xi)$ from eqns (4.58) and (4.59) into eqn (4.60). If eqn (4.60) is true, then

$$(1 + \xi) \sum_{i=0}^{\infty} i(i-1)n_i\xi^{i-2} + m \sum_{i=0}^{\infty} i n_i \xi^{i-1} - \epsilon \sum_{i=0}^{\infty} n_i \xi^i = 0 \tag{4.61}$$

Thus all the separate coefficients of the ξ^i must be identically zero. This requirement for ξ^i gives

$$(i+2)(i+1)n_{i+2} + (i+1)in_{i+1} + m(i+1)n_{i+1} - \epsilon n_i = 0 \tag{4.62}$$

whilst for ξ^{i-1}

$$(i+1)in_{i+1} + i(i-1)n_i + min_i - \epsilon n_{i-1} = 0 \tag{4.63}$$

Subtraction of eqn (4.63) from eqn (4.62) gives the recurrence relationship (4.57), thus demonstrating that if $f(\xi)$ is considered as having the form indicated by eqn (4.58), then eqn (4.60) represents an alternative method of formulating the recurrence relationship (4.57).

The next step is to solve the differential equation (4.60) for $f(\xi)$. If the variables are changed to $x^2 = 4\epsilon(1 + \xi)$ and $y = f(\xi)/x^{1-m}$, the differential equation becomes

$$x^2 \frac{\mathrm{d}^2 y}{\mathrm{d}x^2} + x \frac{\mathrm{d}y}{\mathrm{d}x} - \{(1-m)^2 + x^2\}y = 0 \tag{4.64}$$

This is the standard form of a modified Bessel equation. Of the possible solutions to this equation, Stockmayer accepts the particular solution

$$y = I_\nu(x), \qquad \nu = |1 - m| \tag{4.65}$$

where I_ν is the modified Bessel function of the first kind of order ν. Restriction of the solution to this particular form was presumably on the criterion of physical acceptability subsequently discussed by O'Toole.[8]

Stockmayer gives the following results for the average number of active centres per polymer particle:

$$\begin{aligned}
&\text{(i) if } m \leqslant 1, &\quad \bar{\imath} &= \frac{h}{4}\frac{I_{-m}(h)}{I_{1-m}(h)} \\
&\text{(ii) if } m \geqslant 1, &\quad \bar{\imath} &= -\frac{m-1}{2} + \frac{h}{4}\frac{I_{m-2}(h)}{I_{m-1}(h)}
\end{aligned} \tag{4.66}$$

where $h^2 = 8\epsilon$. (These results are obtained from the fact that $\bar{\imath}$, $= \sum i \cdot n_i / \sum n_i$, is given by $(f'(\xi)/f(\xi))_{\xi=1}$.) If $\epsilon \ll 1$, which corresponds physically to

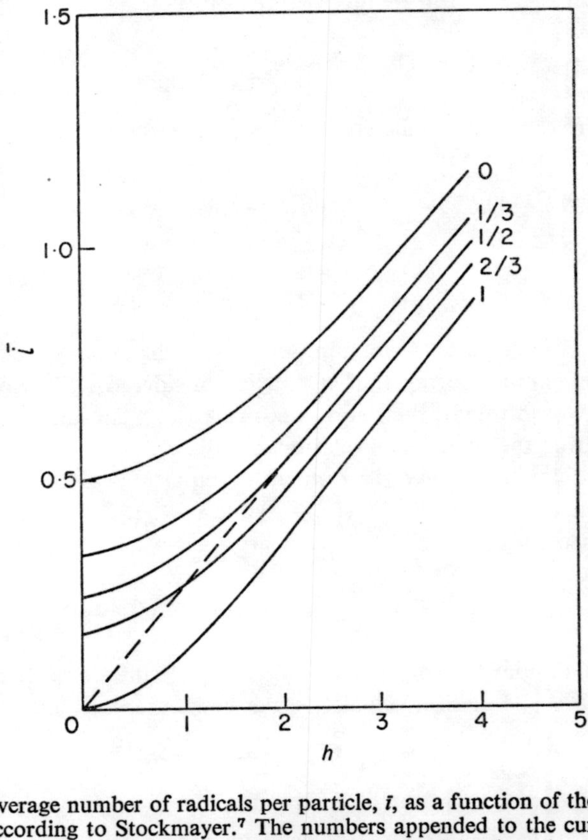

FIG. 4.2. Average number of radicals per particle, $\bar{\imath}$, as a function of the parameters h and m, according to Stockmayer.[7] The numbers appended to the curves give the appropriate values of m. The broken line through the origin gives the prediction for bulk free-radical kinetics.

either a low initiation rate or small particle volume, or both, then series expansion gives

$$\text{(i) if } m \leqslant 1, \qquad \bar{\imath} \cong \frac{1-m}{2} + \frac{\epsilon}{2-m}$$
$$\text{(ii) if } m \geqslant 1, \qquad \bar{\imath} \cong \epsilon/m \tag{4.67}$$

The former of these results reduces to the Smith–Ewart 'Case 2' if ϵ and m are very small. In fact, if the Smith–Ewart assumption is made that radicals cannot leave a locus once they have entered, then $k_0 = 0$ and hence $m = 0$.

The variation of $\bar{\imath}$ with h for various values of m, according to eqs (4.67), is shown in Fig. 4.2. In effect, since $h^2 = 8\epsilon$, they are equivalent to plots of polymerisation rate against square root of initiation rate.

Alexander and Napper,[9] in a most useful review of emulsion polymerisation kinetics, have pointed out that when $m = 0$ (in effect, no loss of radicals by diffusion out of loci) but no restriction is placed upon ϵ, then the rate of polymerisation is proportional to the factor $(h/4)I_0(h)/I_1(h)$, if other things are equal. Alexander and Napper call the ratio $I_0(h)/I_1(h)$ the 'subdivision factor', because, for a given k_t and rate of generation of radicals, it is determined by the ratio v/N, *i.e.* by the extent to which the propagating centres are compartmentalised. The subdivision factor as a function of ϵ and the corresponding value of $\bar{\imath}$ is shown in Fig. 4.3. It appears that the factor differs from unity by less than 10% if $h \geqslant 6$,

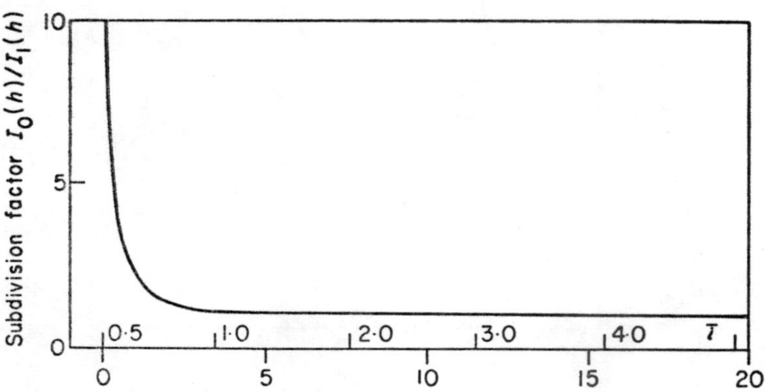

FIG. 4.3. 'Subdivision factor', $I_0(h)/I_1(h)$, as a function of ϵ, according to Alexander and Napper.[9] The values of $\bar{\imath}$ corresponding to those of ϵ are shown on a scale adjacent to that for ϵ.

which corresponds to $\bar{i} \geqslant 1{\cdot}5$, which is relatively small. Alexander and Napper draw the conclusion that the Smith–Ewart 'Case 3' kinetics should be obeyed by systems for which $\bar{i}$ is greater than about 1·5.

O'Toole[8] has criticised, on physical grounds, the particular solution of the Bessel equation adopted by Stockmayer when the rate of diffusion of radicals out of loci is small but non-zero. He attempts to predict the stationary distribution of radicals in particles resulting from a combination of the following processes: (i) acquisition by a zero-order absorption from the aqueous phase or by intraparticle initiation; (ii) a first-order 'desorption'; and (iii) a second-order termination reaction. The recurrence relationship is recast into a form which asserts that the overall rate of conversion of loci containing i radicals into those containing $i + 1$ radicals is zero; so also is the overall rate at which the reverse process occurs. The recurrence equation is then solved to give the probability, P_i, that a single

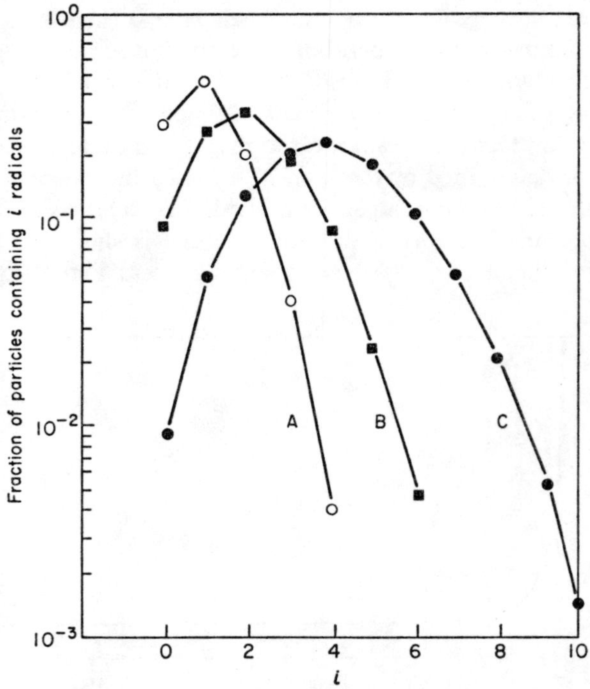

FIG. 4.4. Frequency polygons for distribution of relative frequency of loci containing various numbers of radicals, the distribution being with respect to number of radicals, for case $m = 0$, according to Alexander and Napper.[9] The polygons refer to different values of $\bar{i}$ as follows: A, $\bar{i} = 1$; B, $\bar{i} = 2{\cdot}03$; C, $\bar{i} = 4{\cdot}01$.

locus selected at random contains exactly i radicals. The result which O'Toole gives for P_i is

$$P_i = \frac{h^i 2^{(m-1-3i)/2}}{i!} \cdot \frac{I_{m-1+i}[h/\surd(2)]}{I_{m-1}(h)} \tag{4.68}$$

If $m = 0$, this becomes

$$P_i = \frac{h^i 2^{-(1+3i)/2}}{i!} \cdot \frac{I_{i-1}[h/\surd(2)]}{I_1(h)} \tag{4.69}$$

Alexander and Napper[9] have given typical distributions as predicted by this relationship. These distributions are reproduced in Fig. 4.4. Curves for $\bar{i}$ as a function of h for various values of m, as predicted by the O'Toole theory, are given in Fig. 4.5, together with the corresponding Stockmayer

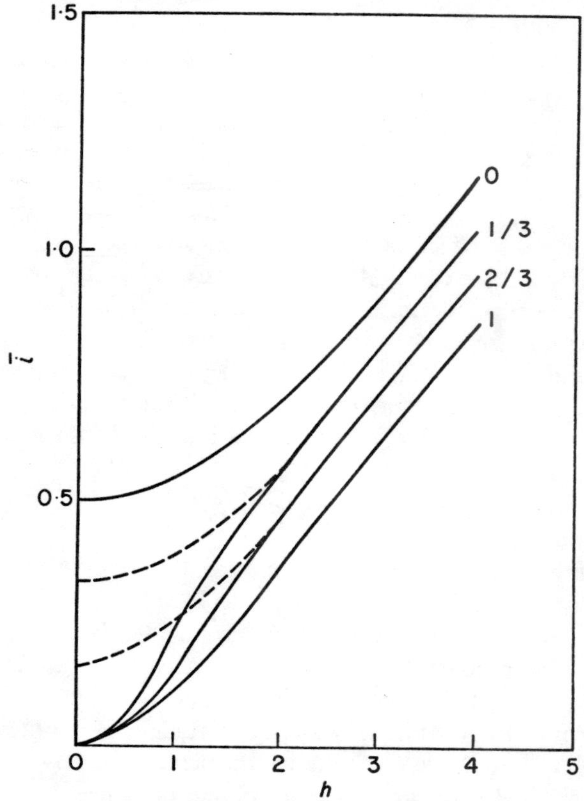

FIG. 4.5. Average number of radicals per particle, $\bar{i}$, as a function of the parameters h and m, according to O'Toole.[8] The numbers appended to the curves give the appropriate values of m. The broken lines represent the predictions of the Stockmayer theory.[7]

curves. It is seen that the two theories predict identical variation of $\bar{\imath}$ with h if $h > 2$; but for $h < 2$, deviations occur unless $m = 0$, the O'Toole theory giving consistently lower average radical populations than the Stockmayer theory.

4.4.2 Napper and Parts treatment of case where $m = 0$[10]

For the particular case of $m = 0$, Napper and Parts have attempted to integrate the expression for the instantaneous rate of polymerisation which is obtained from the Stockmayer theory. By using an expression of the form of eqn (4.1), in which the $\bar{\imath}$ of 1/2 for the Smith–Ewart 'Case 2' is replaced by the $\bar{\imath}$ given by the Stockmayer theory for $m = 0$ (see equation (4.66(i)), it is readily found that the rate of polymerisation can be written as

$$\frac{\mathrm{d}M}{\mathrm{d}t} = k_p[M]N\bar{\imath} = k_p[M]N\cdot\frac{h}{4}\frac{I_0(h)}{I_1(h)} \tag{4.70}$$

Thus the number of molecules of monomer converted per unit volume of aqueous phase in the interval between $t = 0$ and $t = t$ is given by

$$\Delta M = \int_0^t k_p[M]N\cdot\frac{h}{4}\frac{I_0(h)}{I_1(h)}\,\mathrm{d}t \tag{4.71}$$

The problem is to effect the integration and so present ΔM as a function of t, assuming that the reaction occurs exclusively in the stage between the disappearance of micelles and the disappearance of a separate monomer droplet phase. These latter assumptions enable N and $[M]$ to be regarded as constants along with k_p. The following argument seems to be analogous to that given by Napper and Parts, although the notation and most of the presentation are rather different. The integral (4.71) may obviously be re-written as

$$\Delta M = k_p[M]N\int_0^h \frac{h}{4}\frac{I_0(h)}{I_1(h)}\frac{\mathrm{d}t}{\mathrm{d}h}\,\mathrm{d}h \tag{4.72}$$

where the upper limit is the value which h assumes when the time is t. That h is a function of t follows immediately because $h^2 = 8\epsilon = 8v\rho/k_tN$, and v is known to be a function of t; also $h=0$ when $t=0$ because then $v=0$.

The following argument can now be advanced for maintaining that in fact $\{I_0(h)/I_1(h)\}\,\mathrm{d}t/\mathrm{d}h$ is independent of either h or t. The amount of monomer converted in time t, when the particles have volume v, can be expressed in a form which is quite independent of eqn (4.71). The total volume of polymer dispersed in unit volume of aqueous phase is given by $Nv\phi$, where ϕ is the volume fraction of polymer in the particle. (As long as a separate monomer droplet phase is present, ϕ can be taken as constant.) The mass of polymer dispersed in unit volume of aqueous phase at time t is therefore $Nv\phi\sigma$, where σ is the density of the polymer and v is the volume of the particle at time t. Thus

$$\Delta M = \frac{Nv\phi\sigma N_A}{\eta} \tag{4.73}$$

where N_A is the Avogadro number and η is the molecular weight of the monomer. Differentiation with respect to time gives

$$\frac{\mathrm{d}M}{\mathrm{d}t} = \frac{\mathrm{d}(\Delta M)}{\mathrm{d}t} = \frac{N\phi\sigma N_A}{\eta}\frac{\mathrm{d}v}{\mathrm{d}t} \tag{4.74}$$

But $v = (k_t N/8\rho)h^2$, and so $\mathrm{d}v/\mathrm{d}t = (k_t N/4\rho)h\,\mathrm{d}h/\mathrm{d}t$, *i.e.*

$$\frac{\mathrm{d}M}{\mathrm{d}t} = \frac{N\phi\sigma N_A}{\eta}\cdot\frac{k_t N}{\rho}\cdot\frac{h}{4}\frac{\mathrm{d}h}{\mathrm{d}t} \tag{4.75}$$

Comparison of eqns (4.70) and (4.75) shows that, under the conditions stipulated for the problem, $\mathrm{d}h/\mathrm{d}t$ is directly proportional to $I_0(h)/I_1(h)$, and that therefore $\{I_0(h)/I_1(h)\}\,\mathrm{d}t/\mathrm{d}h$ is constant. This in turn implies from eqn (4.72) that the relationship between ΔM and h is of the form

$$\Delta M \propto \int_0^h h\,\mathrm{d}h$$

i.e.

$$\sqrt{(\Delta M)} = \mathrm{K}h \tag{4.76}$$

where K is a composite constant.

Thus far in the argument, it appears that *any* arbitrary function of h could be substituted for $I_0(h)/I_1(h)$, and the conclusion represented by eqn (4.76) would still be reached. However, in order to predict the variation of ΔM with t, it is necessary to obtain the variation of h with t, and this does depend upon the presence of the function $I_0(h)/I_1(h)$ in the expression for $\bar{\imath}$. The above argument has shown that

$$\frac{\mathrm{d}t}{\mathrm{d}h} = \mathrm{K}'\frac{I_1(h)}{I_0(h)} \tag{4.77}$$

where K′ is a second composite constant. It therefore follows that

$$t = \mathrm{K}'\int_0^h \frac{I_1(h)}{I_0(h)}\,\mathrm{d}h \tag{4.78}$$

(The lower limit for h is zero, because $v = 0$ when $t = 0$, and so therefore $h = 0$.) Napper and Parts then evaluated the integral on the right-hand side of eqn (4.78) by a graphical method. Their results for the integral

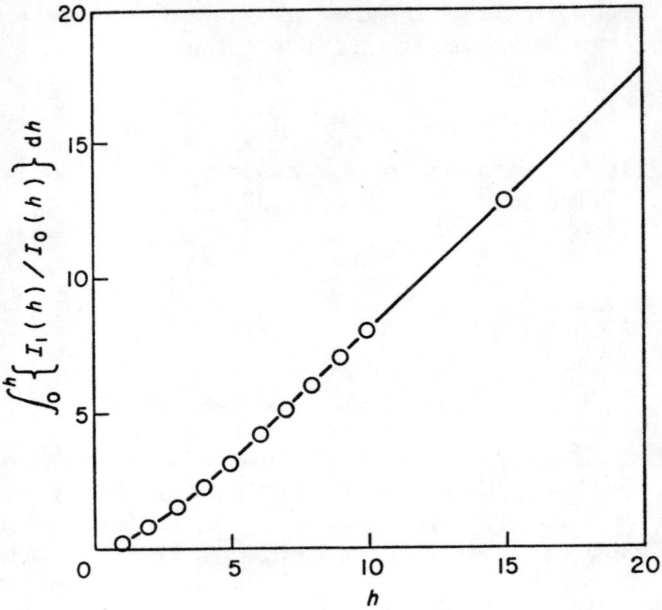

FIG. 4.6. $\int_0^h \{I_1(h)/I_0(h)\}\,dh$ as a function of h, according to Napper and Parts.[10] The integral was evaluated by a numerical procedure.

as a function of h are shown in Fig. 4.6. For values of h between 2 and 20, the variation is almost linear, the actual value of the integral in this range being approximately $0{\cdot}93h - 1{\cdot}35$. Combining this result with eqn (4.76) gives the conclusion that *the square root of the conversion should increase linearly with time*, if $\bar{\imath}$, and therefore h, is sufficiently large for the above approximation to the integral of eqn (4.78) to be valid. Thus the Napper–Parts integration gives what is virtually an alternative solution to the Smith–Ewart 'Case 3'. The Smith–Ewart treatment can be regarded as being an approximation to the above in which the definite integral $\int_0^h \{I_1(h)/I_0(h)\,dh$ is assigned the value h.

4.4.3 Parts, Moore and Watterson's treatment[11,12]

The developments of the Smith–Ewart theory which have been reviewed so far have been largely concerned with the question of the distribution of radicals among actual and potential loci in the system. In particular, a major objective has been a prediction of the mean number of radicals per locus in terms of parameters which have a physical significance for the system. By contrast, the theory of Parts, Moore and Watterson[11] is largely concerned with the calculation of the number of loci present when micellar soap has disappeared.

At any stage in the reaction, the total number, N, of actual and potential loci into which the radicals may enter is regarded as being the sum of the number of micelles, N_m, the number of growing particles, N^0, and the number of dead particles, $N\dagger$. The first assumption made is that the probability of a radical entering a particular type of locus is proportional to the number of such loci. Thus

$$\left.\begin{aligned} -\frac{dN_m}{dt} &= \rho\frac{N_m}{N} + Q(t) \\ \frac{dN^0}{dt} &= \rho\frac{N_m + N\dagger}{N} - \rho\frac{N^0}{N} = \rho\left(1 - \frac{2N^0}{N}\right) \\ \frac{dN\dagger}{dt} &= \rho\frac{N^0}{N} - \rho\frac{N\dagger}{N} = \rho\frac{N^0 - N\dagger}{N} \end{aligned}\right\} \tag{4.79}$$

In these equations, ρ is, as previously, the overall rate at which radicals are acquired, and $Q(t)$ is a function of time which denotes the rate of disappearance of micelles due to their disintegration as a result of adsorption of their constituent molecules on to growing latex particles. $Q(t)$ is thus proportional to the instantaneous rate of increase of the particle–aqueous phase interface. Equations (4.79) yield the simple relationship

$$\frac{dN}{dt} = -Q(t) \tag{4.80}$$

which also follows intuitively, since micelle disintegration is the only one of the postulated processes which affects the total number of actual and potential loci.

Whilst it is admitted that, in reality, $Q(t)$ is a complicated function of time, an initial assumption is made that $Q(t)$ is a constant, κ. Integration then gives

$$N = N_0 - \kappa t \tag{4.81}$$

where N_0 is the number of micelles present at time $t = 0$. Explicit expressions for N_m and N^0 can then be obtained as follows:

$$-\frac{dN_m}{dt} = \rho\frac{N_m}{N} + Q(t) = \frac{\rho N_m}{N_0 - \kappa t} + \kappa \tag{4.82}$$

i.e.

$$\frac{dN_m}{dt} + \frac{\rho N_m}{N_0 - \kappa t} = -\kappa \tag{4.83}$$

This is a linear differential equation of the first order, for which the integrating factor is readily found to be $(N_0 - \kappa t)^{-\rho/\kappa}$. Thus

$$\left[N_m(N_0 - \kappa t)^{-\rho/\kappa}\right]_{t=0}^{t=t} = -\kappa \int_0^t (N_0 - \kappa t)^{-\rho/\kappa}\,\mathrm{d}t$$

$$= \left[\frac{(N_0 - \kappa t)^{1-\rho/\kappa}}{1 - \rho/\kappa}\right]_{t=0}^{t=t} \tag{4.84}$$

i.e.

$$N_m(N_0 - \kappa t)^{-\rho/\kappa} - N_0 \cdot N_0^{-\rho/\kappa} = \frac{\kappa}{\kappa - \rho}\{(N_0 - \kappa t)^{1-\rho/\kappa} - N_0^{1-\rho/\kappa}\} \tag{4.85}$$

Straightforward algebraic manipulation then gives

$$N_m = \frac{N_0 - \kappa t}{\kappa - \rho}\{\kappa - \rho N_0^{(\kappa-\rho)/\kappa}(N_0 - \kappa t)^{(\rho-\kappa)/\kappa}\} \tag{4.86}$$

Similarly, the differential equation for N^0 becomes

$$\frac{\mathrm{d}N^0}{\mathrm{d}t} = \rho\left(1 - \frac{2N^0}{N}\right) = \rho - \frac{2\rho N^0}{N_0 - \kappa t} \tag{4.87}$$

i.e.

$$\frac{\mathrm{d}N^0}{\mathrm{d}t} + \frac{2\rho N^0}{N_0 - \kappa t} = \rho \tag{4.88}$$

The integrating factor for this equation is $(N_0 - kt)^{-2\rho/\kappa}$, and thus

$$\left[N^0(N_0 - \kappa t)^{-2\rho/\kappa}\right]_{t=0}^{t=t} = \rho \int_0^t (N_0 - \kappa t)^{-2\rho/\kappa}\,\mathrm{d}t$$

$$= -\frac{\rho}{\kappa}\left[\frac{(N_0 - \kappa t)^{1-2\rho/\kappa}}{1 - 2\rho/\kappa}\right]_{t=0}^{t=t} \tag{4.89}$$

Insertion of the limits, noting that $N^0 = 0$ at $t = 0$, and rearrangement for N^0 gives

$$N^0 = \frac{\rho(N_0 - \kappa t)}{\kappa - 2\rho}\{(N_0 - \kappa t)^{(2\rho-\kappa)/\kappa} N_0^{(\kappa-2\rho)/\kappa} - 1\} \tag{4.90}$$

N† can be found by difference.

The time, t^*, at which the micelles disappear can be obtained from eqn (4.86) by setting $N_m = 0$. This gives

$$\kappa - \rho N_0^{(\kappa-\rho)/\kappa}(N_0 - \kappa t^*)^{(\rho-\kappa)/\kappa} = 0 \tag{4.91}$$

from which it is found that

$$t^* = \frac{N_0}{\kappa}\left\{1 - \left(\frac{\kappa}{\rho}\right)^{\kappa/(\rho-\kappa)}\right\} \tag{4.92}$$

The total number of remaining active and dead particles then follows from

$$N^* = N_0 - \kappa t^* = N_0\left(\frac{\kappa}{\rho}\right)^{\kappa/(\rho-\kappa)} \tag{4.93}$$

Also, the number of active particles when $t = t^*$ follows from eqn (4.90). The result is

$$N^{0*} = \frac{\rho N_0}{\kappa - 2\rho}\left(\frac{\kappa}{\rho}\right)^{\kappa/(\rho-\kappa)}\left\{\left(\frac{\kappa}{\rho}\right)^{(\kappa-2\rho)/(\kappa-\rho)} - 1\right\} \tag{4.94}$$

From eqns (4.93) and (4.94), it is seen that the fraction of particles which is active at the time when micelles disappear is given by

$$\frac{N^{0*}}{N^*} = \frac{\rho}{\kappa - 2\rho}\left\{\left(\frac{\kappa}{\rho}\right)^{(\kappa-2\rho)/(\kappa-\rho)} - 1\right\} \tag{4.95}$$

Parts, Moore and Watterson point out that, even if the assumption of constant $Q(t)$ does not correspond to the actual situation, eqn (4.95) sets a lower limit for the fraction of particles which is active at the time when micelles disappear. Their argument is that a more realistic assumption for $Q(t)$ would be that it gradually increases with time, so that, for given N_0, ρ and t^*, the actual ratio of micelles to particles is greater than that calculated above. Hence the ratio N^{0*}/N^* given by eqn (4.95) should be smaller than that calculated on any assumption of a gradual increase of $Q(t)$ with t.

Parts, Moore and Watterson then seek to apply their equations to certain published experimental results. From the value of N^0/N at $t = t^*$, it appeared that only about 2% of the particles existing at time t^* had acquired a second radical, this result leading to the prediction that the rate of polymerisation should subsequently fall to about one-half of the value reached at time t^*. This prediction follows from the requirement to maintain consistency with the assumption that the period of constant rate which occurs after locus initiation is complete is due to each particle containing, on average, only half a radical. The theory was therefore rejected as inadequate, since it requires the rate of polymerisation to pass through a maximum at the point where the micelles disappear. This is precisely the difficulty to which attention was drawn in Section 4.2, namely, the implicit prediction of the Smith–Ewart theory that the overall rate of polymerisation should increase progressively throughout the period of locus nucleation, pass through a maximum at about the end of that period, and then fall to a value equal to about one-half of that at the maximum.

The next assumption made is that the rate of entry of radicals, of radius r', from a medium where their concentration is c, into a particle of radius r, where their concentration is zero, is equal to $4\pi c(r + r')$. This expression is approximated to kr, where k is a constant, if $r' \ll r$ and the concentration c is stationary. It is further assumed that the rate of decrease of the total number of particles is proportional to the rate of increase of the surface area of growing latex particles. It is by this mechanism that the micelles are considered to be consumed. The rate of increase in the surface area of a spherical particle which increases its volume at a constant rate μ (whether due to polymerisation, to imbibition of monomer, or to both occurring simultaneously) is equal to $2\mu/r$. If there are N^0 active particles of differing radii, then the rate of increase of total surface area is $2\mu N^0/\bar{r}$, where $\bar{r}$ is the appropriate instantaneous average radius of the growing particles. Given that Θ is the area on the surface of a latex particle which is just covered by the number of soap molecules which are contained in one micelle, then the rate of disappearance of micelles is equal to $2\mu N^0/\Theta\bar{r}$. It is now possible to re-write the differential equations (4.79). The resulting equations were integrated numerically. Again the conclusion was reached that almost all the particles are active at the point where micellar soap disappears. There was no substantial difference in the final conclusion reached on the basis of the two calculations. The similarity was attributed to the fact that the number of micelles initially present is more than three orders of magnitude greater than the number of latex particles which are eventually formed.

It is then stated that agreement with experimental results can be achieved if an assumption equivalent to preferential capture of radicals by the latex particles is introduced. A radical present in the aqueous phase is therefore to be considered as having a different relationship to latent latex particles which have already been formed than it has to the micelles still remaining. The authors realise that it is difficult to justify this difference of relationship in chemical terms. According to Alexander and Napper,[9] possible explanations include (a) that the London dispersion forces of attraction between particles and free radicals are greater than those between micelles and free radicals, because of the larger size of the particles, and (b) that electrostatic interactions influence differently the entry of charge-free radicals into particles and micelles through differences in the surface charge density. Mathematically, this idea is introduced by way of a factor which reduces the probability that a radical will enter a micelle, relative to the probability that it will enter an existing particle. It appears that factors of the order of 10^{-2} are required.

In a second paper, Watterson, Parts and Moore[12] endeavour to use their theory to predict the average degree of polymerisation of the product obtained by emulsion polymerisation. In addition to the usual assumptions of the Harkins theory, they further assume that

(i) supply of monomer by diffusion through the aqueous medium is sufficient to allow each growing radical in each latex particle to react with constant and equal rate; and
(ii) no transfer reactions of any kind occur.

It is inappropriate to consider this aspect of their theory in any detail, if only because it is hard to believe that transfer to polymer could ever be absent in view of the high ratio of polymer to monomer present at the polymerisation locus. However, if a monomer could be found which satisfies this criterion and all the other criteria for the conventional emulsion polymerisation reaction, then presumably comparison between the actual and predicted average degrees of polymerisation would provide a powerful test of the correctness of the underlying assumptions concerning the mechanism of the reaction. It may be noted in passing that although the theory itself is rather complex, the final result for the average lifetime of a growing radical is quite simple, being

$$T = \frac{N}{\rho} \frac{3 + 2\zeta - e^{-2\zeta}}{3 + 2\zeta + e^{-2\zeta}} \tag{4.96}$$

where $\zeta = \rho t/N$, and is a dimensionless variable equal to the average number of radicals acquired per latex particle up to time t. N is the number of latex particles present at the point where the micelles disappear. Given the above assumptions (i) and (ii), the average degree of polymerisation of the product will be proportional to T.

4.4.4 Other contributions

Ugelstad, Mørk and Aasen[13] have pointed out that the O'Toole treatment regards the desorption of a growing radical from a polymerisation locus as involving the destruction of the radical activity. This is kinetically equivalent to a situation in which, in addition to the usual second-order mutual termination reaction, there also occurs within the particles a first-order termination reaction. The effect of this assumption

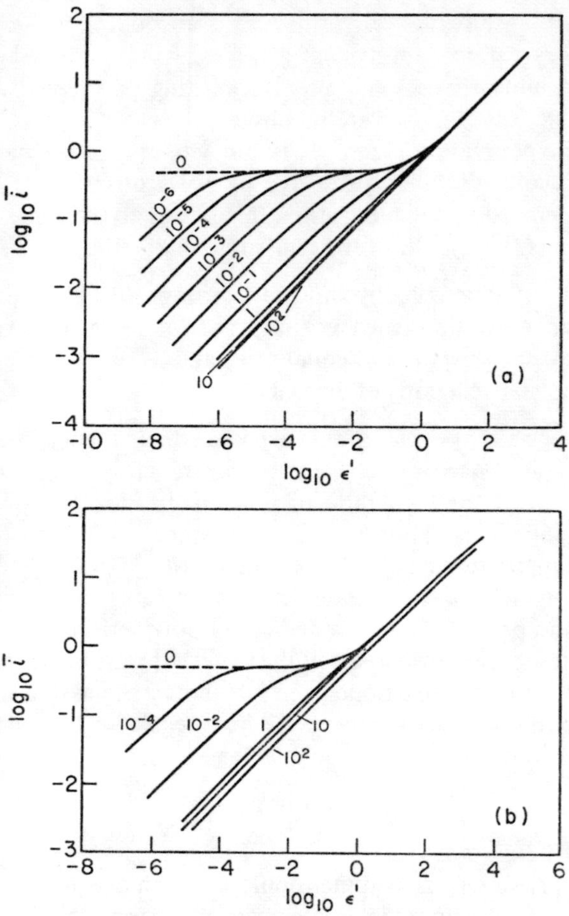

FIG. 4.7.

is to make the rate of absorption of radicals independent of the rate of desorption. However, Ugelstad, Mørk and Aasen consider that, if desorption of radicals from particles occurs, then subsequent reabsorption of those same radicals into particles should be recognised as a possibility and allowed for in the mathematical treatment. They cite the results of Romatowski and Schulz,[14] and of Ley, Hummel and Schneider,[15] as indicating that radicals are both desorbed from and reabsorbed into particles. These results were for the emulsion polymerisation of various monomers initiated by intermittent illumination.

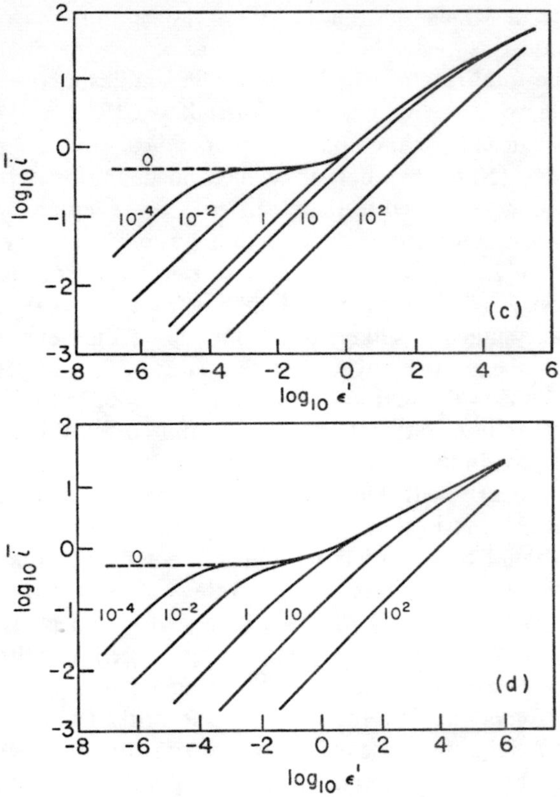

FIG. 4.7—*contd.* Average number of radicals per particle, $\bar{i}$, as a function of the parameters ϵ' and m, according to Ugelstad, Mørk and Aasen.[13] The four graphs refer to different values of the parameter Z as follows: (a) $Z = 0$; (b) $Z = 10^{-4}$; (c) $Z = 10^{-2}$; (d) $Z = 1$. The numbers appended to the curves give the appropriate values of m.

Ugelstad, Mørk and Aasen then give a mathematical treatment based on the assumption that desorbed radicals contribute to the stationary concentration of free radicals in the aqueous phase. Furthermore, their treatment seems to assume also that a desorbed radical is as likely to diffuse into a reaction locus as is a radical which had not previously been subjected to the absorption–desorption process. Allowance is also made for the possibility that the rate coefficient for termination for radicals in the water phase, k_{tW}, may differ from that for radicals in the particles, k_{tP}. Two additional parameters are introduced: (i) $\epsilon' = \rho' v / N k_{tP}$, where

ρ' is the rate at which new 'acquirable' radicals are formed per unit volume of aqueous phase, ϵ being used as before to denote $\rho v/Nk_{tP}$, where ρ is the total rate at which 'acquirable' radicals become available in unit volume of aqueous phase; and (ii) $Z = 2Nk_{tP}k_{tW}/k_a{}^2v$, where k_a is the rate coefficient for the absorption of radicals of all types by polymerisation loci. (Note: Symbols for which no definitions are given here have the meanings assigned to them earlier in this chapter.)

A procedure is outlined for calculating $\bar{\imath}$ as a function of ϵ' and m, given a value for Z. The results for the case $Z = 0$ are given in Fig. 4.7(a). The condition $Z = 0$ is taken as implying that $k_{tW} = 0$, *i.e.* that the radicals are very much less likely to terminate one another in the aqueous phase than in the particles whatever the values of ϵ' and m. However, mathematically $Z = 0$ could also imply that $k_{tP} = 0$, but this would then carry the further implication that $\epsilon' = \infty$, and so for this reason is discounted as a possibility. It is seen from Fig. 4.7(a) that, at low values of $\bar{\imath}$, log $\bar{\imath}$ varies linearly with log ϵ' with a slope of approximately 0·5. This is taken to imply that in this region $\bar{\imath}$, and consequently the rate of polymerisation, should be of order 0·5 with respect to the concentration of initiator in the aqueous phase. At low values of m, the order of $\bar{\imath}$ with respect to ϵ' decreases with increasing values of ϵ' and passes through a region in which the order is approximately zero. The lower the value of m, the lower is the value of ϵ' at which the decrease in order appears. The limiting case for $m = 0$ is represented by the broken line in Fig. 4.7(a); it will be recalled that this case corresponds to a situation in which radicals once absorbed by a locus cannot subsequently diffuse out. At still higher values of ϵ', the order again increases to 0·5, and the value of $\bar{\imath}$ becomes independent of m. It is further seen that, in the case that $Z = 0$, $\bar{\imath}$ is independent of m over the entire range of ϵ' values, and is of order 0·5 with respect to ϵ'.

In Fig. 4.7(b), (c) and (d) are given $\bar{\imath}$ as a function of ϵ' for three non-zero values of Z, namely 10^{-4}, 10^{-2} and 1, respectively. The difference between a value of $\bar{\imath}$ obtained from one of these curves and the value for the same ϵ' and m obtained from Fig. 4.7(a) is taken as indicating the relative contribution of aqueous-phase termination to the overall rate of termination. The effect of aqueous-phase termination is, of course, to reduce the value of $\bar{\imath}$ for any given ϵ' and m. These curves show that aqueous-phase termination represented by $Z = 10^{-4}$ has little effect if $\epsilon' < 10^3$ and $m > 10$, and that these upper limits decrease as Z increases. Thus for $Z = 10^{-2}$, the condition becomes $\epsilon' < 10$, $m < 1$, and for $Z = 1$, $\epsilon' < 10^{-1}$, $m < 10^{-2}$. Ugelstad, Mørk and Aasen have attempted a rough

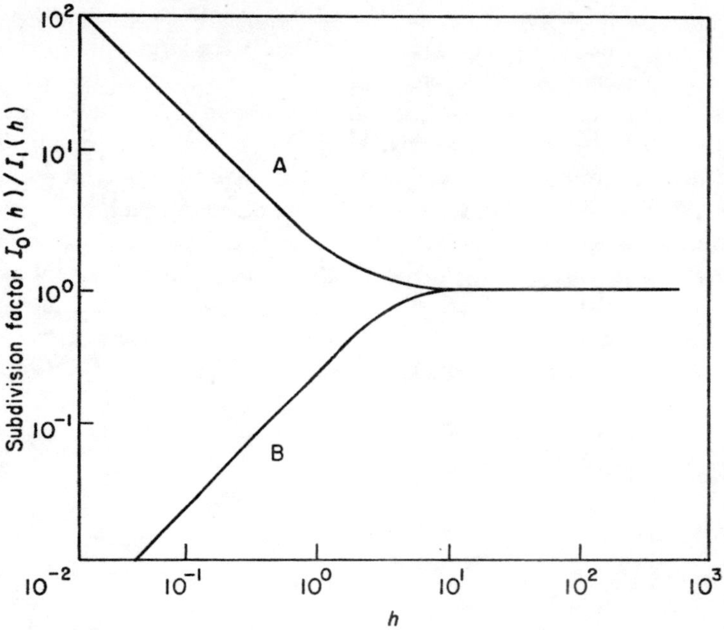

FIG. 4.8. 'Subdivision factor', $I_0(h)/I_1(h)$, as a function of the parameter h, according to van der Hoff.[16,17] The curve A gives the prediction for the situation where radicals are acquired singly, whereas curve B gives that for radicals acquired in pairs.

estimation of Z on the basis that absorption is a simple diffusion process. They conclude that, for most monomers, Z is less than 10^{-4} assuming a reasonable value for N, and that aqueous-phase termination therefore can be neglected up to relatively high values of ϵ' and m.

Van der Hoff[16,17] has published curves showing the variation of the 'subdivision factor', $I_0(h)/I_1(h)$, as a function of h. These curves are reproduced in Fig. 4.8. The upper curve is for the situation where the initiating radicals enter the polymerisation locus singly, as for initiation by water-soluble initiators and also probably for initiation by monomer-soluble initiators activated by water-soluble co-initiators. This curve shows that a polymerisation system will have a subdivision factor of unity, and will therefore obey bulk polymerisation kinetics, if h is larger than about 10. In typical emulsion polymerisations, h is usually less than 1, showing that subdivision factors appreciably greater than unity would be expected.

The lower curve of Fig. 4.8 is for the case where initiation occurs by the decomposition of a monomer-soluble substance into radical pairs within the particles. The subdivision factor is still unity for large h, but for small h the factor decreases with decreasing h. The prediction is therefore that subdivision favours high rate of polymerisation and degree of polymerisation if radicals are acquired singly, but militates against these attributes if radicals are acquired in pairs. Van der Hoff[17] also discusses the variables which influence the value of the parameter h, and notes that high values are favoured by large particles, a high rate of initiation and slow termination.

Reference should also be made to the comprehensive review of the kinetics of emulsion polymerisation which was published by Gerrens in 1959.[18] This contains a thorough discussion of many of the matters raised in this chapter, but is unfortunately available in German only.

REFERENCES

1. W. V. Smith and R. H. Ewart, *J. chem. Phys.*, 1948, **16**, 592.
2. R. N. Haward, *J. Polym. Sci.*, 1949, **4**, 273.
3. E. J. Meehan, *J. Am. chem. Soc.*, 1949, **71**, 628.
4. M. Morton, S. Kaizerman and M. W. Altier, *J. Colloid Sci.*, 1954, **9**, 300.
5. D. C. Blackley, *High Polymer Latices*, Applied Science Publishers, London, 1966, pp. 31–3.
6. P. W. Allen, *J. Colloid Sci.*, 1958, **13**, 483.
7. W. H. Stockmayer, *J. Polym. Sci.*, 1957, **24**, 314.
8. J. T. O'Toole, *J. appl. Polym. Sci.*, 1965, **9**, 1291.
9. A. E. Alexander and D. H. Napper, 'Emulsion polymerisation', in ***Progress in Polymer Science*** (ed. A. D. Jenkins), Pergamon, Oxford, 1971, Vol. 3, pp. 145–97.
10. D. H. Napper and A. G. Parts, *J. Polym. Sci.*, 1962, **61**, 113.
11. A. G. Parts, D. E. Moore and J. G. Watterson, *Makromolek. Chem.*, 1965, **89**, 156.
12. J. G. Watterson, A. G. Parts and D. E. Moore, *Makromolek. Chem.*, 1968, **116**, 1.
13. J. Ugelstad, P. C. Mørk and J. O. Aasen, *J. Polym. Sci.*, Pt A-1, 1967, **5**, 2281.
14. J. Romatowski and G. V. Schulz, *Makromolek. Chem.*, 1965, **85**, 227.
15. G. J. M. Ley, D. O. Hummel and C. Schneider, paper presented at Symposium on New Concepts in Emulsion Polymers, 152nd National Meeting, American Chemical Society, New York, 11–16 Sept. 1966.
16. B. M. E. van der Hoff, *J. Polym. Sci.*, 1958, **33**, 487.
17. B. M. E. van der Hoff, 'Kinetics of emulsion polymerisation', in *Polymerisation and Polycondensation Processes*, Advan. Chem. Ser., American Chemical Society, U.S.A., 1962, no. 34, pp. 6–31.
18. H. Gerrens, *Fortschr. HochpolymForsch.*, 1959, **1**, 234.

CHAPTER 5

THE GARDON THEORY

5.1 INTRODUCTION: QUALITATIVE DESCRIPTION OF MODEL USED

The definitive exposition of the Gardon theory of emulsion polymerisation was published in six papers which appeared in 1968.[1–6] Since that time, summarising papers which present the main features of the theory have appeared in at least two journals.[7,8] The purpose of the theory is to develop quantitatively the consequences of a single, internally-consistent model. The properties to be predicted include (1) the particle size of the latex produced, (2) the conversion–time relationship, (3) the dependence of latex particle size and molecular weight of contained polymer on conversion, and (4) the influences of surface-active substances, of initiator, and of monomer. As will be seen from the following paragraph, the mechanistic model assumed for the reaction system is essentially that proposed by Harkins (for details of which Chapter 3 of this book should be consulted). In effect, much of the Gardon theory is concerned with a re-examination of and recalculation of the Smith–Ewart theory which has been discussed at some length in Chapter 4. Attention is drawn to the fact that the summary of the theory given in this chapter uses a notation which differs in several respects from that used by Gardon.

Concerning the mechanistic model for emulsion polymerisation reactions which is used as a basis for the theory, Gardon[7] states that his calculations are for a system which satisfies the following conditions:

> 'A monomer is mixed with an aqueous solution containing a persulphate initiator and a micellar soap. This mixture is kept at constant temperature until most of the monomer is converted to polymer. Initially the monomer is dissolved in water, absorbed into micelles, and dispersed

in separate droplets. In the beginning of the reaction (interval 1), new latex particles are formed and their number increases. These latex particles absorb monomer and are swollen by it. After all particles are formed, their number is constant in intervals 2 and 3. In intervals 1 and 2 the conversion is sufficiently low for the unreacted monomer to be present in the form of droplets. In interval 3 the conversion is so high that all the monomer droplets disappeared; here the monomer is present only in the monomer-swollen particles and is dissolved in water.

The surface of the micelles and of the latex particles is much larger than the surface of the droplets. Negligible fraction of radicals enter into the droplets and, for all practical purposes, there is no reaction in the droplets. In intervals 1 and 2 the droplets act as an inert reservoir of monomer. Monomer diffuses from the droplets to the locus of polymerisation. The reactor is sufficiently well stirred so that this diffusion does not control the rate of conversion. Instead, the rate is determined by free radical propagation kinetics.'

Thus, it is assumed that all the latex particles subsequently present are formed in interval 1, and that, in the absence of agglomeration, the number of particles is constant in intervals 2 and 3. In intervals 1 and 2, the reaction system consists of three phases: monomer-swollen latex particles, droplets of monomer, and the aqueous phase. In interval 3, the monomer droplets are no longer present, and thus the system contains only two phases, *i.e.* monomer-swollen latex particles and the aqueous phase. Thus in interval 1, the particle number is increasing and monomer droplets are present. In interval 2, the particle number is constant and monomer droplets are present. And in interval 3, the particle number is constant, but no monomer droplets are present.

Further assumptions which emerge as the theory unfolds include the following:

(1) The rate of decomposition of the initiator is slow relative to the rate of polymerisation, so that, for practical purposes, both the initiator concentration and the rate of generation of initiating radicals can be taken as constant.

(2) The absorption of radicals from the aqueous phase into micelles and particles is an irreversible process, *i.e.* the rate of desorption is zero.

(3) The critical micelle concentration of the micelle generator is very small relative to the average concentration present in the aqueous phase as a whole. Thus, for practical purposes, the amount of

monomeric soap dissolved in the aqueous phase can be ignored relative to the total amount present.

(4) As the radical in a particle grows, so the particle absorbs more monomer from the aqueous phase, thus enabling the polymerisation reaction to continue. The monomer droplets which are present during interval 1 merely act as inert reservoirs whose sole function is to keep the aqueous phase saturated with monomer.

(5) As long as a separate monomer droplet phase is present, the ratio of monomer to polymer in a given particle is assumed to be constant and to depend little on the size of the particle. Thus the concentration of the monomer in the reaction loci is taken as independent of conversion whilst monomer droplets are present.

(6) Conventional free-radical polymerisation kinetics are assumed for the reaction occurring within the particles.

5.2 CALCULATION OF FINAL NUMBER OF PARTICLES

Gardon first shows that *each particle absorbs radicals at a rate which is proportional to its surface area.* Radicals are envisaged as being generated at random throughout the aqueous phase, and then as diffusing away from their point of formation. The possibility exists that, as they diffuse from their point of origin, so they add monomer molecules which are dissolved in the aqueous phase. If this process continues long enough, oligomeric radicals will be formed of such a size that they precipitate out of the aqueous phase and so could act as new nuclei for reaction loci. In fact, Gardon believes that this does not happen; the density of micelles and particles is so high that the majority of radicals do not have time to grow to the critical oligomer size for precipitation. (A further discussion of this matter will be found in the next chapter, Section 6.4.) Gardon argues that there must be a critical mean-free-path of diffusion, L^*, which, if it were exceeded, would leave the radical in the aqueous phase for a time sufficient for it to grow to a size exceeding the critical size. Therefore, on average, no radical can move a distance greater than L^* from its point of origin. It is then argued that, on the basis of a reasonable estimate, L^* is considerably greater than the average distance between micelles and particles.

In order to show that particles absorb radicals at a rate proportional to their surface areas, consider a radical which is generated at a point O (see Fig. 5.1 (a)) whose minimum distance away from a spherical particle of radius r is l. Let the length of the tangent from O to the surface of the

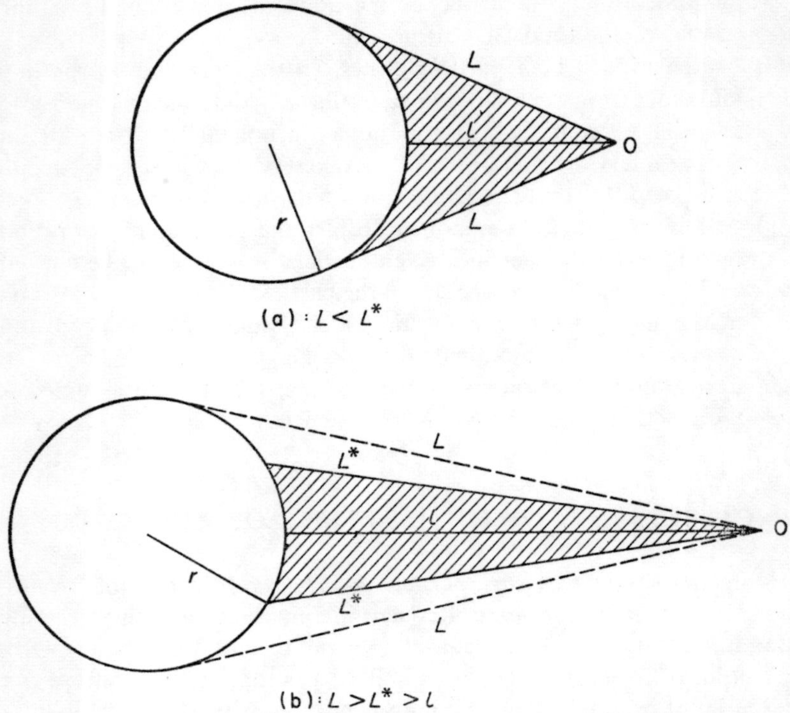

FIG. 5.1. Illustrating Gardon's calculation[1] of the probability that a radical generated at a point O will enter a spherical particle or micelle at a minimum distance l from O. The case where the length of the tangent to the particle is less than the critical mean-free-path of diffusion for the radical is shown at (a). At (b) is shown the case where the tangent is longer than the critical mean-free-path, but l is still less than the critical mean-free-path.

particle be L. Then, so long as $L < L^*$, it is reasonable to assume that the radical generated at O will enter the particle if it moves away from O in a direction within the cone-like volume whose cross-section is shaded in Fig. 5.1 (a). Otherwise the radical will miss the particle. It is then assumed that the probability of the radical entering the particle is the ratio of the volume of the cone-like volume to that of the sphere whose centre is O and whose radius is L. In fact, Gardon states that this is the probability without giving the matter any discussion. It is well known, however, that conclusions concerning probabilities in geometrical situations can depend very much upon the criterion which is adopted for equality of probability. In the present instance, it would seem at least as reasonable to take the required probability as the ratio of the solid angle subtended by the cone-like volume at O to the total solid angle at O. However, the final conclusion regarding the effect of particle radius upon the rate of absorption of radicals would probably be unaltered. As a matter of geometry, Gardon obtains the following result for the probability as defined above:

$$p(r, l) = \frac{l^2 r^2}{4L^3 (r + l)} \tag{5.1}$$

Gardon also considers a second case, illustrated in Fig. 5.1 (b), where $l < L^*$ but $L > L^*$. In this case, not all the directions within the cone-like volume of Fig. 5.1 (a) are available to the radical generated at O, if it is to enter the particle. Only those directions are available which are such that $L' \leqslant L^*$, where L' is the distance from the point O to the surface of the particle. These are the directions contained within a new cone-like volume, whose cross-section is shown shaded in Fig. 5.1 (b). An expression for the probability in this case is also developed. Presumably the number of radicals acquired by particles through the existence of this possibility is small, in view of the conclusions reached regarding the mean-free-path of radicals in relation to the distance between particles and micelles. Gardon then uses these probabilities to calculate the rate of entry of radicals into the particle from all sources of origin whose distances away from the surface of the sphere lie between l and $l + \delta l$. These sources will all lie within a spherical shell whose centre is the centre of the particle and whose radii are $r + l$ and $r + l + \delta l$. The rate of generation of radicals within such a shell is given by the product of its volume, $4\pi (r + l)^2 \delta l$, and the rate of production of effective radicals per unit volume of aqueous phase, which will be denoted by R. Multiplication by the probability that a single radical generated at this distance will enter the particle gives the rate of entry of particles from this shell; and integration over all possible distances from the surface of the particle gives the total rate of entry of radicals.

Gardon obtains a result which, when plausibly approximated in the light of probable conditions prevailing, leads to the conclusion that the total rate of entry of radicals into a particle of radius r is equal to $aL^*R/4$, where a, $= 4\pi r^2$, is the surface area of the particle. The result justifies the conclusion that the rate is proportional to the surface area of the particle. It also follows that the total rate of entry of radicals into all particles and micelles present in unit volume of aqueous phase is given by $AL^*R/4$, where A is the total interfacial area of both particles and micelles per unit volume of aqueous phase. However, this rate can never be greater than R, since it is impossible that more radicals should be absorbed than are generated. Thus, if in fact $AL^* \geqslant 4$, then it follows that the actual rate of absorption of radicals is R, so that all radicals generated in the aqueous phase become absorbed by particles.

The next step is to formulate the differential equation for particle nucleation. This can be done quite simply by noting that, if A_p denotes the aggregate surface area of the monomer–polymer particles contained in unit volume of aqueous phase, then $A - A_p$ denotes that of the micelles. Since the number of particles is only affected by entry of radicals insofar as the radicals enter micelles rather than existing particles, it follows that the rate of nucleation of new particles is proportional to $A - A_p$. This falls to zero when $A_p = A$, *i.e.* when micellar soap has disappeared. It remains at zero thereafter. Thus during the stage where $A_p < A$, we must have that

$$\frac{dN_t}{dt} = R \cdot \frac{A - A_p}{A} = R\left\{1 - \frac{4\pi}{A}\left(\sum n_i r_i^2\right)_t\right\} \tag{5.2}$$

where N_t is the number of particles per unit volume at time t, and n_i is the number of particles of radius r_i, the suffix t denoting that it is the values at the time t which are required. It should be noted that eqn (5.2) is virtually the same as eqn (4.15), being the fundamental equation for the Smith–Ewart 'Case 2' calculation for too few loci. Equation (5.2) here differs from eqn (4.15) in two minor respects: (i) R has been written in place of ρ; and (ii) for A_p has been substituted its value in terms of particle radii on the assumption that the particles are spherical. The writing of R in place of ρ is the mathematical embodiment of the belief that all radicals generated in the aqueous phase are in fact absorbed by particles and micelles.

Gardon next proceeds to consider the rate of growth of a single particle. As is customary in the development of the kinetics of conventional free-radical polymerisation, the rate of disappearance of monomer is written in the form

$$-\frac{d[M]}{dt} = k_p[M]\Sigma[M_i\cdot] \tag{5.3}$$

where k_p is the rate coefficient for propagation, $[M]$ is the molar concentration of monomer *within the reaction locus*, and $\Sigma[M_i\cdot]$ is the molar concentration of radicals of all sizes, also within the reaction locus. The symbol Ω is then introduced to denote the total number of moles of monomer polymerised at time t in the particle. If the volume of the particle at that time is v, then clearly

$$\frac{d\Omega}{dt} = -v\frac{d[M]}{dt} \tag{5.4}$$

It is assumed that, during interval 1, each particle grows as if it contained one radical only. (Note that the assumed number is one, and not one-half. This matter has been discussed in some detail in the previous chapter.) This being so, we must have that

$$N_A v \sum [M_i \cdot] = 1 \tag{5.5}$$

where N_A is the Avogadro number. Also, if the volume fraction of monomer in the particle is denoted by ϕ, and $\overline{V}_1$ is the molar volume of monomer, then

$$\phi = \overline{V}_1[M] \tag{5.6}$$

Combining eqns (5.3)–(5.6) gives

$$\frac{\mathrm{d}\Omega}{\mathrm{d}t} = \frac{k_p \phi}{\overline{V}_1 N_A} \tag{5.7}$$

Let $\overline{V}_2$ denote the molar volume of a monomer unit within the polymer. $\overline{V}_1$ and $\overline{V}_2$ are connected by the relationship

$$\sigma_1 \overline{V}_1 = \sigma_2 \overline{V}_2 \tag{5.8}$$

where σ_1 and σ_2 are the densities of monomer and polymer, respectively.† Then the volume of polymer contained in the particle is $\Omega \overline{V}_2$, and we can write

$$\Omega \overline{V}_2 = \Omega \overline{V}_1 \frac{\sigma_1}{\sigma_2} = v(1 - \phi) = \frac{4\pi r^3}{3}(1 - \phi) \tag{5.9}$$

where r is the radius of the particle at the instant in question. Gardon introduces the symbol K to denote the rate of increase of r^3 with time. It is equivalent to the μ of the Smith–Ewart theory (see Section 4.2). In fact, $K = 3\mu/4\pi$. From eqns (5.9),(5.7) and (5.8) it follows that

$$K = \frac{\mathrm{d}(r^3)}{\mathrm{d}t} = \frac{3\overline{V}_2}{4\pi(1 - \phi)} \frac{\mathrm{d}\Omega}{\mathrm{d}t} = \frac{3k_p}{4\pi N_A} \frac{\sigma_1}{\sigma_2} \frac{\phi}{1 - \phi} \tag{5.10}$$

† It is perhaps unfortunate, and certainly confusing, that, in the literature pertaining to the theory of emulsion polymerisation, the suffixes m and p are used to distinguish both monomer from polymer and micelles from particles. It is pure coincidence that one member of each pair has m as its initial letter and the other p. The convention adopted here is to reserve the suffixes m and p to distinguish micelles from particles, and to use suffixes 1 and 2 to denote monomer (effectively the solvent) and polymer (effectively the solute), respectively. Further complication arises because p is also used as a suffix for rate coefficients to distinguish those for propagation from those for other steps of the polymerisation reaction. However, it is not thought that this additional use of the suffix p will cause confusion in the present context.

Two points are immediately evident from the result embodied in eqn (5.10). In the first place, insofar as ϕ (and therefore the viscosity of the reaction medium) remains constant, k_p probably remains constant. Thus the rate of volume increase of the particle is constant. But, secondly, K can be calculated from quantities determined by experiments other than those concerned with emulsion polymerisation.

From the definition of K, and its lack of dependence upon conversion during interval 1, it follows that the radius at time t of a particle nucleated at time τ is given by

$$r^3 = K(t - \tau) \tag{5.11}$$

if the radius at the time τ is assumed to be negligible relative to r.

The results obtained above can be used to establish a relationship for the variation with time during interval 1 of the total particle surface. During the increment of time between τ and $\tau + \delta\tau$, the number of new particles nucleated will be given by $\delta N_\tau = (\mathrm{d}N_\tau/\mathrm{d}\tau)\,\delta\tau$. Using an argument analogous to that given in Chapter 4 for the Smith–Ewart theory, it follows that the total surface area of particles present at time τ after the start of the reaction is given by

$$A_p(t) = 4\pi K^{2/3} \int_{\tau=0}^{\tau=t} (t - \tau)^{2/3}\,\mathrm{d}N_\tau \tag{5.12}$$

Substituting from eqn (5.2), this gives

$$A_p(t) = 4\pi K^{2/3} R \int_0^t \left\{1 - \frac{4\pi}{A}\left(\sum n_i r_i^2\right)_\tau\right\}(t - \tau)^{2/3}\,\mathrm{d}\tau \tag{5.13}$$

$A_p(t)$ is, of course, equal to $4\pi\,(\Sigma\, n_i r_i^2)_t$. Equation (5.13) is valid over the range of times for which

$$0 \leqslant 4\pi\left(\sum n_i r_i^2\right)_t \leqslant A \tag{5.14}$$

Particle nucleation ceases as soon as $4\pi\,(\Sigma\, n_i r_i^2)_t = A$, there being no soap remaining in micellar form at this point. Gardon solves eqn (5.13) numerically to give the critical value of t at which nucleation ceases. This value is denoted here by t^*. Gardon's result is

$$t^* = 0{\cdot}365\, A^{3/5} R^{-3/5} K^{-2/5} \tag{5.15}$$

The total number of particles, N^*, formed during interval 1 is obtained by integrating eqn (5.2) over the interval from $t = 0$ to $t = t^*$. Gardon's result, obtained numerically, is

$$N = 0{\cdot}208\, A^{3/5} R^{2/5} K^{-2/5} \tag{5.16}$$

It is stated that this is the exact mathematical equivalent of the Smith–Ewart approximation for N^*.

5.3 VARIATION OF CONVERSION WITH TIME DURING INTERVAL 1

Under the assumptions listed in Section 5.1, the amount of polymer formed at any instant is proportional to the total volume of polymer particles present at that instant. Each of the δN_τ particles which is formed during the interval from τ to $\tau + \delta\tau$ has a volume equal to $4\pi K(t - \tau)/3$ at a subsequent time t. The aggregate volume of particles contained in unit volume of aqueous phase at time t is therefore

$$V = \frac{4\pi KR}{3}\int_0^t \left\{1 - \frac{4\pi}{A}\left(\sum n_i r_i^2\right)_\tau\right\}(t - \tau)\mathrm{d}\tau \qquad (5.17)$$

so long as $4\pi\,(\Sigma\, n_i r_i^2)_\tau \leqslant A$. The integration is effected numerically. The final result for conversion of monomer to polymer is expressed in terms of the volume of polymer formed per unit volume of aqueous phase. This latter quantity is denoted here by Π; clearly, the relationship between Π and V is

$$\Pi = (1 - \phi)V \qquad (5.18)$$

Gardon's result for Π is

$$\Pi = 1{\cdot}47\,(1 - \phi)\,KRt^2\,(1 - 1{\cdot}65\,K^{0\cdot62}R^{0\cdot93}A^{-0\cdot93}t^{1\cdot55}) \qquad (5.19)$$

For times shorter than about $t^*/2$, the second term in the bracket is much less than unity, so that

$$\Pi \simeq 1{\cdot}47\,(1 - \phi)\,KRt^2 \qquad (5.20)$$

for $0 \leqslant t \leqslant t^*/2$. Substituting for K from eqn (5.10) gives

$$\Pi \simeq 0{\cdot}351\,\frac{k_p}{N_A}\cdot\frac{\sigma_1}{\sigma_2}\,\phi\,Rt^2 \qquad (5.21)$$

Thus the conclusion is reached that, during the early part of interval 1, the conversion should be proportional to the square of the time, and should also be independent of the amount of soap present. If systems containing different amounts of initiator, but otherwise identical, are compared, then the conversion after a given time should be proportional to the concentration of initiator in the aqueous phase. For the conversion at the end of interval 1, Gardon obtains the value

$$\Pi^* = 0{\cdot}209\,A^{1\cdot2}K^{0\cdot2}R^{-0\cdot2}(1 - \phi) \qquad (5.22)$$

He points out that Π is not a fractional conversion, but an absolute value. The fractional conversion at which nucleation ceases is therefore dependent upon the total amount of monomer present in the system initially.

5.4 GROWTH HISTORY OF INDIVIDUAL PARTICLES DURING INTERVAL 1

During interval 1, the total area of interface between the aqueous phase and the various oil phases (micelles, particles and droplets) is assumed to remain constant at A per unit volume of aqueous phase. The rate of generation of effective radicals per unit volume of aqueous phase is equal to R per second, all of which are absorbed by the oil phases. Thus the rate of entry of radicals into a particle of radius r is given by

$$\frac{\mathrm{d}i}{\mathrm{d}t} = \frac{4\pi r^2}{A} \cdot R \tag{5.23}$$

where i denotes the number of radicals. Consider a particle which is nucleated at time $t = 0$, and whose radius at the instant of nucleation can be taken as zero. At any subsequent time, t, the radius will be $(Kt)^{1/3}$, by eqn (5.11). The growth history of this particle during interval 1 will therefore be described by the differential equation

$$\frac{\mathrm{d}i}{\mathrm{d}t} = 4\pi\left(\frac{R}{A}\right)(Kt)^{2/3} \tag{5.24}$$

Integration over the interval $t = 0$ to $t = t$ gives

$$i = \frac{12\pi}{5} K^{2/3}\left(\frac{R}{A}\right)t^{5/3} \tag{5.25}$$

Gardon now introduces the concept of the average time which must elapse before one radical enters a particle which already contains one growing polymer chain. This time, which is denoted here by $\bar{t}_1$, can be obtained from eqn (5.25) by setting $i = 1$. The result is

$$\bar{t}_1 = \left(\frac{5}{12\pi}\right)^{3/5}\left(\frac{A}{R}\right)^{3/5} K^{-2/5} \tag{5.26}$$

Combining eqns (5.26) and (5.15) gives

$$\frac{\bar{t}_1}{t^*} = \left(\frac{5}{12\pi}\right)^{3/5} \frac{1}{0.365} \tag{5.27}$$

i.e.

$$\bar{t}_1 = 0{\cdot}81\, t^* \tag{5.28}$$

Thus it follows that $\bar{t}_1$ is approximately 80% of the time interval which elapses between the beginning of the reaction and the end of interval 1. Thus only those particles nucleated near the beginning of the reaction are likely to be deactivated by the entrance of a second radical. It also follows that, on average, all those particles which are nucleated in the interval from $t^* - \bar{t}_1$ to t^* are likely to contain no radical other than the one which was responsible for nucleation. Gardon calculates that these particles comprise 67·3% of the number finally present at the end of interval 1.

A similar calculation shows that the average time which elapses between the entry of the second and third radicals, $\bar{t}_2$, is given by

$$\bar{t}_2 = 0{\cdot}486\, t^* \tag{5.29}$$

Clearly, $\bar{t}_1 + \bar{t}_2 > t^*$. Thus it appears that, of the particles present at the end of interval 1, 67·3% should be propagating due to the continuing activity of the radical by whose entry they were nucleated. Of the remaining 32·7%, almost all will be inactive due to the entrance of a second radical. The number which have become re-activated by the entrance of a third radical, or de-activated by the entrance of a fourth, etc., is negligible. Thus it further appears that, at the end of interval 1, the average number of active radicals per particle is 0·673.

5.5 VARIATION OF CONVERSION WITH TIME DURING INTERVAL 2

If the number of particles per unit volume of aqueous phase is N, then the above definitions of Π and Ω give immediately the following relationship between an incremental increase of Ω within a single particle and the corresponding incremental increase of Π:

$$\delta\Pi = N\bar{V}_2\,\delta\Omega = N\bar{V}_1\,\frac{\sigma_1}{\sigma_2}\,\delta\Omega \tag{5.30}$$

Thus

$$\frac{\mathrm{d}\Pi}{\mathrm{d}t} = N\bar{V}_1\,\frac{\sigma_1}{\sigma_2}\,\frac{\mathrm{d}\Omega}{\mathrm{d}t} \tag{5.31}$$

But $\mathrm{d}\Omega/\mathrm{d}t$ is given by eqn (5.7), modified to allow for the possibility that the average number of growing radicals per particle is other than unity. If the number is $\bar{i}$, then eqn (5.5) becomes

$$N_A v \sum [M_i\cdot] = \bar{i} \tag{5.32}$$

Combining eqns (5.4), (5.3) and (5.32) gives the following as the modified version of eqn (5.7):

$$\frac{\mathrm{d}\Omega}{\mathrm{d}t} = \frac{k_p\,\phi}{V_1\,N_A}\,\bar{\imath} \tag{5.33}$$

Substituting eqn (5.33) in eqn (5.31) gives

$$\frac{\mathrm{d}\Pi}{\mathrm{d}t} = N\frac{\sigma_1}{\sigma_2}\frac{k_p\phi}{N_A}\,\bar{\imath} \tag{5.34}$$

In order to apply this entirely general equation to interval 2, it is necessary to consider the assignment of values to N, $\bar{\imath}$ and ϕ. So long as attention is confined to times such that $t \geqslant t^*$, the assumed model gives N as constant. Its value has already been calculated and the result is given in eqn (5.16). Regarding $\bar{\imath}$, the value of 0·5 corresponding to the Smith–Ewart 'Case 2' is assumed. The justification for this value has been adequately discussed in previous chapters of this book. It is sufficient to recall here that the justification depends upon the further assumption that termination occurs very rapidly relative to propagation. Accordingly, no two radicals can coexist within the same particle for other than infinitesimal periods of time. ϕ is taken to be constant for interval 2, and is in principle determinable by experiment. The argument by which this assumption is justified has been summarised in outline in Chapter 4, Section 4.2. This argument is valid so long as unlimited monomer is available for imbibition into the particles. Thus the validity extends for as long as a separate monomer droplet phase is present, that is, for the duration of interval 2.

In the final paper of his series, Gardon[6] examines the assumption of constant ϕ in some detail. For, as he points out, the assumption is at first sight rather arbitrary and implausible unless the polymer within the particle is crosslinked. In fact, two distinct matters have to be considered. In the first place, it is necessary to show that the swelling of a particle of uncrosslinked polymer attains an equilibrium value which is independent of the size of the particle. This is a matter to be decided according to the principles of equilibrium thermodynamics. But, in the second place, it has to be shown that, if such an equilibrium exists in theory, then the diffusion of monomer from droplets through the aqueous phase to the reaction loci is sufficiently fast relative to monomer consumption in the loci for the equilibrium to be achieved and maintained throughout interval 2. Gardon's discussion confirms that ϕ is theoretically a relatively insensitive function of particle size over the range of sizes which are encountered in typical emulsion polymerisation systems. Furthermore, as the particles grow in size during interval 2, their surfaces becomes less densely covered with adsorbed soap molecules, there being now no reservoir of micellar soap to provide additional soap molecules for adsorption. Thus the interfacial free energy at the interface will tend to rise, thereby counteracting any tendency for ϕ to increase with increasing particle size. As noted previously, constancy of ϕ during interval 2 also carries the implication constancy of k_p. Although k_p may vary somewhat throughout the course of a normal bulk polymerisation (although not as much as does k_t), it seems reasonable to suppose that it will not vary significantly if monomer is continuously added at such a rate as to keep the ratio of monomer to polymer constant.

Making the substitution $\bar{i} = 0{\cdot}5$ in eqn (5.34) gives

$$\frac{\mathrm{d}\Pi}{\mathrm{d}t} = 0{\cdot}5\, N \frac{\sigma_1}{\sigma_2} \frac{k_p \phi}{N_A} \tag{5.35}$$

Integration of this equation between the limits $t = t^*$, $\Pi = \Pi^*$ and $t = t$, $\Pi = \Pi$ gives the variation of Π with t over interval 2. The final result is

$$\Pi - \Pi^* = B(t - t^*) \tag{5.36}$$

where B denotes right-hand side of eqn (5.35), which, it will be noted, is a constant throughout interval 2, in view of the assumptions discussed in the previous paragraphs. Equation (5.36) as it stands implies that the conversion–time relationship for interval 2 should be linear. This conclusion in itself is hardly surprising. It is an obvious and immediate consequence of a model for which (a) the number of reaction loci is constant, (b) each locus contains either one or no propagating radical, (c) the number which contain one radical is a fixed proportion of the whole, and (d) the concentration of monomer is constant. The significant feature of the Gardon theory is that it enables the constant B to be calculated from certain fundamental attributes of the system. If the value of N given by eqn (5.16) is substituted into eqn (5.35), and then K substituted from eqn (5.10), it is readily found that

$$B = \frac{\mathrm{d}\Pi}{\mathrm{d}t} = 0{\cdot}104\left(\frac{4\pi}{3}\right)^{2/5}\left\{\frac{A\phi k_p}{N_A} \cdot \frac{\sigma_1}{\sigma_2}\right\}^{3/5} \{R(1-\phi)\}^{2/5} \tag{5.37}$$

In principle, every factor in the expression which forms the right-hand side of this equation is known for a given emulsion polymerisation system, in the sense that each factor can be determined by experiments which require measurements to be made of quantities other than rates of emulsion polymerisation. Therefore, eqn (5.37) provides a basis for predicting the rate of reaction in interval 2 quite independently of any measurements of that rate.

Two further comments are necessary concerning the conversion–time relationship for interval 2. In the first place, eqn (5.36) can be rearranged to give

$$\Pi = Bt + \Pi^* - Bt^* \tag{5.38}$$

This shows that, whilst the conversion–time relationship is linear, it does not in general pass through the origin. If projected back to the starting-point of the reaction, the line will intersect the conversion axis at the point $\Pi^* - Bt^*$. This is positive, zero or negative according as Π^*/t^* is greater than, equal to, or less than B; that is, according as the mean rate during interval 1 is greater than, equal to, or less than the steady rate during interval 2.

The second comment concerns the fact that, having shown that the average number of propagating radicals per particle is 0·673 at the conclusion of interval 1, Gardon then assigns the Smith–Ewart value of 0·5 for interval 2. Thus once again we meet the problem to which reference was made in Chapter 4, Section 4.2, namely, the prediction that the rate of polymerisation at the end of the particle-nucleation stage should exceed the steady rate in the phase which succeeds particle initiation. In common with the Smith–Ewart theory, the Gardon theory predicts that the rate should pass through a maximum at the end of interval 1 and the beginning of interval 2. Concerning this prediction, Gardon merely states that 'this would probably be unnoticeable experimentally'.

5.6 DISTRIBUTION OF PARTICLE SIZES

Gardon states that two factors operate to give a distribution of particle sizes at the end of interval 1. The first arises from the random nature of the entry of radicals into micelles and particles. The times $\bar{t}_1$ and $\bar{t}_2$ introduced in Section 5.4 are averages, the first being for the time which elapses between the entries of the first and second radicals, and the second for the time between the entries of the second and third radicals. In fact, the actual times will be distributed around these averages as means. This implies that some initiated particles will remain active for longer times and some for shorter times than $\bar{t}_1$, and that some particles whose activity has been destroyed by the entrance of a second radical will remain inactive for longer times and some for shorter times than $\bar{t}_2$. Gardon states that the distribution of intervals about $\bar{t}_1$ and about $\bar{t}_2$ will be Poissonian; in fact, for a random entry of radicals it should be exponential. But this does not in any way invalidate Gardon's general argument, because in his calculations he ignores polydispersity arising from this cause; consequently his final estimate of polydispersity is low.

The second factor which gives rise to a distribution of sizes is that particles nucleated at different times have correspondingly different lengths of time during which to grow until the end of interval 1. It is in principle a simple matter to calculate the effect of this factor upon the distribution of particle sizes at the end of interval 1. The detailed mathematics involved are, however, rather complex. Gardon gives the following as his estimates of polydispersity at the end of interval 1, expressed in terms of the ratios of various particle-radius averages:

$$\left.\begin{aligned} \bar{r}_{vs}/\bar{r}_n &= 1{\cdot}03 \\ \bar{r}_w/\bar{r}_n &= 1{\cdot}04 \\ \bar{r}_w/\bar{r}_{vs} &= 1{\cdot}01 \end{aligned}\right\} \qquad (5.39)$$

In these equations: (a) $\bar{r}_n$ denotes the number-average particle radius, *i.e.* $\Sigma\, n_i r_i/N$; (b) $\bar{r}_{vs}$ denotes the volume/surface area-average particle radius, *i.e.* $\Sigma\, n_i r_i^3/\Sigma\, n_i r_i^2$; (c) $\bar{r}_w$ is the mean radius which is obtained by weighting each radius by a factor proportional to the volume of the corresponding particle, *i.e.* $\Sigma\, n_i r_i^4/\Sigma\, n_i r_i^3$. Two comments are necessary concerning these results. First, the ratios of the various radius averages are quite close to unity. Thus it follows that the degree of polydispersity arising from differences in growth times is quite small. But, secondly, it should be noted that the polydispersity due to differences in growth times is independent of the monomer properties, of the mode of initiation, and of the nature or concentration of the micelle generator.

As regards the effect of interval 2 upon the distribution of particle sizes, the prediction is that the distribution becomes narrower throughout the course of interval 2. This prediction is based upon the belief (for which justification is given in Section 5.2) that the rate of growth of the particles is constant irrespective of the size of the particle. It will be recalled that K, which is by definition $\mathrm{d}(r^3)/\mathrm{d}t$ for a single growing particle, is a function of k_p, σ_1, σ_2 and ϕ, and insofar as ϕ and k_p are constant during interval 2, so too is K. Thus over any specified part of interval 2, $\mathrm{d}(r^3)/\mathrm{d}t$ for a given particle is a constant (actually $K/2$) irrespective of size. Thus $r^2\mathrm{d}r/\mathrm{d}t$ is also a constant irrespective of size. Thus $\mathrm{d}r/\mathrm{d}t$ is proportional to r^{-2}, *i.e.* the average rate of increase of particle radius with time is inversely proportional to the square of the radius of the particle at that instant. Thus the small particles should increase in radius more rapidly than the larger, and this should inevitably lead to a sharpening of the particle-size distribution. The conclusion reached is therefore that, although there is a distribution of sizes existing at the close of interval 1, the distribution is narrow and becomes still narrower as the reaction proceeds through interval 2. Thus the theory gives a reasonable explanation for the small spread of particle sizes which is observed in the product of the typical emulsion polymerisation reaction.

5.7 EFFECT OF SLOW TERMINATION WITHIN PARTICLES

It has been emphasised in Section 5.5 that setting the average number of radicals per particle during interval 2 equal to 0·5 implies the assumption that termination within the particles occurs very rapidly relative to propagation. Gardon[3] has given consideration to the consequences of abandoning this assumption whilst, at the same time, retaining all the other features of the Smith–Ewart model. The fact that $\bar{t}_1$ is approximately 80% of t^* suggests that relatively few termination events take place during interval 1. In any case, it appears that the actual extent of conversion during interval 1 is rather low, so that the effect upon the overall conversion–time relationship of slow termination during interval 1 is slight. The situation is otherwise for interval 2. If, as is now assumed, termination between two radicals proceeds at a finite rate, then for a certain time two radicals will coexist within the same particle. Reference to this possibility has already been made in Section 4.2 of Chapter 4. The variation of rate of polymerisation with time at a single reaction locus is illustrated diagrammatically in Fig. 4.1(c) for the case where termination is rapid but not instantaneous. Furthermore, as has also been pointed out in Section 4.2 of Chapter 4, if termination is slow relative to the rate of entry of successive radicals, then there is no reason why a third and even a fourth radical should not have entered before the two coexisting radicals have mutually terminated one another. Thus the possibility exists that a particle can contain any number of radicals. The average number of radicals per particle will exceed 0·5 under these circumstances. Furthermore, this average number will tend to increase with time because the volume of the particles increases with time.

In developing the theory for slow termination, Gardon makes the assumption that the activity of the radicals remains contained within the particles, no matter how many there may be within a given particle. Not only is there no mechanism for the desorption of radicals from particles into the aqueous phase; there is also lacking any mechanism whereby transfer of activity can occur. A further assumption is that, throughout the whole of interval 2, the rate of termination can be specified by a single second-order rate coefficient k_t. This assumption needs some justification because it is well known that, in the case of conventional homogeneous bulk polymerisation, k_t can decrease markedly with conversion, as is evidenced by the Trommsdorff 'gel' effect. The grounds for assuming k_t to be constant are the same as those for assuming k_p to be constant, namely, that if ϕ is constant over interval 2, then both the composition and the viscosity of the reaction loci remain constant. Since k_t (and to a lesser extent k_p) for a given reaction is viscosity-dependent, there is clearly justification for taking it to be independent of conversion throughout interval 2. However, Gardon emphasises that the k_t of his theory is not the conventional low-conversion value usually quoted in the literature. It is the value appropriate to a monomer–polymer mixture of the composition which pertains in the reaction loci during interval 2. To this extent, k_t is not an independent parameter, although it could be determined from bulk polymerisation experiments made over a range of conversions, and its value for the corresponding emulsion polymerisation inferred if the value of ϕ for interval 2 is known.

The calculations are complex. The conclusion reached is that the conversion–time relationship for interval 2 becomes

$$\Pi - \Pi^* = B(t - t^*) + C(t - t^*)^2 \tag{5.40}$$

where B is the steady rate which would occur if termination were infinitely rapid, and C allows for the fact that termination occurs at a finite rate specified by k_t. B is given by eqn (5.37); it depends upon the number of particles and hence upon the amount of micelle generator in the system. C is independent of the number of particles, and depends mainly upon the initiation and termination rates. The explicit expression which Gardon gives for C is

$$C = 0{\cdot}102 \frac{k_p^{\,1\cdot 94}}{k_t^{\,0\cdot 94}} \cdot \frac{\sigma_1}{\sigma_2 N_A} \cdot \frac{\phi^{1\cdot 94}}{(1 - \phi)^{0\cdot 94}} \cdot R \tag{5.41}$$

It will be observed that C is not a function of k_p and k_t separately, but rather of (approximately) the ratio $k_p/\sqrt{k_t}$. Furthermore, C becomes vanishingly small for the case where k_t is very large relative to k_p. It should also be noted that C increases in direct proportion to the rate at which initiating radicals are generated. It is hardly surprising that the effect of slow rate of termination becomes increasingly important as the rate of radical production increases. For an increase in R implies an increase in the frequency with which radicals enter particles, and therefore an increase in the proportion of the reaction time for which two or more radicals can coexist in the same locus.

In principle, k_p can be calculated from B using the right-hand side of eqn (5.35). The ratio k_t/k_p, and hence the value of k_t, can be obtained from B and C. Gardon gives the following expression:

$$\frac{k_t}{k_p} = 0{\cdot}158\left(\frac{RB}{NC}\right)^{1{\cdot}062}\frac{\phi}{1-\phi} \tag{5.42}$$

Gardon also calculates the average number, $\bar{i}$, of radicals per particle throughout interval 2 for the case of slow termination. He finds it to be an increasing function of conversion, the result being

$$\bar{i} = 0{\cdot}5\sqrt{\left(1+\frac{4C\Pi}{B^2}\right)} \tag{5.43}$$

Thus the condition for the attainment of the Smith–Ewart value of 0·5 is $4C\Pi/B^2 \ll 1$. Clearly, $\bar{i}$ has the Smith–Ewart value if C is vanishingly small, *i.e.* if k_t is very large relative to k_p. But assuming that k_p and k_t are fixed, the condition $4C\Pi/B^2 \ll 1$ implies a low initiation rate and a latex of small particle size; for eqn (5.41) shows that, for a given monomer, the only way by which C can be reduced is through reduction in the rate of production of initiating radicals. Furthermore, eqn (5.35) shows that a large N, and hence a small particle size, will favour large B, and therefore small $4C\Pi/B^2$. However, eqn (5.37) shows that B is also an increasing function of R for given A (*i.e.* for a given level of micelle generator). Thus the low initiation rate which favours small C also favours small B. In fact, eqns (5.37) and (5.41), taken together, imply that C/B^2 is proportional to $R^{1/5}$. Hence it follows that whilst low initiation rate favours low $4C\Pi/B^2$, the latter quantity should be rather insensitive to R. It should also be noted that $4C\Pi/B^2$ increases with conversion, so that even if the condition $4C\Pi/B^2 \ll 1$ is fulfilled at the start of interval 2 (where Π is very small), it may not necessarily be true over the whole of interval 2.

As regards the distribution of radicals among particles, Gardon finds that the distribution broadens as reaction proceeds. This is really a consequence of the fact that the rate at which two radicals will terminate falls as the volume of the particle in which they coexist increases. Thus the larger particles will tend to contain more radicals than the smaller ones, and in turn tend to grow more quickly, and thus be able to accommodate, on average, still more radicals. As a consequence, not only does the radical number distribution broaden continuously, but so too does the distribution of growth rates. This in turn leads to a tendency for the particle-size distribution to broaden as the reaction proceeds through interval 2. This tendency opposes, and presumably overcomes, the tendency for the distribution to sharpen when termination occurs rapidly.

5.8 CONVERSION–TIME RELATIONSHIP FOR INTERVAL 3

The transition between interval 2 and interval 3 occurs at the point in the reaction where the monomer droplets just disappear. The value of Π corresponding to this point, which will be denoted by Π^{**}, can be calculated as the point at which the volume of monomer polymerised, plus the volume of monomer imbibed in the polymer formed, plus the volume of monomer dissolved in the aqueous phase, exactly equals the volume of monomer which was initially present in the system. If the volume dissolved in the aqueous phase is negligible, then Π^{**} may be found as follows. Consider a system which initially contained a mass M of monomer and a volume ϑ of aqueous phase. When the conversion is Π, the volume of polymer present is $\Pi\vartheta$, and this must have been formed from an amount of monomer whose volume was $\Pi\vartheta\,\sigma_2/\sigma_1$. The maximum volume of monomer which can be imbibed in a volume $\Pi\vartheta$ of polymer is $\Pi\vartheta\phi/(1-\phi)$. Therefore Π^{**} must be such that

$$\Pi^{**}\,\vartheta\,\frac{\sigma_2}{\sigma_1} + \Pi^{**}\,\vartheta\,\frac{\phi}{1-\phi} = \frac{M}{\sigma_1} \tag{5.44}$$

Rearrangement gives

$$\Pi^{**} = \frac{M}{\vartheta}\cdot\frac{1-\phi}{\sigma_2(1-\phi)+\sigma_1\phi} \tag{5.45}$$

Thus the conversion corresponding to the transition point can readily be calculated if ϕ, σ_1 and σ_2 are known.

During the course of interval 3, polymerisation proceeds at the expense of a fixed amount of monomer within the swollen particles. ϕ no longer remains constant, but decreases with increasing conversion in a calculable manner. A further effect of which quantitative account can readily be taken is shrinkage of the monomer-swollen particles as polymerisation proceeds, σ_2 being greater than σ_1. If, therefore, k_p and k_t remain constant throughout interval 3, the rate of polymerisation would be expected to decrease with increasing conversion, because of monomer depletion. In fact, the reaction would be expected to obey first-order kinetics with respect to monomer concentration. However, in the case of termination, increasing conversion with its accompanying increase in viscosity at the reaction loci is likely to lead to a sharp increase in k_t relative to k_p, and, as has been seen previously, this leads to an increase in rate of polymerisation. Thus the conversion–time curve for interval 3 is determined by the balance of two opposing tendencies: a retardation due to monomer depletion and an acceleration due to the operation of the Trommsdorff 'gel' effect. Gardon states that it is possible that $d\Pi/dt$ actually increases during the early part of interval 3 because the Trommsdorff 'gel' effect more than compensates for the effect of monomer depletion. In the case of rapid termination, as postulated for the Smith–Ewart 'Case 2', monomer depletion will be the sole effect operating if termination remains rapid throughout interval 3 as well as during interval 2.

Gardon makes no attempt to develop an accurate quantitative expression for the conversion–time relationship in interval 3, because of the absense of information concerning the functional relationship between k_t and conversion. It is stated (but no details are given) that calculations were carried out assuming that k_t increases either linearly or quadratically with monomer concentration. The particle volume was assumed to remain constant, *i.e.* any shrinkage due to polymerisation was neglected. The results of these calculations are said to have been of little interest, except that the distribution function for radicals among particles resembles that for interval 2.

5.9 COMPARISONS BETWEEN PREDICTIONS AND EXPERIENCE

Gardon gives extensive comparisons between the various predictions of his theory and the corresponding observations for a wide range of emulsion polymerisation systems. For the details of these comparisons, reference should be made to the original papers. All that can be given here is a summary of his main conclusions.

Concerning interval 1, it appears that the available data confirm the theoretical prediction that the conversion at the cessation of particle nucleation is very small. For most systems, the amount of polymer formed by the end of interval 1 is of the order of 10^{-2} ml of polymer per ml of water. Good qualitative agreement is observed between prediction and observation in respect of the conversion–time relationship during interval 1; the rate of polymerisation increases with time and then, when particle nucleation ceases, it levels off. Gardon finds excellent quantitative agreement between predicted and observed particle sizes. He reports that the difference between theoretical and experimental particle radii was always less than about 20% in the case of the emulsion polymerisation of methyl methacrylate using sodium lauryl sulphate as micelle generator and potassium persulphate as initiator. The concentration of sodium lauryl sulphate was varied over a 60-fold range, and that of potassium persulphate over a 140-fold range. Similar good agreement was found for styrene emulsion polymerisation. The predicted joint dependence of particle number upon the 3/5th power of the soap concentration and the 2/5th power of the initiator concentration was observed for several monomers. Although not discussed in the preceding summary of the theory, Gardon also derives equations by which the molecular weight of the polymer produced can be predicted. He finds good agreement between expected and observed molecular weights, at least as regards order of magnitude. For several monomers, both the rate of polymerisation and the molecular weight of the product vary with initiator and soap concentration in a manner close to the theoretical predictions.

As regards the conversion–time relationship for Interval 2, Gardon finds that there is a large corpus of data which is consistent with the general form of eqn (5.40). For latices of very small particle size, a linear relationship between $\Pi - \Pi^*$ and $t - t^*$ is often observed, whereas, for latices of very large particle size, $\Pi - \Pi^*$ is found to be proportional to $(t - t^*)^2$. This is entirely in accordance with theoretical prediction, bearing in mind the effect which latex particle size would be expected to have upon the relative magnitudes of the coefficients B and C of eqn (5.40). Experimental values of the coefficient B are in good agreement with independent theoretical predictions. Estimates of the ratio k_t/k_p obtained from B and C (see eqn (5.42)) are of the same order of magnitude as independent estimates. Estimates of the average number of radicals per polymer particle show this quantity often to be considerably higher than the Smith–Ewart value of 0·5; values as high as 10 have been found.

The overall impression gained from the Gardon theory is that it confirms the essential correctness of the Harkins theory for those monomers which are sparingly soluble in water and which are readily miscible with their own polymer. It represents a considerable improvement upon and refinement of the Smith–Ewart theory, and yet at the same time confirms the earlier theory as at least a good first approximation to which reality conforms in some cases. The Gardon theory still contains several notable gaps and uncertainties, but doubtless attempts to refine the theory further will subsequently appear in the literature.

REFERENCES

1. J. L. Gardon, *J. Polym. Sci.*, Pt A–1, 1968, **6**, 623.
2. J. L. Gardon, *J. Polym. Sci.*, Pt A–1, 1968, **6**, 643.
3. J. L. Gardon, *J. Polym. Sci.*, Pt A–1, 1968, **6**, 665.
4. J. L. Gardon, *J. Polym. Sci.*, Pt A–1, 1968, **6**, 687.
5. J. L. Gardon, *J. Polym. Sci.*, Pt A–1, 1968, **6**, 2853.
6. J. L. Gardon, *J. Polym. Sci.*, Pt A–1, 1968, **6**, 2859.
7. J. L. Gardon, *Br. Polym. J.*, 1970, **2**, 1.
8. J. L. Gardon, *Rubb. Chem. Technol.*, 1970, **43**, 74.

CHAPTER 6

INITIATION AND INITIATOR SYSTEMS

6.1 GENERAL INTRODUCTION

By initiation is meant the totality of all those processes whereby a propagating centre comes to be present in the polymerisation locus. As thus understood, it can comprise several individual steps. These can include:

(1) formation of radicals within the aqueous phase, by, for instance, chemical reactions involving added initiating substances;
(2) reaction with water molecules to form other radicals;
(3) reaction with monomer molecules dissolved in the aqueous phase, with the formation of oligomeric propagating centres;
(4) transfer of radicals or oligomeric propagating centres from the aqueous phase into the polymerisation loci; and
(5) initiation of propagating centres within the polymerisation loci.

It is the purpose of this chapter to review the known chemistry of the more important of those systems which have been used for the initiation of polymerisation in emulsion. In so doing, it will be convenient to consider, as appropriate, some of the other matters which are listed above, in particular, the problems of understanding the processes whereby a radical species in the aqueous phase becomes transferred to the polymerisation locus.

It is conventional to subdivide initiator systems into two broad classes: (a) *dissociative* initiators, and (b) *redox* initiators. As the name implies, dissociative initiators are those which function by simple dissociation of a molecule or an ion into two radical species, from which species initiation then takes place. For practical purposes, the only important emulsion polymerisation initiators of this type are the inorganic salts of persulphuric acid. The initial step in the initiation process is usually regarded as being the homolytic dissociation of the persulphate ion into two sulphate radical-ions:

$$S_2O_8^{2-} \rightarrow 2SO_4^{\cdot -} \qquad \text{(i)}$$

Redox initiators, on the other hand, are systems comprising two or more substances whose mutual interaction produces free radicals able to initiate polymerisation. The name derives from the fact that these combinations usually depend for their effectiveness upon the interaction between a reducing agent and an oxidising agent. Whilst the distinction between dissociative and redox initiators may at first sight appear to be clear-cut, in reality some blurring of the distinction may occur, because it is generally believed that persulphates in conventional styrene–butadiene emulsion copolymerisation systems in fact function as redox initiators. The other part of the couple is provided by the mercaptan which is primarily added as a chain-transfer agent to reduce the degree of polymerisation of the product. But in spite of this blurring, it is convenient to subdivide this chapter into the two main headings of dissociative and redox initiation systems.

A. DISSOCIATIVE INITIATORS

6.2 KINETICS AND MECHANISM OF PERSULPHATE DECOMPOSITION IN AQUEOUS MEDIA

Studies of the kinetics of the decomposition of metal persulphates in aqueous solution were first made by Levi and Migliorini[1] and later by Green and Masson,[2] and by Kailan and his co-workers.[3–5] There is general agreement among these workers that persulphate decomposition in aqueous solution is first-order with respect to persulphate-ion concentration, and that the reaction is catalysed by hydrogen ions. In the absence of substances which can scavenge the $SO_4^{\cdot -}$ radicals initially formed, the overall reaction in alkaline, neutral and dilute acid solution can be written

$$K_2S_2O_8 + H_2O \rightarrow 2KHSO_4 + \tfrac{1}{2}O_2 \qquad \text{(ii)}$$

Mechanistically, however, the process must take place according to a scheme such as the following:

$$S_2O_8^{2-} \rightarrow 2SO_4^{\cdot -} \qquad \text{(iii)}$$

$$SO_4^{\cdot -} + H_2O \rightarrow HSO_4^{-} + \cdot OH \qquad \text{(iv)}$$

$$2\cdot OH \rightarrow H_2O + \tfrac{1}{2}O_2 \qquad \text{(v)}$$

One of the reaction products is the bisulphate anion and, to the extent that this is able to ionise into hydrogen ions and sulphate ions, and buffers are absent, the pH of the medium is to be expected to drop during the course of the decomposition. In strongly acidic solutions (which are of little interest in conventional emulsion polymerisations), the following reactions are also believed to occur:

$$H_2S_2O_8 + H_2O \rightarrow H_2SO_5 + H_2SO_4 \qquad \text{(vi)}$$

$$H_2SO_5 + H_2O \rightarrow H_2O_2 + H_2SO_4 \qquad \text{(vii)}$$

Of the more recent studies of the kinetics and mechanism of persulphate decomposition which have been published, that of Kolthoff and Miller[6] is of particular interest. Persulphate analyses were made using a polarographic method developed by Kolthoff *et al.*[7] Decompositions were carried out in buffer solutions to ensure that the pH remained constant. The results summarised in Fig. 6.1 show that the reaction is always kinetically of first order with respect to persulphate-ion concentration, and that the rate coefficient is strongly dependent upon pH, especially in acid solution. The results given in Table 6.1 show that the first-order rate coefficient for decomposition in alkaline solution (0·1 mole l^{-1} sodium hydroxide) is independent of ionic strength, but the first order rate coefficient for decomposition in acid solution (0·1 mole l^{-1} perchloric acid) falls with increasing ionic strength. The ionic strength was varied by the addition of sodium perchlorate.

The effect of pH upon the overall first-order rate coefficient for the decomposition of potassium persulphate at 50°C is shown in Fig. 6.2. The rate coefficients were adjusted to a constant ionic strength of 0·4 (to allow comparisons to be made with results obtained at different ionic strengths).

TABLE 6.1

Effect of ionic strength upon the rate of thermal decomposition of 0·01 mole l^{-1} aqueous potassium persulphate at 50°C (according to Kolthoff and Miller[6])

$NaClO_4$ *concentration* (mole l^{-1})	$k_0(\text{min}^{-1}) \times 10^5$†	
	in 0·1 mole l^{-1} NaOH	*in* 0·1 mole l^{-1} $HClO_4$
0	6·05	57·1
0·1	6·02	45·4
0·4	6·04	34·1
0·9	6·05	27·8

† k_0 = rate coefficient for first-order decomposition.

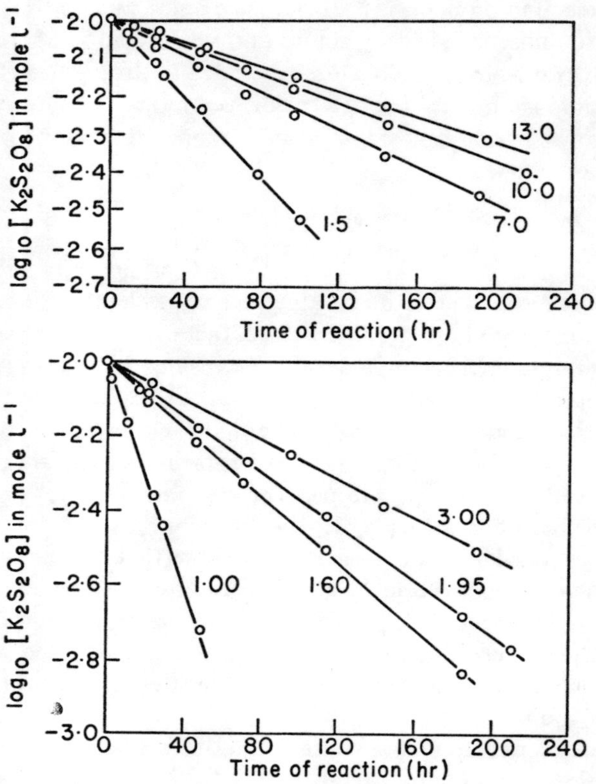

FIG. 6.1. Course of thermal decomposition of potassium persulphate in aqueous solution at 50°C and various hydrogen-ion concentrations, according to Kolthoff and Miller.[6] The initial potassium persulphate concentration was 0·01 mole l^{-1}. The numbers appended to the lines indicate the pH levels.

Clearly, reducing the pH has relatively little effect until the solution is strongly acidic (pH *ca.* 3), when the rate coefficient increases sharply. Kolthoff and Miller state that the variation of overall first-order rate coefficient, k_0, with pH over the range studied can be represented by the equation

$$k_0 = k_1 + k_2[H^+]$$

Thus the overall reaction can be regarded as consisting of two simultaneous reactions: one which is an uncatalysed decomposition, and the other which follows first-order catalysis by hydrogen ions. The effect of temperature upon the decomposition could be interpreted in terms of an activation

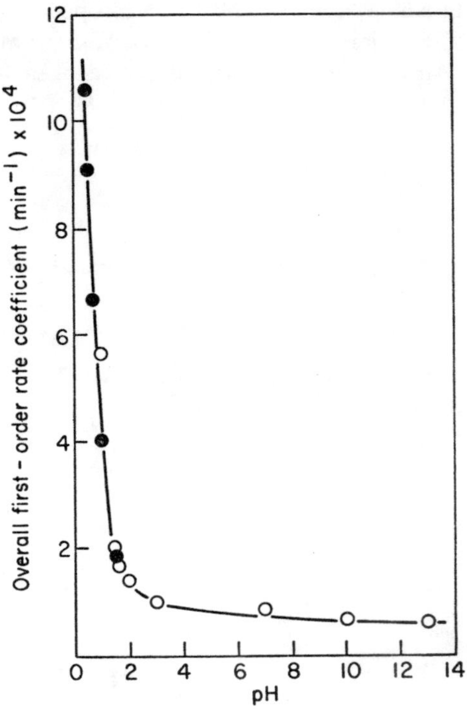

FIG. 6.2. Effect of pH upon overall first-order rate coefficient for thermal decomposition of potassium persulphate in aqueous solution at 50°C, plotted from data given by Kolthoff and Miller.[6] The rate coefficients were adjusted to a constant ionic strength of 0·4. The initial potassium persulphate concentrations were 0·01 mole l^{-1} (to which data points ○ refer) and 0·1 mole l^{-1} (to which data points ● refer).

energy of 33·5 kcal $mole^{-1}$ for the uncatalysed reaction, and of 26·0 kcal $mole^{-1}$ for the acid-catalysed reaction.

The source of the oxygen liberated as a result of persulphate decomposition was investigated by carrying out the decomposition in water which had been enriched with H_2O^{18}. The isotopic composition of the oxygen liberated was assayed by mass spectroscopy. The published results, one set of which is given in Table 6.2, show that the source of the oxygen depends very much upon the pH at which the decomposition occurs. In alkaline and neutral media, the oxygen comes almost exclusively from the water, whereas, in strongly acid solution, most of the oxygen is derived from the persulphate ion.

The mechanism which Kolthoff and Miller propose for the uncatalysed decomposition is essentially that which has been given above in reactions (iii), (iv) and (v). For the reaction catalysed by hydrogen ions, they propose the following:

$$S_2O_8^{2-} + H^+ \rightarrow HS_2O_8^- \rightarrow SO_4 + HSO_4^- \qquad \text{(viii)}$$

$$SO_4 \rightarrow SO_3 + \tfrac{1}{2}O_2 \qquad \text{(ix)}$$

In strong acid, they suggest that the following reaction can also occur:

$$SO_4 + H_2O \rightarrow H_2SO_5 \qquad \text{(x)}$$

A possible mechanism for reaction (viii) could involve the formation of a cyclic transition state for the species $HS_2O_8^-$, with its subsequent decomposition by electron transfer:

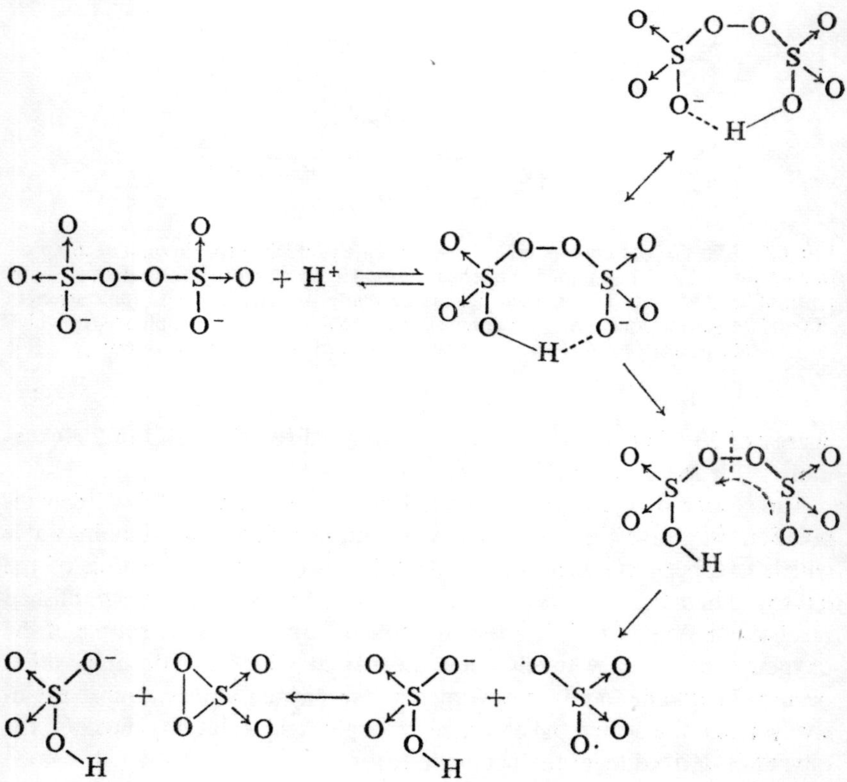

TABLE 6.2

Source of oxygen liberated by decomposition of persulphate at 90°C in water enriched with H_2O^{18} (according to Kolthoff and Miller[6])

pH	%O_2 *from* H_2O	O_2 *from* H_2O : O_2 *from* $S_2O_8^{2-}$
13	100	∞
10	98	49
7·2	96	24
2·1	77	3·3
1·5	67	2·0
1·0	33	0·51
0·3	6	0·06

A possible mechanism for reaction (ix) might then be:

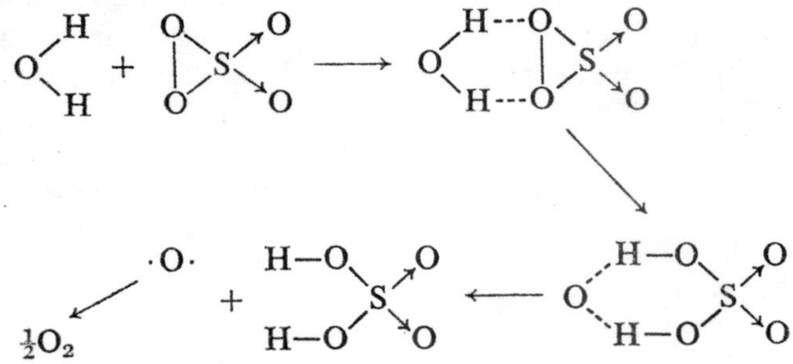

This mechanism is consistent with the observations insofar as SO_3 forming in an aqueous medium would be indistinguishable from H_2SO_4. But it is inconsistent in that it predicts that the liberated oxygen comes from the water. The following alternative mechanism obviates this difficulty:

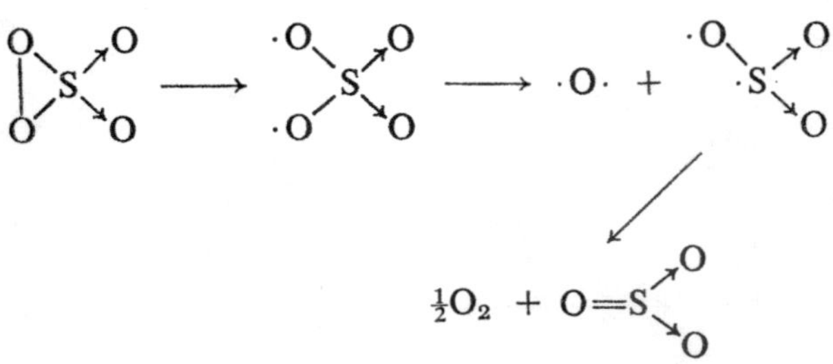

The outline reaction mechanism implied by reactions (iii) to (v) and (viii) to (x) above is obviously consistent with the observed kinetic behaviour if two conditions are fulfilled: (a) that the uncatalysed and acid-catalysed decompositions occur independently of one another, and (b)

that reactions (iii) and (viii) are the respective rate-determining steps. The proposed mechanism is also consistent with the evidence that the oxygen liberated by the uncatalysed reactions comes from the water, whilst that liberated by the acid-catalysed reaction comes from the persulphate. Furthermore, kinetic analysis of the above reaction scheme shows that the ratio of oxygen from water to oxygen from persulphate should be equal to $k_1/k_2\,[H^+]$. In fact, Kolthoff and Miller found reasonable agreement between this ratio and the experimentally-determined ratios at various pH values.

The results of a most thorough investigation into the kinetics of the thermal decomposition of potassium persulphate in aqueous solution has been published by Hakoila.[8] The decomposition was followed by titrating with alkali the bisulphate anions liberated in accordance with reaction (i) The precaution was taken of carefully cleaning the reaction flask and then 'conditioning' it to the reaction mixture. This was done by adding an initial quantity of reaction mixture to the flask, and then renewing it when reaction had proceeded to about 10%. The effect of conditioning in this way was to reduce the extrapolated first-order rate coefficient significantly. The lower values were regarded as more accurately representing the be-

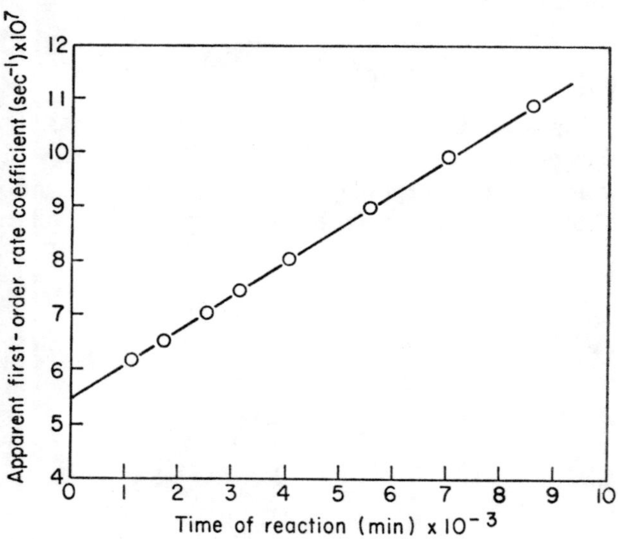

FIG. 6.3. Effect of reaction time upon apparent first-order rate coefficient for thermal decomposition of potassium persulphate in unbuffered aqueous solution at 45°C, plotted from data given by Hakoila.[8] The initial potassium persulphate concentration was 0·2 mole l^{-1}.

TABLE 6.3

Rate coefficients for thermal decomposition of potassium persulphate in pure water at various initial persulphate concentrations and various temperatures (according to Hakoila[8])

	First-order rate coefficient (sec^{-1}) × 10^8			
	Initial concentration of $K_2S_2O_8$			
Temperature	0·0200 mole l^{-1}	0·0500 mole l^{-1}	0·100 mole l^{-1}	0·200 mole l^{-1}
25°C	2·30	1·97	1·70	1·40
35°C	12·0	11·5	10·7	9·9
45°C	66	62	58	54
60°C	578	544	522	499

haviour of the system, because of the susceptibility of the decomposition to catalysing by adventitious impurities.

In unbuffered solutions, the decomposition was found to be autocatalytic, because of the reduction in pH which accompanied the decomposition. This is shown clearly in Fig. 6.3, where the apparent first-order rate coefficient is plotted as a function of reaction time. It is seen that the rate coefficient increases approximately linearly with time, and that a reliable estimate of the initial value can be obtained by extrapolation. Values obtained by such extrapolation are regarded as appropriate to the decomposition in 'pure' water, *i.e.* in water which is pure apart from the presence of the persulphate. Table 6.3 summarises Hakoila's estimates for the rate coefficient for the decomposition of the persulphate ion in pure water at various initial persulphate concentrations and various temperatures. At any given temperature, the rate coefficient decreases with increasing persulphate concentration, the effect being especially marked at higher temperatures. It is regarded as being essentially an ionic strength effect, since the nature of the environment in which the decomposition is occurring alters as the initial concentration of the potassium persulphate increases.

It is of some interest to consider the inferences which can be drawn from these rate coefficients regarding the half-life of persulphate ions at various temperatures. Taking the rate coefficients given in Table 6.3 for an initial persulphate concentration of 0·02 mole l^{-1}, it appears that the respective half-lives at 35°, 45° and 60°C are approximately 1600 h, 292 h and 33 h. These values will not be greatly affected by initial persulphate concentration. It thus appears that, for reaction temperatures up to about 50°C, the half-life is long compared to the usual reaction time for an emulsion polymerisation, but, at temperatures above 50°C, the half-life decreases rapidly, and at 60°C is of the same order of magnitude as the time required for complete conversion in, say, a styrene–butadiene emulsion copolymerisation reaction. However, it must also be borne in mind that, as will appear from Section 6.3, there is likely to be a tendency for other components of the system to accelerate the decomposition of the persulphate, and so reduce its half-life. Interpolation between the rate coefficients shown in Table 6.3 to 50°C (a temperature of particular interest in the application of persulphate initiation) gives a half-life of about 150 h. This is about ten times greater than the longest reaction times normally used at this temperature.

Typical results for the effect of additions of potassium bisulphate and of sulphuric acid upon the rate coefficient extrapolated to zero time are summarised in Fig. 6.4. It is apparent that bisulphate ions have an accelerating effect on the decomposition, as also does sulphuric acid. Furthermore, the variation of rate coefficient with added sulphuric acid is almost linear. Potassium bisulphate is less effective than sulphuric acid in accelerating the reaction.

Hakoila's results for the Arrhenius parameters and thermodynamic activation functions for the thermal decomposition of potassium persul-

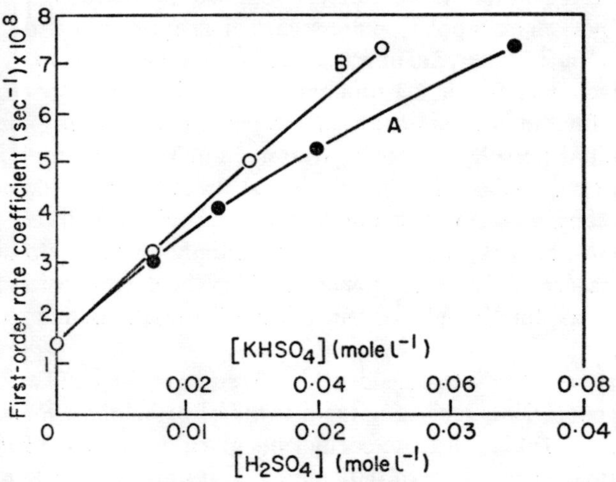

FIG. 6.4. Effect of additions of potassium bisulphate (A) and of sulphuric acid (B) upon the initial first-order rate coefficient for the thermal decomposition of potassium persulphate in aqueous solution at 25°C, according to Hakoila.[8] The initial potassium persulphate concentration was 0·2 mole l^{-1}.

TABLE 6.4

Arrhenius parameters and thermodynamic activation functions for the thermal decomposition of potassium persulphate in neutral solutions (according to Hakoila[8])

	Initial concentration of $K_2S_2O_8$			
	0·0200 mole l^{-1}	0·0500 mole l^{-1}	0·100 mole l^{-1}	0·200 mole l^{-1}
E (kcal mole^{-1})	$31{\cdot}3_9$	$31{\cdot}8_2$	$32{\cdot}3_8$	$33{\cdot}1_4$
$\log_{10} A$ (A = pre-exponential factor)	$15{\cdot}2_0$	$15{\cdot}6_1$	$15{\cdot}9_7$	$16{\cdot}4_5$
$\Delta G\ddagger$ (kcal mole^{-1})	27·1	28·0	28·1	28·2
$\Delta H\ddagger$ (kcal mole^{-1})	30·8	31·2	31·8	32·6
$\Delta S\ddagger$ (cal °K^{-1} mole^{-1})	9·0	10·9	12·5	14·7

phate in neutral solutions are given in Table 6.4. With the exception of entropy of activation, they vary little with initial concentration of potassium persulphate. The activation energy agrees well with that reported by Kolthoff and Miller[6] for the uncatalysed thermal decomposition of potassium persulphate. Hakoila confirms that the effect of initial acidity, whether derived from added sulphuric acid or from added bisulphate ions, is to reduce the activation energy. The lowest value he reports is 25.4 kcal mole^{-1}, and this agrees well with the value of 26.0 kcal mole^{-1} found by Kolthoff and Miller[6] for the acid-catalysed decomposition.

Hakoila has given a detailed kinetic analysis of his results based essentially upon the Kolthoff–Miller scheme of reactions described above. This analysis enabled him to calculate first-order rate coefficients for the thermal decomposition of the hydrogen persulphate anion under various conditions. His results are given in Table 6.5. It is seen that, at any given temperature and level of added potassium bisulphate, the rate coefficient increases markedly as the initial concentration of potassium persulphate is increased. This is interpreted as due to the increase in the ionic strength of the environment. Likewise, the rate coefficient increases as the amount of added potassium bisulphate increases, and this also is interpreted as an ionic strength effect, since the analysis allows for the purely catalytic effect of the hydrogen ions derived from the added bisulphate anions. Rate coefficients at various temperatures for the thermal decomposition of persulphate and hydrogen persulphate anions are shown in Fig. 6.5 as functions of the overall ionic strength of the aqueous medium in which the initial decomposition occurred. Whereas the rate coefficient for the persulphate decomposition falls slightly with increasing ionic strength, that of

TABLE 6.5

Calculated values for the rate coefficient for the thermal decomposition of the hydrogen persulphate anion in aqueous solution at various temperatures and various concentrations of potassium persulphate (according to Hakoila[8])

		First-order rate coefficient (sec^{-1}) × 10^8		
	$KHSO_4$ *concentration*	*Initial concentration of* $K_2S_2O_8$		
Temperature	(mole l^{-1})	0·05 mole l^{-1}	0·1 mole l^{-1}	0·2 mole l^{-1}
25°C	0·02	26·1		
	0·04	28·1	46·2	69·8
	0·06		49·6	74·4
35°C	0·02	107		
	0·04	107	169	223
	0·06		164	233
45°C	0·02	312		
	0·04	351	445	602
	0·06		490	690
60°C	0·02	1790		
	0·04	2150	2550	2890
	0·06		2490	2970

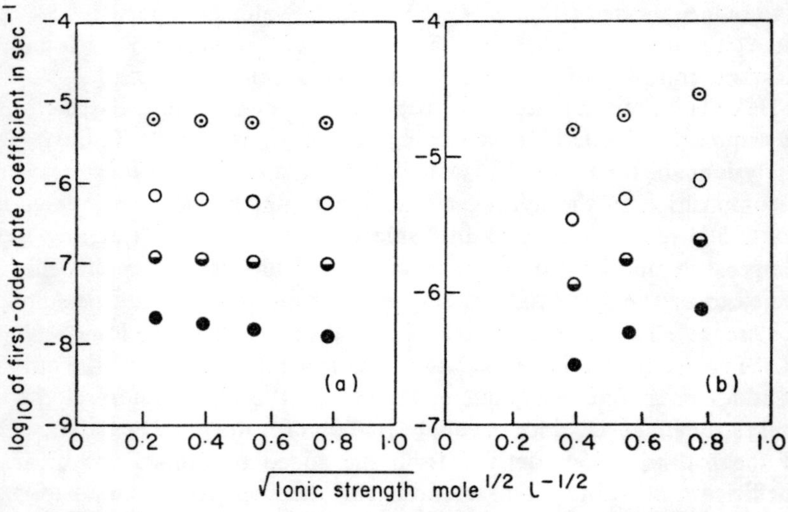

FIG. 6.5. Effect of ionic strength of reaction medium upon rate coefficient at various temperatures for (a) thermal decomposition of persulphate anions in aqueous solution, and (b) thermal decomposition of hydrogen persulphate anions in aqueous solution, according to Hakoila.[8] Significance of points: ⊙ 60°C; ○ 45°C; ◒ 35°C; ● 25°C.

the hydrogen persulphate ion increases markedly. These observations are apparently at variance with those of Kolthoff and Miller[6] (see Table 6.1), but it should be noted that Kolthoff and Miller varied their ionic strengths by adding sodium perchlorate to the reaction medium, whereas Hakoila varied his by adding reactants. As Hakoila remarks, the two sets of results are not necessarily at variance if allowance is made for possible specific ion effects.

6.3 EFFECT OF OTHER SUBSTANCES UPON THERMAL DECOMPOSITION OF PERSULPHATE IN AQUEOUS MEDIA

The studies considered in the previous section have been concerned exclusively with the decomposition of persulphate ions in very simple aqueous media consisting of reactants, products, buffers and certain additives. The situation is more complex in the typical emulsion polymerisation system to the extent that many other substances may be present. The indications from the published literature are that some of these substances may have an effect (usually of acceleration) upon the persulphate decomposition.

Thus Bartlett and Nozaki,[9] studying the emulsion polymerisation of allyl acetate using potassium persulphate as initiator, had occasion to investigate whether the presence of the monomer was likely to have any effect upon the rate of decomposition of the persulphate. For this purpose, they chose to study the effect of a non-polymerising analogue of the monomer, for which ethyl acetate was selected. The results are shown in Fig. 6.6. The decomposition was studied at 80°C in aqueous media maintained alkaline by the presence of 0·45 mole l^{-1} sodium pyrophosphate. Experiments were also carried out to see if oxygen had any effect upon the decomposition; some solutions were exhaustively out-gassed, whilst others were saturated with oxygen. In order to ascertain the effect of ethyl acetate on the reaction, the aqueous solution was saturated with the ester. The results summarised in Fig. 6.6 show that (a) the presence of oxygen has no effect upon the course of the decomposition; (b) the decomposition is accurately first-order both in the absence and in the presence of ethyl acetate; and (c) ethyl acetate has a marked accelerating effect upon the decomposition. Bartlett and Nozaki state that 'It was found that the rate of persulfate decomposition was increased over eight-fold by this change in conditions, the first order rate constant being 0·248 h^{-1} in the absence of ethyl acetate and 2·08 h^{-1} in its presence.' No mechanism for the

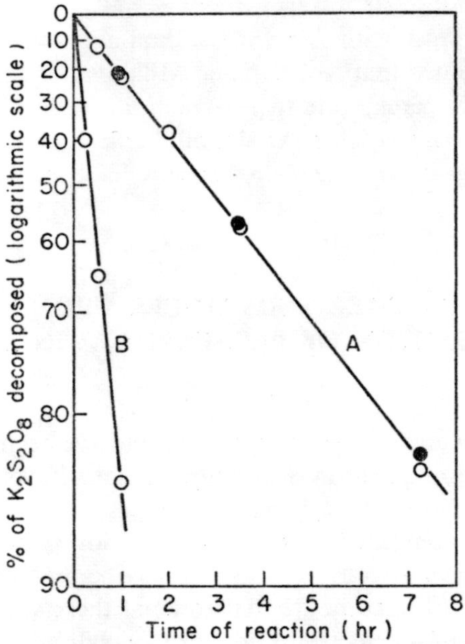

FIG. 6.6. Effect of ethyl acetate upon thermal decomposition of potassium persulphate in aqueous solution at 80°C, according to Bartlett and Nozaki.[9] Line A is for decomposition in the absence of ethyl acetate; line B is for decomposition in the presence of ethyl acetate at saturation concentration. Significance of points: ○ oxygen absent; ● oxygen present. (Note that the vertical scale is logarithmic in percentage of persulphate decomposed, whereas the horizontal time scale is linear. Linearity of data plotted in this way verifies consistency with first-order kinetics.) (The initial potassium persulphate concentrations are not given. The solutions are said to have been kept 'on the alkaline side' with 0·45 mole l^{-1} sodium pyrophosphate.)

accelerated decomposition was given. Bartlett and Nozaki also endeavoured to follow the rate of disappearance of persulphate in the presence of allyl acetate at 80°C, and came to the conclusion that the rate was approximately double its value in the absence of the monomer.

The effect of methanol upon the decomposition of potassium persulphate in aqueous solution has been the subject of a study by Bartlett and Cotman.[10] The temperature of decomposition was 80°C, and the hydrogen-ion concentration was maintained at pH 8 throughout by means of a phosphate buffer. In the absence of methanol, the decomposition was found to be strictly first-order over at least 85% of the reaction. The presence of methanol in a concentration of about 1 mole l^{-1} was observed to cause a

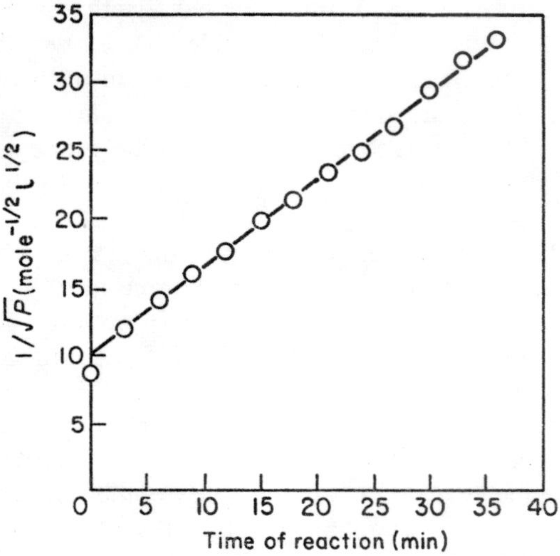

FIG. 6.7. Kinetics of thermal decomposition of potassium persulphate in aqueous methanol solution at 79·8°C, according to Bartlett and Cotman.[10] *P* is the concentration in mole l^{-1} of persulphate remaining after a given time of reaction. The initial potassium persulphate concentration was 0·013 75 mole l^{-1}; that of methanol was 0·488 mole l^{-1}. The pH was 8.

25-fold increase in the rate of decomposition, and to be oxidised to formaldehyde in an amount equivalent to the persulphate which had decomposed. But in addition to its effect in accelerating the decomposition, the presence of the methanol was found to change the order of the reaction from unity to 1·5. This is confirmed by the linearity of a plot of reciprocal square root of amount decomposed against time (see Fig. 6.7). The 1·5-order rate coefficient was observed to vary as the square root of the methanol concentration, so that the conclusions of Bartlett and Cotman concerning the kinetics of the reaction can be summarised by the equation

$$-\frac{d[K_2S_2O_8]}{dt} = k[CH_3OH]^{1/2}\,[K_2S_2O_8]^{3/2}$$

These authors give the following reaction scheme in order to account for their observations:

$$S_2O_8^{2-} + CH_3OH \rightarrow HSO_4^- + SO_4^{\overline{\cdot}} + \cdot CH_2OH \qquad \text{(xi)}$$

$$SO_4^{\overline{\cdot}} + CH_3OH \rightarrow HSO_4^- + \cdot CH_2OH \qquad \text{(xii)}$$

$$S_2O_8^{2-} + \cdot CH_2OH \rightarrow HSO_4^- + SO_4^{\overline{\cdot}} + CH_2O \qquad \text{(xiii)}$$

$$2\cdot CH_2OH \rightarrow CH_3OH + CH_2O \qquad \text{(xiv)}$$

This scheme is consistent with the observed kinetics if the amount of persulphate lost in reaction (xi) is negligible compared with that lost in reaction (xii), *i.e.* the persulphate anion acts as an initiator for a chain reaction which causes its own decomposition. The above scheme is also consistent with formaldehyde as the oxidation product of the methanol, and for its production in an amount equivalent to the persulphate decomposed.

Morris and Parts[11] have undertaken a study of the rate of decomposition of potassium persulphate in the presence of some of the monomers and surface-active substances which may be present in emulsion polymerisation systems. Decompositions were carried out at 50°C, the initial persulphate concentration being approximately 0·1 mole l^{-1}. All the reaction mixtures were maintained at pH 7·0 with a buffer consisting of potassium dihydrogen phosphate and sodium hydroxide. Residual persulphate was determined by a procedure which consisted of adding the persulphate sample to excess arsenious oxide solution and back-titrating the excess with potassium bromate in the presence of potassium bromide and methyl orange. Blank determinations were carried out with all the additives, and none was found to interfere with this analytical procedure. First-order rate coefficients for the decomposition were calculated from plots of the logarithm of residual persulphate against time. In all cases, the plots were linear, but in some instances a distinct break in the line was apparent after which time the points fell on a line of different slope. Two values for the first-order rate coefficient are reported in these cases. The time at which the break in the line occurred was different for the various additives. The results are summarised in Table 6.6. With the exception of the fully-fluorinated soap, the rate of disappearance of persulphate was increased by all the substances investigated. It is interesting to note that the rate of decomposition of persulphate in the presence of sodium hexadecyl sulphate is virtually independent of the soap concentration. Morris and Parts do not accept as correct the interpretation that only molecularly-dissolved sodium alkyl sulphate is capable of catalysing the decomposition. The marked effect which certain common monomers have in accelerating the decomposition of persulphate should be noted; vinyl acetate is particularly effective in this respect.

Kolthoff and Miller[12] have reported upon the interaction between potassium persulphate and carboxylate soaps in aqueous solution. The soaps used were sodium laurate, sodium myristate, sodium palmitate and sodium caprate. The pH was maintained at 10 by the use of a sodium carbonate–sodium bicarbonate buffer. In 10% laurate solution, the rate of

TABLE 6.6

Rate coefficients for the thermal decomposition of potassium persulphate in aqueous solution at 50°C and pH 7·0 in the presence of various additives (according to Morris and Parts[11])

Additive	Concentration (mmole l^{-1})	*First-order rate coefficient* (sec^{-1}) × 10^6	
		Stage I	*Stage II*
None	—	1·4	
Sodium hexadecyl sulphate	0·5 (0·02%)	2·4	1·32
Sodium hexadecyl sulphate	5·0	2·4	
Sodium dodecyl sulphate	0·5	1·6	1·34
A fully-fluorinated soap	? (0·03%)		1·35
Vinyl acetate	50	70	5·2
Vinyl acetate	10	46	3·8
Methyl acrylate	50	7·8	
Methyl acrylate	10	5·8	
Acrylonitrile	50	6·8	2·1
Vinyl acetate and Sodium hexadecyl sulphate	50 5·0	70	5·2
Methyl acrylate and Sodium hexadecyl sulphate	50 5·0	5·7	

disappearance of persulphate was found to be about four times as great as in the buffer solution. In caprate solutions, the rate of disappearance of persulphate was independent of the caprate concentration when the latter was varied from 0·01 to 0·75 mole l^{-1}. Some loss of carboxyl groups occurred in the course of the reaction, and some carbonate was formed. It was also observed that peroxides or hydroperoxides were formed when sodium laurate, myristate and palmitate were oxidised by persulphate in the presence of oxygen. This was taken as evidence of the formation of free-radical intermediates from the soap. The following reactions were proposed:

$$S_2O_8^{2-} + R\underset{\underset{O}{\|}}{C} - O^- \rightarrow SO_4^{\overline{\cdot}} + R\underset{\underset{O}{\|}}{C} - O\cdot + SO_4^{2-} \qquad \text{(xv)}$$

$$SO_4^{\overline{\cdot}} + R\underset{\underset{O}{\|}}{C} - O^- \rightarrow SO_4^{2-} + R\underset{\underset{O}{\|}}{C} - O\cdot \qquad \text{(xvi)}$$

$$R\underset{\underset{O}{\|}}{C} - O\cdot \rightarrow R\cdot + CO_2 \qquad \text{(xvii)}$$

$$R\cdot \rightarrow \,? \qquad \text{(xviii)}$$

Bawn and Margerison[13] have reported on the catalysis of potassium persulphate decomposition by heavy-metal ions. The disappearance of persulphate ions was followed by measuring colorimetrically the rate of disappearance from the system of the stable free radical, diphenylpicrylhydrazyl (DPPH). DPPH is a strongly-coloured substance. It acts as a very efficient scavenger of free radicals, and the reasonable assumption was made that it disappears from the reaction mixture only by interaction with the products of decomposition of the persulphate ions. Thus the rate of disappearance of the radical scavenger was taken to be equal to the rate of decomposition of the persulphate. The fact that the rate of disappearance of DPPH was found to be independent of its concentration was taken as evidence in support of this assumption. Because DPPH is insoluble in water, and potassium persulphate is insoluble in typical organic solvents, it was found necessary to employ a 50:50 ethanol:water mixture as the reaction medium. Bawn and Margerison state that, although quantitative measurements were confined to this particular medium, the rate of consumption of the DPPH was not significantly dependent upon the ethanol: water ratio. They therefore regard conclusions drawn from their study as being applicable to reactions in mixtures of different composition.

The decomposition of the persulphate in the absence of heavy-metal ions was found to be first-order with respect to persulphate concentration. (Note that there should be no drop in pH as reaction occurs, if the radicals produced by the thermal decomposition are completely scavenged; thus complications due to autocatalysis would not be expected.) An activation energy of 28·4 kcal mole^{-1} was reported. The rate of decomposition was found to be unaffected by hydrogen-ion concentration over the range pH 3–7. The hydrogen-ion concentration was varied by the addition of potassium bisulphate to the solution; measurements were restricted to the range pH 3–7, because of the instability of DPPH outside of this range.

In the presence of silver ions (derived from silver nitrate), the rate of consumption of the DPPH was found to obey the relationship

$$-\frac{d}{dt}[DPPH] = (k_1 + k_2[Ag^+])[K_2S_2O_8]$$

where k_1 is the first-order rate coefficient for the uncatalysed reaction (which is of negligible significance relative to the catalysed reaction). The activation energy for the catalysed reaction was found to be 17·9 kcal mole^{-1}. Variation of the hydrogen-ion concentration over the range pH 4–7 (by addition of potassium bisulphate) had no effect upon the rate of consumption of DPPH; neither did the addition of potassium persulphate in 10^{-4} mole l^{-1} concentration. Measurements of silver ion concentration by a colorimetric method using *p*-dimethylaminobenzylidene rhodamine showed that the silver-ion concentration in the solution remained constant throughout the whole period of an experiment. Also, the rate of disappearance of DPPH was again observed to be independent of the initial DPPH concentration, other conditions remaining constant. Therefore the conclusion was drawn that DPPH does not participate in the rate-determining reaction.

Bawn and Margerison propose, for the catalysed decomposition, the following mechanism as being consistent with their observations and with the known facts pertaining to the chemistry of silver ions:

$$Ag^+ + S_2O_8^{2-} \rightarrow Ag^{2+} + SO_4^{\cdot -} + SO_4^{2-} \qquad \text{(xix)}$$

$$Ag^{2+} + OH^- \rightarrow Ag^+ + \cdot OH \qquad \text{(xx)}$$

$$\left.\begin{array}{l} D\cdot + SO_4^{\cdot -} \\ D\cdot + \cdot OH \end{array}\right\} \rightarrow \text{stable products} \qquad \text{(xxi)}$$

where D· denotes the DPPH radical.

Bawn and Margerison also studied the decomposition of potassium persulphate in the presence of cupric ions (derived from copper (II) sulphate). The catalytic effect was much smaller than that observed with silver ions. An appreciable increase in the rate of DPPH was observed only if the cupric-ion concentration was of the same order as that of the persulphate. The kinetics of the reaction were not simple, and the mechanism was not therefore elucidated.

This particular study is of considerable interest in showing that metal ions which can exist in more than one valency state can catalyse the decomposition of persulphate ions. Such ions may enter adventitiously into industrial emulsion polymerisation systems, either as impurities in the water or from defective reactor linings. The result under discussion here suggest that such impurities may give rise to erratic persulphate decomposition rates, and therefore to erratic initiation rates.

6.4 MECHANISM OF INITIATION BY PERSULPHATE IONS

Evans and Baxendale[14] appear to have been the first to suggest that initiation of emulsion polymerisation by persulphate salts involves the decomposition of the anion into two sulphate radical-ions, from which polymerisation then proceeds by a normal radical mechanism. Plausible though this suggestion is, it immediately invites at least three questions:

(1) Is the effective primary initiating radical the sulphate radical-ion, or is it the hydroxyl ion produced by reaction of the sulphate radical-ion with the water of the aqueous phase in accordance with reaction (iv) above?
(2) How does the actual rate of initiation of polymer chains compare with the rate of decomposition of the persulphate ions, *i.e.* what is the initiating efficiency of the sulphate radical-ions?
(3) What is the mechanism by which a radical species passes from the aqueous phase into a micelle or polymer particle?

In the following paragraphs, some consideration will be given to each of these questions in turn.

The principal evidence that the decomposition products of persulphate can act as primary initiating radicals comes from the observation that the polymer molecules prepared using this initiator often contain combined sulphur. Thus Barlett and Nozaki[9] found that polyallyl acetate prepared by emulsion polymerisation using persulphate contains combined sulphur. They regard allyl acetate as an especially favourable monomer to have used for the investigation of this matter, because the polymer chains which form are short and therefore the concentration of terminal groups is relatively high. A sample of polymer was prepared from allyl acetate and an aqueous solution containing 0·366 mole l^{-1} potassium persulphate and 0·45 mole l^{-1} sodium pyrophosphate. The polymer was obtained by extraction with chloroform, removal of allyl acetate and chloroform by evaporation, and then extraction of the polymer with benzene. The molecular weight of the polymer was determined viscometrically. Three types of quantitative estimation were carried out on the polymer: hydrolysis of sulphate by hot hydrochloric acid; destruction oxidation of the polymer and precipitation of the sulphate produced; and estimation of the sodium content. These three methods gave, respectively, 75·1%, 79% and 75% of the number of terminal groups to be expected if there were one such group per polymer molecule. Since most of the growing chains in an allyl acetate polymerisation are probably terminated by a 'degradative chain-transfer' reaction with monomer, it follows that the expected concentration of initiator fragment should correspond to one per polymer molecule.

Smith and Campbell[15] and Smith[16] have shown that polystyrenes prepared by emulsion polymerisation using persulphate as initiator contain sulphur chemically combined in the polymers. Here the expected number of end-groups per polymer molecule is two, since polystyrene terminates by the combination of pairs of propagating radicals. Thus Smith[16] has reported the results of experiments in which initiation was effected by means of S^{35}-labelled potassium persulphate. The polymers isolated were found to be radioactive. Their sulphur contents were estimated by comparing the activity of the purified polymer with that of the latex as a whole. Since the total sulphur content of the latex was known from the ingredients which made up the charge, the sulphur content of the polymer could be calculated from the ratio of the activities. Polymers were prepared at various temperatures using a reaction system which typically comprised 3·00 ml of a 0·55% fatty-acid soap solution, 0·300 ml of a 1·81% S^{35}-labelled potassium persulphate solution, and 2·00 ml of styrene.

After purging with nitrogen, the reaction vessels were sealed and the contents polymerised to virtual completion with appropriate agitation. The method of purification proved to be quite important if reliable results were to be obtained. The procedure finally adopted was to dilute a 1 ml portion of the latex with 100 ml of water, and then flocculate the polymer by addition of hydrochloric acid. After warming, filtering and washing with water and methanol, the polymer was dried on the filter by radiation from an infrared lamp while air was simultaneously sucked through it. It[5] activity was then assayed in this condition. Molecular weights were estimated viscometrically. The results are summarised in Table 6.7, from

TABLE 6.7

Sulphur contents of emulsion polystyrenes polymerised with S^{35}-labelled potassium persulphate (according to Smith[16])

(typical reaction system comprised 2 ml of styrene + 3·3 ml of aqueous phase; % soap and % $K_2S_2O_8$ given in table are presumed to be on the aqueous phase)

Polymerisation temperature (°C)	*g of polymer per mole of sulphur*	*Molecular weight of polymer*	*Atoms of sulphur per polymer molecule*
0·5% soap; 0·165% $K_2S_2O_8$			
30·5	43·5 × 10⁶	20 × 10⁶	0·46
50	4·52 × 10⁶	6·20 × 10⁶	1·4
70	0·62 × 10⁶	1·40 × 10⁶	2·3
90	0·144 × 10⁶	0·274 × 10⁶	1·9
0·5% soap; 0·511% $K_2S_2O_8$			
70	0·33 × 10⁶	0·53 × 10⁶	1·6
2·0% soap; 0·165% $K_2S_2O_8$			
70	1·05 × 10⁶	2·1 × 10⁶	2·0

which it is seen that the number of sulphur atoms per polymer molecule approximates to the expected value of two for polymers prepared at higher temperatures, but is considerably less than two for polymers prepared at 30°C.

Kolthoff, O'Connor and Hansen[17] also investigated the sulphur content of emulsion polystyrene prepared using S^{35}-labelled potassium persulphate. They reported great difficulty in reproducing Smith's results and also their own. Polystyrene prepared using the so-called 'Mutual' recipe (*i.e.* styrene, water, fatty-acid soap, persulphate and mercaptan) apparently contained many more than two sulphur atoms per polymer molecule. But when a solution of the polymer in benzene was washed with water, the radioactivity of the polymer decreased continuously, and prolonged shaking with water brought the number of sulphur atoms down to below two per polymer molecule. Kolthoff, O'Connor and Hansen prepared one series of emulsion polystyrenes using S^{35}-labelled potassium persulphate as the sole initiator, and a further series using persulphate activated by iron(II) pyrophosphate. The effect of the activator was to increase the rate of initiation at any temperature.

Several advantages were claimed to accrue from the use of an activator: at any given temperature, a polymer of much lower molecular weight was obtained than with persulphate alone, and thus the importance of transfer-to-monomer reactions was lessened; also, the relationship between intrinsic viscosity and molecular weight was known with greater certainty than in the case of polymer of very high molecular weight. In addition, it might have been pointed out that polymer of lower molecular weight contains a higher concentration of terminal groups, and end-group analysis is thereby facilitated.

After several preliminary experiments, the following procedure was adopted for isolating the polymer from its latex: The polymer was first precipitated by addition of methanol or ethanol. It was washed continuously with water for 6–8 h, and then with ethanol. It was dried at 50°C in a vacuum oven. The dried polymer was dissolved in benzene (about 100 mg of polymer per 100 ml of benzene). After determination of the viscosity of this solution, and the radioactivity of the polymer, the benzene solution was shaken with an equal volume of water. The measurements were repeated, and the radioactivity of the water layer was also determined. This process was repeated to give the variation of solution viscosity and polymer radioactivity as a function of time of contact with water. A careful check was made to ensure that the sum of polymer radioactivity and radioactivity in the aqueous layer remained constant throughout the extraction process.

TABLE 6.8

Sulphur contents of emulsion polystyrenes polymerised with S^{35}-labelled potassium persulphate (according to Kolthoff, O'Connor and Hansen[17])

(reaction system: styrene 100, water 180, fatty-acid soap 5, $K_2S_2O_8$—variable)

$K_2S_2O_8$ *level* (pphm)	*Polymer-isation temp.* (°C)	*Total time of shaking with water* (min)	*Specific activity* (counts min^{-1} mg^{-1})		*Molecular weight of polymer*	*g of sulphur per g of polymer*	*Atoms of sulphur per molecule of polymer*
			S *in* $K_2S_2O_8$	*polymer*			
0·3	40	0	$2{\cdot}38 \times 10^7$	262	$9{\cdot}0 \times 10^6$	$1{\cdot}1 \times 10^{-5}$	3·1
		20		226	$7{\cdot}5 \times 10^6$	$9{\cdot}5 \times 10^{-6}$	2·2
		40		222	$7{\cdot}2 \times 10^6$	$9{\cdot}3 \times 10^{-6}$	2·1
		100		218	$6{\cdot}9 \times 10^6$	$9{\cdot}2 \times 10^{-6}$	2·0
		120		208	$6{\cdot}8 \times 10^6$	$8{\cdot}7 \times 10^{-6}$	1·9
1·2	50	0	$3{\cdot}14 \times 10^7$	434	$7{\cdot}2 \times 10^6$	$1{\cdot}4 \times 10^{-5}$	3·1
		21		379	$6{\cdot}7 \times 10^6$	$1{\cdot}2 \times 10^{-5}$	2·5
		51		353	$6{\cdot}6 \times 10^6$	$1{\cdot}1 \times 10^{-5}$	2·3
		105		352	$6{\cdot}5 \times 10^6$	$1{\cdot}1 \times 10^{-5}$	2·2
1·2	50	0	$1{\cdot}38 \times 10^7$	—	$5{\cdot}4 \times 10^6$	—	—
		20		118	$5{\cdot}2 \times 10^6$	$8{\cdot}2 \times 10^{-6}$†	1·4
		50		124	$5{\cdot}1 \times 10^6$	$8{\cdot}2 \times 10^{-6}$†	1·4
0·3	80	0	$1{\cdot}49 \times 10^7$	586	$1{\cdot}7 \times 10^6$	$3{\cdot}7 \times 10^{-5}$†	2·1
		30		576	$1{\cdot}3 \times 10^6$	$3{\cdot}6 \times 10^{-5}$†	1·6
		205		572	$1{\cdot}3 \times 10^6$	$3{\cdot}6 \times 10^{-5}$	1·6
		512		487	$1{\cdot}3 \times 10^6$	$3{\cdot}1 \times 10^{-5}$	1·3
		812		419	$1{\cdot}3 \times 10^6$	$2{\cdot}7 \times 10^{-5}$	1·2

† In the original paper, these exponents are given as −5, −5, −6 and −6, respectively; presumably they should have been given as above.

It was also shown that most of the extracted sulphur could be regarded as originating from sulphate and persulphate occluded in the polymer during coagulation. About 15–20% of the extracted sulphur was recovered in a form resembling an alkyl sulphate. Kolthoff, O'Connor and Hansen also reported that after further extraction with water for very long periods of time, there was a continuous, but very slow, loss of sulphur from the polymer, and that this sulphur was recovered mainly as sulphate. This further loss was attributed to very slow hydrolysis of sulphate end-groups combined in the polymer.

Results for polystyrenes prepared using potassium persulphate alone are given in Table 6.8. Results obtained using activated persulphate as the initiator are given in Table 6.9. The latter results give a more conclusive verdict than the former. Kolthoff, O'Connor and Hansen draw the conclusion that, if complications arising from, say, transfer-to-monomer reactions are absent, then the polymer produced contains two sulphur atoms per molecule. This is consistent with initiation by sulphate radical-ions exclusively, and with termination by combination of propagating radicals. Results were also given for polystyrenes prepared in the presence of *m*-dinitrobenzene, which functions as a chain terminator (and therefore as a retarder). These results are summarised in Table 6.10. On the basis of the above considerations, it would now be expected that the number of sulphur atoms per polymer molecule would fall to one, since termination does not now occur principally by combination, but by reaction with the chain terminator. As Table 6.10 shows, there certainly is a reduction in number of sulphur atoms per molecule; in some cases, this number fell to about 0·5.

The overall conclusion which can be drawn from the work of Bartlett and Nozaki, of Smith, and of Kolthoff, O'Connor and Hansen is that, if due allowance is made for transfer reactions and for the termination mechanism, then a substantial proportion of the polymer molecules formed had sulphur-containing moieties as their initiating species. The initiating species has been widely regarded as the sulphate radical-ion, and there is little doubt of the correctness of this view.

Further evidence that sulphate groups become chemically combined with polymer chains when initiation is brought about by potassium persulphate is provided by the work of van den Hul and Vanderhoff.[18] These workers point out that earlier work using S^{35}-labelled potassium persulphate suffered from two defects. The one is of a purely practical nature, and concerns the difficulty of separating from the polymer occluded and adsorbed initiator and its residues. The other is inherent in the method,

TABLE 6.9

Sulphur contents of emulsion polystyrenes polymerised with iron(II)-activated S^{35}-labelled potassium persulphate (according to Kolthoff, O'Connor and Hansen[17])

(reaction system as shown in Table 6.8, with the addition of 1 mmole of iron(II) pyrophosphate activator per 100 g of monomer)

$K_2S_2O_8$ level (pphm)	Polymer-isation temp. (°C)	Total time of shaking with water (min)	Specific activity (counts min^{-1} mg^{-1})		Molecular Weight of polymer	g of sulphur per g of polymer	Atoms of sulphur per molecule of polymer
			S in $K_2S_2O_8$	Polymer			
4·7	5	0	$6{\cdot}42 \times 10^7$	411	$1{\cdot}4 \times 10^6$	$6{\cdot}4 \times 10^{-5}$	2·8
		20		375	$1{\cdot}2 \times 10^6$	$5{\cdot}8 \times 10^{-5}$	2·2
		30		358	$1{\cdot}2 \times 10^6$	$5{\cdot}6 \times 10^{-5}$	2·1
		1470		212	$0{\cdot}8 \times 10^6$	$3{\cdot}3 \times 10^{-5}$	0·9
4·7	5	0	$1{\cdot}01 \times 10^8$	—	$2{\cdot}0 \times 10^6$	—	—
		10		274	$1{\cdot}6 \times 10^6$	$5{\cdot}5 \times 10^{-5}$	2·7
		30		255	$1{\cdot}6 \times 10^6$	$5{\cdot}2 \times 10^{-5}$	2·4
		60		258	$1{\cdot}4 \times 10^6$	$5{\cdot}2 \times 10^{-5}$	2·2
		120		254	$1{\cdot}4 \times 10^6$	$5{\cdot}1 \times 10^{-5}$	2·1
		270		218	$1{\cdot}4 \times 10^6$	$4{\cdot}4 \times 10^{-5}$	1·9
		750		216	$1{\cdot}4 \times 10^6$	$4{\cdot}4 \times 10^{-5}$	1·8
		3210		192	$1{\cdot}2 \times 10^6$	$3{\cdot}9 \times 10^{-5}$	1·5
4·7	25	0	$1{\cdot}01 \times 10^8$	106	$0{\cdot}8 \times 10^6$	10×10^{-5}	2·6
		20		91	$0{\cdot}8 \times 10^6$	$8{\cdot}6 \times 10^{-5}$	2·3
		50		92	$0{\cdot}8 \times 10^6$	$8{\cdot}7 \times 10^{-5}$	2·2
		100		91	$0{\cdot}8 \times 10^6$	$8{\cdot}6 \times 10^{-5}$	2·2
		200		86	$0{\cdot}8 \times 10^6$	$8{\cdot}1 \times 10^{-5}$	2·0
		400		83	$0{\cdot}8 \times 10^6$	$7{\cdot}8 \times 10^{-5}$	2·0
		600		86	$0{\cdot}8 \times 10^6$	$8{\cdot}1 \times 10^{-5}$	2·0

TABLE 6.10

Sulphur contents of emulsion polystyrenes polymerised with S^{35}-labelled potassium persulphate in the presence of *m*-dinitrobenzene as retarder (according to Kolthoff, O'Connor and Hansen[17])

(conditions of preparation: $K_2S_2O_8$ 0·3 pphm; *m*-dinitrobenzene 1 mmole per 100 g of monomer; otherwise reaction system as shown in Table 6.8; temperature 50°C)

% conversion	*Total time of shaking with water* (min)	*Specific activity* (counts min^{-1} mg^{-1})		*Molecular weight of polymer*	*g of sulphur per g of polymer*	*Atoms of sulphur per molecule of polymer*
		S *in* $K_2S_2O_8$	*Polymer*			
45	0	$2{\cdot}18 \times 10^6$	271	$3{\cdot}6 \times 10^5$	$1{\cdot}2 \times 10^{-4}$	1·4
	20		236	$2{\cdot}6 \times 10^5$	$1{\cdot}1 \times 10^{-4}$	0·9
	40		214	$2{\cdot}6 \times 10^5$	$0{\cdot}96 \times 10^{-4}$	0·8
	100		186	$2{\cdot}5 \times 10^5$	$0{\cdot}84 \times 10^{-4}$	1·0
25	0	$1{\cdot}72 \times 10^6$	197	$2{\cdot}8 \times 10^5$	$1{\cdot}2 \times 10^{-4}$	1·0
	32		163	$2{\cdot}5 \times 10^5$	$0{\cdot}99 \times 10^{-4}$	0·7
	65		144	$2{\cdot}5 \times 10^5$	$0{\cdot}87 \times 10^{-4}$	0·7
	100		134	$2{\cdot}5 \times 10^5$	$0{\cdot}81 \times 10^{-4}$	0·6
	307		121	$2{\cdot}5 \times 10^5$	$0{\cdot}73 \times 10^{-4}$	0·6
	507		112	$2{\cdot}5 \times 10^5$	$0{\cdot}68 \times 10^{-4}$	0·5
	707		106	$2{\cdot}5 \times 10^5$	$0{\cdot}64 \times 10^{-4}$	0·5

and concerns its inability to distinguish between sulphur which is present in ionic forms and that which is present in non-ionic forms. In this work, ionic impurities such as soap, residual initiator, initiator residues and buffer were removed from polystyrene latices by treatment with a mixed-bed ion-exchange resin.[19] The process of ion exchange is said to convert anions to their acid form. The latex was then titrated with alkali conductimetrically to give the concentration of sulphate groups present on the surface of the particles. From experiments with a latex prepared in the presence of acrylic acid, and using persulphate as initiator, it was found possible to obtain, from a single titration, estimates of the number of weakly acidic groups (such as carboxyl) and strongly acidic groups (such as sulphate). In fact, for polystyrene latices prepared using a variety of micelle-generating substances (including carboxylates) and persulphate as initiator, it was found that only strongly acidic groups were present after ion exchange. The strongly acidic groups were assumed to be sulphates. The total ionic sulphur in the polymer was estimated by dissolving the polymer in a dioxan–water mixture, subjecting it to ion exchange, and then titrating conductimetrically with alkali. In a few cases, the results for total ionic sulphur were checked against a determination of total sulphur by X-ray fluorescence. Good agreement was reported.

Results for the determination by titration of the surface sulphate groups and the total number of sulphate groups are summarised in Table 6.11. The efficiency of the added persulphate is always less than 100%, in the sense that part may remain undecomposed at the end of the reaction and part may be lost in side reactions (although the products of the side reactions may themselves be efficient initiators). Furthermore, it appears that between 45% and 100% of the sulphate groups combined in the polymer are present on the surface rather than buried in the particles. The contribution of these surface groups to the surface change density is therefore probably significant, as is their contribution to the colloid stability of the latex. As will be seen in Chapter 7, Section 7.12, the stabilising effect of such surface sulphate groups has been put to use in the preparation of synthetic latices by emulsion polymerisation systems which are free from added micelle-forming substances.

The results in Table 6.11 also show that, except for the latex having the largest particles, the total number of sulphate groups per polymer molecule is less than two. (It should perhaps be mentioned in passing that van den Hul and Vanderhoff say number of sulphate *end*-groups per polymer molecule; but, in the present writer's view, they provide no evidence for believing that the sulphate groups are exclusively terminal, although such

TABLE 6.11

Number of sulphate groups per polymer molecule for various polystyrene latices prepared using potassium persulphate as initiator (according to van den Hul and Vanderhoff[18]

(exact details of reaction system not given)

Soap type†	*Average particle diameter (Å) of latex*	$\overline{M}_n$ *by osmometry*	*Number of sulphate groups per polymer molecule* — *Added as persulphate*	*Present on particle surfaces*	*Total sulphate*
A	880	4·3 × 10⁴	3·3	0·98	1·57
A	2340	17·3 × 10⁴	5·2	0·87	1·59
B	1580	8·06 × 10⁴	2·8	1·26	1·21
B	2480	7·5 × 10⁴	2·6	1·02	1·23
C	2540	7·58 × 10⁴	2·6	0·95	0·94
D	1090	10·6 × 10⁴	3·8	0·95	—
D	1870	10·6 × 10⁴	5·5	0·90	—
D	2850	10·1 × 10⁴	3·5	1·03	1·63
D	4470	8·52 × 10⁴	3·1	0·93	2·07

† A: bis-1,3-dimethylbutyl sodium sulphosuccinate.
B: sodium lauryl sulphate.
C: di-2-ethylhexyl sodium sulphosuccinate.
D: potassium oleate.

a belief would certainly be reasonable.) It is recognised that the reason for the deficiency may be the occurrence of side reactions, *e.g.* reaction (iv) above, leading to hydroxyl radicals and the possibility of initiation and termination by those radicals. A consequence would be the presence of combined hydroxyl groups in the polymer.

That hydroxyl groups are in fact present is indicated by the observation that an ion-exchanged latex heated for 6 h at 90°C with potassium persulphate and a catalytic amount of silver nitrate acquires weakly-acidic surface groups. These are presumed to be carboxyl groups, derived by oxidation of the hydroxyl groups. As shown in Fig. 6.8, the balance between sulphate groups and hydroxyl groups depends very much upon the pH of the medium in which the polymerisation is carried out. The number of sulphate groups per polymer molecule passes through a maximum at about pH 8. Van den Hul and Vanderhoff state that the reduction in this number at low pH values correlates well with the belief that the production of hydroxyl radicals increases with decreasing pH. They offer no comment regarding the reduction in the number at high pH also. It may

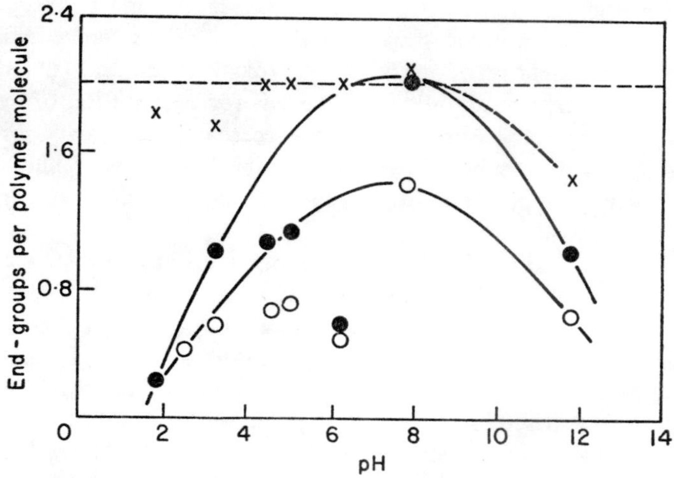

FIG. 6.8. Effect of pH of polymerisation medium upon the number and nature of the end-groups of the emulsion polystyrene obtained using potassium persulphate as initiator, according to van den Hul and Vanderhoff.[18] Significance of points: ○ surface sulphate groups; ● total sulphate groups; × total sulphate plus hydroxyl groups. (Polymerisations were carried out at 90°C. The reaction system contained 0·47 pphm of potassium persulphate; the micelle generator was bis-1,3-dimethylbutyl sodium sulphosuccinate. See text for details of purification of the latices after preparation.)

be that, at low pH, the formation of hydroxyl radicals proceeds by way of a transition state involving hydroxonium ions, whereas at high pH it is through a transition state involving hydroxyl ions. It is perhaps significant that the maximum number of sulphate groups per polymer molecule occurs in approximately that pH range where the sum of the concentrations of hydrogen and hydroxyl ions is a minimum.

It is now necessary to consider the second of the two matters listed above, namely, the comparison between the rate of initiation on the one hand and the rate of production of sulphate radical-ions on the other. Even if it could be shown that all polymer chains are initiated by sulphate radical-ions, this would not, of course, guarantee that all sulphate radical-ions formed in the aqueous phase are effective in initiation. There could be considerable wastage of radicals, for example, by reaction with water—see reaction (iv). That such wastage does occur is evidenced by the drop in pH which accompanies many polymerisations in which persulphate ions are used as initiator, an example of which is given in Fig. 3.8. The obvious explanation for the pH drop is reaction between sulphate radical-ions and water, to form bisulphate ions and hydroxyl radicals. Whilst such a reaction would lead to some sulphate radical-ions being unavailable for initiation, it would not necessarily affect the overall initiation efficiency, because of the formation of one hydroxyl radical for each sulphate radical-ion lost. However, a ready mechanism exists for the reaction of hydroxyl radicals to form non-radical products—see reaction (v)—and on these grounds it would be expected that interaction between sulphate radical-ions and water would lead to overall wastage of radicals in terms of initiation.

The results of Smith[16] for the emulsion polymerisation of styrene, taken together with published values for the rate of decomposition of persulphate ions in aqueous media, provide evidence that the rate of formation of sulphate radical-ions, the rate of combination of sulphur and the rate of initiation of polymer chains are all of the same order of magnitude. The results are summarised in Table 6.12, together with estimates of the rate of formation of polymer particles. It is seen that the rate of production of radicals is roughly within the range from 5×10^{11} to 2×10^{14} per second per millilitre of aqueous phase, depending upon the temperature and the persulphate concentration. From these results it may be concluded that, for styrene emulsion polymerisation at least, most of the polymer chains are initiated by sulphate radical-ions, and that a substantial number of the radicals produced by thermal decomposition of the persulphate are actually used for initiation. It further appears that the rate of formation of polymer particles is of the same order of magnitude at lower polymerisation temperatures, but tends to be one order of magnitude less at higher temperatures.

Regarding the figures given in Table 6.12, the rates of formation of sulphate radical-ions were calculated by the present writer on the basis of rate coefficients for the decomposition of persulphate ions in pure water published by Hakoila.[8] The rate of formation of sulphate radical-ions has been underestimated because no allowance has been made for the possible accelerating influence of other components of the polymerisation system. The rates of combination of sulphur were found from the radioactivity of the polymer produced; the rate of formation of polymer molecules were found from the rate of polymerisation and the molecular weight of the product. The calculation of the rate of formation of particles was effected by more tortuous and controversial procedure. Smith himself admits that his estimates of the initial rate of particle formation cannot be considered as being at all accurate. The procedure was first to estimate the number of particles, N, per unit volume of aqueous phase in the steady state of the reaction from a knowledge of the polymer content and the average particle

TABLE 6.12

Emulsion polymerisation of styrene initiated by potassium persulphate: comparison of quantities dependent upon rate of radical formation (according to Smith[16])

(typical reaction system comprised 2 ml of styrene + 3·3 ml of aqueous phase; % soap and % $K_2S_2O_8$ given in table are presumed to be on the aqueous phase)

Reaction conditions:						
temperature (°C)	30·5	50	70	90	70	70
soap level (%)	0·5	0·5	0·5	0·5	0·5	2·0
$K_2S_2O_8$ *level* (%)*	0·172	0·172	0·172	0·172	0·516	0·172
Rate of combination of sulphur (atoms sec^{-1} (ml aq. ph.)$^{-1}$)	$1·5 \times 10^{11}$	$6·3 \times 10^{12}$	$1·5 \times 10^{14}$	$1·7 \times 10^{15}$	$2·9 \times 10^{14}$	$1·6 \times 10^{14}$
Rate of formation of particles (number sec^{-1} (ml aq. ph.)$^{-1}$)	$2·6 \times 10^{11}$	$4·2 \times 10^{12}$	$3·3 \times 10^{13}$	$1·5 \times 10^{14}$	$4·3 \times 10^{13}$	$2·6 \times 10^{13}$
2× rate of polymer molecule formation (mole sec^{-1} (ml aq. ph.)$^{-1}$)	$6·4 \times 10^{11}$	$8·4 \times 10^{12}$	$8·0 \times 10^{13}$	$1·6 \times 10^{15}$	$2·6 \times 10^{14}$	$1·1 \times 10^{14}$
Rate of formation of radicals by thermal decomposition of $K_2S_2O_8$ (radicals sec^{-1} (ml aq. ph.)$^{-1}$)†	$4·31 \times 10^{11}$	$9·90 \times 10^{12}$	$1·67 \times 10^{14}$	$2·15 \times 10^{15}$	$5·00 \times 10^{14}$	$1·67 \times 10^{14}$

† Calculated by present writer on basis of rate coefficients published by Hakoila.[8] In making these calculations, it has been assumed that the $K_2S_2O_8$ level (marked* in the above table) refers to % on the aqueous phase, and not on the monomer or reaction system as a whole.

diameters (obtained by electron microscopy) of the particles present. The initial rate of particle formation, ρ, was then calculated from N using eqn (4.30), which is given by the Smith–Ewart theory. It will be recalled that this equation reads

$$N = \chi\mu^{-2/5}\varpi^{3/5}\rho^{2/5}S^{3/5}$$

where χ is a constant having a value between 0·37 and 0·53, μ is the rate of increase in volume of the particles, ϖ is the adsorption area of unit mass of soap, and S is the soap content. χ was taken as its lowest value, *i.e.* 0·37; ϖ was taken as 6×10^6 cm^2 g^{-1}; and μ was obtained from an estimate of the average rate of polymerisation per particle and the supposed composition of the particles. The rate of polymerisation per particle could be found from the steady rate of polymerisation divided by N; the composition of the particles was inferred from that of polystyrene particles swollen to equilibrium in the presence of styrene. In passing, it is of interest to note that Smith reported the ratio of the volume of the swollen particle to the mass of polymer present in the particle as 2·6. One further conclusion reached by Smith must be noted, and that is that the rate for the formation of propagating radicals does not appear to depend upon the area of interface between the organic and aqueous phases in the system.

Kolthoff, O'Connor and Hansen[17] re-examined the relationship between the rate of initiation of polymer molecules and the rate of formation of radicals in the aqueous phase for the emulsion polymerisation of styrene initiated by persulphate. They concluded that, as long as a micelle-forming substance is present in the system, the rate of initiation is the same as the rate of formation of sulphate radical-ions. This is the case whether the soap is of the fatty-acid type, or a cationic type such as dodecylammonium chloride. In Fig. 6.9, which is taken from their paper, is collected a number of results for rate of initiation and rate of formation of sulphate radical-ions at various temperatures. It is seen that, within experimental error, both sets of points can be represented by the straight line, from which it may be concluded that no distinction can be made between the two rates over the temperature range studied. However, in the absence of a micelle-forming substance, the rate of initiation was found to be some ten times smaller than in the presence of micelles. Kolthoff, O'Connor and Hansen interpret this as a 'physical', rather than a 'chemical' effect, arising essentially from the reduced concentration of monomer present in the aqueous phase as a whole when no micelle-forming substance is present.

In their study of the emulsion polymerisation of vinyl acetate initiated by potassium persulphate, Dunn and Taylor[20] had occasion to check the

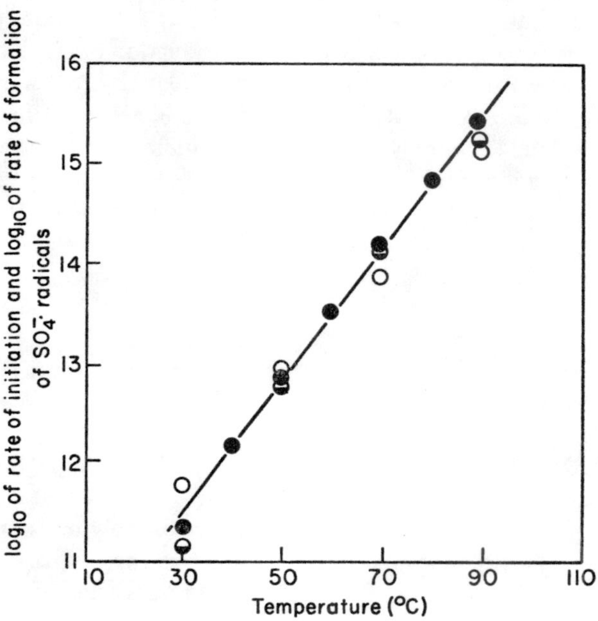

FIG. 6.9. Rate of initiation and rate of formation of sulphate radical-ions as a function of temperature for emulsion polymerisation of styrene initiated with potassium persulphate, according to Kolthoff, O'Connor and Hansen.[17] Significance of points: ○ rate of initiation from rate of polymerisation and molecular weight; ◒ rate of initiation from rate of incorporation of radioactive sulphur; ● rate of formation of sulphate radical-ions by persulphate decomposition. The units of rate of initiation and rate of formation of sulphate radical-ions are radicals per second per millilitre of aqueous phase. (See original paper for primary sources of data.)

initiation efficiency of sulphate radical-ions as it appeared from the inhibition periods induced by the presence of known amounts of either hydroquinone or *p*-benzoquinone. The conclusion reached was that 60% of the radicals formed were effective in initiating polymerisation. The value of the rate coefficient for the thermal decomposition of potassium persulphate in aqueous solution at 60°C was taken from the early work of Green and Masson;[2] this value is slightly higher than that given by Hakoila,[8] but the difference would not greatly affect the conclusion. The relative rates of consumption of hydroquinone and *p*-benzoquinone implied the same rate of initiation if the reasonable assumption was made that a hydroquinone molecule destroys the activity of only one radical, but one *p*-benzoquinone molecule destroys that of two radicals.

The fact that rates of initiation (and, to a lesser extent, rates of particle formation) are generally similar to rates of production of sulphate radical-ions, as estimated from rates of thermal decomposition of persulphate ions, lends support to the view that the rate of initiation is determined solely by the rate of thermal decomposition of the persulphate. It tends to refute the possibility that additional initiation mechanisms are also operative, such as direct interaction with monomer molecules (M) by a reaction such as the following:

$$M + S_2O_8{}^{2-} \rightarrow \cdot MOSO_3{}^{-} + SO_4\overset{-}{\cdot} \qquad \text{(xxii)}$$

That such additional reactions do not occur is consistent with the observation that when discrepancies between rate of initiation and rate of persulphate thermal decomposition are observed, the former is usually smaller than the latter.

Kolthoff, Meehan and Carr[21] have provided additional evidence in support of the view that reactions such as (xxii) contribute little if anything to the rate of initiation in the case of the emulsion polymerisation of allyl acetate. They found that the rate of disappearance of persulphate in aqueous media is almost unaffected by the presence of allyl acetate (in contrast to the pronounced accelerating effect of certain other organic substances, such as methanol, to which reference has been made previously). It is therefore most unlikely that direct interaction between persulphate and monomer occurs. It is also of interest that Kolthoff, Meehan and Carr found the rate of disappearance of persulphate in the presence of methanol and allyl acetate to be almost the same as that in the presence of allyl acetate alone; the pronounced accelerating effect of the methanol is apparently almost completely eliminated by the addition of allyl acetate. Thus it appears that (a) allyl acetate acts as a very efficient scavenger for the radicals which are produced by the thermal decomposition of the persulphate and which could otherwise attack the methanol, and (b) the acceleration of the persulphate decomposition induced by the presence of methanol alone is not due to direct interaction between methanol and the persulphate ions.

It should be noted that, whilst the evidence discussed in the preceding paragraphs tends to eliminate the possibility of direct interaction between monomer and persulphate, it does not preclude the possibility that some initiation may occur by way of a hydrogen-abstraction reaction between monomer and radicals produced directly or indirectly through the decomposition of the persulphate. That combined sulphur is present in the polymer produced shows that not all initiation occurs by this mechanism; but there seems to be no inherent reason why it should not be a contributing mechanism.

Having considered the matter of which radical is the actual polymerisation initiator when persulphates are used, and also the matter of the comparison between the rate of formation of sulphate radical-ions and the rate of initiation, it now remains to discuss the mechanism by which radical species pass from the aqueous phase into micelles or polymer particles. The Harkins theory merely requires that active radical species be acquired by the micelles and particles; it places no restriction upon the mechanism of acquisition. If, as is usual in emulsion polymerisations, the radicals are generated in the aqueous phase and therefore externally to the polymerisation loci, it is a matter of some interest to decide what factors cause a radical or its descendants to pass into a micelle or particle. Even now, the factors which govern this entry are not well understood. The problem is that there is little obvious reason why entities such as the sulphate radical-ion or the hydroxyl radical should pass from the aqueous phase into micelles or particles. In fact, thermodynamic considerations would lead to the belief that partitioning between the polar aqueous phase and the very much less polar micelle or particle phase would be strongly in favour of the former. Furthermore, if the behaviour of sulphate anions can be taken as a guide, even the specific adsorption of the sulphate radical-ion at the particle–aqueous phase interface is probably quite weak,[22] unless the suface density of charges at the interface is high and positive.[23] Thus the problem is especially acute in the case of initiation by sulphate radical-ions when the micelle-generating substance is anionic in character; for here initiation seems to require that a negatively-charged species be adsorbed on to, or should pass through, a negatively-charged double layer. Thus, in such cases, it would be expected that both the hydrophilic character of the radical and its negative charge would militate against entry into micelles or particles by a process such as random diffusion.

For reasons such as these, an alternative point of view has been developed, namely, that initiation of polymerisation occurs within the true aqueous phase by addition of dissolved monomer molecules to the primary free radical. Letting M denote a monomer molecule, the following sequence of reaction can be envisaged:

$$\mathrm{M} + \cdot\mathrm{OSO_3}^- \rightarrow \cdot\mathrm{MOSO_3}^- \qquad \text{(xxiii)}$$

$$\mathrm{M} + \cdot\mathrm{MOSO_3}^- \rightarrow \cdot\mathrm{M_2OSO_3}^- \qquad \text{(xxiv)}$$

$$\vdots \qquad\qquad \vdots$$

$$\mathrm{M} + \cdot\mathrm{M}_n\mathrm{OSO_3}^- \rightarrow \cdot\mathrm{M}_{n+1}\mathrm{OSO_3}^- \qquad \text{(xxv)}$$

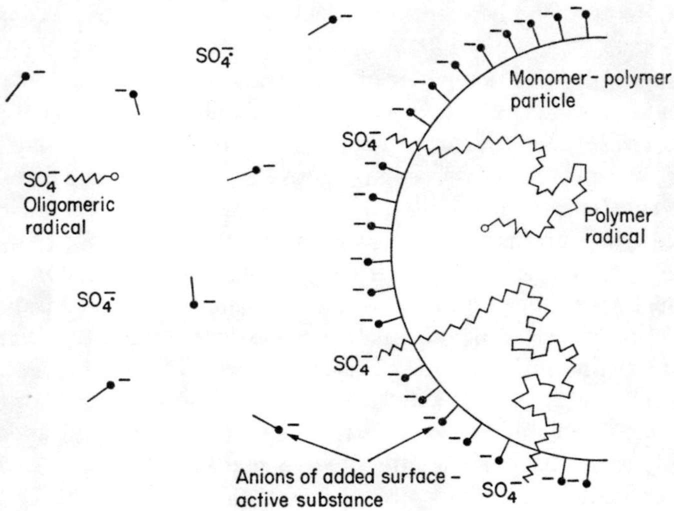

FIG. 6.10. Diagrammatic illustration of mechanism of initiation of polymerisation in an emulsion system by adsorption of oligomeric radical species (after van den Hul and Vanderhoff[18]).

Now as n increases, there is every reason to believe that the oligomeric species $\cdot M_nOSO_3$ will acquire that dual combination of hydrophobic and hydrophilic character which is the differentia of surface-active substances. If such species were present in sufficient numbers (as might well be the case in the polymerisation of relatively water-soluble monomers), then they would tend to form their own micelles, and this tendency would doubtless be encouraged by the mutual termination of radicals within the newly formed micelles, especially if that termination were accompanied by combination rather than by disproportionation. But in any event, as such oligomeric species grow, there will be an increasing tendency for them to be adsorbed at existing micelle and particle surfaces. Furthermore, although only a surface adsorption as far as the polar sulphate anion head-group is concerned, it would be natural for the oligomeric chain bearing the free radical on its end to find its way into the interior of the particles, albeit in the surface layers only. Here, then, is a plausible mechanism whereby the activity of a sulphate radical-ion generated within the aqueous phase can pass into the interior of the micelle without the sulphate anion having to pass through a negatively-charged double layer. It is true that the mechanism does require the adsorption of a negatively-charged ion at a negatively-charged interface, but a definite driving force is now available for the process, namely, the reduction in Gibbs free energy which accompanies the expulsion of the hydrophobic oligomer chain from the aqueous phase.

A diagrammatic representation of these ideas is given in Fig. 6.10, which is taken from a paper by van den Hul and Vanderhoff.[18] In support of the existence of oligomeric free radicals, Alexander and Napper[24] cite the fact that stable latices can be formed from several sparingly-soluble monomers such as vinyl acetate, methyl methacrylate, etc., in the absence of added surface-active substances, but using potassium persulphate as initiator. Extensive dialysis does not seem to lead to desorption of ionogenic groups, suggesting that they are chemically combined.

Some consideration has been given in the literature to the critical degree of polymerisation at which the oligomeric free radical would become insoluble in the aqueous phase and so precipitate as a separate phase if it had not already been adsorbed at an interface. Thus Priest[25] found an upper limit of about 50 for the polymerisation of vinyl acetate. The monomer was polymerised in aqueous solution, free of added surface-active substances, using persulphate anions as the initiator. The reaction was quenched at the point of incipient turbidity. The polymer was isolated and its average degree of polymerisation determined. Napper and Alexander,[26] on the other hand, suggest a critical degree of polymerisation of about 30.

Against the view that entry of radicals into reaction loci occurs by way of the adsorption of oligomers must be set the conclusion which Flory[27] has drawn from certain theoretical calculations, namely, that 'termination in the aqueous phase is utterly negligible, and it may be assumed with confidence that virtually every primary radical enters a polymer particle (or a micelle). Moreover the average lifetime of a chain radical in the aqueous phase (*i.e.* 10^{-5} sec.) is too short for an appreciable expectation of addition of a dissolved monomer molecule by the primary radical prior to its entrance into a polymer particle.'

Flory's argument is as follows. The diffusion current, I, of radicals into a spherical particle of radius r is given by $I = 4\pi r D \cdot \Delta c$, where D is the diffusion coefficient and Δc is the difference between the concentration of the diffusing radicals in the bulk of the external medium and the concentration at the surface of the particle. If the latter is taken as zero, then Δc becomes equal to the stationary concentration of radicals in the aqueous phase. Denoting this concentration by $[\mathrm{R}\cdot]$, it follows that

$$[\mathrm{R}\cdot] = I/4\pi r D$$

Flory then took D as being of the order of $10^{-5}\mathrm{cm}^2\ \mathrm{sec}^{-1}$ and r as about $5 \times 10^{-6}\mathrm{cm}$. I was estimated as being of the order of 10^{-1} per second, on the grounds that there are about ten times more polymer particles in unit volume of aqueous phase than there are radicals produced per second in the same unit volume of aqueous phase. These values give [R·] as being of the order of 10^8 radicals per ml. But radicals are generated at a rate of about 10^{13} ml^{-1} sec^{-1} in a typical system; hence the average lifetime of a radical from generation to capture by a polymer particle or micelle must be of the order of 10^{-5} sec. On the other hand, taking a typical value for the rate coefficient for the mutual termination of radicals, and assuming a stationary concentration of 10^8 radicals per ml, the rate of loss of radicals by termination appears to be many orders of magnitude less than the rate of generation of radicals. On these grounds, termination within the aqueous phase is rejected as being of negligible importance.

The error in Flory's argument probably arises from his assumption that the radicals can diffuse unhindered through a negatively-charged double layer at the same rate as they diffuse through the surrounding aqueous phase. This will almost certainly not be true if the radical is hydrophilic and also negatively charged; an appreciable potential energy barrier, caused by electrostatic repulsion, must be overcome before the radical can enter the micelle or particle. Indeed, one can envisage that many radicals would be 'turned away' by electrostatic repulsions, and only an occasional radical having above-average kinetic energy would be able to penetrate. Thus the time of retention in the aqueous phase is probably much greater than Flory's estimate; termination within the aqueous phase would thus be less negligible than he estimated, and, more relevant to the present discussion, there would be more time for the formation of oligomeric radical species by solution polymerisation.

A conclusion similar to that of Flory was reached by Smith[28] on the basis of similar calculations. His view is summed up by the following quotation: 'Assuming a steady state concentration of 10^8 free radicals per cc. of water solution and a rate of entrance into polymer particles of 10^{13} sec^{-1}, a given free radical will exist in the water solution for only 10^{-5} seconds, which is much too short a time for it to react with one of the styrene molecules dissolved in the water. Also, it is very much too short a time for it to react with another of the free radicals in the water solution. Therefore, substantially none of the free radicals produced in the water will be terminated there but practically all will enter polymer particles; moreover, they will enter them while the free radicals are still small, containing not more than one or two styrene units.'

Roe and Brass[29] have also discussed the problem of the diffusion of free radicals into polymer particles. They regard diffusion as being of only minor importance in bringing radicals to the point of entry into a reaction locus. They consider the dominant process to be the dissociation of persulphate ions into sulphate radical-ions. They do not regard the extended region surrounding a representative particle as being available for free diffusion, because of the presence of other particles within that space, each one of which particles can act as a sink for diffusing radicals. 'We can imagine', say Roe and Brass, 'an initial state in which diffusion from a distance is the dominant process, but it is clear that before long the supply of long range diffusing radicals will be reduced to a negligible level at any point in the interior of the dispersion. With the development of this situation, we are left dependent on local generation as the dominant source of free radicals entering a particle.' They consider their hypothesis to be justified because it provides a satisfactory basis for explaining certain of their observations.

As Dunn[30] has recently emphasised, the problem of electrostatic repulsion between electrical double layer and radical is obviated if initiation occurs by way of hydroxyl ions formed by interaction between sulphate radical-ions and water, rather than by the sulphate radical-ions themselves. Polymerisation might still commence in the aqueous phase, producing surface-active oligomers. But these hydroxyl-terminated species would probably show a stronger tendency to be adsorbed than do ionised sulphate-terminated species. Dunn points out that the results of end-group determinations on emulsion polymers have shown that, under some conditions at least, hydroxyl radicals are responsible for much of the initiation. Also, in those cases where persulphate-initiated polymers contain less than the theoretical expectation of sulphur, it is possible that some of the polymer chains were initiated by hydroxyl radicals.

6.5 EFFECT OF PERSULPHATE CONCENTRATION UPON RATE OF POLYMERISATION

Smith[28] has provided experimental evidence to show that, *provided the number of reaction loci is fixed,* then variation of the persulphate level has no measurable effect upon the rate of polymerisation. This is entirely in accordance with the prediction of the Smith–Ewart theory. The method was to use a particular polystyrene latex as a 'seed' latex, and thus to fix the number of particles. The seed latex was diluted with aqueous potassium persulphate solutions of three different concentrations. Styrene was then added and the rate of polymerisation of the added styrene determined. The

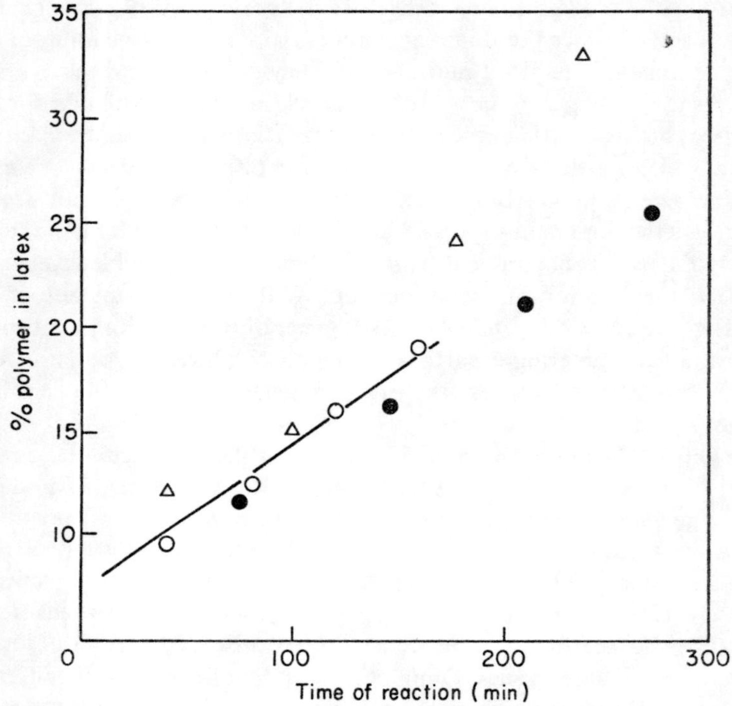

FIG. 6.11. Effect of persulphate concentration upon emulsion polymerisation of styrene in a seed latex at 50°C, according to Smith.[28] The straight line is that obtained by application of the method of least squares to data on two duplicate runs at 'normal' persulphate concentration (*i.e.* 0·175%). The points ○ and △ are for polymerisation with 4 times the normal persulphate concentration; the points ● are for a polymerisation with 1/4 times the normal persulphate concentration.

results are shown in Fig. 6.11. It is clear that, initially at least, the rates obtained from systems in which the persulphate concentration was varied 16-fold could adequately be represented by the line derived for a system containing an intermediate amount of initiator. A tendency towards increased rate with increased conversion was observed in the case of the run at high persulphate concentration; Smith regards this as being possibly due to the formation of new reaction loci whose presence becomes apparent when the persulphate concentration is high.

Whilst it may seem strange at first sight that a 16-fold increase in persulphate concentration should have no effect upon the rate of polymerisation, at least at low conversion, it should be recalled that theoretical

TABLE 6.13

Effect of persulphate concentration upon number of particles produced in emulsion polymerisation of styrene at 50°C (according to Smith[28])

(reaction system: styrene 100, 1% soap solution *ca.* 285; $K_2S_2O_8$—variable)

Persulphate concentration (% on aqueous phase)	0·70	0·175	0·044
Observed relative† number of particles	1·70	1·000	0·62
Predicted relative† number of particles‡	1·74	1·000	0·58

† *i.e.* relative to the number formed when the persulphate concentration was 0·175%.
‡ *i.e.* predicted on the assumption that $N \propto [K_2S_2O_8]^{2/5}$.

considerations suggest that, for a fixed number of loci, only one-half will be active at any one time no matter how many radicals are available, provided that termination is rapid relative to the interval between successive entries of radicals. As a corollary of this, the addition of further initiator can have no effect upon the rate of polymerisation in a conventional emulsion system once the number of loci is fixed. According to the Smith–Ewart theory, the steady rate of polymerisation is affected by the initial initiator concentration only because of its effect upon the number of loci which form during the locus-initiation stage. It will be recalled that the Smith–Ewart prediction is that this number, and therefore the steady polymerisation rate, should be proportional to the 2/5th power of the initiator concentration if initiation is first-order in initiator concentration. The results summarised in Table 6.13, which are due to Smith,[28] show that, for the emulsion polymerisation of styrene initiator by persulphate, the number of particles follows approximately the Smith–Ewart prediction, if the soap concentration is constant.

There are various references in the published literature to the fact that the rate of emulsion polymerisation of styrene initiated by persulphate is approximately proportional to the square root of the initial persulphate concentration. The experimental data do not allow a distinction to be made between this order of reaction and that of 2/5 predicted by the Smith–Ewart theory. The results of Price and Adams[31] summarised in Table 6.14 show that the ratio of k_0, the zero-order rate coefficient with respect to monomer at constant initiator concentration, to the square root of the initial persulphate concentration, is approximately constant. The results of Kolthoff and Dale[32] also confirm approximate proportionality between the rate of polymerisation and the square root of the persulphate concentration in the case of styrene.

TABLE 6.14

Effect of persulphate level upon rate of emulsion polymerisation of styrene at 40°C (according to Price and Adams[31])

(reaction system comprised styrene, water, potassium stearate; exact details not given)

$K_2S_2O_8$ *conc.* mole l^{-1}	k_0 (mole l^{-1} min^{-1})	$k_0/\sqrt{[K_2S_2O_8]}$
$2{\cdot}03 \times 10^{-3}$	0·0101	0·207
$4{\cdot}06 \times 10^{-3}$	0·0130	0·185
$6{\cdot}09 \times 10^{-3}$	0·0199	0·221
$8{\cdot}12 \times 10^{-3}$	0·0218	0·199

† k_0 is the apparent zero-order rate coefficient with respect to styrene at fixed initiator concentration.

In the case of the industrially-important emulsion polymerisation of butadiene and the emulsion copolymerisation of styrene and butadiene, it has for long been recognised that the rate of polymerisation is remarkably insensitive to the level of persulphate initiator in the system. As Bovey *et al.*[33] say, 'Early in the investigation of the standard GR-S recipe it was found that the rate of polymerisation is unaffected when the concentration of persulfate is changed within wide limits.' Industrial experience has repeatedly confirmed the truth of this feature of styrene–butadiene emulsion copolymerisation initiated by potassium persulphate. Typical results are given in Table 6.15 and in Fig. 6.12. The former are due to Haskell and

TABLE 6.15

Effect of persulphate level upon the rate of polymerisation of butadiene and rate of copolymerisation of styrene and butadiene (according to Haskell and Dale[34])

(reaction system: monomers 100, water 180, fatty-acid soap 5, n-C_{12} mercaptan 0·5, persulphate—variable; temperature of polymerisation 50°C)

Persulphate level (pphm)	*% conversion at times shown*				
	Butadiene		*25:75 styrene:butadiene*		
	6 h	12 h	3 h	6 h	12 h
3·0	19	47	14	37	81
1·5	20	51	14	37	81
0·3	24	57	15	37	78
0·15	28	67	17	37	77
0·075	29	61	14	32	73
0·03	27	56	11	27	62

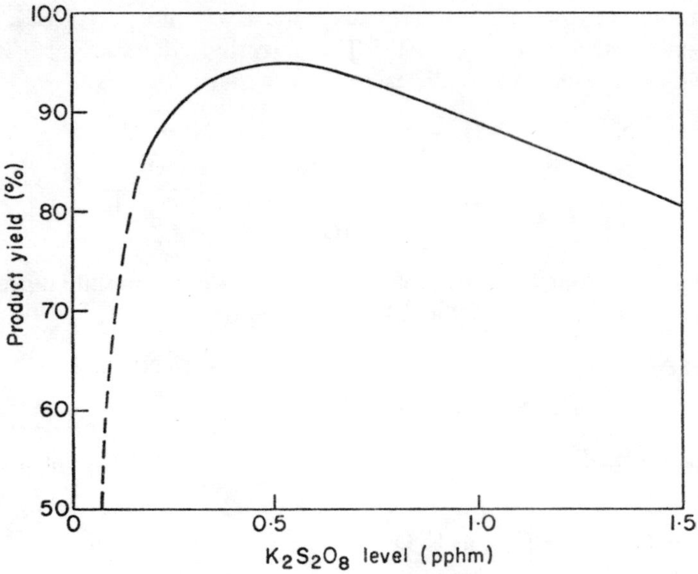

FIG. 6.12. Effect of persulphate level upon product yield in emulsion copolymerisation of styrene and butadiene after 18 h reaction at 40°C, according to Starkweather *et al.*[35] (Reaction system: styrene 25, butadiene 75, water 150, aliphatic (C_{16}) sulphonate 4, sodium naphthalene formaldehyde sulphonate 1, dodecyl mercaptan 0·75, $K_2S_2O_8$—variable, $K_3Fe(CN)_6$ 0·15, excess NaOH 0·15.)

Dale,[34] and the latter is from a paper by Starkweather and his co-workers.[35] The present writer is unaware of any satisfactory explanation of this well-attested insensitivity of rate to persulphate level. Bovey *et al.*[33] comment that 'It seems that persulfate acts not only as an initiator but also as a retarder.' It should be emphasised that, in the case of the emulsion polymerisation of butadiene and the emulsion copolymerisation of styrene and butadiene, the situation is considerably complicated by the fact that persulphates by themselves seem unable to initiate polymerisation. It is necessary for an additional promoting substance to be present if polymerisation is to occur. This substance is usually a mercaptan modifier.

6.6 OTHER DISSOCIATIVE INITIATORS

Whilst inorganic persulphates form by far the most widely used class of dissociative initiator for emulsion polymerisation reactions, various other classes of substance have been proposed. Among these are aromatic diazoamino compounds (I),[36–40] aromatic diazothio ethers (II),[40] and alkali–metal aryl diazoates (III).[41] The diazothio ethers are also known in the literature as diazothiolic esters.

$$\underset{\text{(I)}}{Ar - N = N - NH - Ar'} \qquad \underset{\text{(II)}}{Ar - N = N - S - Ar'}$$

$$\underset{\text{(III)}}{[Ar - N = N - O]^{-}M^{+}}$$

The thermal decomposition of the first two classes of compound is probably adequately represented by the following reactions:

$$Ar - N = N - NH - Ar' \rightarrow Ar\cdot + N_2 + \cdot NHAr' \qquad \text{(xxvi)}$$

$$Ar - N = N - S - Ar' \rightarrow Ar\cdot + N_2 + \cdot SAr' \qquad \text{(xxvii)}$$

In aqueous media, a plausible thermal decomposition mechanism for the diazoate anion would be as follows:

$$[Ar - N = N - O]^{-} + H_2O \rightarrow Ar\cdot + N_2 + \cdot OH + OH^{-} \qquad \text{(xxviii)}$$

With all three classes of compound, reasonable rates of polymerisation can be obtained with styrene and butadiene as monomers in the absence of mercaptan. In this respect they stand in marked contrast to the persulphates. The preparation of these classes of compound is relatively straightforward. The diazoamino compounds can be made by a coupling reaction between an aryl diazonium salt, $ArN_2{}^{+}X^{-}$, and an aromatic amine, $Ar'NH_2$. The diazothio ethers are made by a similar coupling reaction in which the aromatic amine is replaced by an aromatic mercaptan, Ar′SH. The alkali–metal diazoates can be made by adding an appropriate alkali–metal hydroxide to the corresponding aryl diazonium salt.

Reynolds and Cotten[40] have reported the results of experiments carried out to determine the influence of the structure of the diazoamino compound upon its activity as an initiator for the emulsion copolymerisation of styrene and butadiene at 50°C. They found that compounds containing electron-releasing substituents, such as methyl and methoxy, are more active initiators than those containing electron-withdrawing substituents such as halogen. These effects are readily understandable if the decomposition of diazoamino compounds is encouraged by the presence of mutually-repelling dipoles as indicated in structure (IV). The presence of electron-releasing substituents in the aromatic rings will tend to enhance the strength of the dipoles and so encourage decomposition, whereas electron-withdrawing substituents will have the converse effect.

$$\overset{\delta+}{\mathrm{Ar}}-\overset{\delta-}{\mathrm{N}}=\overset{\delta-}{\mathrm{N}}-\mathrm{NH}-\overset{\delta+}{\mathrm{Ar'}}$$

(IV)

(V)

Notwithstanding the activating effect of electron-releasing substituents, relatively large amounts of even the more active diazoamino compounds are required in order to achieve satisfactory rates of polymerisation at 50°C. Thus 4,4′-dimethoxydiazoaminobenzene (V), one of the most active diazoamino initiators investigated, required 10·5 h reaction at 50°C in order to achieve 70% conversion, even though it was present at a level of 2·57 pphm. By comparison, a conventional persulphate-initiated control gave 80% conversion under the same conditions; the reaction time required for 70% conversion was only about 8·5 h.

It appears that the polymers produced using diazoamino compounds as initiator are of rather lower molecular weight than are those obtained using peroxidic catalysts in the absence of added modifier. Thus it seems that these initiators exert a slight modifying effect, as well as an initiating effect. Possibly it is the Ar radicals which mainly initiate the polymerisation, the Ar′NH· radicals functioning as premature chain terminators which prevent the formation of polymer of excessively high molecular weight. However, Reynolds and Cotten conclude that 'the diazoamino compounds cannot be considered modifiers in the sense now used in the synthetic rubber art—*i.e.* agents which lead to the production of benzene-soluble, low-viscosity polymers at relatively high conversions.' Thus in spite of the large amount of 4,4′-dimethoxydiazoaminobenzene used in the experiment referred to in the preceding paragraph, the polymer at 78% conversion contained 67% of benzene-insoluble gel, was extremely tough, and undoubtedly highly crosslinked.

Reynolds and Cotten[40] have shown that the diazothio ethers are not only powerful initiators for the emulsion copolymerisation of styrene and butadiene, but that they also function as efficient modifiers. Again, it seems likely that it is radicals of the type Ar· which initiate the polymerisation, the mercapto radicals of the type Ar′S· functioning as chain terminators. However, it may also be that certain of the by-products of the diazothio ether thermal decomposition may also function as modifiers. In particular, the diaryl disulphide, Ar′SSAr′, which could be formed by the coupling of Ar′S· radicals, would be expected, on general grounds, to display some activity as a modifier.

Reynolds and Cotten have published conversion–time curves for the emulsion copolymerisation of styrene and butadiene at 50°C using as initiator 0·01 moles phm of the products formed by coupling various diazotised aromatic amines with thio-*p*-naphthol. (This mercaptan was found to be especially useful, since it gives relatively stable, crystalline diazothio ethers). The structure of the amines used are indicated as (VI)–(XVI) below. The conversion–time curves show that the initiating activity of the corresponding diazothio ethers increases in the order (VI) to (XVI), except that the compounds derived from (XIII), (XIV) and (XV) have almost identical activity. Thus, as in the case of the diazoamino compounds, the activity tends to be enhanced by the presence of electron-releasing substituents in the ring of the aromatic amine, and to be diminished by the presence of electron-withdrawing substituents. The explanation for these effects is probably the same as that given for the diazoamino compounds. It appears that the effect of substituents in the aromatic ring of the mercaptan is similar.

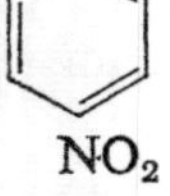

(VI)

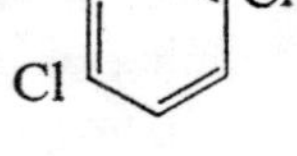

(VII)

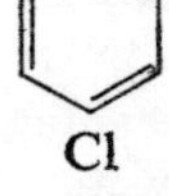

(VIII)

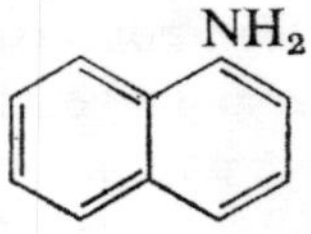

(IX)

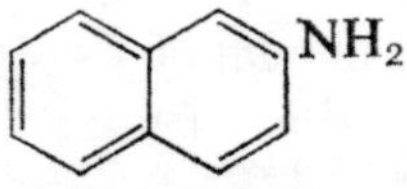

(X)

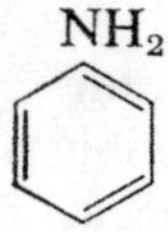

(XI)

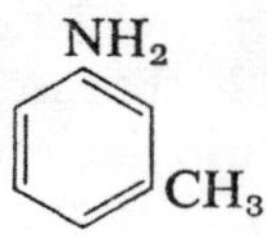

(XII)

(XIII)

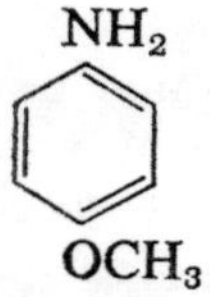

(XIV)

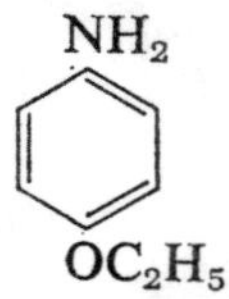

(XV)

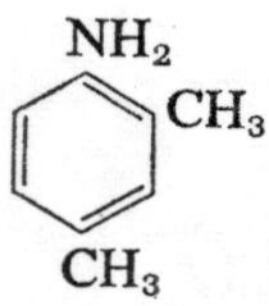

(XVI)

The differences in initiating and modifying activities of the aminoazo compounds and the diazothio ethers are shown clearly by the comparison

(XVII)

between 4,4′dimethoxydiazoaminobenzene (V) (to which reference has already been made above) and the analogous compound 4-(4-methoxybenzenediazomercapto) anisole (XVII), which is prepared by coupling diazotised *p*-anisidine with 4-methoxythiophenol. The diazothio ether gives a conversion of 70% in 4·5 h at 50°C, as compared with the 10·5 h required for the diazoamino compound. Furthermore, the polymer produced is very sticky and clearly of rather low molecular weight. It is completely soluble in benzene. The published conversion–time curves show that, at the levels used in this experiment, this diazothio ether initiates polymerisation more rapidly than does persulphate at conventional levels.

Reynolds and Cotten have reported interesting effects of diazothioether level upon polymerisation rate. In the case of compounds of high activity, the rate was found to increase with increasing amounts of initiator. With compounds of moderate activity, the rate is almost independent of initiator level. And with compounds of low activity, the rate decreases with increasing amount of initiator. It therefore appears that the resultant rate of polymerisation brought about by these compounds is in effect due to the difference between a tendency to initiate and tendency to retard, and that the concentration dependencies of these tendencies may be different in some cases.

Reynolds and Cotten have also shown that water-soluble diazothio ethers can function as efficient initiators and modifiers. The substituent which imparts water-solubility may be on either the amine or the mercaptan component, or on both. The most useful substituents for this purpose are said to be the sulphonate and carboxylate anions, the counterion being an alkali-metal cation. Studies of the effect of structural variations have shown that, in the case of water-soluble compounds, electron-withdrawing deactivating substituents may be present and the compound still be relatively active as an initiator. Indeed, it seems that the presence of deactivating substituents confers the advantage of suppressing an otherwise inconveniently high rate of decomposition in aqueous solution.

A variety of micelle-generating substances can be used satisfactorily in emulsion polymerisation systems initiated by diazothio ethers. In general, any substance which is satisfactory in a persulphate-initiated system is also satisfactory in this system. But, in addition, some surface-active substances which are unsuitable for persulphate-initiated systems can be used with the diazothio ethers. Thus methyl cellulose can be used with these initiators, the products being said to be very stable latices which cannot be coagulated even by strong acid or alkalis. Another interesting feature of initiation by diazothio ethers is the sensitivity of the polymerisation rate to the pH of the aqueous phase. For each compound, there is an optimum pH at which the rate is a maximum. Thus, for the product obtained by coupling diazotised sulphanilic acid and *p*-thiocresol, the optimum pH is in the range 7 to 9, whereas that from diazotised *p*-anisidine and thio-β-naphthol is most effective in carboxylate soap systems at pH levels above 10.

Information on the use of sodium diazoates as initiators for the emulsion copolymerisation of styrene and butadiene at both 122°F and 41°F has been given by Willis *et al.*[41] Preliminary tests showed that the salt derived from *p*-nitroaniline is more active than that derived from aniline. Thus it appears that, in this case, electron-withdrawing substituents have an activating effect. Most of the work reported in this paper is concerned with the production of polymers using a salt of diazotised *p*-nitroaniline. Not only are reasonable rates of polymerisation possible using this initiator, but it is also claimed that vulcanisates from the polymer have better physical properties than do those obtained from persulphate-initiated systems under comparable conditions. It appears that the product of the diazoate-initiated reaction tends to be less branched and crosslinked than that of the persulphate-initiated reaction.

Some interest has also been shown in recent years in the use of water-soluble analogues of azobis*iso*butyronitrile (XVIII) as dissociative initiators for emulsion polymerisation reactions. Suitable water-soluble analogues include 4,4′-azobis-4-cyanovaleric acid (XIX) and its alkali-metal salts, disodium 2,2′-azobis-2-cyanopropane-1-sulphonate (XX), α,α′-azobis*iso*-butyramidinium chloride (XXI), and azobis-(*N*,*N*′-dimethyleneisobutyramidine) (XXII) and its salts with strong acids.

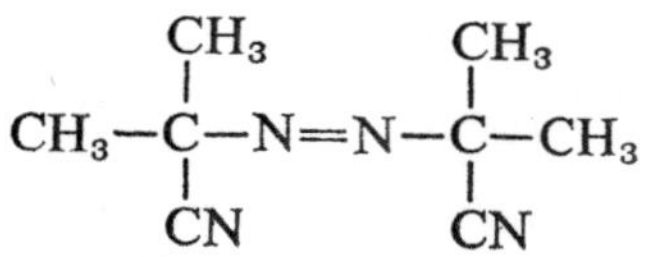

(XVIII)

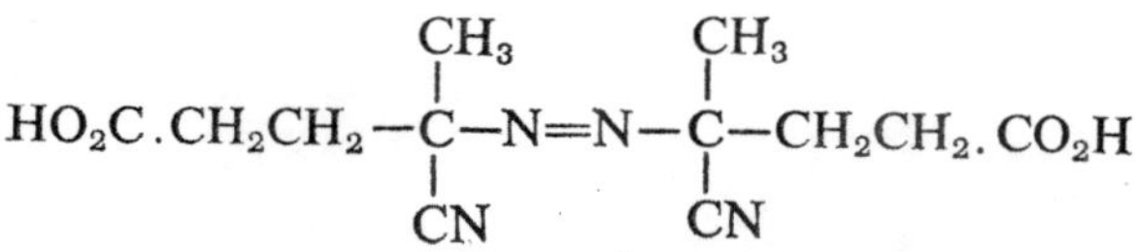

(XIX)

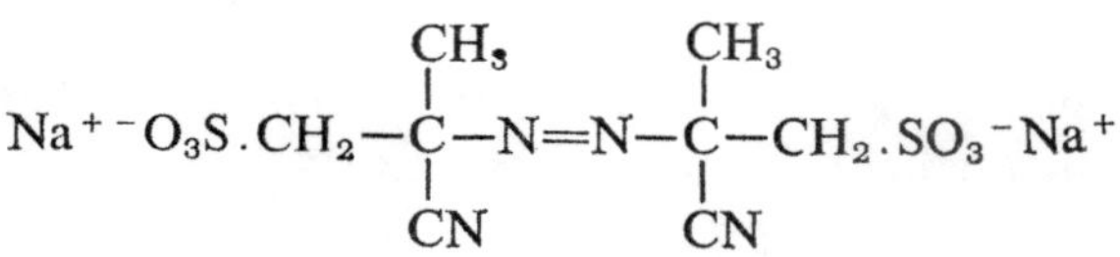

(XX)

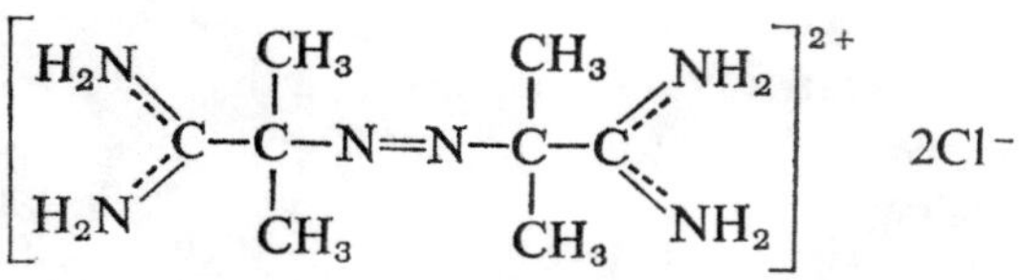

(XXI)

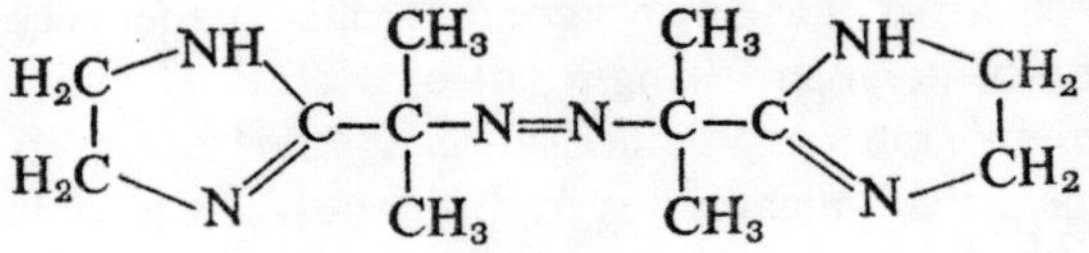

(XXII)

As far as is known, no interest has been shown in this type of compound for the initiation of emulsion polymerisations on the industrial scale. Such interest as has been shown has been confined to fundamental investigations of emulsion polymerisation reactions on the laboratory scale. Azobis*iso*butyronitrile has for many years been used as an initiator for fundamental investigations into the bulk and solution polymerisation of vinyl monomers, its decomposition behaviour being rather simpler than that of, say organic peroxides. Not only is the rate of generation of free radicals strictly first-order with respect to initiator concentration; the rate is also little affected by changes of environment. It is presumed that the water-soluble analogues of azobis*iso*butyronitrile will show similar advantages in emulsion reactions.

The reaction by which these initiators decompose under the influence of heat alone is simple. Nitrogen is eliminated, and two substituted alkyl radicals are produced from the moieties formerly attached to the nitrogen atoms.

B. REDOX INITIATORS

6.7 INTRODUCTION TO REDOX SYSTEMS

As has been pointed out in the general introduction to this chapter (Section 6.1), redox initiators are systems which comprise two or more substances whose mutual interactions produce free radicals which are able to initiate polymerisation. In an extensive review of the initiation of polymerisation processes by redox catalysis, published in 1955, Bacon[42] states that the exploitation of these systems dates from about 1940, when workers in at least three industrial laboratories demonstrated that the rates of polymerisation customarily achieved using peroxy compounds could be greatly increased by using mixtures of oxidising and reducing substances as initiators. He further states that 'The essence of redox polymerisation is that initiation occurs simultaneously with, and in consequence of, oxidation-reduction reactions between a system of catalyst components. The practical result is that polymerisation may occur so rapidly that it becomes feasible to conduct them at low temperatures, or with very small concentrations of initiators, or in equipment designed to function continuously...Redox systems which have been successfully used to initiate polymerisations are numerous, and their number is probably capable of almost indefinite extensions. The great variety of systems so far explored is one of the chief impressions to be gained from a study of the literature.'

The situation is further complicated by the fact that whilst, in certain cases, a metal ion capable of existing in more than one valency state is an essential part of the redox system, in other cases it is added to an existing redox couple as a promotor, thereby giving a three-component system and further scope for multiplicity. Such three-component systems rarely, if ever, effect initiation through the direct intervention of any metal-containing species. Instead, the primary function of the metal ion is to catalyse the interaction between the components of the redox couple. Usually this will occur by way of the establishment of two new redox couples, *i.e.* oxidant plus metal ion in a lower valency state, and reductant plus metal ion in a higher valency state. It can readily be appreciated that the complexity of the system will increase rapidly with the number of components.

The intention of the remainder of this chapter is to discuss some of the redox systems for which published information exists. No attempt has been made to mention all such systems, because of the large number of them. References to further systems are given in the review by Bacon to which attention has been drawn above.

6.8 THE PERSULPHATE–MERCAPTAN SYSTEM

Although mercaptans were first introduced into emulsion polymerisation systems as molecular-weight modifiers, it soon became apparent that they also exert an activating or 'promoting' effect upon persulphate initiators. Indeed, it appears that, as noted previously, in the absence of mercaptan, persulphate-initiated butadiene and styrene–butadiene emulsion polymerisations proceed extremely slowly. On the other hand, persulphate-initiated styrene emulsion polymerisations proceed quite readily, even though mercaptans be absent. Furthermore, the rate of polymerisation in the presence of mercaptan is little affected by the mercaptan concentration. Thus Kolthoff and Dale[32] reported that varying the level of *n*-dodecyl mercaptan in a persulphate-initiated styrene emulsion polymerisation over the range from 0 to 2 pphm had only a 'slight effect' upon the rate of conversion. In fact, when 2 pphm was used, a slight decrease in the rate of conversion was observed; this was attributed to the purely diluent effect of the mercaptan.

The promoting effect of various mercaptans upon the persulphate-initiated emulsion copolymerisation of styrene and butadiene is illustrated in Fig. 6.13, which is taken from a paper by Starkweather and his co-workers.[35] That a marked promoting effect should have been observed with the system used for these particular experiments is surprising, since the persulphate was already activated by the addition of a small amount of potassium ferricyanide. Kolthoff and Harris[43] have confirmed that the level of mercaptan which is required to achieve the maximum promoting effect in the styrene–butadiene copolymerisation system is very small. Thus they observed that 0·0005 pphm of commercial dodecyl mercaptan had sufficient promoting effect to bring about 70% conversion in 24 h at 50°C,

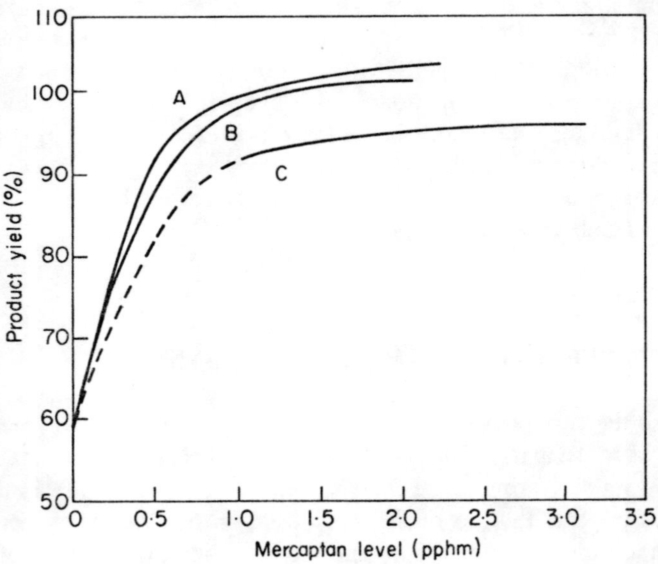

FIG. 6.13. Effect of mercaptan level and type upon product yield in emulsion copolymerisation of styrene and butadiene after 20 h reaction at 40°C, according to Starkweather *et al.*[35] Curve A is for dodecyl mercaptan, curve B for 'No. 30 white oil mercaptan', and curve C for pinene mercaptan. (Reaction system: styrene 25, butadiene 75, water 150, oleic acid 4, sodium naphthalene formaldehyde sulphonate 1, mercaptan—variable, $K_2S_2O_8$ 1, $K_3Fe(CN)_6$ 0·15, excess NaOH 0·75.)

TABLE 6.16

Effect of level of commercial dodecyl mercaptan on conversion in styrene–butadiene emulsion copolymerisation (according to Kolthoff and Harris[43])

(reaction system: styrene 25, butadiene 75, water 180, fatty-acid soap 5, mercaptan—variable, $K_2S_2O_8$ 0·3)

Mercaptan level (pphm)	% *conversion after* 12 h *at* 50°C
0·2	81·0
0·3	80·1
0·4	79·3
0·5	79·4
0·6	79·2
0·7	79·4
0·8	78·1
0·9	79·1
1·0	79·0

whilst 0·005 pphm mercaptan gave 70% conversion after 12 h. In the absence of mercaptan, the conversion after 12 h was less than 2%.

The results summarised in Table 6.16 show that, in the case of activation by commercial dodecyl mercaptan, no increase in conversion after a fixed period was observed with increasing amounts of mercaptan over the range

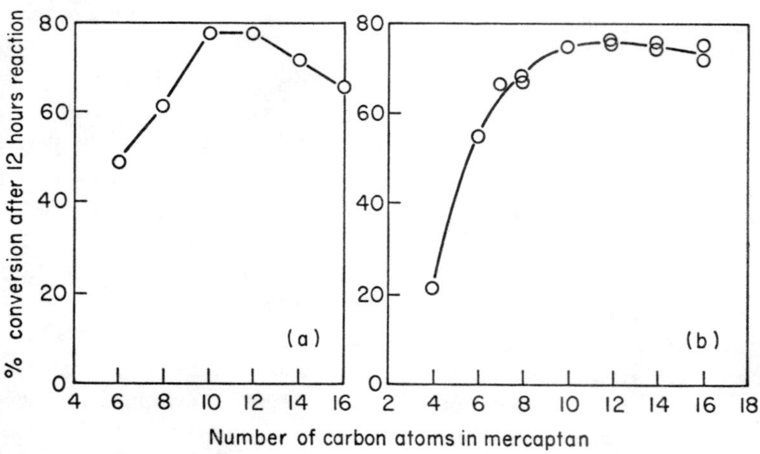

FIG. 6.14. Effect of number of carbon atoms in an aliphatic mercaptan upon effectiveness as promoter for persulphate initiation of styrene–butadiene emulsion copolymerisation at 50°C, according to Kolthoff and Harris.[43] Graph (a) is for primary mercaptans, graph (b) for tertiary mercaptans. (Reaction system: styrene 25, butadiene 75, water 180, fatty-acid soap 5, $K_2S_2O_8$ 0·3, plus *either* primary mercaptan 0·5 *or* tertiary mercaptan 0·3.)

from 0.2 to 1 pphm. Indeed, Kolthoff and Harris state that the rate of conversion is little affected by mercaptan concentration over the range from 0·02 to 2 pphm. *t*-dodecyl mercaptan at very low concentration was found to be a poorer promoter than the corresponding primary mercaptan; thus 0·01 pphm of the tertiary mercaptan was observed to be only about as effective as 0·0005 pphm of commercial dodecyl mercaptan. The effect of mercaptan chain length upon promoting efficiency is shown in Fig. 6.14 (a) for primary aliphatic mercaptans and in Fig. 6.14(b) for tertiary aliphatic mercaptans. In general, the rate of conversion reaches a maximum for aliphatic mercaptans when about 12 carbon atoms are present. With less than 10 carbon atoms, the rate of conversion falls off rapidly with decreasing mercaptan molecular weight. With more than 12 carbon atoms, the rate of conversion falls off slightly with increasing molecular weight. The rates of conversion with primary and tertiary dodecyl mercaptans are very similar, the conversions obtained after 12 h at 50°C being slightly less for the tertiary than for the primary mercaptan.

Kolthoff and Harris found it convenient to classify mercaptans which are poorer promoters than *n*-dodecyl mercaptans into two groups. Goup I comprises those which are poor promoters *per se*. A typical example is hexanedithiol. Mercaptans in this class do not interfere with polymerisation, but are merely poor promoters when used by themselves. A normal or near-normal rate of conversion is observed in the presence of these promoters if a good promoter is added. Group II comprises those mercaptans which are able to interfere with polymerisation, such as *t*-butyl mercaptan. Not only are they ineffective as promoters in themselves, but are able effectively to suppress the rate of polymerisation in the presence of good promoters. This effect is associated with their enhanced water-solubility, presumably allowing them to interact in the aqueous phase with the radicals produced by the persulphate. It is perhaps significant that Blackley and McGilvray[44] have recently observed severe retardation of the persulphate-initiated emulsion polymerisation of styrene and butadiene when mercaptoacetic and mercaptopropionic acids are added to the system. These acids are quite water-soluble, and would certainly be expected to partition in such a way that considerable proportions would be present in the aqueous phase. The retarding effect was observed whether a conventional promoting mercaptan was present or not.

Kolthoff and Harris have also provided evidence to suggest that a slight advantage may be gained, especially at high conversions, by using a mixture of two mercaptans, one of which is consumed at a lesser rate than the other. In this way, the effect of mercaptan depletion can be minimised.

Results are given which suggest that a blend of 0·3 pphm of *n*-dodecyl mercaptan plus 0·1 pphm of primary tetradecyl mercaptan is slightly more effective than 0.3 pphm of *n*-dodecyl mercaptan if the conversion exceeds about 80%.

The promoting effect of mercaptans is usually explained as being due to the formation of a redox couple with the persulphate. The following are the reactions conventionally given:

$$S_2O_8^{2-} \rightarrow SO_4^{\overline{\cdot}}$$

$$SO_4^{\overline{\cdot}} + HSR \rightarrow HSO_4^{-} + \cdot SR \qquad \text{(xxix)}$$

$$M + \cdot SR \rightarrow \cdot MSR, \text{ etc.} \qquad \text{(xxx)}$$

It is certainly reasonable to suppose that radicals such as $\cdot SR$ or $\cdot M_nSR$ could more easily enter negatively-charged micelles and particles than could $SO_4^{\overline{\cdot}}$ or $\cdot M_nSO_4^{-}$. In support of this mechanism may be cited the finding of Kolthoff and Miller[12] that mercaptans containing from 8 to 12 carbon atoms solubilised in aqueous solutions of saturated fatty-acid soaps are oxidised by persulphate ions to disulphides. The latter are most easily envisaged as being formed by the combination of mercaptyl radicals thus:

$$2 \cdot SR \rightarrow RSSR \qquad \text{(xxxi)}$$

The rate of disappearance of persulphate in soap solutions was found to be unaffected by the presence of the mercaptan. The rate of disappearance of mercaptan was observed to be independent of the initial mercaptan concentration, and, in the case of normal mercaptans, equal to the rate of disappearance of persulphate at mercaptan concentrations above 0·008 mole l^{-1}. (In the case of tertiary mercaptans, the rate of oxidation in soap solution was found to be less than the rate of disappearance of persulphate). In all cases reported, oxidised normal mercaptans were recovered quantitatively in the form of the corresponding disulphide. For normal mercaptans, the overall stoichiometry of the reaction appeared to be consistent with the equation

$$S_2O_8^{2-} + 2RSH \rightarrow 2HSO_4^{-} + RSSR \qquad \text{(xxxii)}$$

In order to account for the disappearance of equivalent amounts of mercaptan and persulphate in the presence of fatty-acid soap, Kolthoff and Dale[32] assume that any carboxyl radicals formed by reactions (xv) and (xvi) above can abstract a hydrogen atom from the mercaptan as follows:

$$R\underset{\underset{O}{\|}}{C}{-}O\cdot + R'SH + OH^{-} \rightarrow R\underset{\underset{O}{\|}}{C}{-}O^{-} + R'S\cdot + H_2O \qquad \text{(xxxiii)}$$

TABLE 6.17

Persulphate–sulphur contents of emulsion polystyrenes polymerised with S^{35}-labelled potassium persulphate in presence of *n*-dodecyl mercaptan (according to Kolthoff, O'Connor and Hansen[17])

(reaction system: styrene 100, water 180, fatty-acid soap 5, $K_2S_2O_8$ 1·2; *n*-dodecyl mercaptan—variable, temperature 50°C)

n-Dodecyl mercaptan level (pphm)	% *conversion*	*Total time of shaking with water* (min)	*Specific activity* (min^{-1} mg^{-1}) S *in* $K_2S_2O_8$	*Specific activity* (min^{-1} mg^{-1}) *Polymer*	*Molecular weight of Polymer*	g *of* S_2O_8 *sulphur per* g *of polymer*	*Atoms of* S_2O_8*-sulphur per mole of polymer*
0	84	0	$3·14 \times 10^7$	434	$7·2 \times 10^5$	$1·4 \times 10^{-5}$	3·1
		21		379	$6·7 \times 10^5$	$1·2 \times 10^{-5}$	2·5
		51		353	$6·6 \times 10^5$	$1·1 \times 10^{-5}$	2·3
		105		352	$6·5 \times 10^5$	$1·1 \times 10^{-5}$	2·2
0	67	0	$1·38 \times 10^7$	—	$5·4 \times 10^5$	—	—
		20		118	$5·2 \times 10^5$	$8·2 \times 10^{-6}$	1·4
		50		124	$5·1 \times 10^5$	$8·2 \times 10^{-6}$	1·4
0·25	68	0	$1·38 \times 10^7$	325	$1·8 \times 10^5$	$10·2 \times 10^{-6}$	$5·7 \times 10^{-2}$
		26		234	$1·8 \times 10^5$	$7·4 \times 10^{-6}$	$4·1 \times 10^{-2}$
		70		242	$1·8 \times 10^5$	$7·5 \times 10^{-6}$	$4·3 \times 10^{-2}$
		105		223	$1·8 \times 10^5$	$7·0 \times 10^{-6}$	$3·9 \times 10^{-2}$
0·5	67	0	$1·38 \times 10^7$	120	$3·2 \times 10^5$	$8·3 \times 10^{-6}$	$8·3 \times 10^{-2}$
		20		110	$3·2 \times 10^5$	$7·6 \times 10^{-6}$	$7·6 \times 10^{-2}$
		50		112	$3·2 \times 10^5$	$7·8 \times 10^{-6}$	$7·8 \times 10^{-2}$
0·5	81	0	$1·38 \times 10^7$	352	$0·9 \times 10^5$	$11·1 \times 10^{-6}$	$3·2 \times 10^{-2}$
		21		239	$0·9 \times 10^5$	$7·5 \times 10^{-6}$	$2·2 \times 10^{-2}$
		65		238	$0·9 \times 10^5$	$7·5 \times 10^{-6}$	$2·2 \times 10^{-2}$
		105		238	$0·9 \times 10^5$	$7·5 \times 10^{-6}$	$2·2 \times 10^{-2}$
1·0	81	0	$1·38 \times 10^7$	348	$0·4 \times 10^5$	$10·9 \times 10^{-6}$	$1·3 \times 10^{-2}$
		21		251	$0·4 \times 10^5$	$7·9 \times 10^{-6}$	$1·0 \times 10^{-2}$
		51		243	$0·4 \times 10^5$	$7·6 \times 10^{-6}$	$0·9 \times 10^{-2}$
		105		233	$0·4 \times 10^5$	$7·3 \times 10^{-6}$	$0·9 \times 10^{-2}$

However, Kolthoff, O'Connor and Hansen[17] dissent from the view that such reactions as these are important in the case of styrene emulsion polymerisation. Their conclusion is that 'the mercaptan acts solely as a chain transfer agent in the emulsion polymerization of styrene with persulfate as activator, and that any chemical reaction between persulfate and mercaptan does not play a role in the polymerization'. They draw this conclusion from the observation that the amount of persulphate sulphur per unit mass of polymer is independent of the amount of mercaptan in the system. The persulphate was estimated by radioactivity assay of polymers prepared using S^{35}-labelled potassium persulphate. Their results are summarised in Table 6.17.

Whatever may be the precise mechanism by which mercaptans exert their promoting effect upon persulphates, it remains to be explained why persulphates by themselves are ineffective as initiators for the emulsion polymerisation of butadiene, and the emulsion polymerisation of styrene and butadiene unless the styrene is present in excess. As far as the present writer is aware, no adequate explanation of this phenomenon has ever been given. The problem is especially puzzling in the case of the styrene–butadiene copolymerisation, since the emulsion polymerisation of styrene is readily achieved by initiation with persulphate alone.

6.9 ACTIVATION OF PERSULPHATE BY SULPHITES AND RELATED COMPOUNDS

The persulphate–bisulphite redox system has been used industrially to effect the emulsion polymerisation of acrylonitrile. Berry and Peterson[45] have suggested that the primary step in this redox system is as follows:

$$S_2O_8^{2-} + HSO_3^- \rightarrow SO_4^{2-} + SO_4^-\cdot + HSO_3\cdot \qquad \text{(xxxiv)}$$

If initiation occurs from both sulphate radical-ions and bisulphite radicals, then polymers will be obtained which contain both sulphate and sulphonate end-groups. It has also been suggested[46] that the following side reactions can occur:

$$SO_4^-\cdot + HSO_3^- \rightarrow SO_4^{2-} + HSO_3\cdot \qquad \text{(xxxv)}$$

$$2HSO_3\cdot \rightarrow H_2S_2O_6 \qquad \text{(xxxvi)}$$

The first of these is a radical-transfer reaction, which would not affect the total radical concentration; the second would lead to a loss of radicals. If reaction (xxxvi) were slow relative to reactions (xxxiv) and (xxxv), then the polymer produced would contain mainly sulphonate end-groups.

Tsuda[47] has reported the results of a study whose purpose was to determine the relative numbers of end-groups present in acrylonitrile polymers prepared using this redox system. Polymerisations were carried out at 25°C without agitation. The pH was maintained at 3 by the addition of small amounts of sulphuric acid. Experiments were carried out on systems containing (a) S^{35}-labelled sodium bisulphite plus unlabelled potassium persulphate, and (b) unlabelled sodium bisulphite plus S^{35}-labelled potassium persulphate. Polymerisation was stopped at low conversions ($<10\%$). The polymers were purified by precipitation in water from solution in dimethylformamide. Tsuda claims that radioactivity measurements showed the polymer to have been completely purified by this treatment. Concentrations of the respective types of end-group were determined by comparing the activity of the polymer with that of an inactive polymer to which had been added known amounts of the labelled initiator. It appears that because of difficulties in attaining reproducibility, it was sometimes necessary to determine the concentration of both types of end-groups on a single polymer. A complex isotope-dilution procedure was used for this purpose. This involved reducing all the sulphur-bearing end-groups to hydrogen sulphide, absorbing in zinc acetate–acetic acid solution, and then converting to methylene blue by the addition of *p*-aminodimethylaniline and iron(III) chloride. The total amount of methylene blue was estimated from an optical density measurement. The methylene blue was then adsorbed on to a known amount of active carbon and the radioactivity determined. In order to calibrate the method, polyacrylonitriles were prepared using as initiators (a) S^{35}-labelled sodium bisulphite plus iron(III) salt at 25°C, and (b) labelled potassium persulphate on its own at 60°C.

Tsuda observed that, under all the conditions investigated, sulphonate is the major end-group, even if the bisulphite concentration is very much less than that of the persulphate. He also found that the higher the persulphate concentration, the higher is the proportion of sulphate end-groups; and the higher the concentration of bisulphite, the higher the proportion of sulphonate end-groups. Furthermore, the proportion of sulphate end-groups increases as the monomer concentration increases. Tsuda established the following empirical relationship between the ratio of end-groups in the polymer, the ratio of components in the initiator, and the initial monomer concentration, [M]:

$$\frac{[\sim SO_3^-]}{[\sim SO_4^-]} = 1 + 0{\cdot}27\left\{\frac{[NaHSO_3]}{[M]\,[K_2S_2O_8]}\right\}^{1\cdot5}$$

The experimentally-observed relationship between the ratio of end-groups in the polymer and the ratio of components in the initiator is shown in Fig. 6.15, together with the line given by the above equation. In this graph, $\log_{10}\{[\sim SO_3^-]/[\sim SO_4^-] - 1\}$ has been plotted against

$$\log_{10}\{[NaHSO_3]/([M]\,[K_2S_2O_8])\}.$$

The above equation requires a linear relationship of slope 1·5. No theoretical explanation for the above relationship is given by Tsuda.

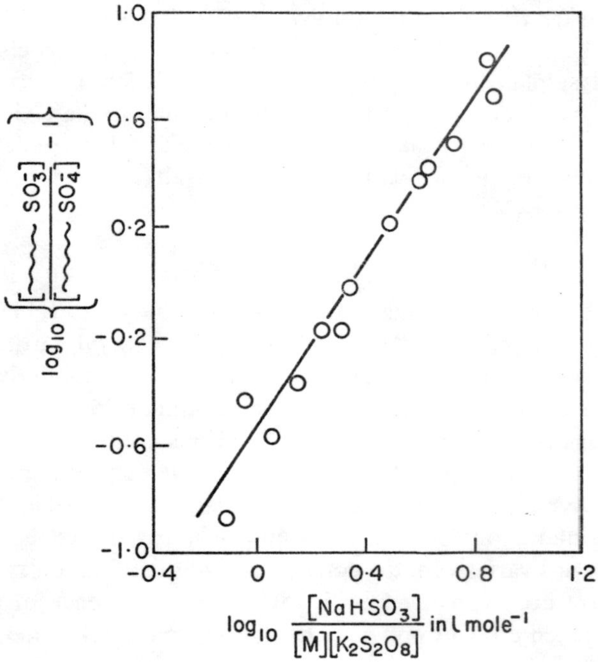

FIG. 6.15. Relationship between the ratio of sulphonate to sulphate end-groups in polyacrylonitrile produced by emulsion polymerisation, and the ratio of bisulphite to persulphate in the initiation system, according to Tsuda.[47] (Full details of the reaction systems are given in the original paper. The temperature of polymerisation was 25°C; the initial pH was 3.)

Berry and Peterson[45] have reported upon the sulphur contents of polytetrafluoroethylenes prepared by emulsion polymerisation using various combinations of potassium persulphate, sodium bisulphite and iron(II) sulphate or iron as initiator. They observed that the polymer was always non-radioactive when radioactive sulphur was incorporated in the persulphate, whether the persulphate was used alone, in combination with iron, or in combination with bisulphite and iron(II) sulphate. By contrast, the polymer was always radioactive when the bisulphite component was labelled with radioactive sulphur. Furthermore, polymers containing sulphur of bisulphite origin were found to retain their sulphur content when heated at 65°C for 3 h as a dispersion in water at a pH between 2 and 3. The conclusions drawn were that the polymers contained no sulphur derived from the persulphate, and that the sulphur derived from the bisulphite was probably present as hydrolysis-resistant sulphonate groups.

Regarding the absence of persulphate-derived sulphur from the polymers it was suggested that either the sulphate radical-ion is ineffective in initiating the emulsion polymerisation of tetrafluoroethylene, or else the perfluoroalkyl sulphate group formed during initiation is very prone to hydrolysis. No evidence is given in support of either of these views, but on balance the latter seems the more likely. Berry and Petersen suggest the following reactions for activation by iron(II) salts:

$$S_2O_8^{2-} + Fe^{2+} \rightarrow SO_4^{2-} + SO_4^{\overline{\cdot}} + Fe^{3+} \qquad \text{(xxxvii)}$$

$$HSO_3^- + Fe^{3+} \rightarrow HSO_3\cdot + Fe^{2+} \qquad \text{(xxxviii)}$$

Reference may also be made to the recent study of Peebles[48] of the persulphate–bisulphite initiated copolymerisation of acrylonitrile and vinyl acetate. A theoretical molecular weight distribution was derived for an acrylonitrile homopolymer forming in a continuous-feed reactor under homogeneous conditions. The derived equations are then compared with data obtained on acrylonitrile–vinyl acetate copolymers prepared by emulsion polymerisation. It was reported that, on average, each polymer molecule contains one sulphonate group, whereas the concentration of sulphate groups varies with the composition of the initiation system. This variation has been explained by invoking the occurrence of a transfer reaction between growing polymer chains and the bisulphite ion activator:

$$\sim\sim\cdot + HSO_3^- \rightarrow \sim\sim H + SO_3^{\overline{\cdot}} \qquad \text{(xxxix)}$$

This reaction regenerates bisulphite-derived initiating species; their concentration is therefore enhanced relative to that of persulphate-derived initiating species.

A related initiation system is that comprising persulphates and metabisulphites (pyrosulphites). The latter compounds have the composition $M_2^IS_2O_5$, and are derived from the corresponding bisulphite by loss of water. Of the two structures (XXIII) and (XXIV) which can be written for metabisulphite ion, (XXIV) appears to be preferred. As regards the

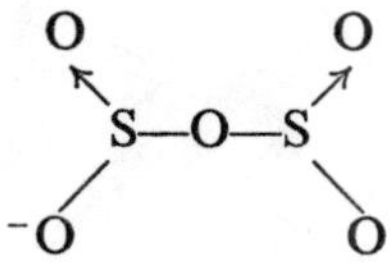

(XXIII)

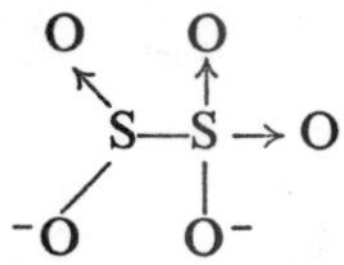

(XXIV)

reactions which occur between persulphates and metabisulphites in aqueous solution, it is presumed that the latter act as a source of bisulphite ions (formed by hydrolysis), and that the reactions then proceed as for the persulphate–bisulphite system. Hay and his co-workers[49] have reported

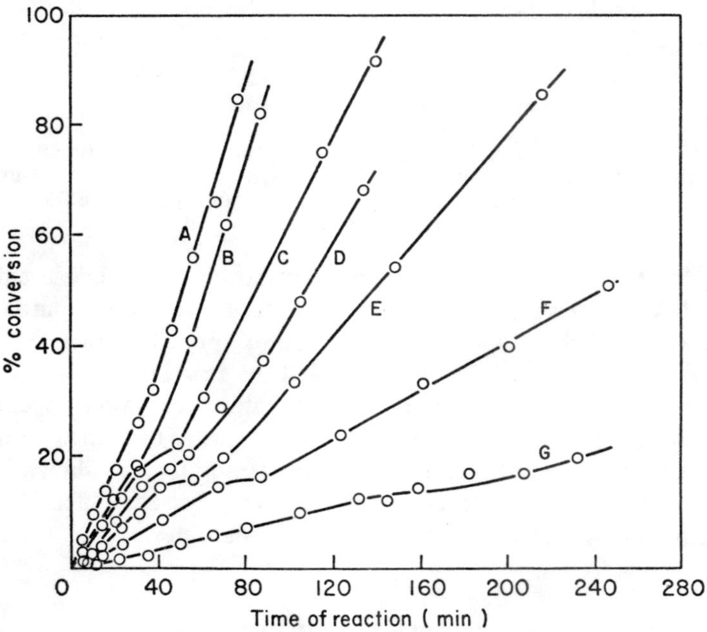

FIG. 6.16. Effect of initiator level upon conversion–time relationship for emulsion polymerisation of vinylidene chloride at 32°C, according to Hay *et al.*[49] (Reaction systems: vinylidene chloride 100, sodium lauryl sulphate 2, $(NH_4)_2S_2O_8$ x, $Na_2S_2O_5$ x, water 118 − 2x. The various curves refer to different values of x as follows: A, 0·60; B, 0·45; C, 0·30; D, 0·225; E, 0·15; F, 0·075; G, 0·0375.)

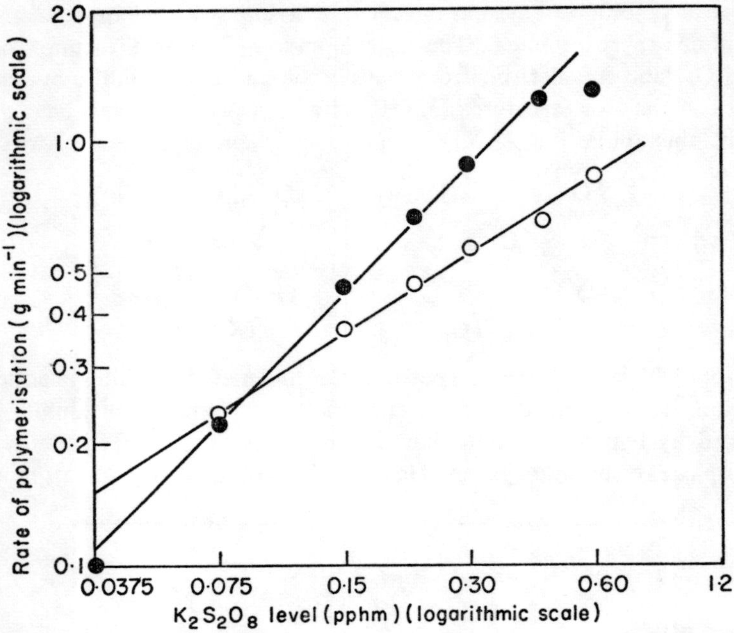

FIG. 6.17. Effect of initiator level upon initial polymerisation rate (points ○) and final polymerisation rate (points ●) for emulsion polymerisation of vinylidene chloride at 32°C, according to Hay *et al.*[49] (See caption to Fig. 6.16 for details of reaction systems.)

upon the kinetics of the emulsion polymerisation of vinylidene chloride initiated by equal weights of ammonium persulphate and sodium metabisulphite. The effect of initiator concentration upon the conversion–time curve is shown in Fig. 6.16. It is seen that the reaction consists of three distinct stages: an initial stage of constant rate, a transition stage, and a final stage of constant rate which tends to be greater than the initial rate. The initiator concentration affects both the steady rates as shown in Fig. 6.17, but the orders of reaction for the two rates are different, being respectively 0·60 and 1·0 for the initial and final rates.

6.10 THE CHLORATE–BISULPHITE SYSTEM

Combinations of chlorates and bisulphites are apparently important for the emulsion polymerisation of vinyl chloride.[50,51] Firsching and Rosen[51] have reported the result of a study of this system, the main objective of which was to determine the amounts of initiator fragments which become chemically combined in the resultant polymer. Polymerisations were carried out at 50°C. The standard recipe was as follows:

Vinyl chloride	19 g
Water	100 ml
Sodium lauryl sulphate (Duponol ME)	0·5 ml
$NaHSO_3$	0·15 g
$KClO_3$	0·028 g

Using S^{35}-labelled sodium bisulphite and unlabelled potassium chlorate, they found that, with the standard level of initiator, about 0·3 atom of sulphur was combined with each polymer molecule, whereas when the initiator concentration was doubled, about 0·5 atom of sulphur became chemically combined. The amount of combined sulphur was not significantly affected by the presence or absence of a surface-active compound. Using Cl^{36}-labelled potassium chlorate and unlabelled sodium bisulphite, they found that, with the standard level of initiator, about 0·08 atom of chlorate-derived chlorine per polymer molecule was chemically incorporated. Changing the initiator level brought about a corresponding change in the amount of chlorate-derived chlorine. Experiments were also carried out in which the sodium bisulphite was replaced by its adduct with ethyl acrylate. No polymer was formed. The conclusion was drawn that an adduct of sodium bisulphite and the monomer is not an intermediate in the initiation process.

The reactions by which the chlorate–sulphite system effects initiation appear to be complex. The following possibilities have been suggested by Firsching and Rosen:

$$SO_3^{2-} + ClO_3^- \rightarrow [O_3SOClO_2]^{3-} \rightarrow SO_4^{2-} + ClO_2^- \qquad \text{(xl)}$$

$$[O_3SOClO_2]^{3-} + 2H^+ + CHCl{:}CH_2 \rightarrow \dot{C}HCl\cdot CH_2SO_3^- + ClO_2 + H_2O \qquad \text{(xli)}$$

Hypochlorite ions can also be envisaged as being formed by a continuation of reaction (xl):

$$SO_3^{2-} + ClO_2^- \rightarrow [O_3SOClO]^{3-} \rightarrow SO_4^{2-} + ClO^- \qquad \text{(xlii)}$$

Hypochlorite ions could then react with sulphite ions to produce free radicals as follows:

$$ClO^- + H^+ \rightleftharpoons HOCl \qquad \text{(xliii)}$$

$$SO_3^{2-} + HOCl \rightarrow SO_4^{2-} + HO\cdot + Cl\cdot \qquad \text{(xliv)}$$

or

$$SO_3^{2-} + HOCl \rightarrow SO_4^{\dot{-}} + HO^- + Cl\cdot \qquad \text{(xlv)}$$

Firsching and Rosen provide evidence to suggest that reactions between sulphite and hypochlorite ions are relatively unimportant for the initiation of polymerisation. The mechanism by which chlorate-derived chlorine enters the polymer could be termination by the chlorine radicals produced in reactions (xliv) and (xlv). An alternative origin could be transfer with molecular chlorine which might be formed by combination of the chlorine radicals, or by reaction between hypochlorous acid and the hydrochloric acid which could in turn be formed by further reduction of the hypochlorous acid by the sulphite anion.

The use of the chlorate–sulphite redox system for the initiation of the emulsion polymerisation of acrylonitrile has been described by Thomas, Gleason and Mino.[52]

6.11 THE HYDROGEN PEROXIDE–IRON SYSTEM

Combinations of hydrogen peroxide and iron(II) salts have for many years been used as an oxidising agent in preparative and analytical chemistry. These combinations are known generically as 'Fenton's reagent'. The mechanism of the interaction between hydrogen peroxide and iron(II) salts has been extensively studied. The principal features of this interaction have been summarised by Medalia and Kolthoff[53] as follows:

(1) In aqueous solutions containing only iron(II) salts, hydrogen peroxide, and stable inorganic salts and acids, the stoichiometric reaction

$$H_2O_2 + 2Fe^{2+} + 2H^+ \rightarrow 2Fe^{3+} + 2H_2O \qquad \text{(xlvi)}$$

is followed if the iron(II) is present in excess over the peroxide.

(2) When the peroxide is in large excess over the iron(II), less iron is oxidised than would correspond to the hydrogen peroxide consumed if reaction (xlvi) were followed. The balance of the hydrogen peroxide is decomposed to oxygen and water.

(3) In the presence of many organic compounds, as well as of certain easily-oxidised inorganic compounds, such as the iodide ion, part of the hydrogen peroxide oxidises the added compound.

The following mechanism was proposed by Haber and Weiss[54] for the interaction:

$$H_2O_2 + Fe^{2+} \rightarrow HO^- + HO\cdot + Fe^{3+} \qquad \text{(xlvii)}$$

$$HO\cdot + H_2O_2 \rightarrow H_2O + HO_2\cdot \qquad \text{(xlviii)}$$

$$HO_2\cdot + H_2O_2 \rightarrow O_2 + H_2O + HO\cdot \qquad \text{(xlix)}$$

$$HO\cdot + Fe^{2+} \rightarrow Fe^{3+} + HO^- \qquad \text{(l)}$$

This is a chain reaction for which reaction (xlvii) is the initiation, reactions (xlviii) and (xlix) are the propagation, and reaction (l) is the termination. According to this mechanism, the iron(II) ions are effective in both initiation and termination. Evidently when iron(II) ions are present in similar concentration to the peroxide, termination occurs so rapidly that the effective kinetic chain length is unity.

Baxendale, Evans and Park[55] studied the polymerisation of vinyl compounds in aqueous solution initiated by the hydrogen peroxide–iron(II) redox system in the light of the Haber–Weiss mechanism. They concluded that the monomer competes with the iron(II) ions for the hydroxyl radicals produced in reaction (xlvii), and that, when the monomer is present in sufficient excess, reaction (l) is completely suppressed. The number of iron(II) ions oxidised per molecule of peroxide consumed falls from a maximum of two to a minimum of one when the monomer concentration is increased from zero. Methyl acrylate is a rather less efficient scavenger for hydroxyl radicals than is acrylonitrile. According to this view, the initiating species for the polymerisation is the hydroxyl radical. Confirmation of this conclusion is provided by the evidence of Evans[56] concerning the concentration of hydroxyl groups in polystyrene prepared using this system as initiator. Hydroxyl groups were determined by reaction with methylmagnesium iodide, and were found to correspond to approximately two per polymer molecule. This would be the expected value for polystyrene if initiation were exclusively by hydroxyl ions, since, in this case, termination occurs by combination rather than by disproportionation.

Occasional references to the use of hydrogen peroxide in combination with iron(III) salts occur in the literature. Haber and Weiss[54] proposed the following as the initial and rate-determining step of the reaction:

$$HO_2^- + Fe^{3+} \rightarrow HO_2\cdot + Fe^{2+} \qquad \text{(li)}$$

This is an electron-transfer step involving a hydroperoxide anion. The resulting hydroperoxy radical could then disappear by the reverse of this reaction, by reaction with hydrogen peroxide as in reaction (xlix), or by reaction with iron(III) ions as follows:

$$HO_2\cdot + Fe^{3+} \rightarrow H^+ + O_2 + Fe^{2+} \qquad \text{(lii)}$$

The iron(II) ions produced in reaction (li) could also react with hydrogen peroxide by the reactions discussed above. Many other reactions have been proposed to account for the interaction between iron(II) salts and hydrogen peroxide. For a review of these, the reader is referred to the paper by Medalia and Kolthoff.[53]

Marvel *et al.*[57] have reported upon the emulsion copolymerisation of styrene and butadiene at 50°C, initiated by hydrogen peroxide and iron salts. Both iron(II) and iron(III) salts were used and found to be equivalent in their activities. The activity of both types of salt is increased by complex-formation with pyrophosphate ions. Polymerisation does not commence for about an hour, during which time the hydrogen peroxide completely disappears and some organic peroxide is formed. Although these observations suggest that hydrogen peroxide is not directly responsible for initiating the polymerisation, the hydrogen peroxide is nevertheless essential to the subsequent reaction, since systems charged without the hydrogen peroxide fail to polymerise at all. Marvel and his co-workers made the plausible suggestion that it is the organic peroxides formed during the initial induction period which are the true polymerisation initiators. The results given in this paper are of further interest in that they show clearly the advantage of having the iron complexed, in this case with pyrophosphate ions. Thus, whereas in their standard recipe, hydrogen peroxide and iron(II) sulphate gave 41% conversion after 4 h at 50°C, the addition of sodium pyrophosphate increased the yield of 78%. The corresponding conversions for hydrogen peroxide and iron(III) sulphate were 9% and 80%. No discussion is given concerning the reason for the higher yields in the presence of the pyrophosphate ions. A plausible explanation is that, by sequestration of the iron ions, the rate of production of radicals is controlled in such a way that wastage by mutual interaction is much less than occurs in the absence of the sequestering agent.

6.12 HYDROPEROXIDE–IRON SYSTEMS

Initiation systems comprising organic hydroperoxides and iron(II) salts have been extensively used for the low-temperature emulsion copolymerisation of styrene and butadiene. The radical-forming reaction is usually given as

$$ROOH + Fe^{2+} \rightarrow RO\cdot + {}^{-}OH + Fe^{3+} \qquad \text{(liii)}$$

initiation occurring by way of the alkoxy radical. The system is capable of great diversification because of (a) the wide range of organic hydroperoxides which is available; (b) the possibility of controlling the availability of iron(II) ions by using performed complexes or else very insoluble iron(II) salts; and (c) the possibility of adding other components to the system, such as reducing sugars. It should be noted that the hydroperoxide will generally partition mainly in the non-aqueous phases of the system, whereas the iron compound will be confined to the aqueous phase. This feature provides additional control over the rate at which initiating radicals are produced through limitation of the rate at which the two components of the couple are brought together.

If a reducing agent is also added to the system, then the function of the iron(II) compound can be viewed as that of a catalyst which promotes interaction between hydroperoxide and reducing agent. The relationships between the several components of the system can be displayed as shown in Fig. 6.18. The hydroperoxide oxidises the iron(II) ions in accordance with reaction (liii), thereby producing alkoxy radicals which initiate polymerisation. The iron(III) ions also produced in turn oxidise the reducing agent, being themselves reduced to iron(II) ions once again.

A typical recipe for the emulsion polymerisation of styrene and butadiene using a hydroperoxide–iron–reducing sugar system is the following,

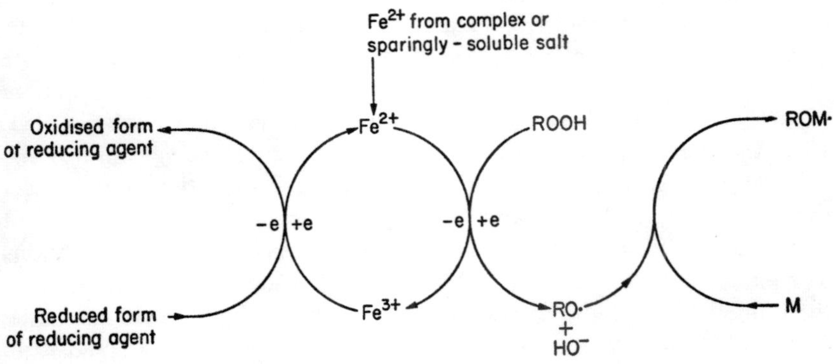

FIG. 6.18. Illustrating the interactions between the various components of a hydroperoxide–iron–reducing agent initiation system.

given by Shearon, McKenzie and Samuels.[58] The levels of the various ingredients refer to parts by weight:

Styrene	29
Butadiene	71
Water	180
Sodium soap of disproportionated rosin acid (Dresinate 731)	4·7
t-C_{12} mercaptans	0·2
Cumene hydroperoxide	0·15
Dextrose	2·5
$FeSO_4.12H_2O$ ($7H_2O$?)	0·1
$Na_4P_2O_7$ ($12H_2O$?)	0·6
Polymerisation temperature	41°F
Reaction time	20–22 h
Conversion	60%

The sodium pyrophosphate in this recipe is used to complex the iron(II) ions. It has been found necessary to adopt a carefully-controlled procedure for preparing the complex if consistent results are to be obtained. Shearon, McKenzie and Samuels[58] give the following instructions: 'The sodium pyrophosphate is added slowly to water heated to between 120° and 130°F. After all the pyrophosphate is in solution the glucose is added and dissolved. The temperature of the resulting solution is raised to boiling and maintained at 212°F for 30 minutes. It is then cooled to 80° to 90°F and the ferrous sulfate is dissolved as the final step.'

The critical importance of the preparative procedure for the solution containing the complexed iron and the reducing sugar has been emphasised by Pryor, Harrington and Druesedow[59] as follows: 'The quality of the iron pyrophosphate–dextrose activator suspension is one of the most important factors in cold GR-S reaction rate and uniformity. The activator is prepared by a special procedure. A dextrose–potassium pyrophosphate solution is heated, aged a short time, and partially cooled, ferrous sulfate solution is mixed in, and the mixture is again aged and finally cooled. However, the exact temperatures and times of heating, ageing, and cooling, and the turbulence during mixing proved to be very critical with respect to activator activity, as judged by early development work. It was apparent that these conditions could be most readily controlled to exact specifications in equipment for continuous activator preparation.' These authors give the following procedure as that originally adopted by the B.F. Goodrich Chemical Co. at their Port Neches plant: 'Dissolve dextrose and potassium pyrophosphate in 10 parts of water at about 90°F. Dissolve ferrous sulfate in 1 part of water adjusted to pH 3·0 with sulfuric acid. Pump dextrose–

pyrophosphate solution to continuous heat exchanger, heating to 220°F in about 1 minute. Age dextrose–pyrophosphate solution in continuous holdup tank 4 minutes at 220°F. Cool to 140°F in continuous heat exchanger. Add ferrous sulfate solution stream in continuous mixer. Cool activator suspension to room temperature in continuous heat exchanger to run to storage tank.' Subsequently it was found that polymerisation times could be reduced by making various amendments to this procedure, including, *inter alia*, ageing the dextrose–pyrophosphate solution for a longer time at a lower temperature.

A wide range of organic hydroperoxides has been proposed for use in the hydroperoxide–iron system. The following are the names and structures of a few of the more common: *t*-butyl hydroperoxide (XXV), *iso*propylbenzene hydroperoxide (cumene hydroperoxide) (XXVI), di*iso*propylbenzene hydroperoxide (XXVII), *p*-methane hydroperoxide (XXVIII), methylcyclohexyl hydroperoxide (XXIX), *p*-cymene hydroperoxide (XXX) and tetralin hydroperoxide (XXXI).

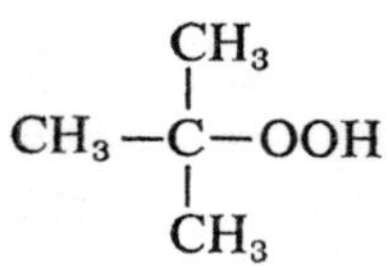
(XXV)

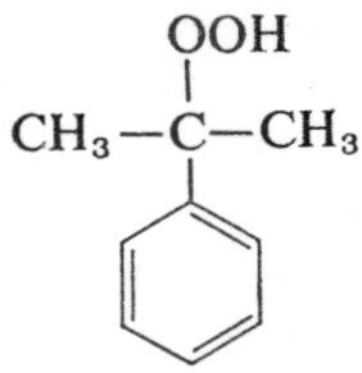
(XXVI)

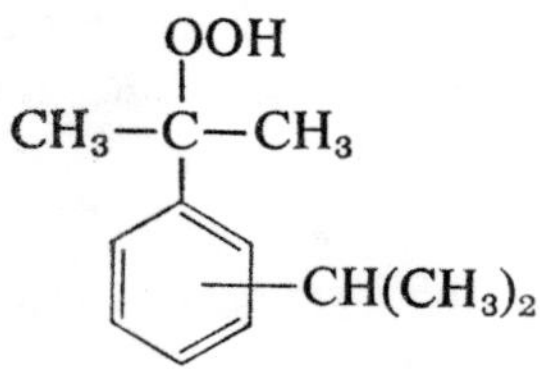
(XXVII)

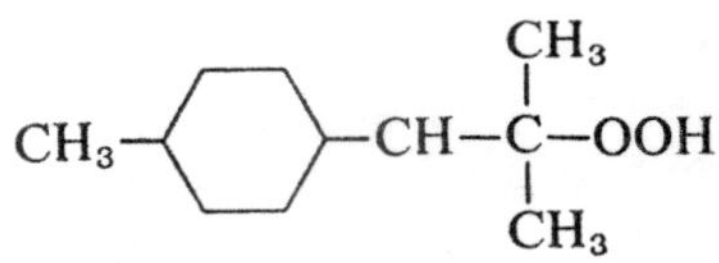
(XXVIII)

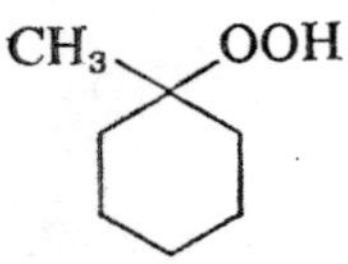
(XXIX)

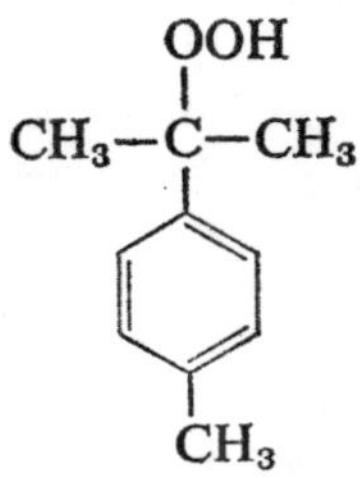
(XXX)

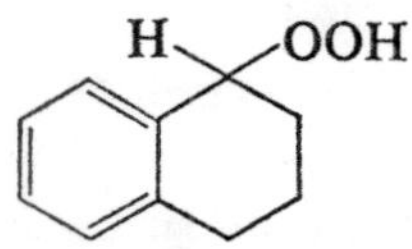
(XXXI)

Several detailed studies of the hydroperoxide–iron system have been published. Thus Wicklatz, Kennedy and Reynolds[60] reported on the effectiveness of a wide range of hydroperoxides in the following standard recipe (the so-called 'custom recipe'), all part being by weight:

Styrene	28
Butadiene	72
Water	180
Potassium soap of disproportionated rosin acid (Dresinate 214)	4·7
Mixed tertiary mercaptans (C_{12}, C_{14} and C_{16})	0·25
Hydroperoxide	variable
Dextrose	1·0
$Fe_2SO_4.7H_2O$	0·14
$K_4P_2O_7$	0·165
KCl	0·5
KOH	0·037

Polymerisation was effected at 5°C. Reactions were carried out on the beverage-bottle scale. In view of the complexity of the initiation system and the critical nature of the method of bringing the reagents together, it is though desirable to quote the procedure as given by Wicklatz, Kennedy and Reynolds:

'(a) The Dresinate 214 is dissolved in slightly less than 160 parts of water. After boiling it for ten minutes and allowing it to cool to room temperature, the solution is adjusted to pH 10 with a concentrated solution of potassium hydroxide. Sufficient water is added to bring the total number of parts of water in making the solution to 160.

(b) The potassium chloride is dissolved in the Dresinate solution and the mixture is placed in the reaction bottle.

(c) The dextrose and 0·037 part of potassium hydroxide are dissolved together in ten parts of water and heated at 70°C, for 30 minutes. After allowing it to cool to room temperature this solution is added to the Dresinate 214–potassium chloride solution in the reaction bottle.

(d) A solution of MTM in 90% of the charge of styrene is added to the bottle.

(e) The air space above the liquid in the bottle is flushed with nitrogen; the bottle is corked and cooled in a salt-ice bath.

(f) When if has cooled sufficiently, the bottle is charged with cold, liquid butadiene and capped with a self-sealing Crown cap. It is stored in a refrigerator until steps (g) and (h) can be carried out.

(g) The ferrous sulfate is dissolved in five parts of water and a solution of the pyrophosphate, also made with five parts of water, is added to it. After it is heated at 60°C for twenty minutes, the mixture is cooled quickly to room temperature. This mixture is called the activator solution and it is important to prevent its oxidation by the atmosphere in the course of its preparation. The activator solution is most conveniently set aside for future use in a six-ounce, Crown-capped bottle which has been swept out previously with nitrogen. The seal is made with a perforated self-sealing Crown cap. After the bottle is capped the contents are put under 25 p.s.i.g. pressure with oxygen-free nitrogen.

(h) The initiator solution is made by dissolving the required amount of hydroperoxide ... in 10% of the charge of styrene.

(i) When the activator and initiator solutions have been prepared, the bottle which had been charged previously with emulsifier solution MTM, and monomers is transferred from the refrigerator to a polymerisation bath which is maintained at +5°C. The bottle is allowed to rotate in the bath for 10–15 minutes in order to allow the contents to reach thermal equilibrium with the bath. It is then removed and the initiator solution is injected by means of a hypodermic syringe. The bottle is shaken quickly and 10 parts of activator solution is immediately injected. The activator solution is transferred from its storage bottle to the reaction bottle by means of a hypodermic syringe. After the activator solution has been injected, the reaction bottle is immediately returned to the polymerisation bath and the time at which it is replaced is taken as zero time for the purpose of observing the rate of polymerisation.'

As a result of experiments conducted in this way, Wicklatz, Kennedy and Reynolds found that the effectiveness of a hydroperoxide as an initiator for low-temperature polymerisation depends very much upon its structure. Furthermore, for each hydroperoxide there is an optimum mole ratio of hydroperoxide:iron. The values of these ratios as given by Wicklatz, Kennedy and Reynolds are listed in Table 6.18. There appears to be a definite trend towards a higher optimum mole ratio as the molecular weight of the hydroperoxide increases.

The relative effectiveness of the various hydroperoxides investigated is

TABLE 6.18

Optimum mole ratio hydroperoxide: iron for various hydroperoxides in the 'custom recipe' at 5°C (according to Wicklatz, Kennedy and Reynolds[60])

Hydroperoxide	*Molecular weight*	*Mole ratio*
2,3-Dimethylbutane hydroperoxide	118	0·75
Methylcyclohexane hydroperoxide	130	0·75
Cumene hydroperoxide	152	0·75
2,2,5-Trimethylhexane hydroperoxide	160	0·75
1,2,3,4-Tetrahydronaphthalene hydroperoxide	164	1·5
sec-Butylbenzene hydroperoxide	166	0·75
p-Cymene hydroperoxide	166	1·0
Aliphatic alkylate hydroperoxide	174	1·0
1-methyl-1,2,3,4-Tetrahydronaphthalene hydroperoxide	178	1·0
5-Phenylpentene-2-hydroperoxide	178	1·5
Chloro*iso*propylbenzene hydroperoxide	186·5	1·0
Cyclohexylbenzene hydroperoxide	192	0·75
Di*iso*propylbenzene hydroperoxide	194	0·75
*iso*propyl-1,2,3,4-tetrahydronaphthalene hydroperoxide	207	1·5
t-Butyl*iso*propylbenzene hydroperoxide	208	1·0
Di*iso*propyltoluene hydroperoxide	208	2·0
1,2,3,4,4a,9,10,10a-Octahydrophenanthrene hydroperoxide	218	1·0
5-(4-*iso*propylphenyl)-2-pentene hydroperoxide	220	2·0
(1-Methylbutyl)-*iso*propylbenzene hydroperoxide	222	2·0
Chlorodi*iso*propylbenzene hydroperoxide	228·5	0·75
Tri*iso*propylbenzene hydroperoxide	236	1·0–1·5
1,2-Diphenylbutane hydroperoxide	242	1·5
Di-*t*-butyl*iso*propylbenzene hydroperoxide	264	1·0
(1-Methylhendecyl)-toluene hydroperoxide	292	2·5
1,2-Bis-(dimethylphenyl)-butane hydroperoxide	298	1·5
(1-Methylhendecyl)-*iso*propylbenzene hydroperoxide	320	2·0

summarised in Table 6.19. In each case, the mole ratio of hydroperoxide to iron was at the optimum. Some correlations between performance and structure emerge from these results. Wicklatz, Kennedy and Reynolds believe that the concentration of hydroperoxide in the aqueous phase of the system is an important factor. For hydroperoxides having the general structure (XXXII), the order of increasing effectiveness for different R

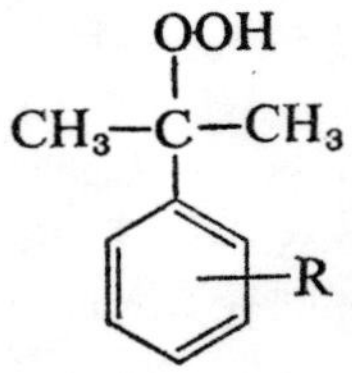

(XXXII)

is as follows: $H < CH_3(CH_2)_2\overset{|}{C}H(CH_3) < CH_3 < CH_3(CH_2)_9\overset{|}{C}H(CH_3) < (CH_3)_2\overset{|}{C}H < Cl < (CH_3)_3C—$. It also appears that substituting increasing numbers of alkyl groups in the benzene ring of *iso*propylbenzene hydroperoxide has an activating effect. Thus, for hydroperoxides having the general structure (XXXIII), the activity increases in the order $n = 0, 1, 2$ when R is both $(CH_3)_2\overset{|}{C}H$ and $(CH_3)_2\overset{|}{C}(CH_3)$; furthermore, the *sec*-butyl substituent has a greater effect than the *iso*propyl. Again, comparison of

TABLE 6.19

Order of effectiveness of various hydroperoxides as initiators in the 'custom recipe' at 5°C (according to Wicklatz, Kennedy and Reynolds[60])

Hydroperoxide	*Rate of conversion* ($\%h^{-1}$)†
Chlorodi*iso*propylbenzene hydroperoxide	15·8
Di-*t*-butyl*iso*propylbenzene hydroperoxide	14·7
Cyclohexylbenzene hydroperoxide	14·3
1,2,3,4,4a,9,10,10a-Octahydrophenanthrene hydroperoxide	13·2
t-Butyl*iso*propylbenzene hydroperoxide	13·1
Chloro*iso*propylbenzene hydroperoxide	11·1
Tri*iso*propylbenzene hydroperoxide	10·2
(1-Methylhendecyl)-toluene hydroperoxide	9·8
Di*iso*propyltoluene hydroperoxide	9·4
*Iso*propyl-1,2,3,4-tetrahydronaphthalene hydroperoxide	9·1
o-,*m*-,*p*-Di*iso*propylbenzene hydroperoxide	8·8
(1-Methylhendecyl)-*iso*propylbenzene hydroperoxide	8·7
m-Di*iso*propylbenzene hydroperoxide	8·6
1-Methyl-1,2,3,4-tetrahydronaphthalene hydroperoxide	8·3
5-(4-*Iso*propylphenyl)-2-pentene hydroperoxide	8·2
1,2-Bis-(dimethylphenyl)-butane hydroperoxide	8·0
1,2-Diphenylbutane hydroperoxide	7·7
2,2,5-Trimethylhexane hydroperoxide	7·7
p-Cymene hydroperoxide	7·5
o-,*m*-,*p*-Cymene hydroperoxide	7·3
Aliphatic alkylate hydroperoxide	6·5
(1-Methylbutyl)-*iso*propylbenzene hydroperoxide	5·9
Cymene hydroperoxide	5·8
Methylcyclohexane hydroperoxide	4·6
1,2,3,4-Tetrahydronaphthalene hydroperoxide	4·0
2,3-Dimethylbutane hydroperoxide	3·3
sec-Butylbenzene hydroperoxide	2·2
5-Phenylpentene-2-hydroperoxide	<2·0

† Average based upon time to reach 60% conversion.

chloro*iso*propylbenzene hydroperoxide with *iso*propylbenzene hydroperoxide, and of chlorodi*iso*propylbenzene hydroperoxide with di*iso*propylbenzene hydroperoxide, shows that halogen substitution in the benzene

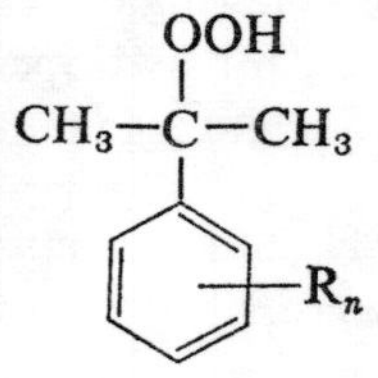

(XXXIII)

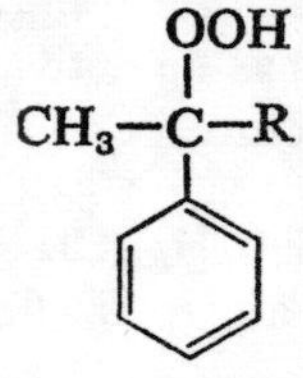

(XXXIV)

ring has an activating effect. But the comparison between *iso*propylbenzene hydroperoxide and *sec*-butylbenzene hydroperoxide indicates that increasing the size of the alkyl group in structure (XXXIV) has a deactivating effect. Typical conversion–time curves for the 'custom recipe' at 5°C with various hydroperoxides are shown in Fig. 6.19. These curves are remarkably linear up to high conversions, and induction periods appear to be absent.

The suitability of various hydroperoxides as initiators for styrene–butadiene emulsion polymerisations to be run at sub-zero temperatures (actually − 10°C) has been studied by Fryling and Follett.[61] The formulation used was as follows, all parts being by weight:

Styrene	30
Butadiene	70
Water	192
Methanol	48
Potassium soap of disproportionated rosin acid (Dresinate 214)	3·5
Potassium fatty-acid soap (K-SF flakes)	1·5
Mixed tertiary mercaptans	0·25
Hydroperoxide	(1·0 mmole)
$FeSO_4.7H_2O$	0·278 (1·0 mmole)
$Na_4P_2O_7.10H_2O$	0·446 (1·0 mmole)
KCl	variable

The effect of varying the level of initiator, keeping the mole ratios of hydroperoxide:iron:pyrophosphate constant at 1:1:1 is summarised in Table 6.20. The hydroperoxide was di*iso*propylbenzene hydroperoxide and the level of potassium chloride 0·4 pphm. It is clear that no advantage can be gained by increasing the initiator level above about 1·0 mmole parts per

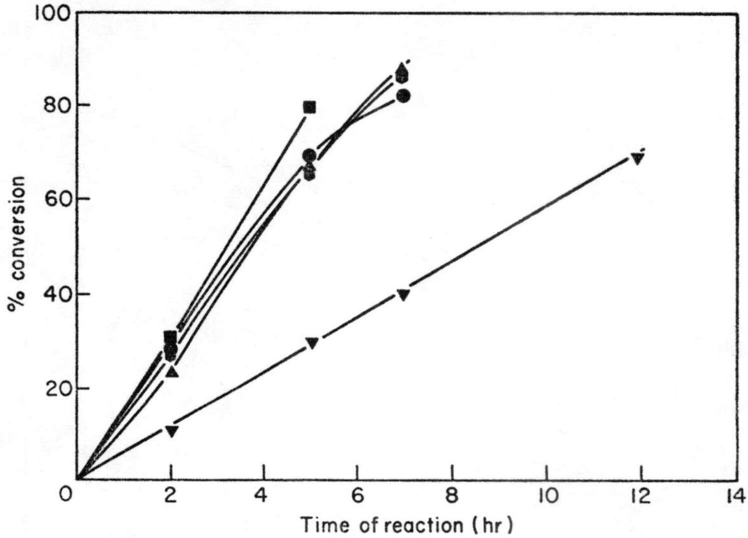

FIG. 6.19. Typical conversion–time curves for the 'custom recipe' reacted at 5°C, with various hydroperoxides as oxidising component of the redox couple, according to Wicklatz, Kennedy and Reynolds.[60] Significance of points: ■ chlorodi*iso*propylbenzene hydroperoxide; ● cyclohexylbenzene hydroperoxide; ▲ 1, 2, 3, 4, 4a, 9, 10, 10a-octahydrophenanthrene hydroperoxide; ⬢ *t*-butyl*iso*propylbenzene hydroperoxide; ▼ cumene hydroperoxide. (See text for details of reaction system.)

100 parts of monomer. Fryling and Follett regarded 0·3 as a convenient level for the rest of their work, as the polymerisation rate was satisfactory and the lower iron content of the recipe could minimise the possible deleterious effects of this element in the polymer subsequently produced.

The effect of varying the mole ratio of hydroperoxide to iron is shown in Fig. 6.20. The hydroperoxide was again di*iso*propylbenzene hydroperoxide, and the level of potassium chloride was 0·25 pphm. The levels of iron and pyrophosphate in the system were kept constant at 0·3 mmole phm, and the hydroperoxide level was varied in such a way that the mole ratio of hydroperoxide to iron varied from 0.75 to 2·0. These results illustrate the fact that an optimum mole ratio does exist; in this case it is at about 1·0, although the course of the reaction is not greatly affected by variation over the range 0·75 to 1·5.

Of the various hydroperoxides studied by Fryling and Follett in recipes balanced to give optimum rates, tri*iso*propylbenzene hydroperoxide appeared to be the most effective, with *t*-butyl*iso*propylbenzene

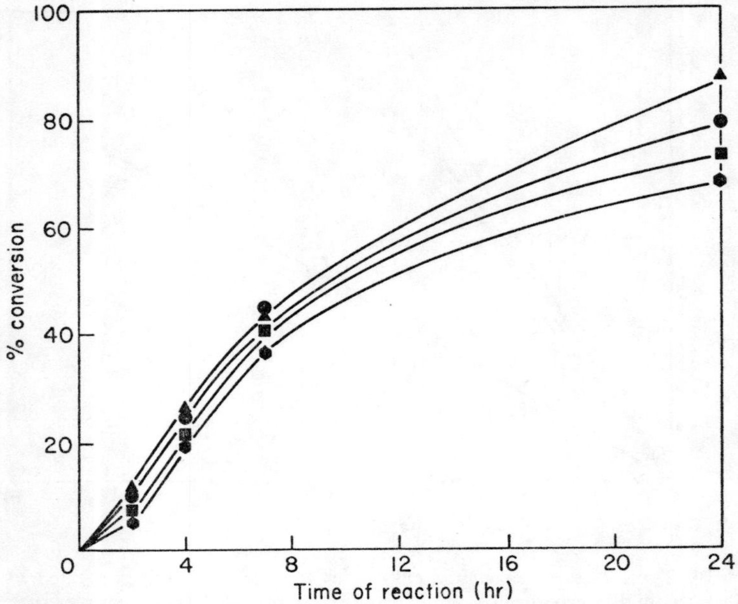

FIG. 6.20. Effect of di*iso*propylbenzene hydroperoxide:iron mole ratio upon course of emulsion copolymerisation of styrene and butadiene at −10°C, plotted from data given by Fryling and Follett.[61] Significance of points: ● 0·75 mole ratio HP:Fe; ▲ 1·0 mole ratio HP:Fe; ■ 1·5 mole ratio HP:Fe; ⬢ 2·0 mole ratio HP:Fe. (See text for details of reaction system.)

TABLE 6.20

Conversion as a function of time for emulsion copolymerisation of styrene and butadiene at −10°C using as initiator di*iso*propylbenzene hydroperoxide, iron(II) sulphate and sodium pyrophosphate in 1:1:1 mole ratio (according to Fryling and Follet[61])

(reaction system: styrene 30, butadiene 70, water 192, methanol 48, potassium soap of disproportionated rosin acid 3·5, potassium fatty-acid soap 1·5, mixed tertiary mercaptans 0·25, di*iso*propylbenzene hydroperoxide—variable, $FeSO_4.7H_2O$—variable, $Na_4P_2O_7.1OH_2O$—variable, KCl 0·4)

Initiator level (mmoles phm)	*% Conversion after times shown* 2 h	4 h	7 h	24 h
0·1	0	2	5	16
0·2	5	13	23	61
0·3	8	19	36	82
0·5	13	30	58	95
0·7	13	31	60	95
1·0	17	38	68	97
1·5	13	28	51	92

hydroperoxide and cyclohexylbenzene hydroperoxide almost as effective. All were found to be markedly more effective than cumene hydroperoxide, as also was di*iso*propylbenzene hydroperoxide. The hydroperoxides of tetralin and methylcyclohexane were found to be poor as initiators in this system. Fryling and Follett summarized their conclusions regarding the correlation between hydroperoxide structure and its performance in their recipe as follows: 'The hydroperoxides most successful in the emulsion copolymerisation of butadiene and styrene contained the $-C{=}\overset{|}{C}-\underset{*}{\overset{|}{C}}-$ allyl skeleton structure partially incorporated in a benzene ring and a side chain which contained a tertiary hydrogen replaced by the hydroperoxide group, and the most reactive of the oxidants had more than three but less than fifteen carbon atoms substituted on the benzene ring.'

They also observed an inverse correlation between the rate of polymerisation and the tendency of the hydroperoxide to partition in the aqueous phase. Their results are given in Table 6.21, together with the compositions of the systems used to simulate polymerisation reaction mixtures. Measurements were made at -10°C and at $+5$°C, the system for the lower temperature containing methanol as an antifreeze. Furthermore, measurements were made in the absence and in the presence of normal quantities of soap. Thus it is possible to infer from these results the effects of the following variables upon partitioning between the aqueous and hydrocarbon phases: (a) hydroperoxide structure, (b) temperature/presence or absence of methanol, and (c) presence or absence of soap. As would be expected, in the majority of cases the partitioning is strongly in favour of the hydrocarbon phase. Again, as expected, the larger is the alkyl substituent in the benzene ring, the more favourable is the partitioning. Temperature and presence or absence of methanol appear to have little effect upon the general trends; in fact the results for the $+5$°C methanol-free system are surprisingly similar to those for the -10°C system which contained methanol. It is evident that the soap in those systems where it was present exerted a strong solubilising effect; this is particularly marked in the case of cumene hydroperoxide.

The most surprising feature of these results is the apparent inverse correlation between initiating activity and solubility in the aqueous phase. Other things being equal, a direct correlation would have been expected. However, on the basis of the results given by Fryling and Follett, it is impossible to distinguish effects due to differences in solubility from those due to differences in reactivity.

TABLE 6.21

Distribution of hydroperoxide between aqueous and hydrocarbon phases at −10°C and +5°C (according to Fryling and Follett[61])

(1) *Simulated reaction systems (parts by weight)*

	−10°C	+5°C
Styrene	30	30
Butadiene	70	70
Water	192	192
Methanol	48	0
Soap	5·0 or 0	5·0 or 0
Hydroperoxide	0·18–0·21	0·18–0·21

(2) *Distribution of hydroperoxide between phases*

				% Hydroperoxide in	
Hydroperoxide	*Soap*	*Soap level* (pphm)	*Temp.* (°C)	*Aqueous phase*	*Hydro-carbon phase*
Cumene hydroperoxide	none	—	−10	17	83
Di*iso*propylbenzene hydroperoxide	none	—	−10	10	90
Tri*iso*propylbenzene hydroperoxide	none	—	−10	6	94
t-Butyl*iso*propylbenzene hydroperoxide	none	—	−10	2	98
Dodecyl*iso*propylbenzene hydroperoxide	none	—	−10	2	98
Cumene hydroperoxide	K laurate	5·0	−10	40	60
Di*iso*propylbenzene hydroperoxide	K laurate	5·0	−10	21	79
Tri*iso*propylbenzene hydroperoxide	KRA/KFA†	3·5/1·5	−10	16	84
t-Butyl*iso*propylbenzene hydroperoxide	KRA/KFA†	3·5/1·5	−10	10	90
Dodecyl*iso*propylbenzene hydroperoxide	KRA/KFA†	3·5/1·5	−10	10	90
Cumene hydroperoxide	none	—	+5	10	90
Di*iso*propylbenzene hydroperoxide	none	—	+5	8	92
Tri*iso*propylbenzene hydroperoxide	none	—	+5	4	96
t-Butyl*iso*propylbenzene hydroperoxide	none	—	+5	2	98
Dodecyl*iso*propylbenzene hydroperoxide	none	—	+5	3	97
Cumene hydroperoxide	K laurate	5·0	+5	39	61
Di*iso*propylbenzene hydroperoxide	K laurate	5·0	+5	17	83
Tri*iso*propylbenzene hydroperoxide	KRA/KFA†	3·5/1·5	+5	15	85
t-Butyl*iso*propylbenzene hydroperoxide	KRA/KFA†	3·5/1·5	+5	9	91
Dodecyl*iso*propylbenzene hydroperoxide	KRA/KFA†	3·5/1·5	+5	12	88

† KRA = potassium soap of disproportionated rosin acid.
KFA = potassium fatty-acid soap.

A detailed study of the behaviour of styrene–butadiene emulsion copolymerisation systems containing cumene hydroperoxide and reacted at 40°C has been reported by Vandenberg and Hulse.[62] Polymerisation was observed to occur much more rapidly than in equivalent persulphate-initiated systems. Thus in a typical recipe, 72% conversion occurred within 2 h, this corresponding to a sixteen-fold increase in rate as compared with that found in a system initiated with potassium persulphate. The following features of the reaction are of special interest:

(1) The initial rate of polymerisation is not very dependent upon the cumene hydroperoxide level. Thus the rate increases by only about 30% as the hydroperoxide level is increased from 0.04 to 0.4 pphm, and then falls off by an equal amount as the hydroperoxide level is further increased to 1·4 pphm. However, with less than about 0·1 pphm of peroxide, the polymerisation either stops or slows down before a high conversion is reached. This was shown to be due to hydroperoxide depletion, as the addition of more hydroperoxide after the reaction had ceased was observed to cause it to proceed at the original rate.

(2) The polymerisation rate can be controlled by adjusting the level of reducing sugar (fructose) in the system. Thus, decreasing the fructose level from the usual 0·5 pphm to 0·02 pphm decreases the polymerisation rate approximately fourfold without affecting the linearity of the conversion–time curve. On the other hand, increasing the fructose level to 4·0 pphm increases the initial polymerisation rate only slowly. In this case, the major effect is to slow down the polymerisation before a high conversion is reached. The decrease in rate occurs at a lower conversion at the higher level of fructose, and was attributed to hydroperoxide depletion. It appears that the optimum amount of hydroperoxide depends upon the fructose level: it is greater or less than 0·17 pphm according as the fructose level is greater or less than 0·5 pphm.

(3) Mercaptans are unnecessary for initiation by the cumene hydroperoxide redox system, although they were observed to minimise the retardation period found in systems containing insufficient amounts of iron.

(4) The effect of temperature upon the cumene hydroperoxide redox system is similar to that found generally in vinyl polymerisation reactions. Increasing the temperature from 40° to 50°C doubles the initial polymerisation rate. However, after 50% conversion, the rate

falls off very markedly. This was attributed to hydroperoxide depletion, and was taken as indicating that the decomposition of the hydroperoxide is more greatly accelerated by increase in temperature than is the overall polymerisation reaction. Limitation of the polymerisation rate by diffusion of monomers from droplets to reaction loci was suggested as a possible underlying cause.

Kolthoff and Medalia[63] have studied the emulsion copolymerisation of styrene and butadiene at 30°C using the cumene hydroperoxide–iron–reducing sugar system. The iron was added either as iron(II) sulphate or iron(III) nitrate. The iron(III) salt was found to give slightly lower polymerisation rates than the iron(II) salt, but the system was considered simpler and more amenable to interpretation. The assumption made was that the iron(III) was reduced by the sugar to iron(II). On the basis of experiments carried out on systems from which sugar was omitted, it was concluded that the reaction between iron(II) and cumene hydroperoxide is

TABLE 6.22

Effect of sugar type upon rate of emulsion copolymerisation of styrene and butadiene initiated by hydroperoxide–iron–sugar system at low temperature (according to Fryling *et al.*[66])

(reaction system: styrene 28, butadiene 72, water 180, fatty-acid soap 1·2, rosin-acid soap 3·5, mixed tertiary mercaptans 0·40, cumene hydroperoxide 0·15, sugar 1·0, $FeSO_4.7H_2O$ 0·10, $Na_4P_2O_7.10H_2O$ 1·00; temperature 15°C)

Sugar	*% conversion after* 8 h
Sorbose	71
D-Ribose	65
Levulose	63
Invert sugar	52
D(+)-Xylose	39
Melibiose hydrate	24
Cellobiose	22
L(+)-Arabinose	21
β-Lactose	18
D(+)-Galactose	17
D(+)-Mannose	14

Note: Fryling *et al.*[66] also give results for two other reductants, namely, benzoin and aldol. The reported conversions after 8 h are 35% and 12%, respectively. In each case, the level of the reductant was 1 pphm.

too rapid to bring about significant polymerisation; the systems are, in effect, iron(III) systems, whether iron(II) or iron(III) is charged.

A study of various sugars in the cumene hydroperoxide–iron–sugar system has been published by Kolthoff, Medalia and Youse.[64] Again, experiments were confined to the emulsion copolymerisation of styrene and butadiene. Polymerisations were carried out at 0° and at 30°C. At 0°C, much higher rates were observed with fructose than with glucose, but the effectiveness of glucose could be much improved by boiling the glucose with alkali. The consequent improvement was attributed not to conversion to fructose, but to the formation of degradation products. One of the possible degradation products, dihydroxyacetone, was found to be a much more effective reducing agent for polymerisations at 0°C than either fructose or boiled glucose. Of the possible degradation products, dihydroxyacetone was found to be a much more effective reducing agent for polymerisations at 0°C than either fructose or boiled glucose. A detailed study of the use of dihydrooxyacetone was subsequently published by Kolthoff and Medalia.[65] They have shown that various crude reaction mixtures can be used in place of pure dihydroxyacetone. A typical example is that obtained from glycol by oxidising with a mixture of hydrogen peroxide and an iron(II) salt. Further information upon the effect of changing the sugar type is given in a paper by Fryling *et al.*[66] Their observations are summarised in Table 6.22.

The general dissatisfaction with the iron–pyrophosphate activator system can be judged from the large number of alternatives which have been proposed for it. Note has already been taken of the need for following a carefully-devised set procedure when preparing the activator solution. The problems stem partly from the fact that the iron complex is sensitive to oxidation by air. But superimposed upon the indeterminacy of the chemical species present are the vagaries associated with the preparation of colloidal suspensions, because the complex is only sparingly soluble in water. Thus, for example, Fryling *et al.*,[66] referring to the difficulties associated with the use of iron(II) sulphate–pyrophosphate activators, state that 'If ferrous sulfate is used, the activator solution must be carefully protected from atmospheric oxidation and prepared very shortly before using. To be practical for use on a large scale, such a solution must be stabilized and it is a matter of importance that stabilization can be achieved by inclusion of the reducing sugar in the activator solution.'

A significant advance in eliminating the uncertainties associated with the preparation of the activator solution was the discovery that the water-soluble iron(II) complexes of ethylenediamine tetraäcetic acid or its salts are powerful activators for hydroperoxide-initiated emulsion polymerisation. Thus Mitchell, Spolsky and Williams[67] used this complex with sugar as a regenerant and obtained 60% conversion of styrene and butadiene in about 17 h at 5°C. Considerably less iron was used in this system than in the more usual pyrophosphate system.

There has also been considerable interest in the possibility of using alternative reductants in place of sugars, or, at least, of diminishing the amount of sugar in the recipe. Thus, Neklutin, Westerhoff and Howland[68] have enumerated the disadvantages of sugar-containing initiation systems as follows:

'(1) Necessity of a complicated activator make-up procedure in order to obtain good rates of polymerisation.
(2) Instability of the final latex on storage because of a continued oxidation of sugar resulting in a drop in pH and prefloc formation if the pH is not continuously adjusted.
(3) Difficulty in arresting the polymerisation with known water-soluble shortstopping agents, which resulted in the use of oil-soluble materials that are more difficult to introduce into the latex.'

A further disadvantage of systems containing conventional levels of reducing sugar is mentioned by Smith *et al.*[69] This concerns the difficulty of obtaining satisfactory dried films from the lattices produced using such systems, the presence of the sugar causing the films to remain sticky and hygroscopic. Neklutin, Westerhoff and Howland[68] then proceed to describe the characteristics of various sugar-free hydroperoxide iron(II) systems. The activators studied were an iron(II) sulphate–potassium pyrophosphate complex, for which a carefully controlled preparative procedure was still required; an iron(II) sulphide suspension; an iron(II) soap which was prepared *in situ* and which worked well only if the soap was of the fatty-acid type; and an iron(II) silicate colloidal dispersion which worked best with soaps of the rosin-acid type, and which was claimed to result in the simplest and most economical iron-containing recipe hitherto developed. Results are also presented for initiation by a hydroperoxide activated by an amine, but consideration of this system is deferred until later in this chapter (see Section 6.13).

In a paper which appeared some months before that of Neklutin, Westerhoff and Howland, Kolthoff and Medalia[63] reported that the reaction between iron(II) ions and cumene hydroperoxide can be moderated to a suitably slow rate in a sugar-free system if the iron(II) is introduced as a suspension of a very sparingly-soluble iron(II) salt. By this means, the concentration of iron(II) ions is kept extremely small, and the rate-determining step for the formation of radicals is the rate of release of iron(II) ions from the salt. Very good rates of conversion were observed in recipes for which the glucose had been replaced by sodium sulphide, and it was surmised that the effective reducing agent was the very sparingly-soluble iron(II) sulphide which was precipitated. In a later, more detailed study of this system, Kolthoff and Medalia[70] gave the following as the order of addition of the various ingredients: sodium sulphide, sodium pyrophosphate, iron(II) sulphate, rosin-acid soap, cumene hydroperoxide and mercaptan dissolved in the styrene, and butadiene. Concerning the appearance of the system, they say, 'On the addition of ferrous sulfate, a black precipitate of ferrous sulfide is formed. After polymerization starts, the color of the emulsion is dark blue and this color fades as the reaction proceeds. On reaching about 80% conversion, the latex becomes colorless. Even a colored latex becomes colorless on sampling; that is, on exposure to air.' The following alternative methods of mixing the ingredients were investigated:

A. All ingredients except iron(II) sulphate charged at polymerisation temperature. Charge shaken vigorously; iron(II) solution injected and charge again shaken.
B. Aqueous solutions of potassium sulphate, sodium sulphide, iron(II) sulphate and soap added in that order; then hydroperoxide, mercaptan and monomers.
C. Aqueous solutions of sodium sulphide, iron(II) sulphate, potassium sulphate and soap added in that order; then hydroperoxide, mercaptan and monomers.

Methods B and C gave rates of conversion nearly as high as those obtained with the recommended method of mixing described previously, whereas method A (in which potassium sulphate was also included) gave a rate only about two-thirds as great as that given by the recommended method. A charge prepared according to method A, but with 1 part of potassium pyrophosphate rather than potassium sulphate, gave conversions approximately equal to those obtained with the recommended method. These observations illustrate the complexities of the system. The use of preformed, washed suspensions of iron(II) sulphide was also investigated. All the procedures investigated using the preformed sulphide gave low and poorly-reproducible rates of conversion.

Regarding the mechanism for the initiation reaction, the gradual disappearance of the colour of the iron(II) sulphide suggests that the sulphide ions also act as a reductant, either for the hydroperoxide directly or for the iron(III) ions which are formed from the hydroperoxide and the iron(II) ions. Reducing sugars may also be included in hydroperoxide–iron(II) sulphide systems, but the advantage of such inclusion seem doubtful.

Yet a further variant of the hydroperoxide–iron system has been investigated by Kolthoff and Meechan.[71] The iron(III) complex of ethylenediamine tetraäcetic acid, formed by interaction between iron(III) nitrate and disodium dihydrogen ethylenediamine tetraäcetate, is reduced by hydrazine to the corresponding iron(II) complex. This in turn reacts with the hydroperoxide to produce the radicals which initiate polymerisation. The rate of conversion can be varied within wide limits by changing the concentrations of iron(III), EDTA, or hydrazine. With a given concentration of iron(III), the maximum rate is found with a molar concentration of ethylenediamine tetraäcetic acid slightly smaller than that of the iron(III). The excess of iron(III) over ethylenediamine tetraäcetic acid is kept in aqueous suspension as a pyrophosphate complex. The addition of potassium pyrophosphate up to 1 pphm increases the rate of polymerisation and yields a normal latex which is free from pre-coagulum. In the absence of potassium pyrophosphate, a viscous latex prone to pre-coagulation is obtained. In yet another variant of this system, sodium sulphide has been proposed as a replacement for the hydrazine. Thus Kolthoff and Meehan[72] have claimed that satisfactory rates of polymerisation can be achieved using a combination of hydroperoxide, the iron(III) complex of ethylenediamine tetraäcetic acid and sodium sulphide. Mercaptan is not necessary for initiation. Added oxygen was found to have but a slight effect which could be largely eliminated by the addition of sodium dithionite. A system was developed for which 70–75% conversion in 12 h at −5°C was claimed.

Kolthoff and Meehan[73] have also shown that the iron(II) component of a hydroperoxide–iron(II) couple can be added as sodium nitroprusside. The characteristics of the system depend upon whether or not excess alkali is present. In a fatty-acid soap recipe without added alkali, the presence of a primary mercaptan is essential for initiation by the hydroperoxide–nitroprusside combination; without mercaptan, or if tertiary mercaptan is present, little polymerisation occurs. The reaction can be promoted by the addition of hydroxylamine, but several reducing agents of comparable strength which were investigated do not have this effect. The addition of alkali to raise the pH to between 11 and 12 causes a marked increase in the polymerisation rate. At such pH levels, hydroxylamine has no effect upon the rate, and mercaptan is not essential for initiation; in most instances, relatively minor differences in rate were found between systems in which a mercaptan was present and those from which it was absent. The following reactions were proposed to account for initiation in systems to which excess alkali had been added:

$$[Fe^{II}(CN)_5NO]^{2-} + 2OH^- \rightleftharpoons [Fe^{II}(CN)_5NO_2]^{4-} + H_2O \qquad \text{(liv)}$$

$$ROOH + [Fe^{II}(CN)_5NO_2]^{4-} \rightarrow RO\cdot + OH^- + [Fe^{III}(CN)_5NO_2]^{3-} \qquad \text{(lv)}$$

The first step, which is a known reaction of nitroprusside ions in alkaline media, involves the formation of pentacyanate–nitrito iron(II) ions. These are then presumed to be oxidised to pentacyanatonitrito iron(III) ions by interaction with the hydroperoxide. Polymerisation is assumed to be initiated by the alkoxy radicals produced as a consequence of this oxidation. In the case of the systems to which excess alkali has not been added, it was suggested that initiation proceeds by way of oxidation of the pentacyanatonitrito iron(II) ion to the iron(III) analogue by interaction with the hydroperoxide, followed by reaction with the mercaptan to produce initiating radicals thus:

$$RSH + [Fe^{III}(CN)_5NO_2]^{3-} \rightarrow RS\cdot + H^+ + [Fe^{II}(CN)_5NO_2]^{4-} \qquad \text{(lvi)}$$

This proposal is consistent with the observation that polymerisation does not occur in the absence of the nitroprusside, so that the cause of the polymerisation cannot be direct interaction between mercaptan and hydroperoxide. Furthermore, in the absence of hydroperoxide, the combination of nitroprusside and primary mercaptan does not initiate polymerisation. The chemistry underlying the promoting effect of hydroxylamine in this system is obscure; Kolthoff and Meehan suggest that the hydroxylamine may react with the nitroprusside to give a product which then reacts with the hydroperoxide to form pentacyanatonitrito iron(III) ions.

In a further paper,[74] Kolthoff and Meehan describe the use of sodium dithionite as a reductant in conjunction with a hydroperoxide, the iron(III) complex of ethylene diamine tetraäcetic acid, and potassium pyrophosphate. The addition of the latter results in a latex which is of normal fluidity and free from pre-coagulation up to 80% conversion. The level of sodium dithionite in relation to that of hydroperoxide is apparently rather critical if low conversions are to be avoided due to exhaustion of hydroperoxide.

The use of sodium formaldehyde sulphoxylate as a reducing agent for the hydroperoxide–iron(II/III) ethylenediamine tetraäcetate system has been described in some detail by Brown *et al.*[75] in a paper which appeared in 1954. This initiation system has subsequently been widely used for the production of synthetic diene rubbers by low-temperature emulsion polymerisation. The principal advantages claimed for this combination as compared with earlier systems are ease of preparation, low cost, reproducibility, and responsiveness of polymerisation rate to initiator level.

The effects of varying the type and level of iron-chelating agent in this initiation system have been reported in some detail by Prince and Spitz.[76] Andersen and Proctor[77] have reported the results of an investigation into the kinetic aspects of this combination. Evidence is offered in support of the view that iron(III) ethylene diamine tetraäcetate and sodium formaldehyde sulphoxylate interact to give a free radical which Andersen and Proctor tentatively suggest has the structure $(HCHO)SO_2^{\cdot}$.

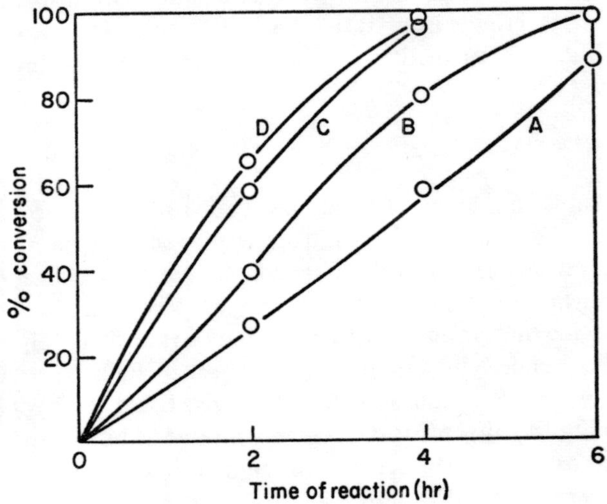

FIG. 6.21. Effect of varying level of iron(II) salt upon course of styrene–butadiene emulsion copolymerisation at 5°C initiated by hydroperoxide–iron–sodium formaldehyde sulphoxylate system, according to Brown *et al.*[75] (Reaction system: styrene 30, butadiene 70, water 180, potassium soap of disproportionated rosin acid 4·5, mixed tertiary mercaptans 0·3, di*iso*propylbenzene hydroperoxide 0·15, sodium formaldehyde sulphoxylate dihydrate 0·10, $FeSO_4.7H_2O$—variable, tetrasodium ethylenediamine tetraäcetate ?, KCl 0·3. The letters appended to the curves denote the levels of $FeSO_4.7H_2O$ as follows: A, 0·02; B, 0·03; C, 0·05; D, 0·07.)

The rate of polymerisation with this initiation system depends very much upon the amount of iron. Typical results taken from the paper by Brown *et al.*[75] are shown in Fig. 6.21. Brown *et al.* state that, although it might have been possible to obtain higher rates of polymerisation using higher levels of iron, the purpose of this work was in fact to develop a low-iron system in which the available iron rapidly interchanges its valency state. The level of sodium formaldehyde sulphoxylate is also important, as is illustrated in Fig. 6.22. Several other aldehyde and ketone sulphoxylates were evaluated by Brown *et al.*;[75] although most of them were effective none was appreciably better than the commercially-available sodium formaldehyde sulphoxylate. Investigation of the effect of the order in which the several ingredients are charged showed that, in general, with fatty-acid soap, the order is not critical except that the soap should not be added to uncomplexed iron as this causes precipitation of an insoluble

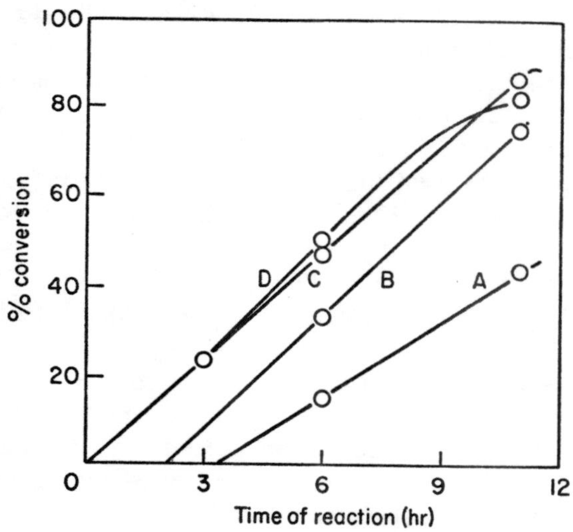

FIG. 6.22. Effect of varying level of sodium formaldehyde sulphoxylate upon course of styrene–butadiene emulsion copolymerisation at 5°C initiated by hydroperoxide–iron–sodium formaldehyde sulphoxylate system, according to Brown *et al.*[75] (Reaction system: styrene 30, butadiene 70, water 180, potassium soap of disproportionated rosin acid 4·5, mixed tertiary mercaptans 0·3, di*iso*propylbenzene hydroperoxide 0·04, sodium formaldehyde sulphoxylate dihydrate—variable, $FeSO_4.7H_2O$ 0·015, tetrasodium ethylenediamine tetraäcetate ?, KCl 0·3. The letters appended to the curves denote the levels of sodium formaldehyde sulphoxylate dihydrate as follows: A, 0·02; B, 0·04; C, 0·06; D, 0·08.)

iron soap. With rosin-acid soaps, however, very slow rates of polymerisation were obtained in all cases in which the iron–ethylenediamine tetraäcetic acid complex was not formed before mixture of either component with the soap.

6.13 THE HYDROPEROXIDE–POLYAMINE SYSTEM

In the preceding paragraph, reference was made to the desire to reduce the level of iron in an activated hydroperoxide initiation system. The presence of appreciable quantities of iron in the eventual polymer is undesirable for at least two reasons: (a) it may cause discoloration, and (b) it may catalyse oxidative degradation of the polymer. For these reasons, endeavours have been made to develop alternative activation systems for hydroperoxides which are nominally iron-free. Of the several systems which have been investigated, the one which has received most attention is that which uses aliphatic polyamines as activator.

Whitby *et al.*[78] noted that many aliphatic and alicyclic amines are capable of decomposing organic peroxides vigorously. In consequence, they undertook a preliminary screening of a large number of amines as activator for the emulsion polymerisation of styrene initiated by a peroxide catalyst. In fact, most of the peroxide-decomposing amines and amino compounds examined were found to have no activating effect in this system. The polyalkylenepolyamines were, however, observed to have a powerful activating effect in the emulsion polymerisation of styrene although only a mild effect in the bulk polymerisation of this monomer. Hydroperoxides of the type of cumene hydroperoxide were found to be particularly susceptible to activation by polyalkylenepolyamines. Whitby and his co-workers have also given extensive information on the comparative effectiveness of a wide range of polyalkylenepolyamines. The polyethylenepolyamines from diethylenetriamine to nonaëthylenedecamine were investigated, and the conclusion reached that the effectiveness reaches a maximum in tetraëthylenepentamine and pentaëthylenehexamine, and then gradually decreases as the molecular size is further increased. Dipropylenetriamine and tripropylenetetramine were found to be more effective than the corresponding ethylenepolyamines. As examples of the effectiveness of the polyamines in activating low-temperature polymerisation, the following may be cited: (a) in a soap emulsion at 10°C with 0.21 pphm of cumene hydroperoxide and 0·2 pphm of tetraëthylenepentamine as initiator, styrene was polymerised to 73% conversion in 1 h; (b) a 30:70 styrene–butadiene mixture in the same system gave 70% of polymer in 6 h.

As a result of an extensive investigation into correlations between the structure of the polyalkylenepolyamine and its ability to function as an activator in hydroperoxide-initiated emulsion polymerisations, Whitby and his co-workers came to the conclusion that two structural requirements must be fulfilled if successful activation is to occur:

(1) The amine must contain, within the same molecule, amino groups of different degrees of substitution; in particular, it appears that the amino groups must be either primary and secondary, or primary and tertiary. As an example illustrating this conclusion, it was found that ethylenediamine (XXXV) is a very weak activator, its symmetrical *N,N'*-dialkyl substitution derivatives (XXXVI) have no activating power, but its *N*-monoalkyl substitution derivatives (XXXVII) are effective activators.

$$\underset{\text{(XXXV)}}{H_2NCH_2CH_2NH_2} \qquad \underset{\text{(XXXVI)}}{RHN\,CH_2CH_2NHR}$$

$$\underset{\text{(XXXVII)}}{H_2NCH_2CH_2HNR}$$

(2) The amino groups must be separated by not more than two carbon atoms. Thus tri(β-ethylamino)amine (XXXVIII), with two carbon atoms between the nitrogen atoms, was observed to activate, whereas tri(γ-aminopropyl)amine (XXXIX), with three, did not. Likewise, triethylenetetramine (XL) activated, whereas 1,3-bis(3'-aminopropylamino)propane (XLI) did not.

$$\underset{\text{(XXXVIII)}}{N(CH_2CH_2NH_2)_3} \qquad \underset{\text{(XXXIX)}}{N(CH_2CH_2CH_2NH_2)_3}$$

$$\underset{\text{(XL)}}{H(HNCH_2CH_2)_3NH_2} \qquad \underset{\text{(XLI)}}{H(HNCH_2CH_2CH_2)_3NH_2}$$

Further information on the use of amine activators has been given by Spolsky and Williams.[79] They confirmed that tetraëthylenepentamine can be used as the sole activator for cumene hydroperoxide, but found that its effectiveness is enhanced by the addition of sugar. In this particular system, the effectiveness of the polyamine was diminished by adding iron(II) ions, but, under other circumstances, the activating effect of polyamines was found to be enhanced by iron(II) ions. Spolsky and Williams do not agree with the contention of Whitby and his co-workers[78] that the ability of a polyamine to activate is associated with the presence of primary and secondary amino groups. They suggest instead that the effectiveness is nearly proportional to the amine content of the polyethylenepolyamine, and that primary and secondary amines are equivalent in their abilities to activate.

In a further study, Embree, Spolsky and Williams[80] found that, for the emulsion copolymerisation of styrene and butadiene, a polyamine mixture comprising 80% of diethylenetriamine and 20% of tetraëthylenepentamine is best for uniform rates of conversion. The polymerisation rate can be adjusted by adding an iron(II) salt or an iron-sequestering agent. Similar amines were found to be ineffective for the activation of acrylonitrile–butadiene emulsion copolymerisation. This was attributed to cyanoethylation of the amine, with loss of reducing power. This can be prevented or slowed down by the addition of digested D-glucose or other source of enediol to the amine solution.

Neklutin, Westerhoff and Howland[68] have confirmed that the emulsion copolymerisation of styrene and butadiene can be brought about by the use of hydroperoxides and polyamines. Of the polyamines evaluated, tetraëthylenepentamine gave the fastest rates. But diethylenetriamine was also regarded as a favourable possibility, in view of the then prevailing conditions of supply, cost and uniformity. According to Brown *et al.*,[75] the hydroperoxide–polyamine system suffers from two main disadvantages in relation to styrene–butadiene emulsion copolymerisation: (a) the amines which are available commercially do not give satisfactory rates of polymerisation with rosin-acid soaps, except under conditions that are difficult to attain in large-scale operations; and (b) polyamines have been found to contribute significantly to the discoloration of articles made directly from the polymer latices.

It appears that little is known regarding the precise chemistry of the hydroperoxide–polyamine system. It is possible to write a plausible radical-generating reaction as follows:

$$\mathrm{ROOH + R'NH_2 \rightarrow RO\cdot + R'NH\cdot + H_2O} \qquad \text{(lvii)}$$

but the present writer is unaware of any systematic explanation of the differences in promoting activity displayed by various amines. There is also some doubt as to whether, in practice, hydroperoxide–polyamine systems are truly iron-free, for it had been claimed that the rigorous purification of the components from all traces of iron results in a reduction of the effectiveness of the combination. It may be that, in practice, it is minute traces of an iron–polyamine complex which interact with the hydroperoxide. If so, this could explain the fact that these combinations are rather sensitive to changes of pH; for the cause could then be variable hydrolysis of the iron–polyamine complex. The present writer is inclined to the view that salt formation between the amine and hydroperoxide components may provide the initial pathway for the reaction. It is at least possible to envisage the formation of a species such as $[ROO]^-[R'NH_3]^+$. Furthermore, if the $[R'NH_3]^+$ has complexed within it minute traces of iron, such salt formation would inevitably bring the iron into close proximity with the hydroperoxide entity, thereby enabling the iron to exert a catalytic influence at an effective concentration which is considerably higher than that prevailing in the system as a whole. The formation of such salts would certainly explain why the behaviour of hydroperoxide–amine pairs in ionising media such as water can differ markedly from their behaviour in non-polar media.

6.14 OTHER SYSTEMS

Marvel *et al.*[81] have published an investigation into a redox system comprising dibenzoyl peroxide and iron(II) pyrophosphate. The conclusion was reached that, for the emulsion copolymerisation of 25 parts of styrene and 75 parts of butadiene in 200 parts of water and 5 parts of fatty-acid soap, the following are the optimum levels of the ingredients essential for the initiation of polymerisation:

Dibenzoyl peroxide	0·25 pphm
$FeSO_4.7H_2O$	0·50 pphm
$Na_4P_2O_7.12H_2O$	3·0 pphm

At very low levels of iron, the addition of a reducing agent, such as fructose, is beneficial. A similar system has been investigated by Johnson and Bebb.[82]

The use of potassium chromate as the oxidising component of redox couple for the emulsion copolymerisation of styrene and butadiene has been investigated by Kolthoff and Meehan.[83] A variety of reducing agents was used with it, of which arsenic(II) oxide seemed the most promising. The results summarised in Table 6.23 show the effect of the levels of chromate and arsenic upon the course of the reaction at 30°C. Experiments with systems containing 5 pphm of potassium myristate and 0·3 pphm of potassium chromate gave the highest rates of polymerisation at 20° and 30°C when the level of arsenic(III) oxide was 0·07 pphm. Polymerisation occurs very rapidly under these conditions; at 30°C, the rate is about 20% per hour. Other fatty-acid soaps or sodium alkyl sulphates can be used in place of potassium myristate, but the rates tend to be rather less. Mercaptans were observed to have a slight accelerating effect, but the presence of

TABLE 6.23

Effect of levels of potassium chromate and arsenic(III) oxide upon course of emulsion copolymerisation of styrene and butadiene at 30°C (according to Kolthoff and Meehan[83])

(reaction system: styrene 25, butadiene 75, water 180, sodium fatty-acid soap 5, dodecyl mercaptan 0·5, K_2CrO_4—variable, As_2O_3—variable)

K_2CrO_4 *level* (pphm)	As_2O_3 *level* (pphm)	*% conversion at times indicated*			
		3 h	6 h	9 h	12 h
0·1	0·3	7	16	23	31
0·3	0·01	6	19	25	30
0·3	0·025	9	23	36	46
0·3	0·05	14	39	66	74
0·3	0·07	19	52	—	—
0·3	0·1	17	51	81	—
0·3	0·3	11	23	39	60
0·6	0·05	17	53	80	88
0·6	0·1	19	56	83	96
0·6	0·2	20	57	88	100

mercaptan is not necessary for the initiation of polymerisation. Arsenic(III) sulphide can also be used as the reducing component, but the activity tends to vary between different suspensions. Potassium antimonyl tartrate in combination with potassium chromate gives rapid polymerisation with fatty-acid soap at 40°C. Tin(II) is much less effective than arsenic(III) or antimony(III). The following inorganic reducing agents were reported to be inefficient as polymerisation initiators in combination with chromate: iron(II), thallium(I), dithionite, hydrazine, hydroxylamine and sulphide.

Interaction between persulphate ions and iron(II) ions provides a very rapid means for generating free radicals. Bacon[84] has shown that this combination can be used to initiate the polymerisation of acrylonitrile. Fordham and Williams[85] have investigated the mechanism of the interaction, and found the observed features to be consistent with the following bimolecular reaction:

$$Fe^{2+} + S_2O_8^{2-} \rightarrow Fe^{3+} + SO_4^{\overline{\cdot}} + SO_4^{2-} \qquad \text{(lviii)}$$

Thus the reaction is kinetically of second order (*i.e.* first-order in each of the concentrations of persulphate ions and iron(II) ions). Furthermore, the variation of the rate coefficient with the ionic strength of the reaction medium is such as to suggest that the reaction is between two oppositely-charged divalent ions. The reaction is not inhibited by the presence of α,α'-dipyridyl (a complexing agent for iron), provided the amount does not exceed two molecules of dipyridyl per atom of iron. If sufficient dipyridyl is present to form the tri-dipyridyl iron complex, then the reactivity of the couple is destroyed.

The use of silver nitrate as a promoter for butadiene emulsion polymerisations and styrene–butadiene emulsion copolymerisations initiated by either hydrogen peroxide or potassium persulphate has been reported by Rainard.[86] Undoubtedly a redox couple is formed between the silver ions and the peroxide initiator. Silver nitrate tends to retard the emulsion polymerisation of butadiene alone initiated by hydrogen peroxide, but to increase the rate of polymerisation when styrene is present as a comonomer. In the latter case, the rate passes through a maximum as the silver ion concentration is increased. Similar effects are observed when the peroxide component is potassium persulphate.

A recent paper by Shukla and Misra[87] has described the use of a redox couple formed between the permanganate ion and thiourea for initiating the polymerisation of acrylamide in aqueous media. An acidic medium is required. The reactions responsible for effecting initiation are given as follows:

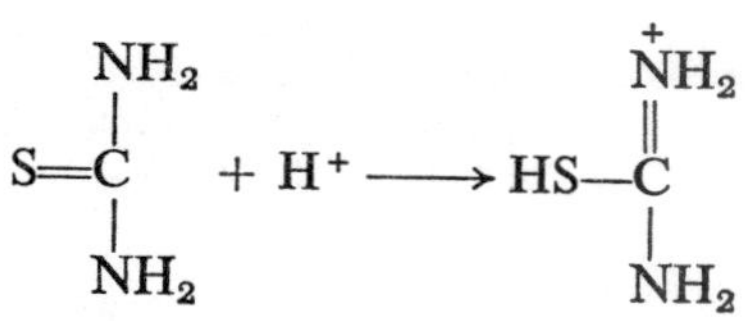

(lix)

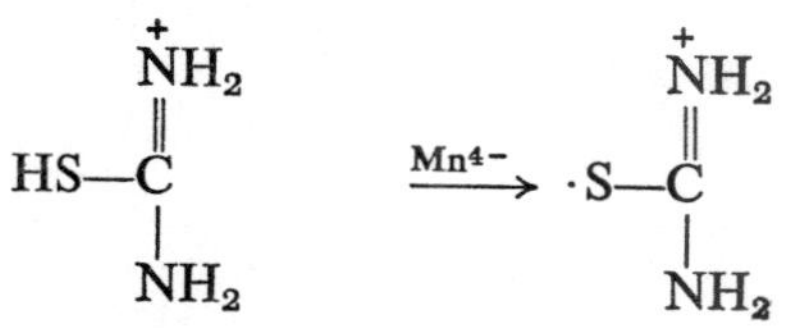

(lx)

For polymerisation carried out at 30°C, the rate was found to be almost first-order with respect to permanganate concentration over the range from $0{\cdot}5 \times 10^{-2}$ to $1{\cdot}4 \times 10^{-2}$ mole l^{-1} and to be independent of the thiourea concentration. The rate is also first-order with respect to acid concentration over the range from $2{\cdot}85 \times 10^{-2}$ to $11{\cdot}4 \times 10^{-2}$ mole l^{-1} and this finding is consistent with reaction (lx) above.

REFERENCES

1. M. G. Levi and E. Migliorini, *Gazz. chim. ital.*, 1906, **36II**, 599.
2. L. Green and O. Masson, *J. chem. Soc.*, 1910, **97**, 2083.
3. A. Kailan and L. Olbrich, *Wien. Akad. Ber.*, 1926, **IIb**, **135**, 423.
4. A. Kailan and L. Olbrich, *Monatsh.*, 1926, **47**, 449.
5. A. Kailan and E. Leisek, *Monatsh.*, 1928, **50**, 403.
6. I. M. Kolthoff and I. K. Miller, *J. Am. chem. Soc.*, 1951, **73**, 3055.
7. I. M. Kolthoff, L. S. Guss, D. R. May and A. I. Medalia, *J. Polym. Sci.*, 1946, **1**, 340.
8. E. Hakoila, *Annales Universitatis Turkuensis*, Ser. A, 1963, **66**, 7.
9. P. D. Bartlett and K. Nozaki, *J. Polym. Sci.*, 1948, **3**, 216.
10. P. D. Bartlett and J. D. Cotman, *J. Am. chem. Soc.*, 1949, **71**, 1419.
11. C. E. M. Morris and A. G. Parts, *Makromolek. Chem.*, 1968, **119**, 212.
12. I. M. Kolthoff and I. K. Miller, *J. Am. chem. Soc.*, 1951, **73**, 5118.
13. C. E. H. Bawn and D. Margerison, *Trans. Faraday Soc.*, 1955, **51**, 925.
14. M. G. Evans and J. H. Baxendale, *Trans. Faraday Soc.*, 1946, **42**, 195.
15. W. V. Smith and H. N. Campbell, *J. Chem. Phys.*, 1947, **15**, 338.
16. W. V. Smith, *J. Am. chem. Soc.*, 1949, **71**, 4077.
17. I. M. Kolthoff, P. R. O'Connor and J. L. Hansen, *J. Polym. Sci.*, 1955, **15**, 459.
18. H. J. van den Hul and J. W. Vanderhoff, *Br. Polym. J.*, 1970, **2**, 121.
19. H. J. van den Hul and J. W. Vanderhoff, *J. Colloid and Interface Sci.*, 1968, **28**, 336.
20. A. S. Dunn and P. A. Taylor, *Makromolek. Chem.*, 1965, **83**, 207.
21. I. M. Kolthoff, E. J. Meehan and E. M. Carr, *J. Am. chem. Soc.*, 1593, **75**, 1439.
22. J. P. Friend and A. E. Alexander, *J. Polym. Sci.*, Pt A-1, 1968, **6**, 1833.
23. D. C. Grahame, *Chem. Rev.*, 1947, **41**, 441.
24. A. E. Alexander and D. H. Napper, 'Emulsion polymerisation', in *Progress in Polymer Science* (ed. A. D. Jenkins), Pergamon, Oxford, 1971, Vol. 3, pp. 145–97.
25. W. J. Priest, *J. phys. chem., Ithaca*, 1952, **56**, 1077.
26. D. H. Napper and A. E. Alexander, *J. Polym. Sci.*, 1962, **61**, 127.
27. P. J. Flory, *Principles of Polymer Chemistry*, Cornell University Press, Ithaca, New York, 1953, pp. 208–9.
28. W. V. Smith, *J. Am. chem. Soc.*, 1948, **70**, 3695.
29. C. P. Roe and P. D. Brass, *J. Polym. Sci.*, 1957, **24**, 401.
30. A. S. Dunn, *Chem. Ind.*, 1971, 1406.
31. C. C. Price and C. E. Adams, *J. Am. chem. Soc.*, 1945, **67**, 1674.
32. I. M. Kolthoff and W. J. Dale, *J. Am. chem. Soc.*, 1945, **67**, 1672.
33. F. A. Bovey, I. M. Kolthoff, A. I. Medalia and E. J. Meehan, *Emulsion Polymerisation*, Interscience, New York, 1955, p. 344.
34. B. Haskell and W. J. Dale, unpublished experiments, University of Minnesota, 1944, quoted by F. A. Bovey, I. M. Kolthoff, A. I. Medalia and E. J. Meehan, *Emulsion Polymerisation*, Interscience, New York, 1955, p. 345.
35. H. W. Starkweather *et al.*, *Ind. Engng Chem.*, 1947, **39**, 210.
36. W. D. Wolfe/Wingfoot Corp., U.S. Pat. 2,235,625 (1941).
37. C. F. Fryling/B. F. Goodrich Co., U.S. Pat. 2,313,233 (1943).
38. B. S. Garvey/B. F. Goodrich Co., U.S. Pat. 2,376,963 (1945).
39. H. W. Walker/E. I. du Pont de Nemours & Co., U.S. Pat. 2,382,684 (1945).
40. W. B. Reynolds and E. W. Cotten, *Ind. Engng Chem.*, 1950, **42**, 1905.
41. J. M. Willis, G. Alliger, B. L. Johnson and W. M. Otto, *Ind. Engng Chem.*, 1953, **45**, 1316.
42. R. G. R. Bacon, *Q. Rev. chem. Soc.*, 1955, **9**, 287.

43. I. M. Kolthoff and W. E. Harris, *J. Polym. Sci.*, 1947, **2**, 41.
44. D. C. Blackley and A. P. McGilvray, unpublished results; A. P. McGilvray, Ph.D.Thesis, C.N.A.A., 1973.
45. K. L. Berry and J. H. Peterson, *J. Am. chem. Soc.*, 1951, **73**, 5195.
46. M. Katayama and T. Ogoshi, *Chemy high Polym.* (*Tokyo*), 1956, **13**, 6.
47. Y. Tsuda, *J. appl. Polym. Sci.*, 1961, **5**, 104.
48. L. H. Peebles, *J. appl. Polym. Sci.*, 1973, **17**, 113.
49. P. M. Hay, *et al.*, *J. appl. Polym. Sci.*, 1961, **5**, 23.
50. A. Hill/Diamond Alkali Company, U.S. Pat. 2,673,192 (1954).
51. F. H. Firsching and I. Rosen *J. Polym. Sci.*, 1959, **36**, 305.
52. W. M. Thomas, E. H. Gleason and G. Mino, *J. Polym. Sci.*, 1957, **24**, 43.
53. A. I. Medalia and I. M. Kolthoff, *J. Polym. Sci.*, 1949, **4**, 377.
54. F. Haber and J. Weiss, *Proc. R. Soc.*, 1934, **A147**, 332.
55. J. H. Baxendale, M. G. Evans and G. S. Park, *Trans. Faraday Soc.*, 1946, **42**, 155.
56. M. G. Evans, *J. chem. Soc.*, 1947, 266.
57. C. S. Marvel, R. Deanin, C. J. Clous, M. B. Wyld and R. L. Seitz, *J. Polym. Sci.*, 1948, **3**, 350.
58. W. H. Shearon, J. P. McKenzie and M. E. Samuels, *Ind. Engng Chem.*, 1948, **40**, 769.
59. B. C. Pryor, E. W. Harrington and D. Druesedow, *Ind. Engng Chem.*, 1953, **45**, 1311.
60. J. E. Wicklatz, T. J. Kennedy and W. B. Reynolds, *J. Polym. Sci.*, 1951, **6**, 45.
61. C. F. Fryling and A. E. Follett, *J. Polym. Sci.*, 1951, **6**, 59.
62. E. J. Vandenberg and G. E. Hulse, *Ind. Engng Chem.*, 1948, **40**, 932.
63. I. M. Kolthoff and A. I. Medalia, *J. Polym. Sci.*, 1950, **5**, 391.
64. I. M. Kolthoff, A. I. Medalia and M. Youse, *J. Polym. Sci.*, 1951, **6**, 93.
65. I. M. Kolthoff and A. I. Medalia, *J. Polym. Sci.*, 1951, **6**, 189.
66. C. F. Fryling, S. H. Landes, W. M. St John and C. A. Uraneck, *Ind. Engng Chem.*, 1949, **41**, 986.
67. J. M. Mitchell, R. Spolsky and H. L. Williams, *Ind. Engng Chem.*, 1949, **41**, 1592.
68. V. C. Neklutin, C. B. Westerhoff and L. H. Howland, *Ind. Engng Chem.*, 1951, **43**, 1246.
69. H. S. Smith, H. G. Werner, J. C. Madigan and L. H. Howland, *Ind. Engng Chem.*, 1949, **41**, 1584.
70. I. M. Kolthoff and A. I. Medalia, *J. Polym. Sci.*, 1951, **6**, 209.
71. I. M. Kolthoff and E. J. Meehan, *J. Polym. Sci.*, 1952, **9**, 343.
72. I. M. Kolthoff and E. J. Meehan, *J. Polym. Sci.*, 1952, **9**, 433.
73. I. M. Kolthoff and E. J. Meehan, *J. Polym. Sci.*, 1953, **11**, 71.
74. I. M. Kolthoff and E. J. Meehan, *J. appl. Polym. Sci.*, 1959, **1**, 200.
75. R. W. Brown, C. V. Bawn, E. B. Hansen and L. H. Howland, *Ind. Engng Chem.*, 1954, **46**, 1073.
76. A. K. Prince and R. D. Spitz, *Ind. Engng Chem.*, 1960, **52**, 235.
77. H. M. Andersen and S. I. Proctor, *J. Polym. Sci.*, Pt A, 1965, **3**, 2343.
78. G. S. Whitby, N. Wellman, V. W. Floutz and H. L. Stephens, *Ind. Engng Chem.*, 1950, **42**, 445.
79. R. Spolsky and H. L. Williams, *Ind. Engng Chem.*, 1950, **42**, 1847.
80. W. H. Embree, R. Spolsky and H. L. Williams, *Ind. Engng Chem.*, 1951, **43**, 2553,
81. C. S. Marvel, R. Deanin, C. G. Overberger and B. M. Kuhn, *J. Polym. Sci.*, 1948. **3**, 128.
82. P. H. Johnson and R. L. Bebb, *J. Polym. Sci.*, 1948, **3**, 389.

83. I. M. Kolthoff and E. J. Meehan, *J. Polym. Sci.*, 1952, **9**, 327.
84. R. G. R. Bacon, *Trans. Faraday Soc.*, 1946, **42**, 140.
85. J. W. L. Fordham and H. L. Williams, *J. Am. Chem. Soc.*, 1951, **73**, 4855.
86. L. W. Rainard, *J. Polym. Sci.*, 1947, **2**, 16.
87. J. S. Shukla and D. C. Misra, *Makromolek. Chem.*, 1973, **173**, 91.

CHAPTER 7

MICELLE-GENERATING SUBSTANCES

7.1 INTRODUCTION

This chapter is concerned with those substances which are added to emulsion polymerisation systems ostensibly for the primary purpose of generating micelles within the aqueous phase. The substances under consideration here are also known collectively by various other names, of which 'emulsifier', 'surfactant' and 'soap' are perhaps the most common.

It is important to realise that, in practice, the micelle generator may be required to perform certain other ancillary functions. Thus it is usually required to assist with the emulsification of the monomer, and to stabilise the growing latex particles, and also to stabilise the particles of the final latex. Although auxiliary surface-active substances may be added, either at the beginning of the reaction or during its course, in order to perform those ancillary functions, nevertheless in many cases it falls to the micelle-generating substance to perform them. In fact, it is rather surprising that a single substance can often be found which does meet all these requirements; if this had not been the case, then in all probability the discovery of the process of emulsion polymerisation would have been considerably delayed. It is a corollary of these considerations that, if no surface-active substances are added other than the initial micelle generator, then it is the latter which determines the colloidal properties of the latex which is eventually produced, notably the response of the latex to the various destabilising influences to which it may be subjected.

The plan of this present chapter is as follows. It is first necessary to consider in more detail than hitherto the phenomena of micelle formation in aqueous media and of the solubilisation of hydrophobic liquids by aqueous solutions of surface-active substances. This is followed by a discussion of solution polymerisation in aqueous solutions of micelle generators. A selection of typical results for emulsion polymerisation in the presence of each, in turn, of the various types of micelle-generating substance is then given and discussed. Classification of the substances is by way of the chemical nature of the surface-active moiety, *i.e.* as anionic, cationic and non-ionogenic. Anionics are conveniently subdivided into carboxylates, sulphonates and sulphates; and the carboxylates are further conveniently subdivided into fatty-acid, rosin-acid and other types. The chapter concludes with consideration of a few topics which are of current interest, namely, the use of oligomeric and polymerisable micelle generators, the preparation of latices of relatively uniform particle size, and, finally, the preparation of so-called 'emulsifier-free' latices.

7.2 FACTORS AFFECTING MICELLE FORMATION

The concern here is chiefly with the factors which affect (a) the concentration at which micelles first appear in an aqueous solution in substantial numbers (the so-called 'critical micelle concentration'), and (b) the size of the micelles which form. The size of micelles is usually expressed in terms of the so-called 'aggregation number', that is, the average number of 'monomeric' surface-active ions or molecules which are present in each micelle. Also of relevance to ionic micelle generators is the so-called 'fractional charge' of the micelle, that is, its actual effective charge expressed as a fraction of what it would have been if each ion in the micelle had contributed a full integral electronic charge. The reason for the effective charge being less than the product of the aggregation number and the charge on each ion is usually given as strong counterion adsorption. The latter effectively neutralises part of the real charge carried by the micelle.

Of the factors which affect the characteristics enumerated in the preceding paragraph, two are of great importance, namely, (1) the chemical structure of the surface-active substance, and (2) the concentration of simple electrolytes in the aqueous solution. A comprehensive review of the phenomenon of micelle formation in aqueous solutions has recently been published by Shinoda.[1] The remarks contained in this section are largely based upon that review; readers who desire original literature references for the various values of micelle characteristics given here as being taken from Shinoda's review should consult the review itself.

TABLE 7.1

Critical micelle concentrations for various long-chain carboxylates (according to Shinoda[1] and to Harkins[2])

Carboxylate	*Temperature* (°C)	*Critical micelle concentration* (mole l^{-1})	*Reference*
n-$C_7H_{15}CO_2K$	26	0·035†	(2)
n-$C_9H_{19}CO_2K$	26	0·095	(2)
n-$C_{11}H_{23}CO_2K$	26	0·024	(2)
n-$C_{13}H_{27}CO_2K$	26	0·006	(2)
n-$C_{17}H_{35}CO_2K$	55	0·000 45	(1)
cis-$CH_3(CH_2)_7CH{:}CH(CH_2)_7CO_2K$	50	0·001 2	(1)
trans-$CH_3(CH_2)_7CH{:}CH(CH_2)_7CO_2K$	50	0·001 5	(1)
$CH_3(CH_2)_7(CHOH)_2(CH_2)_7CO_2K$	55	0·008	(1)
cis-$CH_3(CH_2)_5CH(OH)CH_2CH{:}CH(CH_2)_7CO_2K$	55	0·003 6	(1)
K abietate	25	0·012	(1)
K dehydroabietate	25	0·027	(1)

† Sample not specially purified.

Some results for critical micelle concentration of various ionic substances are given in Tables 7.1 to 7.4. In Table 7.1, the entries of which are taken partly from the review by Shinoda and partly from a paper by Harkins,[2] are given values of this concentration for a range of long-chain carboxylates. It is clear that the concentration drops sharply as the chain length of the carboxylate increases. In fact, the relationship between critical micelle concentration and chain length is approximately logarithmic, being of the form $\log_{10}$ (cmc) $= A - Bn$ where A and B are constants and n is the number of carbon atoms in the hydrophobic chain. The value of B is approximately $\log_{10}2$, and thus the critical micelle concentration is approximately halved for each additional carbon atom in the chain. The insertion of a double bond in the chain increases the critical micelle concentration considerably; the effect is virtually the same for both *cis* and *trans* double bonds. As expected, the concentration is increased by the presence of hydrophilic substituents such as hydroxyl.

Table 7.2 gives values of the critical micelle concentration for a range of sodium alkylbenzenesulphonates. Again a logarithmic dependence upon chain length of the alkyl group is observed. The benzene ring itself is approximately equivalent to the presence of an additional three n-alkyl carbon atoms. Changing the position of the alkyl substituent in the benzene

TABLE 7.2

Critical micelle concentrations for various sodium alkylbenzenesulphonates (according to Shinoda[1])

Alkyl substituent	*Temperature* (°C)	*Critical micelle concentration* (mole l^{-1})
p-Hexyl	75	0·037
p-Heptyl	75	0·021
o-Octyl	55	0·019
p-Octyl	60–25	0·015–0·012
p-Octyl	35	0·0147
p-Nonyl	75	0·006 5
p-Decyl	50	0·003 81
o-Dodecyl	55	0·001 7
o-Dodecyl	30	0·001 59
m-Dodecyl	30	0·001 46
p-Dodecyl	30	0·001 19
p-Dodecyl	60	0·001 2
p-Tetradecyl	75	0·000 66

ring has relatively little effect upon the concentration. The effect of the hydrophilic head-group upon the critical micelle concentration of ionic micelle generators is summarised in Table 7.3, from which it is seen that the effect is relatively slight.

For substances containing saturated straight-chain hydrocarbon hydrophobic groups, the dominant effect is that of chain length. The logarithmic relationship between critical micelle concentration and chain length is usually observed regardless of the nature of the polar head-group. Values for the parameters of the logarithmic relationship are summarised in Table 7.4 from which is evident the remarkable similarity of these values for diverse types of surface-active molecule. From this table is also evident

TABLE 7.3

Effect of hydrophilic group upon critical micelle concentration (according to Shinoda[1])

Molecule	*Temperature* (°C)	*Critical micelle concentration* (mole l^{-1})
$C_{12}H_{25}CO_2K$	25	0·012 5
$C_{12}H_{25}SO_3K$	25	0·009
$C_{12}H_{25}SO_3Na$	25	0·008 1
$C_{12}H_{25}NH_2Cl$	30	0·014
$C_{12}H_{25}N(CH_3)_3Cl$	30	0·016–0·020
$C_{12}H_{25}N(CH_3)_3Br$	25	0·016

TABLE 7.4

Parameters of the relationship $\log_{10}$ (cmc) $= A - Bn$ for various types of surface-active molecule (according to Shinoda[1])

Molecular type	*Temperature* (°C)	*A*†	*B*‡
RCO_2K	25	1·63	0·290
RCO_2K	45	1·74	0·292
RSO_3Na	40	1·59	0·294
RSO_3Na	50	1·63	0·294
RSO_4Na	45	1·42	0·295
RNH_3Cl	45	1·79	0·296
$RN(CH_3)_3Br$	60	1·77	0·292
$RCH(CO_2K)_2$	25	1·54	0·220
Na 2-*n*-alkylbenzenesulphonate	55	—	0·292

† Note that these values of *A* give critical micelle concentration in mole l^{-1}, *n* being the number of carbon atoms in the hydrocarbon chain of the molecule.
‡ The value of *B* can be usefully compared with that of $\log_{10} 2$, *i.e.* 0·301.

the general closeness of the parameter *B* to the value $\log_{10} 2$; only in the case of the dicarboxylate series does *B* deviate seriously from this value. In so far as changing the counterion does not produce an insoluble compound which therefore precipitates from aqueous solution, it is found that the critical micelle concentration does not depend markedly upon the nature of the counterion. Nor do small changes in temperature appear to have a profound effect.

A selection of results for aggregation number and fractional charge is given in Table 7.5. Regarding the fractional charge values, it is seen that these are considerably less than unity, suggesting that strong counterion adsorption must occur to a considerable extent. However, comparison of the results for lithium and sodium dodecyl sulphates leads one to question to what extent counterion adsorption really does determine the fractional charge value; for the smaller size of the lithium ion relative to the sodium ion would be expected to result in a stronger adsorption tendency and therefore a smaller fractional charge, rather than a larger one as indicated in Table 7.5. Concerning aggregation numbers, it is seen that these are generally within the range 50–100. The value given for sodium dibutylbenzenesulphonate seems to be exceptionally high. The aggregation numbers for magnesium alkylsulphonates seem to be roughly double those for the corresponding sodium alkylsulphonates.

The addition of simple inert electrolytes, such as sodium chloride, has an effect upon all three of the properties under discussion. The effects of

TABLE 7.5

Properties of micelles formed from typical micelle generators in aqueous solution (according to Shinoda[1])

Micelle generator	*Aggregation number*	*Fractional charge*
$C_{11}H_{23}CO_2K$	50	
$C_{10}H_{21}SO_3Na$	40	
$C_{12}H_{25}SO_3Na$	54	
$(C_{10}H_{21}SO_3)_2Mg$	104	
$(C_{12}H_{25}SO_3)_2Mg$	107	
$C_{10}H_{21}SO_4Na$	50	
$C_{12}H_{25}SO_4Na$	62	0·168
$C_{12}H_{25}SO_4Li$	63	0·225
Na dihexylsulphosuccinate	26	
Na dioctylsulphosuccinate	48	
Na di*iso*hexylsulphosuccinate		0·336
Na di*iso*octylsulphosuccinate		0·39
Na dibutylbenzenesulphonate	229	
Na *m*-dodecylbenzenesulphonate	57	
Na *p*-dodecylbenzenesulphonate	24	

potassium chloride, potassium sulphate, sodium pyrophosphate and sodium sulphate upon the critical micelle concentration of potassium laurate and sodium dodecyl sulphate are shown in Fig. 7.1 (a), which is taken from a paper by Corrin and Harkins.[3] It is seen that the relationship between the critical micelle concentration and the electrolyte concentration, expressed in gram equivalents per unit volume, is independent of electrolyte type for a given surface-active substance. The addition of small amounts of electrolyte causes a marked reduction in the critical micelle concentration, but the effect becomes progressively less as larger quantities are added. Ultimately, a stage is reached where the addition of further electrolyte has little effect. Similar effects are observed with cationic surface-active substances—see Fig. 7.1 (b). Over the range of electrolyte additions for which data are available, there is a linear relationship between the logarithm of the critical micelle concentration and the logarithm of the electrolyte concentration. This fact is illustrated in Fig. 7.2 (a) and (b) for anionic and cationic micelle generators respectively. Also illustrated in Fig. 7.1 (a) is the observation that the addition of a non-electrolyte (urea) has almost no effect upon critical micelle concentration.

The results of Stigter and Mysels[4] for the effect of sodium chloride upon the aggregation number and micellar charge of sodium dodecyl sulphate

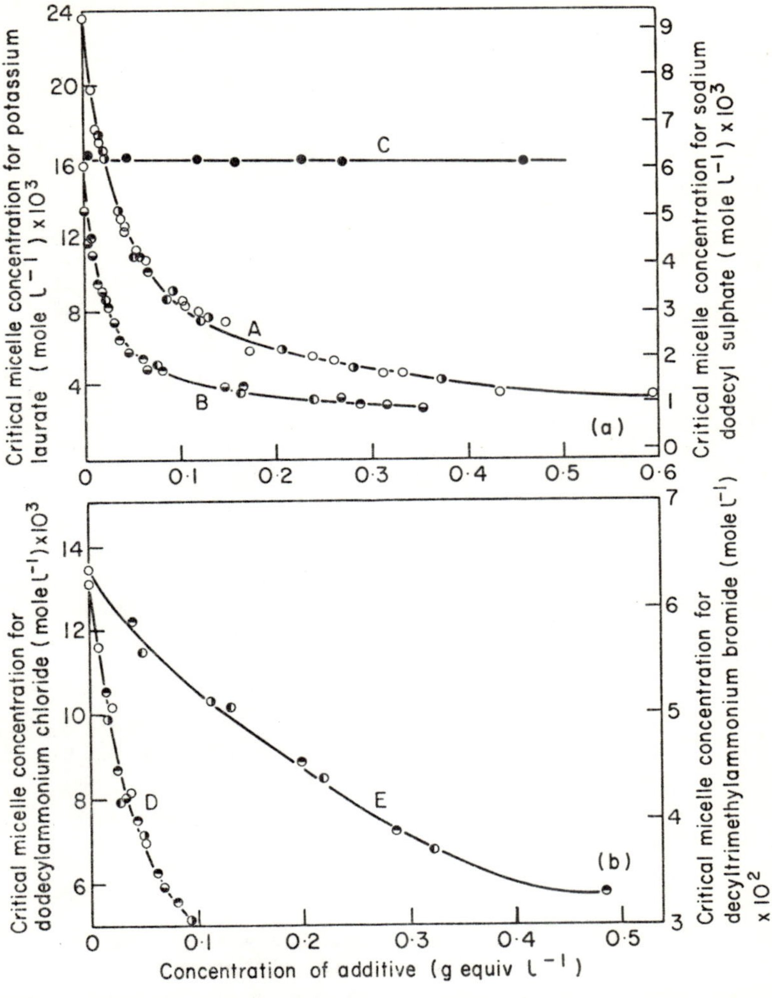

FIG. 7.1. (a) Effect of simple electrolyte upon critical micelle concentration of potassium laurate (A) and sodium dodecyl sulphate (B), according to Corrin and Harkins.[3] Results are also given for the effect of urea upon the critical micelle concentration of sodium dodecyl sulphate (C). Significance of points: ○ KCl; ◒ NaCl; ◑ K_2SO_4; ◐ $Na_4P_2O_7$; ◓ Na_2SO_4; ● urea. (b) Effect of simple electrolytes upon critical micelle concentration of dodecylammonium chloride (D) and decyltrimethylammonium bromide (E) according to Corrin and Harkins.[3] Significance of points: ◐, ◑ NaCl; ◒, ◓ $BaCl_2$; ○ $LaCl_3$.

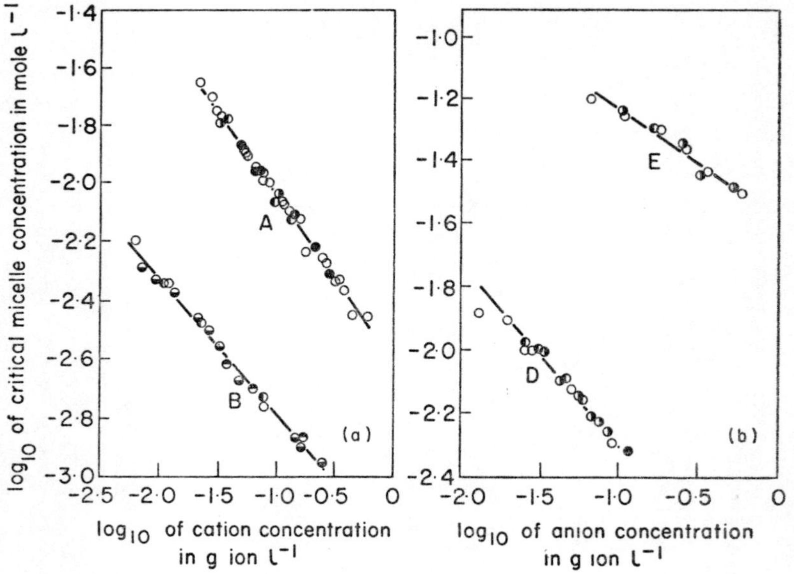

FIG. 7.2. Data of Fig. 7.1 re-plotted on double logarithmic scales, according to Corrin and Harkins.[3] See caption to Fig. 7.1 for significance of letters A to E appended to lines. Significance of points: (a) ○ KCl; ◒ NaCl; ◑ K_2SO_4; ◓ Na_2SO_4; ◐ $Na_4P_2O_7$; (b) ○ NaCl; ◑ $BaCl_2$; ◐ $LaCl_3$.

are summarised in Table 7.6. It is seen that the addition of the electrolyte promotes the formation of rather larger micelles of higher fractional charge.

Hobbs[5] has given a theory which purports to explain the effects of simple electrolytes upon critical micelle concentration and upon micelle size. The logarithmic relationship between critical micelle concentration and electrolyte concentration is successfully derived.

TABLE 7.6

Effect of sodium concentration upon aggregation number and micellar charge for sodium sulphate (according to Stigter and Mysels[4])

NaCl concentration (mole l^{-1})	*Aggregation number*	*Fractional charge*
0	80	0·287
0·01	89	0·281
0·03	100	0·285
0·05	105	0·295
0·10	112	0·324

Some values for the critical micelle concentration of various non-ionogenic polyethylene oxide adducts are collected in Tables 7.7 and 7.8. The information contained in these tables is taken from a second review by Shinoda[6] which is concerned with the properties of aqueous solutions of non-ionogenic surface-active substances. Table 7.7 shows the effect of increasing the length of the hydrophilic (polyethylene oxide) part of the molecule. It will be seen that the critical micelle concentration is little affected by this variable. The trend is for the critical concentration to increase as the number of ethylene oxide units per molecule increases. Table 7.8 shows the effect of increasing the length of the hydrophobic part of the molecule for adducts of ethylene oxide and fatty alcohols. As in the case of ionic surface-active substances, the critical micelle concentration decreases sharply with increasing hydrocarbon chain length; as in the former case, the relationship appears to be such that the logarithm of the critical micelle concentration varies linearly with hydrocarbon chain length. As would be expected, there is an upper limit to the chain length, beyond which the compound is insufficiently soluble in water for the critical micelle concentration to be attained.

TABLE 7.7

Effect of hydrophilic chain upon the critical micelle concentration of various polyethylene oxide adducts in aqueous solution (according to Shinoda[6])

Compound	*Temperature* (°C)	*Critical micelle concentration* (mole l^{-1})
n-$R_{10}O(CH_2CH_2O)_4H$	25	$6{\cdot}8 \times 10^{-4}$
n-$R_{10}O(CH_2CH_2O)_5H$	25	$8{\cdot}1 \times 10^{-4}$
n-$R_{10}O(CH_2CH_2O)_8H$	25	$9{\cdot}0 \times 10^{-4}$
n-$R_{12}O(CH_2CH_2O)_4H$	25	$0{\cdot}4 \times 10^{-4}$
n-$R_{12}O(CH_2CH_2O)_5H$	23	$0{\cdot}57 \times 10^{-4}$
n-$R_{12}O(CH_2CH_2O)_7H$	23	$0{\cdot}8 \times 10^{-4}$
n-$R_{12}O(CH_2CH_2O)_9H$	23	$1{\cdot}0 \times 10^{-4}$
n-$R_{12}O(CH_2CH_2O)_{12}H$	23	$1{\cdot}4 \times 10^{-4}$
i-$R_8C_6H_4O(CH_2CH_2O)H$	25	†$0{\cdot}5 \times 10^{-4}$
i-$R_8C_6H_4O(CH_2CH_2O)_2H$	25	†$0{\cdot}75 \times 10^{-4}$
i-$R_8C_6H_4O(CH_2CH_2O)_3H$	25	†$1{\cdot}0 \times 10^{-4}$
i-$R_8C_6H_4O(CH_2CH_2O)_4H$	25	†$1{\cdot}4 \times 10^{-4}$
i-$R_8C_6H_4O(CH_2CH_2O)_5H$	25	$1{\cdot}8 \times 10^{-4}$
i-$R_8C_6H_4O(CH_2CH_2O)_7H$	25	$2{\cdot}7 \times 10^{-4}$
i-$R_8C_6H_4O(CH_2CH_2O)_8H$	25	$2{\cdot}9 \times 10^{-4}$
i-$R_8C_6H_4O(CH_2CH_2O)_9H$	25	$3{\cdot}1 \times 10^{-4}$
i-$R_8C_6H_4O(CH_2CH_2O)_{10}H$	25	$3{\cdot}5 \times 10^{-4}$

† These values are for solubility, and not for critical micelle concentration.

TABLE 7.8

Effect of hydrophobic chain length upon the critical micelle concentrations of polyethylene oxide adducts in aqueous solution (according to Shinoda[6])

Compound	*Temperature* (°C)	*Critical micelle concentration* (mole l^{-1})
$R_4O(CH_2CH_2O)_6H$	20	0·796
$R_6O(CH_2CH_2O)_6H$	20	0·074
$R_8O(CH_2CH_2O)_6H$	25	0·009 8
$R_{10}O(CH_2CH_2O)_6H$	23·5	0·000 92
$R_{12}O(CH_2CH_2O)_6H$	25	0·000 087
$R_{16}O(CH_2CH_2O)_6H$	25	0·000 001

From the results discussed above, it is possible to form a picture of the situation which exists in the aqueous phase of a typical emulsion polymerisation system before the reaction begins. The typical system contains of the order of 25–30 grams of micelle generator per litre of aqueous phase. Molecular weights of micelle generators are typically about 300, and therefore the molarity of the aqueous phase with respect to micelle generator is of the order of 10^{-1}. This is greatly in excess of the critical micelle concentration, and therefore most of the micelle generator is in fact present as micelles. In each millilitre of aqueous phase, there will be approximately 5×10^{19} molecules of micelle generator. If an aggregation number of between 50 and 100 is assumed, this gives the concentration of micelles as being between 5×10^{17} and 1×10^{18} per millilitre. Each micelle will carry an effective charge equal to between 5 and 10 electron units. Addition of simple electrolytes will cause the critical micelle concentration to fall even lower, but this will not bring about a significant change in the proportion of the micelle-forming substance which is present as micelles. It will, however, tend to make the micelles larger and to increase their effective ionic charge.

7.3 SOLUBILISATION OF HYDROPHOBIC LIQUIDS BY MICELLES

Extensive reviews of the phenomenon of solubilisation by aqueous solutions of micelle-forming substances have been published by Klevens[7] and by McBain and Hutchinson.[8] Tables 7.9–7.12, and Figs. 7.3–7.5 illustrate the main features of the solubilisation of the type of molecule which is of interest in the study of emulsion polymerisation. In considering the results summarised in these tables and figures, it is perhaps helpful to realise that, apart from temperature, the effects of the following variables upon the extent of solubilisation are especially relevant for application to emulsion polymerisation studies: (a) soap type, (b) soap concentration, (c) nature of solubilised substance, and (d) concentration of simple electrolytes.

The results given in Table 7.9 are from a paper by McBain and Richards[9] in which were reported the results of a study of the solubilisation of various organic liquids in aqueous solutions of three soaps: two anionic carboxylates and a cationic amine soap. The method used by these workers for measuring the extent of solubilisation depends upon the observation that, as a water-insoluble liquid is added in increasing amounts to an aqueous solution of a surface-active substance, so there comes a point at which the turbidity of the solution rises sharply. This is interpreted as the point at which both true aqueous phase and micelles are saturated and the excess organic liquid then appears as emulsified droplets. As McBain and Richards state, 'The solubilized material is in solution in the sense that it is not present as suspended or protected particles or emulsified droplets, but is incorporated in the colloidal particles of the detergent itself. It is therefore in solution in the same sense that the soap itself is in solution ... As soon as the saturation value for solubilization is exceeded, any excess of the solubilized material appears as suspended particles or droplets of emulsion and causes a steep rise in turbidity.'

Referring now to the results given in Table 7.9, it is seen that the solubilising power of the three soaps studied decreases in the order dodecylammonium chloride > sodium oleate > potassium laurate. The difference between sodium oleate and potassium laurate is due mainly to the difference in chain length of the hydrophobic part of the molecule, rather than to the difference in counterion. It also appears that the tendency of paraffin hydrocarbons to be solubilised decreases rapidly as the number of carbon atoms increases; the same generalisation is true for the alkylbenzenes. It is interesting to note that benzene shows rather more tendency to be solubilised into the carboxylate micelles than does *n*-hexane, but rather less tendency to be solubilised into the micelles of the cationic soap.

Data for the solubilisation of ethylbenzene (a convenient model for styrene) in aqueous solutions of various potassium *n*-alkanoates are given in Table 7.10. Several interesting features are illustrated by these results. Thus the solubilising power of a soap increases quite sharply as both its concentration and chain length are increased. When the extents of

TABLE 7.9

Solubilisation of various organic liquids by 0·1 mole l^{-1} aqueous solutions of sodium oleate, potassium laurate and dodecylammonium chloride at 25°C (according to McBain and Richards[9])
(values corrected for solubility in water)

Liquid	*Sodium oleate*		*Potassium laurate*		*Dodecylammonium chloride*	
	g solubilised per 100 ml *of solution*	*Molecules per molecule of soap*	*g solubilised per* 100 ml *of solution*	*Molecules per molecule of soap*	*g solubilised per* 100 ml *of solution*	*Molecules per molecule of soap*
n-Hexane	0·398	0·46	0·156	0·18	0·64	0·75
n-Heptane	0·340	0·34	0·117	0·12	0·54	0·54
n-Octane	0·210	0·18	0·096	0·08	0·33	0·29
n-Nonane	0·142	0·11	0·071	0·06	0·29	0·22
n-Decane	0·047	0·05	0·045	0·03	0·18	0·13
n-Dodecane	0·015	0·01	0·008	0·005	0·11	0·06
n-Tetradecane	0·000	0·00	0·000	0·00	0·02	0·01
Benzene	0·594	0·76	0·226	0·29	0·51	0·65
Toluene	0·466	0·51	0·116	0·13	0·45	0·49
Ethylbenzene	0·419	0·40	0·214	0·20	0·40	0·38
p-Xylene	0·383	0·36	0·210	0·20	0·36	0·34
p-Cymene	0·342	0·26	0·103	0·08	0·26	0·19

TABLE 7.10

Solubilisation of ethylbenzene in aqueous solutions of various potassium *n*-alkyl carboxylates at 25°C
(according to Klevens[7] values corrected for solubility in water)

Solution	*Soap concentration* [m. (kg *of solution*)$^{-1}$]	*Moles of micellar soap per* kg *of solution*	*Molecules of dissolved ethylbenzene per* kg *of solution*	*Molecules of ethylbenzene per molecule of soap*	*Molecules of ethylbenzene per molecule of micellar soap*	*Molecules of ethylbenzene per micelle of* 150 *molecules*
—	0	0	0·0016	—	—	—
Potassium octanoate	0·30	0	0·0012	0·004	—	—
	0·48	0·085	0·012	0·025	0·141	21
	0·662	0·267	0·033	0·048	0·124	19
	0·827	0·432	0·066	0·80	0·152	22
Potassium decanoate	0·10	0·005	0·0014	0·014	—	—
	0·232	0·137	0·027	0·116	0·197	30
	0·435	0·340	0·067	0·154	0·197	30
	0·500	0·405	0·087	0·174	0·214	33
	0·717	0·622	0·145	0·202	0·233	35
Potassium dodecanoate	0·042	0·017	0·007	0·166	0·411	62
	0·195	0·170	0·062	0·318	0·364	54
	0·396	0·371	0·151	0·382	0·407	61
	0·500	0·475	0·212	0·424	0·446	67
	0·603	0·578	0·273	0·452	0·472	71
	0·628	0·603	0·291	0·463	0·482	72
	0·860	0·835	0·435	0·506	0·522	78
Potassium tetradecanoate	0·096	0·090	0·054	0·563	0·600	90
	0·242	0·236	0·176	0·728	0·745	107
	0·347	0·341	0·272	0·784	0·798	120
	0·432	0·426	0·348	0·807	0·817	123
	0·500	0·494	0·427	0·855	0·866	130
	0·566	0·560	0·492	0·872	0·888	133
Potassium hexadecanoate	0·070	0·068	0·074	1·06	1·09	159
	0·154	0·152	0·176	1·14	1·15	173
	0·228	0·226	0·302	1·32	1·33	190
	0·292	0·290	0·430	1·47	1·48	220

solubilisation are converted to molecules of solubilised substances per molecule of micellar soap, it is seen that this value is, in general, reasonably constant for a given soap at all soap concentrations. The value does, however, increase rapidly with the number of carbon atoms in the hydrophobic portion of the soap molecule, being of the order of 0·15 for the C_8 carboxylate and of the order of 1·0–1·5 for the C_{16} carboxylate. This dependence upon chain length of the hydrophobic part of the molecule can be readily understood, in that the longer the chain length, the greater will be the volume of the interior of the micelle, and therefore the greater will be its capacity of solubilise. A further point of interest which is apparent from he results given in Table 7.10 is that, in the absence of micelles, the

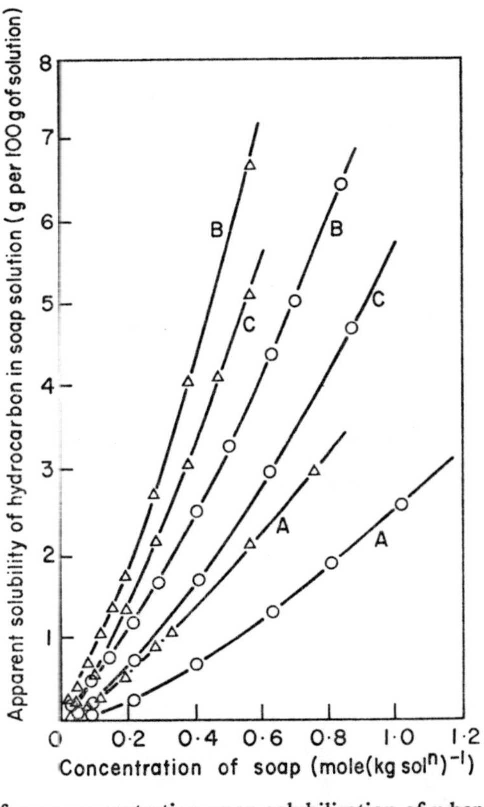

FIG. 7.3. Effect of soap concentration upon solubilisation of *n*-heptane (A), benzene (B) and ethylbenzene (C) in aqueous solutions of potassium laurate (points ○) and potassium myristate (points △) at 25°C, according to Stearns *et al.*[10]

solubility of a hydrocarbon liquid in a soap solution is virtually the same as its solubility in water.

Fig. 7.3 shows the results of Stearns *et al.*[10] for the solubilisation of three hydrocarbons, *n*-heptane, benzene and ethylbenzene, in potassium laurate and potassium myristate solutions. The equilibrium extents of solubilisation were again determined by the turbidimetric method. Concerning this method, Stearns and his co-workers state that it 'works well if there is a large difference between the refractive index of the oil and that of the soap solution. It becomes less and less sensitive as the refractive index of the oil approaches that of the soap solution. When the refractive index of the oil equals that of the soap solution, the solution no longer appears cloudy when excess oil is present, since there is no light scattering by the emulsion drops present.'

The results summarised in Fig. 7.3 show that the amount of hydrocarbon solubilised does not increase linearly with the amount of soap in the solution. Furthermore, these results confirm that potassium myristate (C_{14}) is a more effective solubiliser than is potassium laurate (C_{12}), and that *n*-heptane shows considerably less tendency to be solubilised than does benzene. Substitution of an alkyl group into the benzene ring reduces the tendency to be solubilised, and the larger is the alkyl group, the larger is the reduction in solubilising tendency. The results summarised in Fig. 7.4 which are also due to Stearns and his co-workers,[10] illustrate a progressive fall in tendency to be solubilised for the series benzene, ethyl benzene, propylbenzene. It is also of interest that *iso*propylbenzene shows a stronger tendency to be solubilised than does *n*-propylbenzene, being approximately equivalent to ethylbenzene in this respect. Stearns and his co-workers have also provided convincing evidence that the increase in solubility which accompanies an increase in the soap concentration is associated with the presence of increasing numbers of micelles, and so is naturally interpreted as being due to imbibition within the micelles. The solubility of 2-nitrodiphenylamine, a yellow solid, was determined in dilute solutions of potassium laurate. It was found that, within the limits of experimental uncertainty, the solubility remained constant until the concentration of soap had reached a level which corresponded to the critical micelle concentration as determined by independent experiments.

The effect of simple electrolytes is to increase considerably the extent of solubilisation. The results of Stearns *et al.*[10] for ethylbenzene solubilised in 0.1 mole l^{-1} potassium myristate solution are summarised in Fig. 7·5. Stearns and his co-workers have provided evidence to suggest that, at low soap concentrations (*i.e.* near the critical value), the effect of the electrolyte

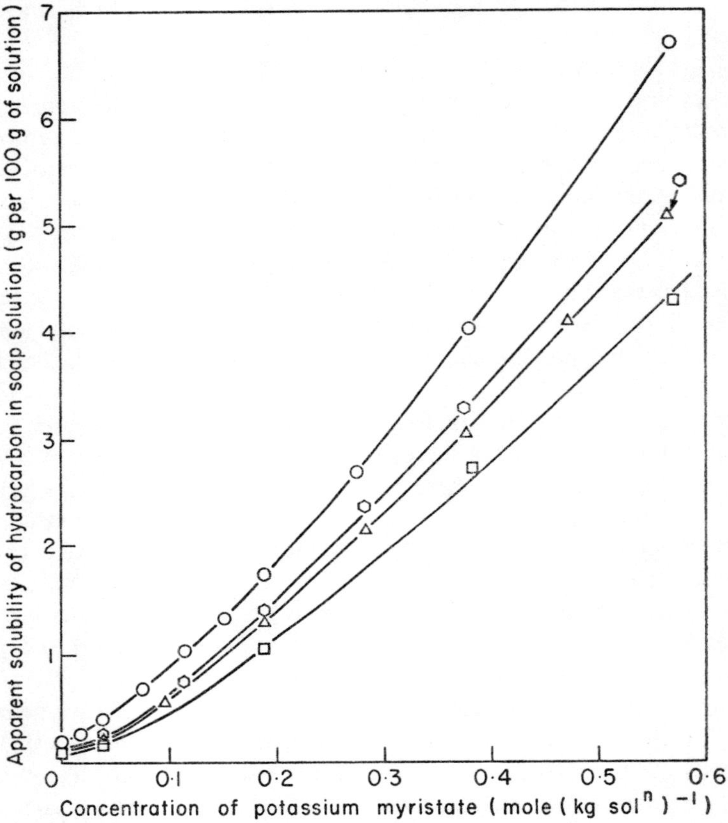

FIG. 7.4. Solubilisation of benzene (points ○), ethylbenzene (points △), *n*-propylbenzene (points □) and *iso*propylbenzene (points ⬡) in aqueous solutions of potassium myristate at 25°C, plotted from data given by Stearns *et al.*[10]

is essentially to make more micellar soap available through the lowering of the critical micelle concentration. At higher soap concentrations, however, they could not account for the increased solubilisation solely in terms of the decrease in critical micelle concentration and the consequent increase in the amount of micellar soap. It may well be that the clue to the understanding of these observations is to be found in the increase in micelle size which, as has been noted in Section 7.2, accompanies the addition of simple electrolytes to aqueous solutions of micelle-forming substances.

The foregoing remarks have been concerned exclusively with the solu-

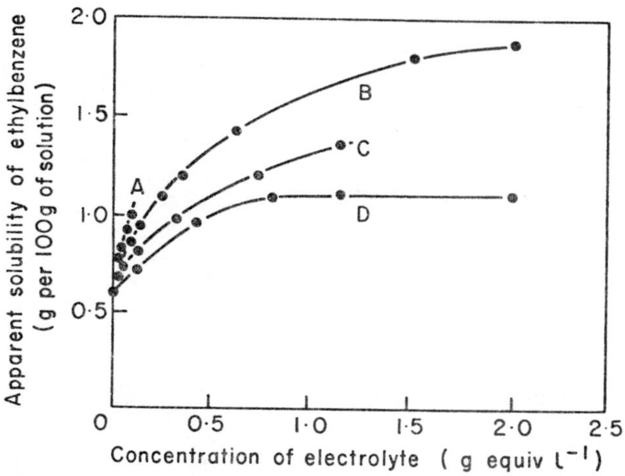

FIG. 7.5. Effect of various simple electrolytes upon the solubilisation of ethylbenzene in 0·1 mole l^{-1} potassium myristate at 25°C, according to Stearns *et al.*[10] The curves summarise data for four electrolytes as follows: A, NaCl; B, KCl; C, K_2SO_4; D, $K_4Fe(CN)_6$.

bilisation of hydrocarbon liquids. For the purposes of understanding emulsion polymerisation reactions, it is clearly desirable to have information upon the solubilisation of more polar compounds, since these would serve as models for the more polar monomers such as vinyl acetate and acrylonitrile. Of course, the more polar liquids are more soluble in pure water than are the hydrocarbons. However, in the results published by McBain and Richards[9] and given in Table 7.11, allowance has been made for enhanced water-solubility. Comparison with the results given in Table 7.9 shows that, whereas for the hydrocarbons the order of decreasing solubilising power for the three soaps is dodecylammonium chloride > sodium oleate > potassium laurate, for the more polar compounds the order is sodium oleate > dodecylammonium chloride > potassium laurate. Correlations between the structure of a compound and its tendency to be solubilised are difficult to draw. The marked tendency for ethers, ketones and esters to be solubilised is not unexpected. Comparing *n*-octane, octylamine and *n*-octyl alcohol, it is seen that, rather surprisingly, the order of decreasing tendency to be solubilised in the two carboxylates is *n*-octyl alcohol > *n*-octane ≃ octylamine. In the cationic soap, the tendency of octylamine to be solubilised is similar to that of the corresponding alcohol.

In Table 7.12 are collected results given by Harkins and Oppenheimer,[11] for the solubility of various alcohols, an amine and a mercaptan in a range

TABLE 7.11

Solubilisation of various polar organic liquids by 0·1 mole l^{-1} aqueous solutions of sodium oleate, potassium laurate and dodecylammonium chloride at 25°C (according to McBain and Richards[9])
(values corrected for solubility in water)

Compound	*Sodium oleate*		*Potassium laurate*		*Dodecylammonium chloride*	
	g solubilised per 100 ml *of solution*	*Molecules per molecule of soap*	*g solubilised per* 100 ml *of solution*	*Molecules per molecule of soap*	*g solubilised per* 100 ml *of solution*	*Molecules per molecule of soap*
Methyl *t*-butyl ether	1·936	2·20	1·460	1·66	1·80	2·05
Methyl *iso*butyl ketone	1·820	1·82	1·200	1·20	1·78	1·78
Amyl acetate	1·990	1·71	1·038	0·89	1·69	1·45
*Iso*propyl *t*-butyl ether	0·858	0·73	0·166	0·14	0·62	0·53
Octylamine	0·088	0·07	0·088	0·07	0·17	0·13
n-Octyl alcohol	0·764	0·59	0·385	0·29	0·24	0·18
2-Ethylhexanol	0·610	0·47	0·083	0·064	0·47	0·36
Lauryl alcohol	0·206	0·13	0·049	0·03	0·08	0·05
Oleic acid	0·137	0·05	0·051	0·018	0·07	0·02
Tributyrin	1·124	0·37	0·327	0·11	0·68	0·22
Triolein	0·000	0·00	—	—	0·00	—

TABLE 7.12

Solubility of various *n*-primary alcohols, an amine, and a mercaptan in 0·3 mole l^{-1} aqueous solutions of various soaps (according to Harkins and Oppenheimer[11])

Compound	*Solubility of compound* (moles per kg *of solution*) *in solution of soap indicated*							
	$C_7H_{15}CO_2K$	$C_8H_{17}CO_2K$	$C_9H_{19}CO_2K$	$C_{10}H_{21}CO_2K$	$C_{11}H_{23}CO_2K$	$C_{12}H_{25}CO_2K$	$C_{13}H_{27}CO_2K$	$C_{12}H_{25}SO_4Na$
$C_7H_{15}OH$	0·099	0·139	0·188	0·201	0·201	0·212	0·226	0·260
$C_8H_{17}OH$	0·023	0·095	0·127	0·148	0·146	0·156	0·158	0·171
$C_9H_{19}OH$	0·004	0·076	0·096 5	0·116 5	0·126	0·132	0·140	0·144
$C_{10}H_{21}OH$	0·003	0·063	0·092 3	0·104	0·111	0·122	0·136	0·138
$C_{11}H_{23}OH$	0·003	0·072	0·086 5	0·107	0·107 5	0·116	0·114	0·125
$C_{12}H_{25}OH$†	0·002	0·056	0·084 0	0·087 0	0·082	0·077	0·034	0·123
$C_{10}H_{21}NH_2$	0·004	0·102	0·11	0·113	0·113	0·113	0·119	0·110
$C_7H_{15}SH$	0·000 7	0·001 4	0·010 2	0·017 1	0·039 3	0·059 1	0·090	0·092 5

† Values deduced by Klevens[7] from graphs in original reference.

of soap solutions. For the alcohols, the chain length of both alcohol and soap has been varied systematically. As expected, the tendency for solubilisation to occur increases with increasing chain length of the soap, and decreases with decreasing chain length of the alcohol. In the case of the higher alcohols, there appears to be a tendency for the solubilising ability of the soap to pass through a maximum as its chain length is increased. The amine tends to be solubilised to a rather smaller extent than does the corresponding alcohol. The mercaptan dissolves to a far more limited extent than does the alcohol containing the same number of carbon atoms.

Further evidence that the phenomenon of solubilisation is due to imbibition of the added substance within micelles already present in the aqueous phase has come from the study of micelle structure by the method of X-ray diffraction. Thus Hughes, Sawyer and Vinograd[12] have reported that the addition of a hydrocarbon liquid to a soap solution causes the size of the micelles to increase. The increase in size is proportional to the ratio of hydrocarbon to soap, and the effect of a given additive depends markedly upon its chemical structure. More extensive information on the application of X-ray diffraction methods to micelles is given in two papers by Harkins, Mattoon and Corrin.[13,14] The second of these papers is concerned specifically with the effects of solubilisation upon the dimensions of micelles as revealed by X-ray methods.

The information given in the preceding paragraphs enables us to supplement the generalised picture given in Section 7.2 of the situation pertaining in the aqueous phase of a typical emulsion polymerisation system before the reaction begins. It will be recalled that the conclusion was reached that the concentration of micelles will lie between about 5×10^{17} and 1×10^{18} per millilitre of aqueous phases. It is now possible to conclude that each micelle will contain between about 10 and 50 solubilised monomer molecules, depending upon the nature of the micelle-generating substance. Two further observations are immediately obvious:

(1) The number of monomer molecules in each micelle is very much less than the degree of polymerisation of the typical polymer eventually produced. Therefore if polymerisation starts in the micelles, there must be an influx of monomer from elsewhere as polymerisation proceeds.

(2) The actual amount of monomer present in the micelles is a very small proportion of the whole. The upper limit is about 10^{-4} moles of monomer solubilised per millilitre of aqueous phase, this corres-

ponding to 10^{18} micelles per millilitre each containing about 50 molecules. By contrast, the total monomer content of a typical emulsion polymerisation system is not usually less than the equivalent of 5×10^{-3} moles per millilitre of aqueous phase, and may be rather more than this. Thus, at most, it would be expected that only about 2% of the monomer is present in the micelles at the start of the reaction.

7.4 POLYMERISATION IN SOLUTIONS OF SURFACE-ACTIVE SUBSTANCES

In an endeavour to gain further insight into the mechanism of the emulsion polymerisation of styrene, Bovey and Kolthoff[15] investigated its rate of polymerisation in both pure water and in aqueous soap solutions, there being no excess of styrene present as a separate phase. It was hoped that measurements of the polymerisation rate in pure water would throw some light on the role of the true aqueous phase in a normal emulsion polymerisation. Polymerisations were carried out under nitrogen. Potassium persulphate was used as initiator. Residual styrene was determined by amperometric titration with potassium bromate. Residual styrene as a function of time of reaction is shown in Fig. 7.6 (a) for polymerisation in pure water, and in Fig. 7.6 (b) for polymerisation in 0·093 mole l^{-1} aqueous dodecylammonium chloride. These data are replotted in Figs. 7.7 (a) and (b), respectively, in order to demonstrate that, over much of its course, the reaction is kinetically of second order with respect to styrene concentration.

Bovey and Kolthoff also found that in both pure water and aqueous dodecylammonium hydrochloride, the rate of polymerisation is also approximately proportional to the square root of the persulphate concentration. Furthermore, for the polymer produced in dodecylammonium chloride solution, the degrees of polymerisation were found to be approximately inversely proportional to the square root of the persulphate concentration. These are all features which are characteristic of a 'normal' bulk or solution polymerisation in which termination is bimolecular with respect to growing radicals. Comparing polymerisation in both water and aqueous dodecylammonium chloride, Bovey and Kolthoff observed that the rate coefficients for the latter were only about 70% of those for the former. They also made measurements of the rates of polymerisation in aqueous solutions of various fatty-acid soaps, and found that the rates

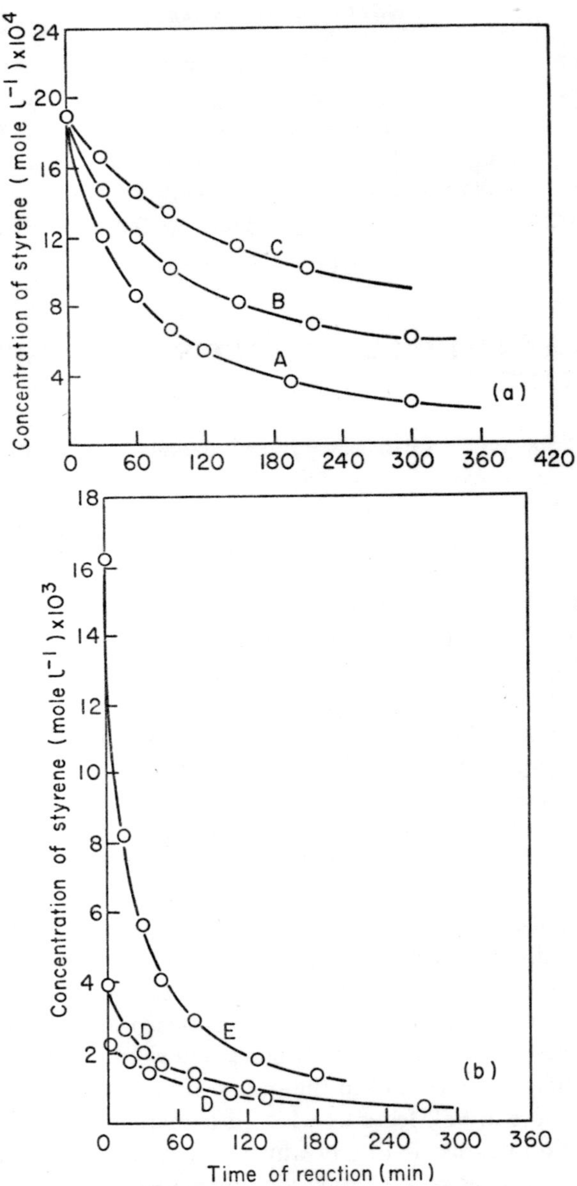

FIG. 7.6. Persulphate-initiated solution polymerisation of styrene at 50°C in (a) water and (b) 0·093 mole l^{-1} aqueous dodecylammonium chloride solution, according to Bovey and Kolthoff.[15] The letters appended to the curves distinguish data for different concentrations of potassium persulphate as follows: A, 6·2 × 10^{-3} mole l^{-1}; B, 1·54 × 10^{-3} mole l^{-1}; C, 3·85 × 10^{-4} mole l^{-1}; D, 6·2 × 10^{-3} mole l^{-1}; E, 1·5 × 10^{-3} mole l^{-1}. (Note that the two curves labelled D show the effect of varying the initial monomer concentration whilst keeping the concentrations of initiator and soap constant.)

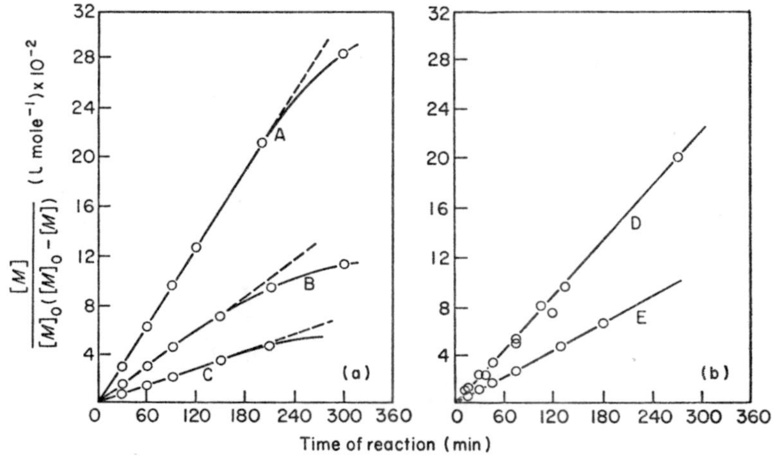

FIG. 7.7. Data of Fig. 7.6 re-plotted to demonstrate extent to which the solution polymerisation of styrene in aqueous media is kinetically of second order with respect to monomer concentration. $[M]$ denotes the styrene concentration after an arbitrary time of reaction, and $[M]_0$ denotes the initial value. (A linear variation of $[M]/\{[M]_0([M]_0 - [M])\}$ with time is consistent with second-order kinetics.) See caption to Fig. 7.6 for significance of letters A to E appended to curves.

were considerably faster than in aqueous dodecylammonium chloride solution under otherwise equivalent conditions. Whereas the presence of the dodecylammonium chloride retarded the polymerisation compared to that in pure water, the presence of a fatty-acid soap, such as potassium palmitate, increased the rate markedly. It should be noted that the concentrations of soaps used in these experiments were well in excess of the critical values for micelle formation.

These results appear to indicate that, in water alone, the polymerisation proceeds as a 'normal' solution polymerisation, and in aqueous dodecylammonium chloride it proceeds as a solution polymerisation in which the soap acts as a mild retarder. In the aqueous solutions of fatty acids, however, it appears that the micelles are able to exert an accelerating effect upon the reaction.

Further information on the polymerisation of styrene in aqueous soap solution is given in a paper by Frilette and Hohenstein,[16] to which reference has been made earlier in this book. Other papers on the subject of polymerisation in aqueous solution include those of Bacon,[17] of Baxendale *et al.*[18,19] and of Morgan.[20] The work of Baxendale and co-workers has been considered in a previous chapter of this book.

7.5 FATTY-ACID SOAPS

A large amount of information has been published concerning the effect of various fatty-acid soaps in systems for the emulsion copolymerisation of styrene and butadiene at about 50°C using potassium persulphate as initiator. The formulation for these studies has often been the so-called 'standard GR-S recipe'. It is as follows:

Styrene	25
Butadiene	75
Water	180
Soap	5·0
n-Dodecyl mercaptan	0·5
Potassium persulphate	0·3

The proportions refer to parts by weight. The effect of varying the chain length of saturated fatty-acid soaps has been reported by Carr *et al.*[21] Their results are summarised in Fig. 7.8. In this diagram, conversion is given as a function of time for systems containing, respectively, caprate, laurate, myristate, palmitate and stearate soaps. It is not clear whether sodium or potassium soaps were used. From these results, it is seen that there is no detectable difference between the behaviour of palmitate and stearate. But as the chain length is reduced below that of palmitate, the

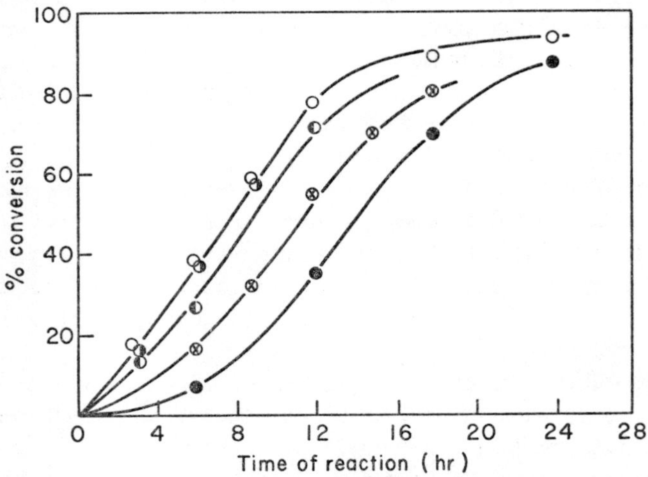

FIG. 7.8. Effect of varying chain length of fatty-acid soap upon conversion–time relationship for standard GR-S system reacted at 50°C, according to Carr *et al.*[21] Significance of points: ● caprate; ⊗ laurate; ◐ myristate; ◑ palmitate; ○ stearate.

initial rate of polymerisation becomes progressively lower. However, after about 20% conversion, all the soaps give roughly the same rate of conversion. It seems reasonable to attribute the initial differences to a combination of the following two factors: (a) the proportion of soap present in micellar form will fall with decreasing chain length of the soap, because of the rise in the critical micelle concentration; (b) the micelles from the soaps with shorter chains will solubilise fewer monomer molecules than those with longer chains. However, once the reaction is underway, it seems reasonable that the soaps should be able to support a fixed number of reaction loci regardless of their chain length. Superimposed upon these considerations is the fact that each of the reaction systems contained 5 parts *by weight* of soap; the actual number of molecules of micelle generator present therefore diminished as the chain length increased. Depending upon the chain length, each additional carbon atom reduces the number of molecules by between 5% and 10%.

Kolthoff, Meehan and Carr[22] have given extensive results which show the effect upon the conversion–time relationship of varying the level of soap in the standard GR-S recipe at 50°C. Two typical sets of data are shown in Fig. 7.9. At (a) are given the curves for a pure soap, potassium laurate, whereas at (b) are given curves for a commercial soap, 'Rubber Reserve' soap, which comprised a mixture of sodium stearate, sodium

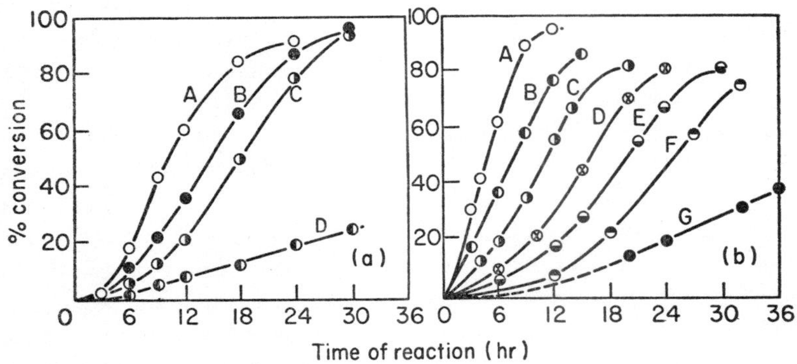

FIG. 7.9. (a) Effect of amount of potassium laurate upon conversion–time relationship for standard GR-S system reacted at 50°C, according to Kolthoff, Meehan and Carr.[22] Letters appended to curves refer to level of soap as follows: A, 5 pphm; B, 5/2 pphm; C, 5/4 pphm; D, 5/8 pphm. (b) Effect of amount of Rubber Reserve soap (sodium stearate, palmitate and oleate) upon conversion–time relationship for standard GR-S recipe reacted at 50°C, according to Kolthoff, Meehan and Carr.[22] Letters appended to curves refer to level of soap as follows: A, 10 pphm; B, 5 pphm; C, 5/2 pphm; D, 5/4 pphm; E, 5/8 pphm; F, 5/16 pphm; G, 5/32 pphm.

palmitate and sodium oleate. Qualitatively-similar features are observed with all fatty-acid soaps. The initial rate of conversion becomes less as the amount of soap is decreased, the initial rate being approximately proportional to the amount of soap. At conversions beyond about 25%, the amount of soap ceases to have an effect on the course of the reaction, and the subsequent rates are approximately equal, regardless of the initial amount of soap. (Actually, more precisely, the rate decreases slightly with decreasing initial amounts of soap.)

In the case of potassium laurate, the initial rate with 5/8 pphm of soap is much less than would be expected by extrapolation of the initial rates for systems containing 5, 5/2 and 5/4 pphm of soap. For a system containing 5/8 pphm of soap, the molar concentration of soap in the aqueous phase is 0·015 mole l^{-1}. This is considerably below the value of 0·024 mole l^{-1} for the critical micelle concentration of potassium laurate (see Table 7.1), even allowing for the effect of the potassium persulphate in depressing the critical micelle concentration through its action as an electrolyte. These considerations clearly suggest that the low rate of polymerisation observed in the case of 5/8 pphm of potassium laurate might be due to the absence of soap in micellar form. But in accepting this explanation, it should be noted that, although micelles can be presumed to have been essentially absent from the system containing 5/8 pphm of potassium laurate, the observed polymerisation rate was certainly not zero.

A similar effect was observed by Kolthoff *et al.*[22] with sodium caprate. In this case, since the critical micelle concentration is higher (about 0·1 mole l^{-1}, see Table 7.1), the sharp decrease in initial rate occurs at higher soap levels (between 5 and 5/2 pphm of soap).

Some confirmation of the essential correctness of this explanation for the sharp fall in rate as the soap level is decreased comes from the observation of Kolthoff *et al.*[22] that the addition of a simple electrolyte, potassium chloride, to a system containing 5/32 pphm of potassium myristate caused the initial rate to be increased several-fold. In the absence of the electrolyte, the critical micelle concentration was not attained, and in consequence the initial rate was low. Addition of the electrolyte lowered the critical micelle concentration so that it was attained, and therefore part of the soap was present in micellar form. These observations are certainly consistent with the view that the presence of surface-active molecules in micellar form enhances the rate of polymerisation, even if the presence of such micelles is not essential to the occurrence of polymerisation.

From time to time, the suggestion has been made that, by adding relatively large quantities of a simple electrolyte, relatively inefficient soaps of shorter chain length (*e.g.* caprylate or caprate) can be made into efficient micelle generators. Theoretically, the possibility exists of producing by this method a latex which could be drastically destabilised by merely adding water sufficient to ensure that the critical micelle concentration in the diluted electrolyte solution is not exceeded. It is not known if this idea has ever been explored practically. A further reference to this phenomenon will be found in Section 9.3 of Chapter 9.

Kolthoff, Meehan and Carr[22] have provided additional apparently-convincing evidence of the importance of solubilising power in determining the initial rate of polymerisation in the standard GR-S polymerisation system. As a measure of solubilising power, they have taken the solubility of dimethylaminoazobenzene in an aqueous solution of the soap. They have shown that the solubility of styrene in such solutions is directly proportional to the solubility of dimethylaminoazobenzene, and so have concluded that the solubility of the latter substance can be taken as a measure of the tendency to solubilise the type of monomers present in the

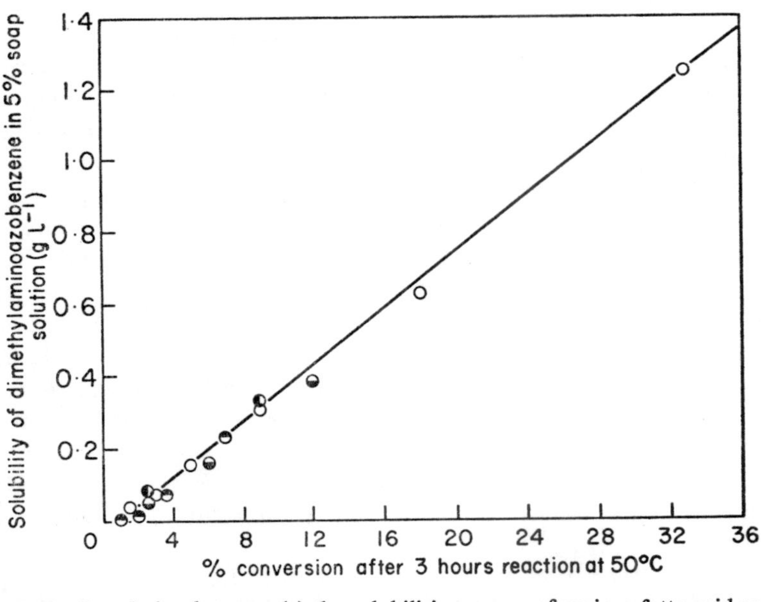

FIG. 7.10. Correlation between (a) the solubilising power of various fatty-acid soaps for dimethylaminoazobenzene and (b) the initial rate of polymerisation in the standard GR-S system reacted at 50°C using these soaps, according to Kolthoff, Meehan and Carr.[22] Significance of points: ○ Rubber Reserve soap; ◒ potassium myristate; ◓ potassium laurate; ◐ sodium caprate.

GR-S system. As a measure of rate of polymerisation, they took the conversion after 3 h reaction. The time of 3 h was chosen arbitrarily to represent the initial stage of the reaction. They argue that, if a longer time had been chosen, the effect of the polymer particles would have become important, whereas if a shorter time had been selected, the experimental uncertainty for the conversions would have been greater. The correlation is shown in Fig. 7.10, from which it is evident that a surprisingly good linear relationship exists between solubilising power and conversion after 3 h. The intercept of the line at zero solubilising power corresponds to about 0.8% conversion in 3 h; this is said by Kolthoff, Meehan and Carr to be of the correct order of magnitude for the polymerisation rate in the absence of soap, or in the presence of soap in non-micellar form.

Smith[23] has given data for the effect of soap level upon the number of particles formed in the course of a styrene emulsion polymerisation. Potassium persulphate (0·175% on the aqueous phase) was used as the initiator, and the amount of styrene used was such as to give a latex which contained 35% of polystyrene. The type of soap used is not specified, but it is presumed from other work reported by Smith that it was of the saturated fatty-acid type. The polymerisation temperature was 50°C. The reaction was carried to 90–95% conversion. The number of particles formed was obtained from the average polymer content of the latex and the average volume of the polystyrene particles. The latter were obtained from particle-size determination by electron microscopy. The results are shown in Fig. 7.11, where the number of particles formed per gram is plotted against the soap concentration in the initial aqueous phase, both scales being logarithmic. It will be recalled (see Chapter 4) that the Smith–Ewart theory predicts direct proportionality between the number of particles formed per gram of soap and the reciprocal 2/5th power of the soap concentration. On logarithmic scales, the relationship should therefore be linear and of slope $-2/5$. A line of this slope is superimposed upon Fig. 7.11. It can be seen that, insofar as the rather scattered data permit any quantitative conclusions to be drawn, a line of slope $-2/5$ can be said to give a not unreasonable representation. It can therefore be claimed that these results are not inconsistent with the requirements of the Smith–Ewart theory.

Further information on the effect of level of fatty-acid soap upon the number and size of particles formed by emulsion polymerisation has been given in a series of papers by Morton, Salatiello and Landfield.[24–26] Some results are summarised in Tables 7.13 and 7.14. In the former are given results for the persulphate-initiated emulsion polymerisation of butadiene

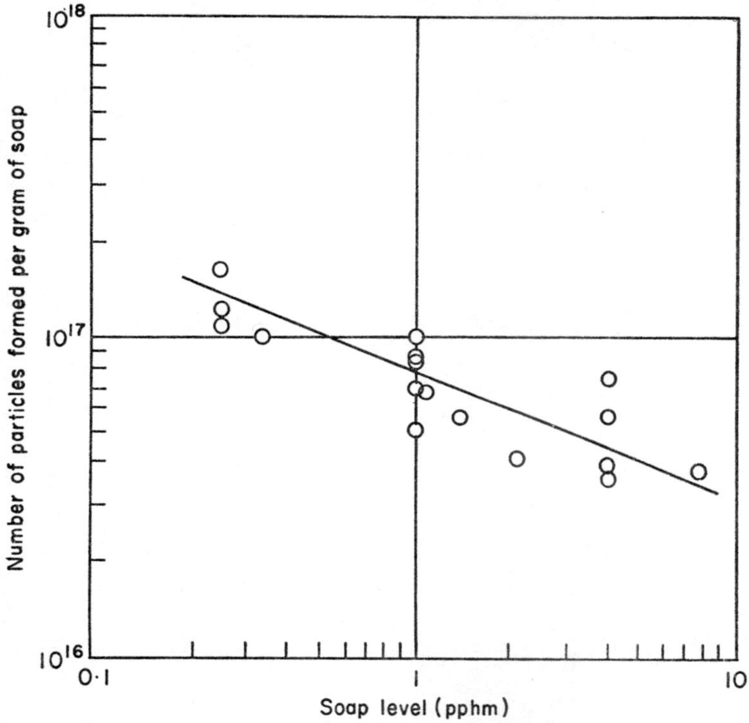

FIG. 7.11. Effect of soap level upon the number of particles formed per unit weight of soap in persulphate-initiated emulsion polymerisation of styrene at 50°C, according to Smith.[25] (See text for details of reaction system and procedure.)

at 50°C. It can be seen that, although the number of particles increases with increasing soap level, the rate of polymerisation per particle does not show much change over a fourfold change of soap level. Morton, Salatiello and Landfield state that the number of particles formed per gram of soap is more consistent with an exponent of $-0{\cdot}5$ than with exponent of $-2/5$ predicted by the Smith–Ewart theory. It is doubtful, however, whether there are here sufficient data to allow discrimination between an exponent of $-0{\cdot}4$ and one of $-0{\cdot}5$. Other significant points which are illustrated by the results given in Table 7.13 are the following: (a) the use of higher levels of soap results not only in a faster polymerisation but also in the formation of a latex of smaller particle size; and (b) comparing the number of particles per millilitre of aqueous phase at lower and higher conversions, it is seen that there is evidence of particle agglomeration and coalescence as the reaction proceeds.

TABLE 7.13

Effect of soap level upon number and size of particles formed in emulsion polymerisation of butadiene at 50°C (according to Morton, Salatiello and Landfield[24])

(reaction system: butadiene 100, water 180, potassium Rubber Reserve soap—variable, mercaptan 0·18, $K_2S_2O_8$ 0·3)

Soap level (pphm)	1·4		2·8		5·6
Rate of polymerisation (% per hour)	3·74		4·83		7·66
% Conversion	32·6	77·0	29·4	65·0	62·9
Particle diameter (Å)	490	790	440	660	590
Particle diameter (Å) (calculated)	360	580	330	480	430
Number of particles per ml of aqueous phase	$7·0 \times 10^{15}$	$4·4 \times 10^{15}$	$9·2 \times 10^{15}$	$6·7 \times 10^{15}$	$8·8 \times 10^{15}$
Number of particles per gram of soap	$5·0 \times 10^{17}$	$3·2 \times 10^{17}$	$3·3 \times 10^{17}$	$2·4 \times 10^{17}$	$1·6 \times 10^{17}$
Rate of polymerisation per particle (g sec^{-1})	—	$1·31 \times 10^{-21}$	—	$1·11 \times 10^{-21}$	$1·34 \times 10^{-21}$

TABLE 7.14

Effect of soap level upon number and size of particles formed in emulsion polymerisation of styrene initiated by sulphite-activated persulphate, and by a hydroperoxide–polyamine redox couple (according to Morton, Salatiello and Landfield[25])

(reaction systems: (A) styrene 100, water 180, potassium Rubber Reserve soap—variable, $K_2S_2O_8$ 0·3, Na_2SO_3 0·05; (B) styrene 100, water 180, potassium Rubber Reserve soap—variable, cumene hydroperoxide 0·1, triethylenetetramine 0·1, sodium ethylenediamine tetraäcetate 0·05, KCl 1·0)

Reaction system	A	A		B		B	
Soap level (pphm)	5	1		5		2	
Initial initiator charge	0·3	0·3		0·1		0·1	
Injected initiator charge	—	—		—		0·1	
Initiation temperature	40	40		5		5	
Propagation temperature (°C)	40	40	50	5	15	5	15
% Conversion	71·6	52·4	62·1	60·6	81·1	67·0	60·6
Rate of polymerisation (% per hour)	34·5	13·3	26·5	13·5	23·3	9·2	14·3
Particle diameter (Å)	1260	1640	1520	910	890	1150	1130
Particle diameter (Å) (calculated)	930	1200	1120	670	660	840	830
Number of particles per ml of aqueous phase	$0{\cdot}90\times10^{15}$	$0{\cdot}30\times10^{15}$	$0{\cdot}43\times10^{15}$	$2{\cdot}0\times10^{15}$	$2{\cdot}9\times10^{15}$	$1{\cdot}18\times10^{15}$	$1{\cdot}05\times10^{15}$
Rate of polymerisation per unit volume of aqueous phase (g sec^{-1}ml^{-1})	$5{\cdot}3\times10^{-5}$	$2{\cdot}1\times10^{-5}$	$4{\cdot}1\times10^{-5}$	$2{\cdot}1\times10^{-5}$	$3{\cdot}6\times10^{-5}$	$1{\cdot}42\times10^{-5}$	$2{\cdot}20\times10^{-5}$
Rate of polymerisation per particle (g sec^{-1})	$5{\cdot}8\times10^{-20}$	$6{\cdot}5\times10^{-20}$	$9{\cdot}2\times10^{-20}$	$1{\cdot}05\times10^{-20}$	$1{\cdot}24\times10^{-20}$	$1{\cdot}2\times10^{-20}$	$2{\cdot}1\times10^{-20}$

Table 7.14 gives similar information for the emulsion polymerisation of styrene initiated by both sulphite-activated persulphate and by a hydroperoxide–polyamine redox couple. (The sulphite activator was introduced in order to eliminate induction periods.) The number of particles produced per unit volume of aqueous phase by a given weight of soap is considerably less than for butadiene, but the rate of polymerisation per particle is much greater. As expected, the rate per particle is greater at 40° and 50°C than at 5° and 15°C. But, again, for a given temperature and initiation system, the rate per particle is almost independent of soap level. It is again evident that a higher level of soap produces a final latex of smaller particle size.

It appears from these results that the concentration of particles present in the final latex produced by reaction systems of the types under consideration is of the order of from 10^{15} to 10^{16} per millilitre of aqueous phase. Assuming the number of micelles in the initial aqueous phase to have been of the order of 10^{18} per millilitre, this means that the eventual number of particles (and therefore presumably of reaction loci) is between two and three orders of magnitude less than the initial number of micelles. Thus the probability of any given micelle becoming a latex particle is very small. Furthermore, at 50°C and the persulphate level used in the standard GR-S recipe, the rate of production of radicals is about 10^{13} per second per millilitre of aqueous phase. If each radical were to create one polymerisation locus, this would imply that the period of locus nucleation should last between 10^2 and 10^3 seconds, that is, from about 2 to 20 minutes.

Considerable interest has been shown in the effect upon polymerisation rate of adventitious impurities in the fatty-acid soap used. Particular attention has been paid to the effects of the soaps derived from unsaturated acids. Thus Wilson and Pfau[27] found that linoleate (I) and linolenate (II) soaps retard the copolymerisation of styrene and butadiene in the standard

$$CH_3(CH_2)_4CH{:}CHCH_2CH{:}CH(CH_2)_7CO_2^-$$

(I)

$$CH_3CH_3CH{:}CHCH_2CH{:}CHCH_2CH{:}CH(CH_2)_7CO_2^-$$

(II)

GRS polymerisation system. The replacement of up to 10% of sodium palmitate by sodium linoleate was found to result in a reduction of the conversion after 12 h of 1·4% for each 1% of sodium palmitate replaced. Linolenate soaps were observed to exert approximately three times as great a retarding effect as the same amount of linoleate soap. When the two double bonds in linoleic acid were isomerised so as to become conjugated, the ability to retard was lost. The general conclusion was reached that soaps containing one or more activated methylene groups of the

$$\text{—CH=CH—CH}_2\text{—CH=CH—} \qquad \text{(III)}$$

type(III), such as linoleates and linolenates, are retarders. This view was confirmed by the observation that 1,4-pentadiene retards emulsion polymerisation to an extent similar to that shown by linolenates. Presumably the retarding action is brought about by the susceptibility of the allylic methylene hydrogen to abstraction by a growing free radical, together with stabilisation of the resulting soap radical by resonance as follows:

$$\text{—CH=CH—}\dot{\text{C}}\text{H—CH=CH—}$$

$$\text{—CH=CH—CH=CH—}\dot{\text{C}}\text{H—} \leftrightarrow \text{—CH=CH—}\dot{\text{C}}\text{H—CH=CH—} \leftrightarrow \text{—}\dot{\text{C}}\text{H—CH=CH—CH=CH—}$$

TABLE 7.15

Effect of partial replacement of sodium palmitate by (a) sodium linoleate and (b) sodium linolenate upon conversion in the standard GR-S system after 12 h at 50°C (according to Wilson and Pfau[27])

% sodium palmitate	*% sodium linoleate or sodium linolenate*	*% conversion*	
		Linoleate	*Linolenate*†
100	0	81·5	79·5
99	1		75
98	2	79	73
97	3		67
96	4	76·5	67
95	5		63
94	6	73·5	61
93	7		48
92	8	71	54
91	1		
90	10	69	43,51
100‡		79	

† The linolenate was a mixture of 80% linolenate and 20% linoleate.
‡ The total amount of sodium palmitate was reduced by 10%. Comparison of this result with those for the 90/10 blends shows that the reduction in conversion in the latter cases was not due merely to reduction in the amount of sodium palmitate, but also to the retarding influence of the unsaturated soap.

The situation is probably analogous to that of the so-called 'degradative chain transfer' which occurs during the free-radical polymerisation of allylic compounds. Some of the observations of Wilson and Pfau on the effect of linoleate and linolenate soaps are summarised in Table 7.15.

These observations have been confirmed and supplemented by Carr *et al.*[28] In Fig. 7.12 are reproduced conversion–time curves for standard GR-S systems containing potassium palmitate and a mixture of potassium palmitate and sodium linoleate. These curves show clearly that the linoleate functions as a retarder rather than as an inhibitor. There is no induction period, and the rate of polymerisation is consistently lower than that of the control. The curve for a system in which part of the potassium palmitate was replaced by water shows again that the effect of the linoleate is not merely to dilute the palmitate.

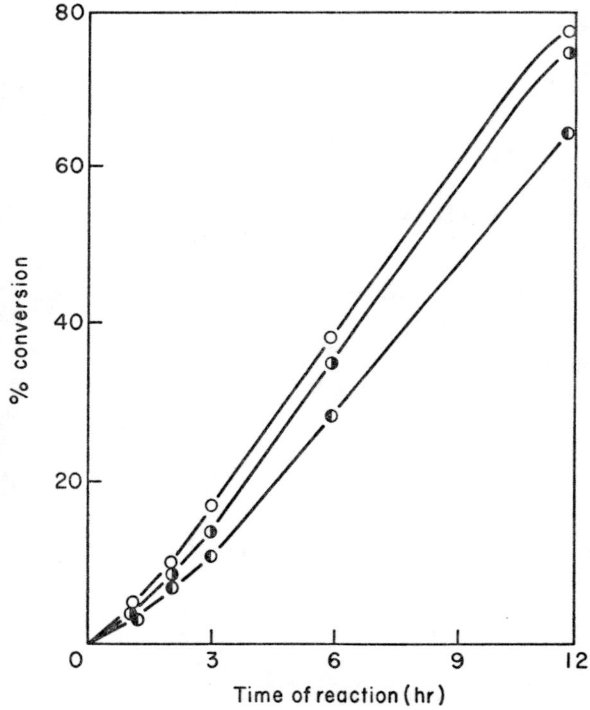

FIG. 7.12. Effect of partial replacement of potassium palmitate by sodium linoleate on conversion–time relationship for standard GR-S system reacted at 50°C, according to Carr *et al.*[28] Significance of points: ○ potassium palmitate; ◑ 90 parts of potassium palmitate + 10 parts of water; ◐ 91·3 parts of potassium palmitate + 8·7 parts of sodium linoleate.

TABLE 7.16

Effect of hydrogenation upon performance of tallow soaps in the standard GR-S system at 50°C (according to Carr *et al.*[28])

Sample	*Iodine value*†	*Thiocyanogen* value*	*Difference*	*% conversion after* 12 h
A	51·5	46·8	4·7	72
A	41·1	38·4	2·7	76
A	32·6	30·9	1·7	77
A	20·8	19·2	1·6	78
B	59·1	50·9	8·2	67
B	39·7	37·2	2·5	76
B	31·0	29·4	1·6	77
B	22·2	20·7	1·5	78
Control	—	—	—	77

* See footnote in text for the significance of iodine and thiocyanogen values.

Carr and his co-workers have further shown that linoleate soaps exert a retarding effect when tertiary mercaptans are used as modifiers, as well as when primary mercaptans are used. They have also shown that potassium palmitate, potassium stearate and sodium oleate all give the same conversion after 12 h at 50°C, so that the one double bond in the oleate has no retarding effect. Furthermore, the retarding effect of poly-unsaturated soaps can be eliminated by hydrogenation. The results given in Table 7.16† are for soaps prepared from two samples of tallow which had been subjected to varying degrees of hydrogenation at 150°C using a nickel catalyst. As these results show, full hydrogenation is by no means necessary in order to attain a rate of polymerisation equal to that of the control. Carr and his co-workers state that 'Apparently hydrogenation to an iodine value of approximately 40 is sufficient to give a conversion comparable to that of the control. Selective hydrogenation of linoleate and linolenate in the fats occurs. Once these substances are eliminated, no further hydrogenation is necessary, and the fats yield soaps with which normal rates of conversion are obtained.'

Carr *et al.*[28] have also reported on the effect of traces of lecithin and tocopherol upon the performance of a fatty-acid soap in the standard GR-S system. Both these substances are possible impurities which might be

† Reference to Table 7.16 will show that the soap acids are characterised by both iodine value and thiocyanogen value. The former is a measure of total unsaturation, and the latter a measure of unsaturation which is present as linoleic and linolenic compounds.

TABLE 7.17

Effect of addition of lecithin and tocopherols to the standard GR-S system containing Rubber Reserve soap (according to Carr *et al.*[28])

Type and amount of additive (parts per 100 parts of soap)		*% conversion after following times at 50°C*		
		3 h	6 h	12 h
0		16	37	78
Lecithin	0·2	15	36	77
Lecithin	0·4	16	37	78
Tocopherol	0·01	15	35	77
Tocopherol	0·1	10	32	72

found in a commercial fatty-acid soap. The results are given in Table 7.17. The conclusion is that lecithin and tocopherol, in the amounts in which they may occur in commercial soaps, do not affect the rate of polymerisation. Even excessive amounts of lecithin do not affect conversion, but excessive amounts of tocopherol decrease conversion by imparting an induction period. However, in the latter case, the retardation proceeds normally after the induction period is completed. Thus tocopherol acts as an inhibitor rather than as a retarder.

It must be emphasised that the foregoing remarks concerning the effects of soap unsaturation refer exclusively to performance in styrene–butadiene emulsion copolymerisation systems. That the effects observed in one system cannot necessarily be used to make inferences concerning behaviour in another is illustrated by the fact that Hopff and Fákla[41] have recently shown that sodium oleate seriously retards the emulsion polymerisation of vinyl chloride at 40°C relative to, say, sodium stearate. They have further reported that, whilst sodium linoleate has an even more marked retarding effect, its effect is not greatly different from that shown by sodium oleate.

Few results seem to be available showing the effect of changing the counterion of a fatty-acid soap upon its performance in an emulsion polymerisation recipe. Kuznetsov and Lebedev[29] have published results which indicate that, as far as the particle size of the final latex is concerned, it makes little difference whether one uses the sodium, potassium or ammonium soap of a fatty acid. But if the lithium soap is used, then the final particle size is approximately double that given by the other soaps.

An extensive study of the use of lithium soaps for controlling the particle size of styrene–butadiene copolymer latices obtained by emulsion polymerisation has been published by Serafini and Bobalek.[30] Results are presented to show that, although lithium soaps form micelles, the solubility of the soap in water only just exceeds the critical micelle concentration. This appears to be especially the case for lithium stearate. The critical micelle concentrations reported for lithium oleate and stearate are $1{\cdot}0 \times 10^{-3}$ mole l^{-1} and $2{\cdot}7 \times 10^{-4}$ mole l^{-1} respectively. It is shown that uniformity of particle size can be attained by restricting the development of new particles to the early stages of the reaction by using lithium soaps which have a limited capacity for micellar nucleation, even when used in quantities sufficient to stabilise the resultant latex. The solubility properties of lithium stearate in particular are such as to provide a self-regulating mechanism for control of micellar nucleation of new polymer particles. Control of the number of soap micelles, and thereby of the number of particles nucleated, the rate of polymerisation, and the ultimate particle size of the latex formed, can be effected by varying the concentration of added lithium ions.

Serafini and Bobalek emphasise that this approach to the control of uniformity and size of latex particles is applicable only to systems in which micelles are believed to be the main sites where new particles originate. When one or more of the monomers is appreciably water-soluble, as in the case of, say, styrene–acrylonitrile copolymerisation, then nucleation occurs in both micelles and in the true aqueous phase. This complicates the problem of control of particle size. Serafini and Bobalek give no information on the rates of polymerisation which can be achieved using lithium soaps. It is suspected that, as a consequence of rather low soap solubility, these would be rather low.

The use of synthetic branched-chain carboxylate soaps for the emulsion copolymerisation of styrene and butadiene is the subject of a paper by Duck, Waterman and La Heij.[31] These acids are prepared by reaction of mono-olefins with either formic acid or carbon monoxide and water. They are fully saturated with the carboxyl functionality attached mainly to a quaternary carbon atom. The branched-chain structure renders their aqueous alkali soap solutions less susceptible to gelation. Their soaps have been considered especially as replacements for rosin-acid soaps in low-temperature systems, where the separation of soap gel on cooling can be a problem. The evaluation of these soaps was carried out in a low-temperature 29/71 styrene–butadiene system, in which initiation was effected by a combination of *p*-methane hydroperoxide, a complexed iron salt, and sodium formaldehyde sulphoxylate. These workers found that, as the molecular size of the synthetic soap is varied, the highest rate of polymerisation is obtained using the soap derived from the C_{15}–C_{19} acid fraction.

With the soaps from the C_{26} and the C_{11}–C_{14} fractions, the polymerisation rates are still high but are lower than those obtained with the C_{15}–C_{19} fraction. The C_9–C_{11} acid fraction gives soaps having very poor efficiency in these emulsion polymerisation systems. Duck and his co-workers concluded that the optimum molecular size corresponds to a soap containing about 17 carbon atoms. They draw attention to the fact that the number of carbon atoms in the abietic acid structure is 20.

Table 7.18 gives comparative figures for the various structural groups present in both the synthetic acids and in abietic acid. The general similarities between the two structures are brought out in this table. Evidence is also presented by Duck, Waterman and La Heij to show that the sodium soaps of the C_{15}–C_{19} synthetic acids retain their water-solubility at low temperatures much better than does, say, sodium stearate, dilute solutions of which tend to gel rapidly below about 34°C. By comparison, a solution of the C_{15}–C_{19} synthetic acids was found to remain fluid and to show no tendency to gel even down to 5°C. The low-temperature solubility behaviour of the synthetic soaps is said to be better even than that of the rosin-acid soaps. Another advantage claimed for the synthetic acids is that, due to their branched structure, their calcium and magnesium salts are more soluble than are those of, say, stearic acid. They should therefore show less tendency to be precipitated from hard water than do conventional fatty-acid soaps from natural sources.

TABLE 7.18

Comparison of structural group analysis of C_{15}–C_{19} synthetic carboxylic acid and abietic acid (according to Duck, Waterman and La Heij[31])

	Number of groups per molecule	
Group	*Synthetic acids*†	*Disproportionated rosin acids*‡
CO_2H	1·0	1·0
CH_3	3·4	4·0
CH_2	9·3	7·0
CH	0·2	4·3
Quaternary C	2·8	3·6
C atoms	16·7	20
Molecular weight	284	302

† Experimentally determined.

‡ Calculated assuming composition to be an equimolar mixture of dehydroabietic, dihydroabietic and tetrahydroabietic acids.

The effect upon polymerisation rate of varying the level of synthetic C_{15}–C_{19} soap is shown in Fig. 7.13. At (a) rate is plotted as a function of soap level, and points for a similar system containing a rosin-acid soap are included for comparison. It is seen that, for a given level of soap, the C_{15}–C_{19} synthetic soap gives a higher rate of conversion than does the rosin-acid soap. Thus it should be possible to achieve a given rate of polymerisation

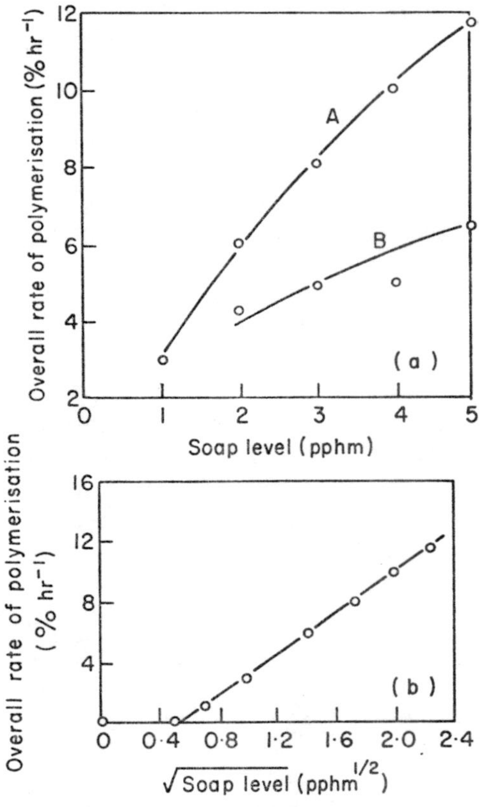

FIG. 7.13. (a) Effect of soap level upon rate of polymerisation in low-temperature (5°C) hydroperoxide–polyamine-initiated styrene–butadiene emulsion copolymerisation systems containing (A) C_{15}–C_{19} synthetic carboxylate sodium soap, and (B) potassium soap of disproportionated rosin acid, according to Duck, Waterman and La Heij.[31] (b) Data for C_{15}–C_{19} synthetic carboxylate sodium soap plotted to show rate of polymerisation as a linear function of square root of soap level. (Reaction system: styrene 29, butadiene 71, water 200, soap—variable, non-ionogenic soap 0·02, *t*-dodecyl mercaptan 0·20–0·23, *p*-menthane hydroperoxide 0·036, sodium formaldehyde sulphoxylate dihydrate 0·03, $FeSO_4.7H_2O$ 0·012, ethylenediamine tetraäctic acid 0·025, $Na_3PO_4.12H_2O$ 0·70.)

using a lower level of the synthetic soap than of the rosin-acid soap. In Fig. 7.13 (b), the results for the synthetic soap are plotted to show that the rate of polymerisation increases linearly with the square root of the soap level. It also shows that no polymerisation is observed below about 0·4 pphm of soap. This is presumably the minimum amount of soap required to saturate the water and also the monomer–aqueous phase interface, before micelle formation can commence. It is of interest to note that this is one of the few instances recorded in the published literature of the complete absence of polymerisation when the soap is present in an amount lower than that presumably required for micelle formation.

Duck, Waterman and La Heij also report that they have carried out polymerisations with the potassium, ammonium and lithium soaps of the C_{15}–C_{19} synthetic acids, as well as with the sodium soap. The polymerisation rates observed with the four soaps under the same conditions were virtually equal. The latices produced all had similar high surface free energies (59–63 ergs cm^{-2}), from which it was concluded that they all contained small particles of roughly the same size.

In concluding this discussion of the behaviour of fatty-acid soaps in emulsion polymerisation systems, it is appropriate to point out that the soap used can have a profound effect upon the properties of the latex which is produced. This is illustrated by the results of Kuznetsov and Lebedev[32] for the flow properties of styrene–butadiene copolymer latices prepared using potassium alkanoates of increasing chain lengths. Curves for viscosity as a function of volume fraction of contained polymer are given in Fig. 7.14. From this diagram, it is seen that the viscosity at any given polymer content increases progressively as the chain length of the soap is increased. The authors attributed this to an increase in the effective hydrodynamic size of the latex particles as the size of the soap molecule is increased. Results for the effect of chain length of soap upon the viscosity of polyvinyl chloride latices made using sodium fatty-acid soaps have been reported by Hopff and Fákla.[41] They have found that, instead of a progressive rise with increasing chain length, the viscosity is relatively low with the caprate, is rather higher with laurate, falls very slightly with myristate and palmitate, and rises to the highest value with stearate.

A further example of the way in which the choice of fatty-acid soap can affect the properties of the latex eventually produced is to be found in the ammonium and amine soaps. As far as performance during the actual emulsion polymerisation is concerned, they appear in general to be little different from the corresponding alkali–metal soap. But in the final latex they are fugitive when the latex dries down to a film. As water is lost from

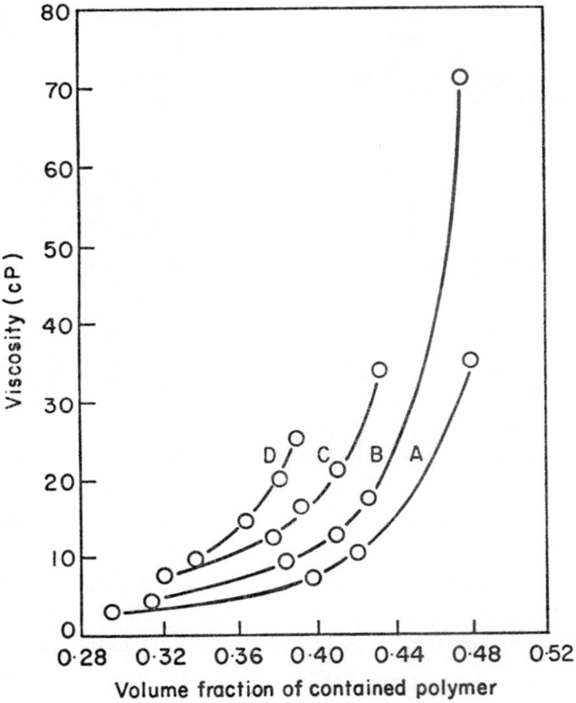

FIG. 7.14. Effect of chain length of fatty-acid soap upon relationship between low-shear viscosity and polymer content for styrene–butadiene copolymer latices produced by emulsion polymerisation, according to Kuznetsov and Lebedev.[32] Curves A, B, C and D are respectively for potassium laurate, potassium myristate, potassium palmitate and potassium stearate.

the latex by evaporation, so too is ammonia or the amine, especially if the temperature is raised. Ultimately only the free fatty acid is left. The latex film is therefore less sensitive to water than if an alkali–metal soap had been used. It is, however, still sensitive to alkaline solutions. Such soaps are sometimes known as *volatile-base* soaps. Ammonia is the most common volatile base, but is rather too volatile for some purposes. Morpholine and triethanolamine are less volatile alternatives.

7.6 ROSIN-ACID SOAPS

Rosin (also known as *colophony*) is a resinous substance which is obtained from certain pine trees, either by the solvent extraction of wood stumps, or by wounding the living tree and collecting the exudate (this is the so-called *turpentining* process). The product of the first process is known as *wood rosin;* the exudate obtained by the second process consists of a mixture of rosin and turpentine, and its rosin fraction is known as *gum rosin.*

Rosin is a translucent, brittle substance at room temperature. It is graded by colour, which may vary from yellow to brown. It is insoluble in water, but dissolves readily in organic solvents. It also dissolves in strong alkalis to give solutions of rosin-acid soaps or *resinates.*

In chemical composition, rosin comprises some 90% of the so-called *resin acids* and about 10% of non-acidic materials. The term 'resin acid' is used to denote the individual acids, and thus to distinguish them from the mixture, which is known as *rosin acid.* The non-acidic fraction includes the esters of rosin and fatty acids, together with various hydrocarbons, including terpenes. The acid fraction of rosin is a complex mixture of acidic materials, the main components having the molecular formula $C_{19}H_{29}CO_2H$ and containing two carbon–carbon double bonds per molecule. These main components fall into two broad chemical types, the abietic acid type, in which the two double bonds are conjugated, and the pimaric acid type, in which they are not. In the first group are abietic acid (IV), levopimaric acid (V), neoabietic acid (VI), dehydroabietic acid (VII), dihydroabietic acid (VIII) and tetrahydroabietic acid (IX). In the second group are dextropimaric acid (X) and *iso*dextropimaric acid (XI). A review of the chemistry of rosin and its derivation has been published by Stone-

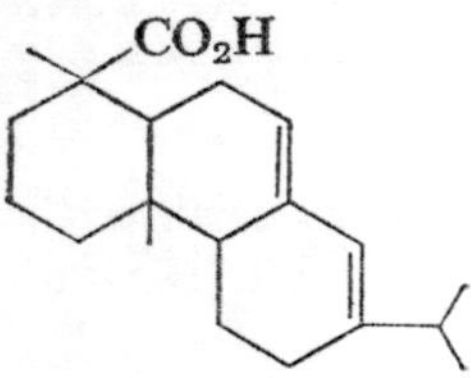

(IV)

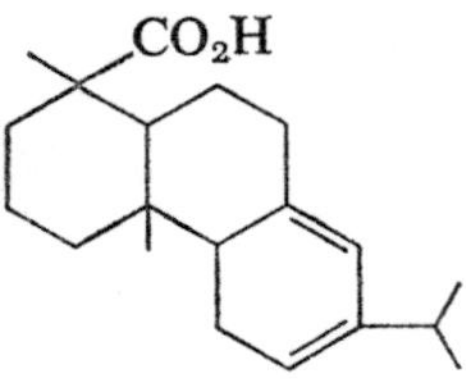

(V)

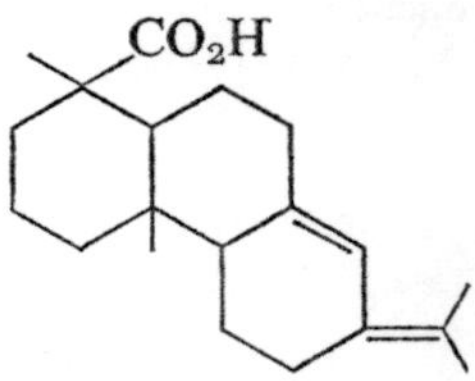

(VI)

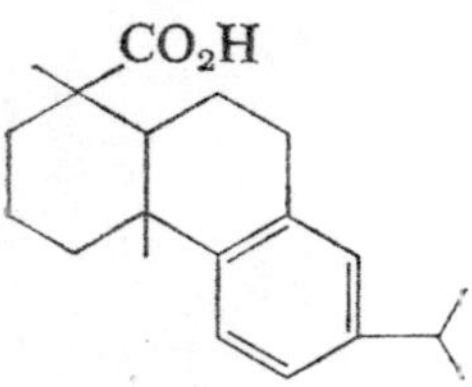

(VII)

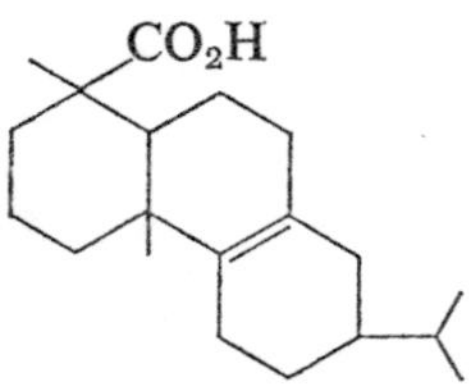

(VIII)

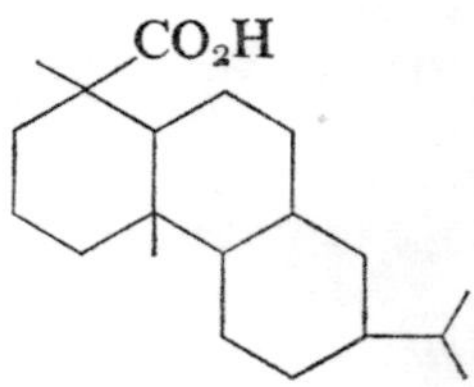

(IX)

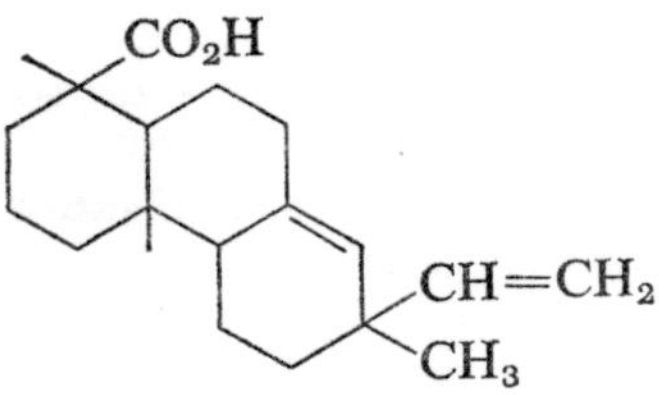

(X)

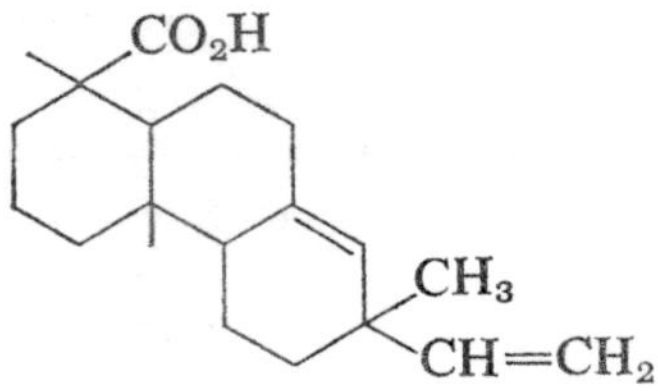

(XI)

cipher and Turner.[33] Studies of the behaviour of rosin-acid soaps in styrene–butadiene emulsion copolymerisation recipes have been published by Hays, Drake and Pratt,[34] and by Azorlosa.[35] In practice, rosin-acid soaps are mainly of interest in systems for the low-temperature copolymerisation of styrene and butadiene. They show much less tendency to form gels when aqueous solutions are cooled, and this is clearly a considerable advantage in expediting reaction at low temperatures.

It has been found that commercial rosins and rosin derivatives are themselves unsatisfactory for use in styrene–butadiene emulsion copolymerisation systems. Hays, Drake and Pratt[34] have shown that the low rates of conversion attained are mainly due to the presence of two types of detrimental substance: acids of the abietic type, and phenolic inhibitors. Their results for the conversions obtained in a persulphate-initiated system using various types of pure resin acid are summarised in Table 7.19. It is clear that the abietic-type acid, as exemplified by abietic and levopimaric acids, have a detrimental effect upon the yield, whereas the pimaric type, as exemplified by *iso*dextropimaric acid, have no such detrimental effect. These results also show that both hydrogenation and dehydrogenation of abietic acid convert it to a satisfactory material. Hays, Drake and Pratt conclude that the inhibiting effect of abietic-type acids is due to the presence of conjugated carbon–carbon double bonds. These must be destroyed if inhibitory effects are to be eliminated. Hydrogenation to either the di- or tetra-hydro acid obviously fulfils this requirement. So too does dehydrogenation, since it creates a third carbon–carbon double bond, and is accompanied by rearrangement to form an aromatic system (see structure (VII) above).

These observations on the relative effectiveness of soaps derived from the various resin acids hace been confirmed by Azorlosa.[35] He has published conversion–time curves for persulphate-initiated styrene–butadiene emulsion copolymerisation systems containing the various acids. These curves are shown in Fig. 7.15 (a), from which is evident on the one hand the strong inhibiting and retarding effect shown by the abietic-acid type

TABLE 7.19

Effectiveness of pure resin-acid soaps in persulphate-initiated emulsion copolymerisation of styrene and butadiene (according to Hays, Drake and Pratt[34])

(reaction system: styrene 25, butadiene 75, water 180, resin acid 5†, dodecyl mercaptan 0·5, potasium persulphate 0·3)

Acid	*% conversion after* 14 h *at* 50°C
Abietic	trace
Dehydroabietic	80
Tetrahydroabietic	82
Dihydroabietic	71
Disproportionated rosin (refined)	78
10% Levopimaric + 90% disproportionated rosin	40
10%*iso*dextropimaric + 90% dehydroabietic	80
Dihydrodextropimaric	79

† Whilst most of the work on resin acids reported in the literature was carried out using alkali resin-acid soaps, the work of Hays, Drake and Pratt was *apparently* carried out using the acids themselves. Their recipe quite specifically states 'rosin or resin acid', and contained no added alkali with which a soap could be formed. Again, they state that 'Polymerization tests on various pure *resin acids* ... showed that ...' (present writer's italics). On the other hand, their paper speaks in various places of disproportionated rosin *soap*, and, indeed, these words appear in the title to the paper. By contrast, Azorlosa[35] was most careful to point out that he used the sodium soaps derived from the various resin acids.

soaps, and, on the other, the virtual equivalence of the pimaric types, the hydrogenated abietates and the dehydrogenated abietates. In Fig. 7.15 (b) are shown the results of Azorlosa for the addition of small amounts of sodium abietate to an effective soap, sodium dehydroabietate. As might be expected from previous results, the addition of even small amounts of sodium abietate has the effect of markedly reducing the yield of polymer after any given time of reaction. Examination of the conversion–time curves shows that increasing the amount of abietate soap decreases the initial rate of polymerisation without changing the slope of the steady-rate portion of the curve appreciably. For this reason, Azorlosa regards sodium abietate as an inhibitor rather than a retarder. The relationship between the amount of sodium abietate and the conversion after 14 h reaction at 50°C is shown in Fig. 7.15 (c).

Whilst the conjugated double bonds of the abietic-type acids can be eliminated by both hydrogenation and dehydrogenation, a more convenient method is by *disproportionation.* According to Hays, Drake and Pratt,[34] this modification can be brought about by heating with a palladium catalyst. The effect is to bring about simultaneous hydrogenation of some

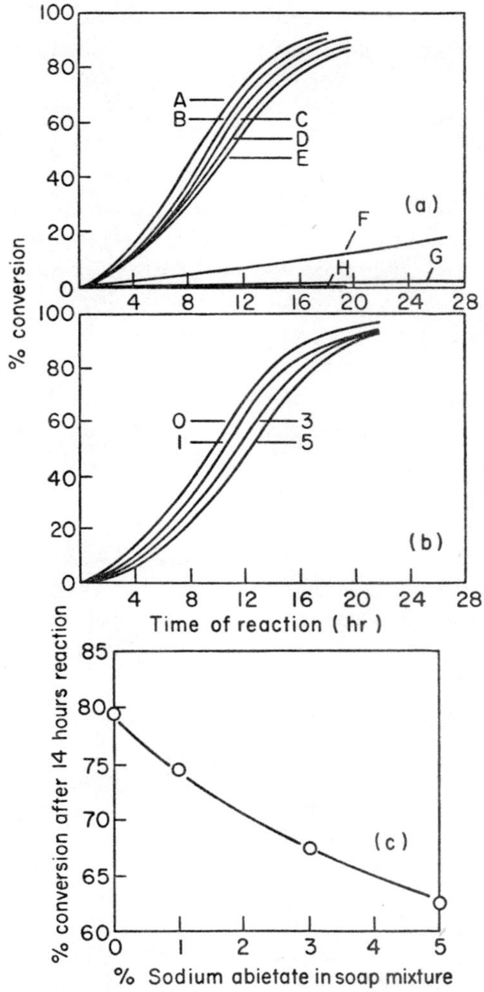

FIG. 7.15. (a) Conversion–time curves for persulphate-initiated emulsion copolymerisation of styrene and butadiene at 50°C using various sodium resin–acid soaps, according to Azorlosa.[35] The resin-acid soaps were as follows: tetrahydroabietate (A), dehydroabietate (B), dextropimarate (C), dihydroabietate (D), *iso*dextropimarate (E), abietate (F), *neo*abietate (G), levopimarate (H). (b) Conversion–time curves for persulphate-initiated emulsion copolymerisation of styrene and butadiene at 50°C using as soap sodium dehydroabietate containing various amounts of sodium abietate, according to Azorlosa.[35] The number appended to each curve gives the percentage of sodium abietate in the soap mixture. (c) Effect of sodium abietate upon conversion after 14 h at 50°C in persulphate-initiated emulsion copolymerisation of styrene and butadiene using sodium dehydroabietate as main soap, according to Azorlosa.[35] (Reaction system in each case: styrene 25, butadiene 75, water 180, sodium resinate 5, dodecyl mercaptan 0·5, $K_2S_2O_8$ 0·3.)

abietic acid molecules and dehydrogenation of others. It is apparently possible, by this means, completely to convert the abietic-type acids to the satisfactory hydro- and dehydro-acids. The pimaric-type acids probably also undergo disproportionation, but since neither they nor their disproportionation products are detrimental in styrene–butadiene emulsion copolymerisation systems, it appears that all the constituents of disproportionated rosin-acid soap should perform satisfactorily in this type of system. It is indeed clear from Fig. 7.15 (b) and (c) that the effectiveness of the disproportionation reaction as a means of obtaining a modified rosin soap satisfactory for use in emulsion polymerisation systems depends entirely upon the feasibility of removing virtually completely all acids of the abietic type.

Hays, Drake and Pratt have also shown that, in addition to retarding and inhibiting acids of the abietic type, rosin acids contain phenolic substances which function as inhibitors of emulsion polymerisation. Thus even when a sample of wood rosin has been disproportionated in such a manner as to be completely free from abietic-type acids, it does not usually give a satisfactory conversion rate unless it is further refined. Listed as possible methods of refining are (a) adsorption, (b) distillation, (c) crystallisation, and (d) solvent extraction. It is stated that the effectiveness of the resultant soap depends upon both the method of refining and, to some extent, the individual sample.

Various other methods have been proposed for eliminating the conjugated double bonds of the abietic-type acids. Thus Feldon, McKennon and Lawrence[36] have evaluated the use of the disodium soap of the ethyl ester of maleopimaric acid (XII) in the persulphate-initiated emulsion

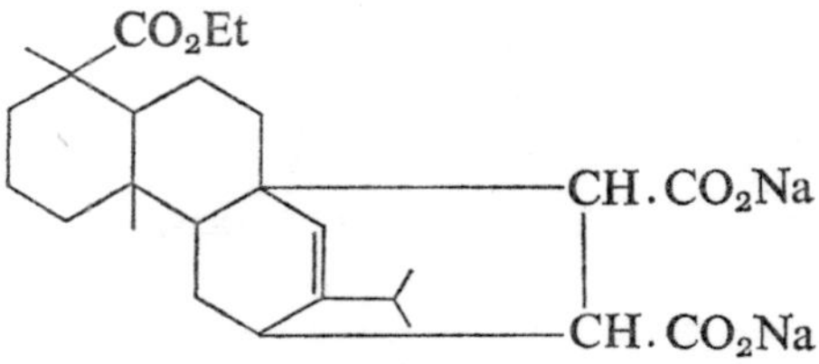

(XII)

copolymerisation of styrene and butadiene at 122°F. Maleopimaric acid is made by carying out a Diels–Alder reaction between maleic anhydride and abietic-type acids. As can be seen from the structure, the conjugated double-bonds have been eliminated by this process. The conclusion was reached that, using disodium ethyl maleopimarate, 16·4 h reaction is required to achieve 72% conversion as compared with 14.3 h for the same system containing disproportionated rosin-acid soap. The viscosity of the latices obtained using the ethyl maleopimarate soap was found to be similar to that of the control latices using disproportionated rosin-acid soap, whereas the chemical stability of the former was observed to be greater in two cases out of three.

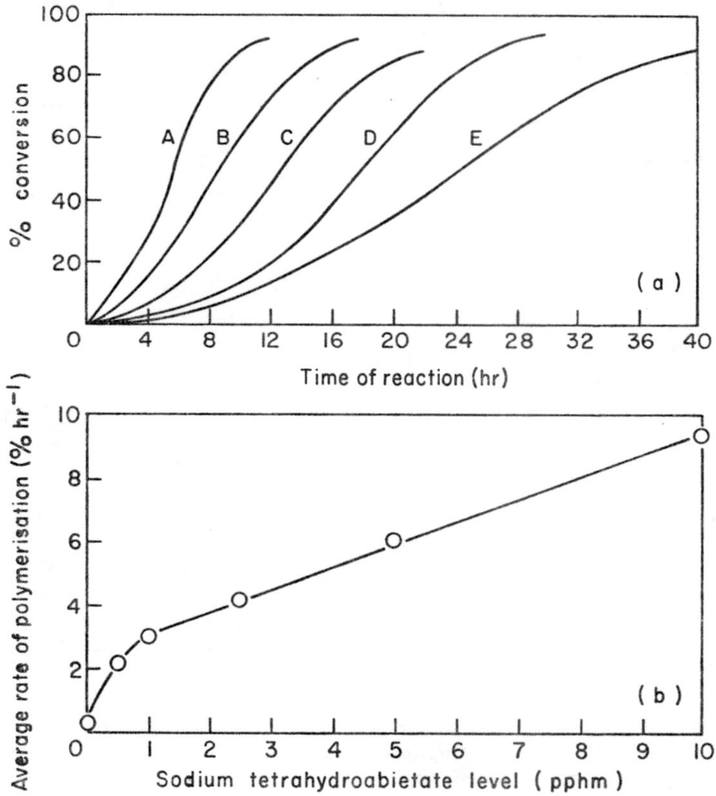

FIG. 7.16. (a) Effect of amount of sodium tetrahydroabietate upon conversion–time relationship for persulphate-initiated emulsion copolymerisation of styrene and butadiene at 50°C, according to Azorlosa.[35] Letters appended to curve refer to level of soap as follows: A, 10 pphm; B, 5 pphm; C, 2·5 pphm; D, 1 pphm; E. 0·5 pphm. (b) Relationship between average rate of polymerisation and level of sodium tetrahydroabietate in persulphate-initiated emulsion copolymerisation of styrene and butadiene at 50°C, according to Azorlosa.[35] (Reaction system: styrene 25, butadiene 75, water 180, sodium tetrahydroabietate—variable, dodecyl mercaptan 0·5, $K_2S_2O_8$ 0·3.)

The effect of the level of a resin-acid soap, sodium tetrahydroabietate, upon the conversion–time relationship for the persulphate-initiated emulsion copolymerisation of styrene and butadiene at 50°C has been reported by Azorlosa.[35] His results are shown in Fig. 7.16 (a). As in the case of the fatty-acid soaps (see Fig. 7.9), the conversion after a given time of reaction falls with decreasing amount of soap. But, unlike the fatty-acid soaps, the level of resin-acid soap affects the steady rate of polymerisation as well as the initial stage of the reaction. Plotting average rate of polymerisation (the nature of this average is unspecified) against the level of sodium tetrahydroabietate—Fig. 7.16 (b)—shows that the relationship is linear over the range from 1 to 10 pphm of soap. Below 1 part of soap, the rate was observed to fall off rapidly. This was interpreted as being due to the presence of insufficient soap to form micelles. (The critical micelle concentration of potassium dehydroabietate at 25.8°C has been reported as being between 2·5 and 3·2 × 10^{-2} mole l^{-1};[37] this corresponds to about 2 pphm of soap in a system of the standard GR-S type.) It should be noted that, according to these results, the relationship between polymerisation rate and soap level for rosin-acid soaps differs somewhat from that shown by the synthetic C_{15}–C_{19} fatty-acid soaps—see Fig. 7.13 (b)—which have been proposed by Duck, Waterman and La Heij[31] as alternatives for rosin-acid soaps.

For carrying out the emulsion polymerisation of certain monomers at

TABLE 7.20

Effect of varying ratio of rosin-acid soap to fatty-acid soap upon conversion–time relationship for styrene–butadiene emulsion copolymerisation at −10°C, initiated by a hydroperoxide–iron redox couple (according to Fryling and Follett[38])

(reaction system: styrene 30, butadiene 70, water 192, methanol 48, total soap 5, mixed tertiary mercaptans 0·25, di*iso*propylbenzene hydroperoxide 0·3 mmole, $FeSO_4.7H_2O$ 0·083 4, $Na_4P_2O_7.10H_2O$ 0·134, KCl 0·25)

Level of potassium soap of disproportionated rosin acid (pphm)	*Level of potassium fatty-acid soap* (pphm)	*% conversion after times shown*		
		4 h	7 h	24 h
5·0	0·0	0	6	16
4·0	1·0	14	25	60
3·5	1·5	21	40	83
2·5	2·5	27	49	86
1·5	3·5	26	49	86
1·0	4·0	37	63	92
0·0	5·0	33	55	92

low temperature, it is quite common practice to use mixtures of rosin-acid and fatty-acid soaps. Table 7.20 summarises the results of Fryling and Follett[38] concerning the effect of varying the ratio of disproportionated rosin-acid soap to fatty-acid soap upon the conversion–time relationship for styrene–butadiene emulsion copolymerisation at −10°C. The initiation system comprised a hydroperoxide–iron redox couple. Potassium soaps were used. It is seen that the rate of conversion is low when exclusively rosin-acid soap is used, but rises quite sharply as rosin-acid soap is replaced by fatty-acid soap. Fryling and Follett regarded a blend of 3·5 pphm of rosin-acid and 1·5 pphm of fatty-acid soap as giving a satisfactory conversion rate. They also observed that replacement of the potassium rosin-acid soap in this blend ratio by the corresponding sodium soap leads to extreme retardation of the reaction.

Fryling and Follett[38] have also published results which illustrate the effect of pH upon the conversion–time relationship for an emulsion copolymerisation of styrene and butadiene at −10°C, using as soap the 3·5/1·5 blend of disproportionated rosin-acid soap and fatty-acid soap. Their results are given in Table 7.21, from which it is seen that the maximum rate of conversion occurs at about pH 10.5, which corresponds almost exactly to 100% neutralisation. Over-neutralisation of the soap leads to marked

TABLE 7.21

Effect of pH upon conversion–time relationship for styrene–butadiene emulsion copolymerisation at −10°C, initiated by a hydroperoxide–iron redox couple (according to Fryling and Follett[38])

(reaction system: styrene 30, butadiene 70, water 192, methanol 48, potassium soap of disproportionated rosin acid 3·5, potassium fatty-acid soap 1·5, mixed tertiary mercaptans 0·25, di*iso*propylbenzene hydroperoxide 0·3 mmole, $FeSO_4.7H_2O$ 0·083 4, $Na_4P_2O_7.10H_2O$ 0·134, KCl 0·25, KOH—variable)

pH	KOH (ml *of* 0·1*N solution per* 100 g *of monomers*)	*% neutralisation of carboxylic acids*		*% conversion after times shown*			
		Rosin acid	*Total acid*	2 h	4 h	7 h	24 h
9·0	0·000	90	93	10	21	31	61
9·5	0·345	94	96	13	26	40	62
10·0	0·691	97	98	14	27	44	74
10·5	0·862	99	99+	15	29	46	77
11·0	1·250	103	102	13	25	38	69
11·5	1·730	108	105	15	23	29	53
12·0	2·580	118	113	12	16	19	22

retardation of the reaction. These results might be regarded as atypical, in that the system was polymerised at a sub-zero temperature, and contained methanol as an antifreeze. It is therefore of interest to note that Starkweather and his co-workers[39] have published curves for methanol-free oleate/resinate systems which show the conversion passing through a maximum as the excess alkali content of the system is increased. Whilst it is easy to understand that hydrolysis of the soap, with consequent loss of activity, will occur at low pH values, it is not so clear why the rate of polymerisation diminishes at high pH values. It may be that the excess alkali is functioning as an electrolyte, for it is known that an increase in the ionic strength of the aqueous phase can cause a reduction in the rate of polymerisation (see Chapter 9, Section 9.2). In this connection, it is of interest to note some early results by Dogadkin[40] concerning the effect of pH upon yield for the emulsion copolymerisation of butadiene using a fatty-acid soap (ammonium oleate). Apparently the system did not contain an added initiator. These results, which are reproduced in Table 7.22 show clearly that the yield increases with increasing neutralisation of the fatty acid, and, indeed, appears to increase even though the point of complete neutralisation is exceeded. However, in comparing these results with those of Fryling and Follett, it should be noted that the latter neutralised their acids with a strong alkali (potassium hydroxide), whereas Dogadkin used a weak alkali (ammonium hydroxide). This difference is reflected in the actual pH levels recorded in the two cases.

TABLE 7.22

Effect of pH upon yield obtained in butadiene emulsion polymerisation† (according to Dogadkin[40])

(reaction system: butadiene 50 ml, aqueous phase 50 ml, oleic acid 1·5 ml)

Composition of aqueous phase		*Alkali excess over the saponification value (normality)*	pH *of latex after reaction*	% *conversion after* 5 days
Water	*0·5% aqueous ammonium hydroxide*			
37·6	8·4	−0·02	—	32·00
34·3	11·7	−0·01	8·65	40·88
32·7	13·3	−0·00	8·90	50·88
27·6	18·4	+0·01	9·10	47·80
24·3	21·7	+0·02	9·05	50·10
17·6	28·4	+0·04	9·35	101·60

† Temperature unspecified.

7.7 SULPHATES AND SULPHONATES

Surface-active sulphates and sulphonates find considerable application in emulsion polymerisation systems, especially in those cases where it is desired to carry out the polymerisation in acid medium (in which, of course, fatty-acid and rosin-acid soaps are unstable). Such micelle generators may also be used where it is desired to confer upon the final latex the property of colloid stability towards the destabilising influence of either acids or heavy-metal ions.

Hopff and Fákla[41] have recently described the effect of the structure of sulphates and sulphonates upon performance in the emulsion polymerisation of vinyl chloride at 40°C. The ammonium persulphate–sodium bisulphite redox couple was used as initiator. They reported that the sodium salts of primary alkyl sulphates give good results. The influence of chain length over the range C_{12} to C_{18} is not significant. The polymers of highest molecular weight are obtained using sodium cetyl sulphate. Conversion–time curves are shown in Fig. 7.17. Unsaturation causes retardation similar to that observed when sodium oleate was compared with, say, sodium stearate in their investigation of fatty-acid soaps. (It will be recalled that Hopff and Fákla found severe retardation when sodium

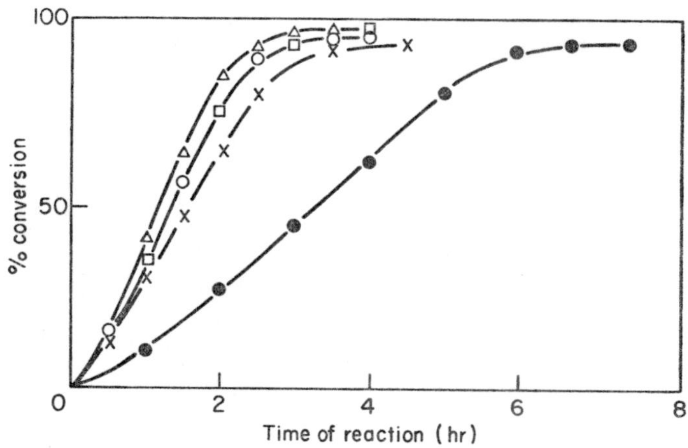

FIG. 7.17. Effect of chain length of sodium *n*-alkyl sulphate upon conversion–time relationship for emulsion polymerisation of vinyl chloride at 40°C, according to Hopff and Fákla.[41] Significance of points: ● C_{10}; × C_{12}; △ C_{14}; ○ C_{16}; □ C_{18}. (Reaction system: vinyl chloride 100, water 250, micelle generator 2, $(NH_4)_2S_2O_8$ 0·5, $NaHSO_3$ 0·5; pH adjusted to 10.)

TABLE 7.23

Comparison of sodium *n*-alkyl sulphates and sodium *n*-alkylsulphonates as micelle generators for emulsion polymerisation of vinyl chloride at 40°C (according to Hopff and Fákla[41])

(reaction system: vinyl chloride 100, water 250, micelle generator 2, ammonium persulphate 0·5, sodium bisulphite 0·5; pH 10)

		Sodium n-alkyl sulphate				*Sodium n-alkylsulphonate*			
Alkyl group	*No. of carbon atoms*	*Reaction time* (h)	*%* *conversion*	*Surface free energy of latex* (erg cm^{-2})	*Molecular weight of polymer*†	*Reaction time* (h)	*%* *conversion*	*Surface free energy of latex* (erg cm^{-2})	*Molecular weight of polymer*†
Capryl	C_{10}	7·5	92·8	27·83	9·899 × 10^4	6	92·8	27·12	9·02 × 10^4
Lauryl	C_{12}	4·5	93·4	25·91	11·363 × 10^4	4	94·0	24·80	10·72 × 10^4
Myristyl	C_{14}	3·5	95·1	26·40	12·430 × 10^4	3	95·4	25·13	10·86 × 10^4
Cetyl	C_{16}	4·0	94·5	30·12	13·319 × 10^4	2·5	95·8	26·75	11·14 × 10^4
Stearyl	C_{18}	4·0	96·0	34·80	13·089 × 10^4	3	96·7	30·40	12·35 × 10^4
Oleyl	C_{18}	17·0	90·3	37·71	7·757 × 10^4	15·5	91·0	32·10	6·54 × 10^4

† From intrinsic viscosity measurements.

oleate was used, as well as when fatty-acid soaps containing multiple unsaturation, *e.g.* sodium linoleate, were used.) Sulphates of alkylphenol polyglycol ethers were found to be relatively ineffective.

The sodium salts of alkylsulphonic acids give higher rates of polymerisation than do the corresponding alkyl sulphates, but also give polymers of lower molecular weights. Some comparative data are summarised in Table 7.23. The yield increases slightly with increasing chain length. Unsaturation causes the same degree of retardation as is observed with sodium oleate.

The salts of secondary alkyl sulphates were found to be much less active than the corresponding straight-chain derivatives. Sulphonates of succinic esters were found to be effective; optimum performance was observed with the dioctyl ester. Studies on the alkylbenzenesulphonates showed that, if the alkyl group is short, then the benzene ring must contain two alkyl

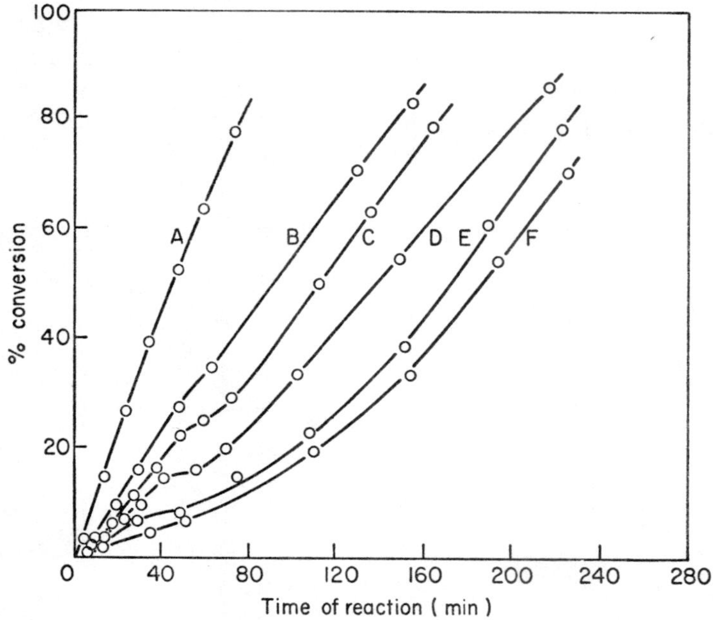

FIG. 7.18. Effect of level of micelle generator upon conversion–time relationship for emulsion polymerisation of vinylidene chloride at 32°C, according to Hay *et al.*[42] (Reaction systems: vinylidene chloride 100, water 117·7, sodium lauryl sulphate—variable, $(NH_4)_2S_2O_8$ 0·15, $Na_2S_2O_5$ 0·15. The letters appended to the curves refer to different levels of soap as follows: A, 10 pphm; B, 5 pphm; C, 3 pphm; D, 2 pphm; E, 1 pphm, F, 0·5 pphm.)

substituents if the substance is to be effective. The highest molecular weights were obtained with *n*-dodecylbenzenesulphonate. Of the alkyl naphthalenesulphonates studied, sodium di*iso*butylnaphthalensulphonate, when pure, gave the best results. The commercial material was found to be ineffective; this was attributed to the presence of impurities.

Results for the emulsion polymerisation of vinylidene chloride at 32°C using sodium lauryl sulphate as soap and an ammonium persulphate–sodium metabisulphite redox couple as initiator have been published by Hay and his co-workers.[42] The effect of soap level upon the conversion–time curve is shown in Fig. 7.18. It is apparent that the reaction proceeds in three reasonably well-defined stages. The rate of the initial stage increases as the level of sodium lauryl sulphate increases, but the rate attained in the final stage does not depend very much upon the soap level. In some respects, this behaviour compares well with that observed when the level of soap is varied in the emulsion copolymerisation of styrene and butadiene (see Fig. 7.9). The effect of sodium lauryl sulphate level upon initial rate is summarised in Fig. 7.19, from which it is seen that a linear relationship exists between the logarithm of the rate and that of the soap level. The slope of this relationship corresponds to an order of reaction of 0·6, *i.e.* to exactly that predicted by the Smith–Ewart theory for the eventual *steady* rate of reaction. By contrast, the results of Hay and co-workers

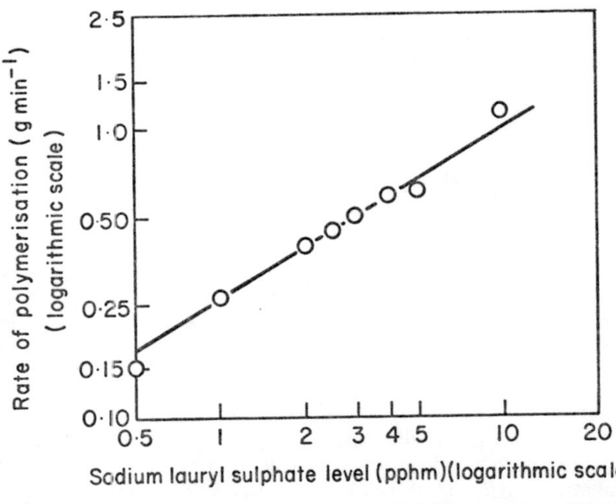

FIG. 7.19. Effect of level of micelle generator upon initial rate of emulsion polymerisation of vinylidene chloride at 32°C, according to Hay *et al.*[42] (See caption to Fig. 7.18 for details of reaction systems.)

indicate that the final rate of reaction is approximately of zero order with respect to soap level.

Hay and his co-workers have also published results for the percentage conversion at which foaming ceases within the system. Initially, sufficient soap was present in all systems to ensure that considerable foaming occurred. The foam was observed to persist during the early part of the reaction and then to subside rather suddenly. A relationship of direct proportionality was found to exist between the conversion at which foaming ceased and the level of sodium lauryl sulphate in the system (see Fig. 7.20). It is then stated that 'These data agree with a 1·2-power dependence on surfactant concentration, a relationship derived from the basic Smith–Ewart theory, if it is assumed that the time at which foaming stops is coincident with the disappearance of micelles.' But no demonstration of this 1·2-power dependence is given. What these authors possibly had in mind was that, according to the Smith–Ewart theory, (a) the time interval over which locus initiation occurs is proportional to $[\text{soap}]^{0.6}$, and (b) the final steady rate of reaction is also proportional to $[\text{soap}]^{0.6}$. Therefore, if the average rate during the period of locus initiation is also proportional to $[\text{soap}]^{0.6}$, these considerations give the conversion at the conclusion of locus initiation as proportional to $[\text{soap}]^{0.6} \times [\text{soap}]^{0.6}$, *i.e.* to $[\text{soap}]^{1.2}$. However, as far as the Smith–Ewart theory itself is concerned, conclusion (a) above certainly applies to the locus-initiation stage, but

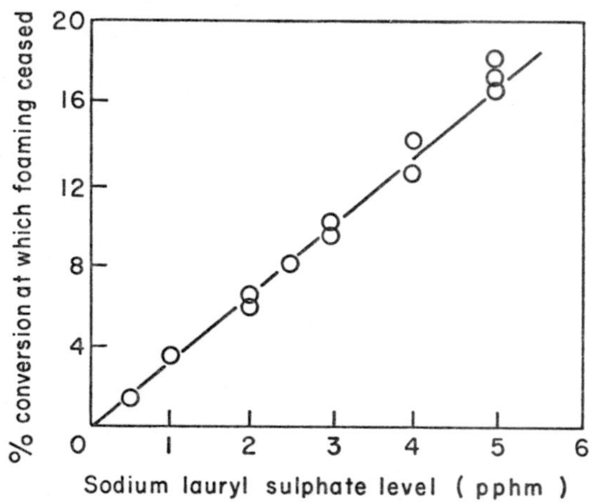

FIG. 7.20. Effect of level of micelle generator upon conversion at which foaming was observed to cease during the emulsion polymerisation of vinylidene chloride at 32°C, according to Hay *et al.*[42] (See caption to Fig. 7.18 for details of reaction systems.)

conclusion (b) applies only to the steady rate of polymerisation after locus initiation is complete.

Schulze, Tucker and Crouch[43] have reported upon the use of alkylaryl-sulphonate micelle generators in low-temperature (41°F) styrene–butadiene emulsion copolymerisations. They have shown that, if the sulphonates are specially treated to remove sodium sulphate and unsulphonated oils, then performance is very satisfactory on both laboratory and pilot-plant scale. The micelle generators used were the sodium salts of alkylbenzenesulphonic acids and alkyltoluenesulphonic acids having aliphatic side-chains of from about ten to about fifteen carbon atoms. The effect of the chain length of the alkyl group upon performance in a system initiated by a cumene hydroperoxide–iron redox couple is summarised in Table 7.24, from which it appears that optimum performance occurs when the side-chain contains about twelve carbon atoms. It should also be noted that the performance of most of the sulphonates was superior to that of the rosin-acid soap control. The effect of sulphonate level upon the conversion after 13 h at 41°F is shown in Fig. 7.21. Schulze, Tucker and Crouch draw attention to the fact that, in contrast to more conventional styrene–butadiene emulsion copolymerisation systems, very satisfactory rates of polymerisation can be obtained with the sulphonate level as low as 1·5 pphm. Furthermore,

TABLE 7.24

Effect of chain length of alkyl group upon performances of sodium alkyltoluenesulphonates in a low-temperature styrene–butadiene emulsion copolymerisation system (according to Schulze, Tucker and Crouch[43])

(reaction system: styrene 28, butadiene 72, water 180, micelle generator 4·7, sodium naphthalene formaldehyde sulphonate 0·10, mixed tertiary mercaptans, 0·24, cumene hydroperoxide 0·10, dextrose 1·0, $FeSO_4.7H_2O$ 0·4, $K_4P_2O_7$ 0·177, KCl 0·50, KOH 0·10)

Boiling range of olefin used for alkylation (°F)	*Average formula of sulphonate side-chain*	% *conversion after* 16 h *at* 41°F
320–350	$C_{10}H_{21}$	65·6
350–375	$C_{11}H_{23}$	70·0
375–400	$C_{11}H_{23}$ to $C_{12}H_{25}$	72·2
400–425	$C_{12}H_{25}$	66·9
425–450	$C_{13}H_{27}$	57·5
450–475	$C_{14}H_{29}$	67·8
475–500	$C_{15}H_{31}$	58·1
Control†	—	61·0

† Recipe containing same weight of sodium soap of disproportionated rosin acid.

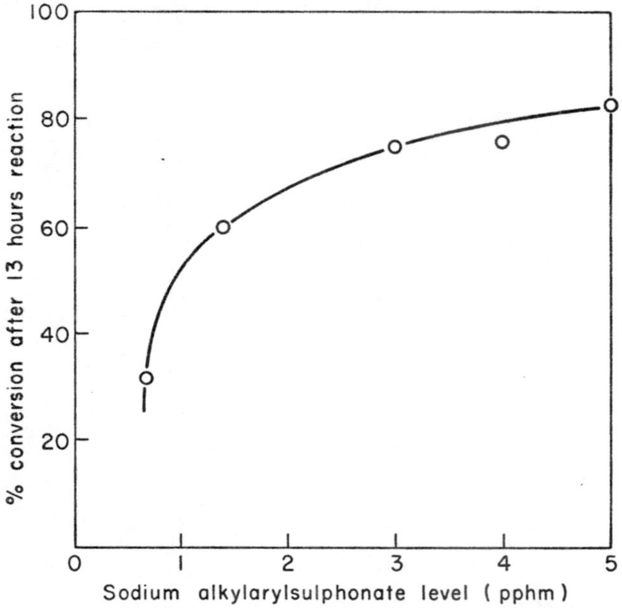

FIG. 7.21. Effect of level of sodium alkylarylsulphonate upon conversion in low-temperature (41°F) emulsion copolymerisation of styrene and butadiene, according to Schulze, Tucker and Crouch.[43] (Reaction system believed to be: styrene 28, butadiene 72, water 180, sodium alkylarylsulphonate—variable, mixed tertiary mercaptans 0·24, cumene hydroperoxide 0·01, dextrose 1·0, $FeSO_4.7H_2O$ 0·08, $K_4P_2O_7$ 0·101, KOH 0·03.)

latices prepared using this or even lower levels of sulphonate are said to have been fluid and stable, showing no apparent tendency towards premature coagulation or gelation.

The importance of impurities, notably electrolytes such as sodium sulphate, upon the rate of polymerisation in recipes containing sulphates and sulphonates is well known. Sensitivity to impurities, and possibly to small changes in chemical type, is especially marked in the case of the alkylarylsulphonates. The present writer is not aware that any satisfactory explanation of this sensitivity has ever been proposed, nor, indeed, of the development of any rational procedures for mitigating its effects. Blackley and McGilvray[44] have observed that the conversion–time curve for persulphate-initiated styrene–butadiene copolymerisations using sodium alkyl sulphates and alkylsulphonates can depend very much upon the source (and therefore, presumably, the purity) of the micelle-generating substance.

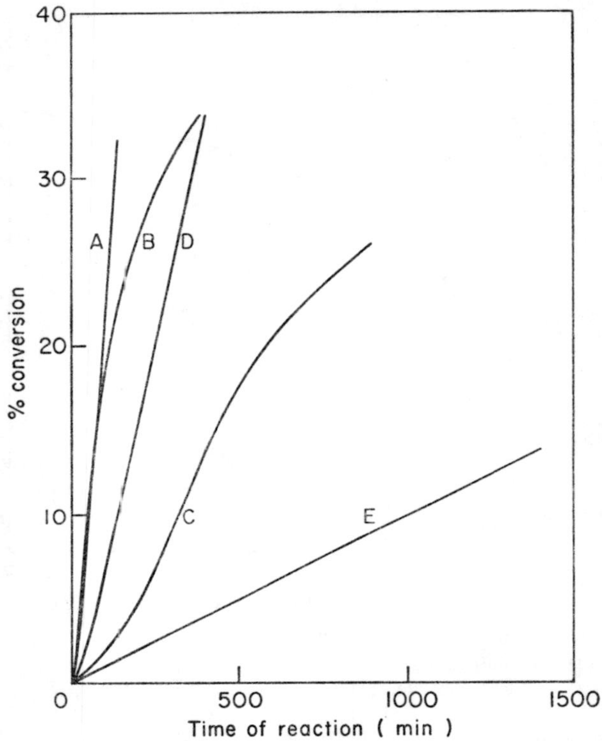

FIG. 7.22. Conversion–time curves for persulphate-initiated emulsion copolymerisation of styrene and butadiene at 60°C, using sodium dodecyl sulphates (A, B and C) and sodium dodecylsulphonates (D and E) from different sources, according to Blackley and McGilvray.[44] The experimental points are omitted for clarity. (Reaction system: styrene 50, butadiene 50, water 180, sodium dodecyl sulphate or dodecylsulphonate 5, *n*-dodecyl mercaptan 0·5, $K_2S_2O_8$ 0·8.)

Some typical results are shown in Fig. 7.22, from which the serious magnitude of the variations is evidence. These observations accord well with industrial experience.

7.8 CATIONIC SOAPS

Cationic micelle generators have usually been used in emulsion polymerisation systems when it has been desired to produce a latex whose particles carry positive instead of negative charges. Such latices have been proposed from time to time for the impregnation of fibrous substances which normally carry negative charges and so tend to repel the particles of an anionic latex; but cationic latices have not in fact found widespread application. Thus interest in the use of cationic micelle generators has been limited. In the main, the types which have been used are the alkylammonium salts, such as dodecylammonium chloride, and the quaternary alkylammonium salts, such as cetyltrimethylammonium bromide and cetylpyridinium bromide.

The use of cationic soaps in high- and low-temperature styrene–butadiene emulsion copolymerisations has been briefly described by Howland *et al.*[45] Two typical formulations using dodecylammonium salts as micelle generator are as follows:

	High-temperature formulation	*Low-temperature formulation*
Styrene	50	30
Butadiene	50	70
Water	120	180
Dodecylammonium chloride	5	—
Dodecylammonium acetate	—	3
t-Hexadecyl mercaptan	0·35	—
Mixed tertiary mercaptans	—	0·15
t-Butyl hydroperoxide	0·30	—
Di*iso*butyl hydroperoxide	—	0·05
$FeCl_2$	—	0·05
$AlCl_3.6H_2O$	0·36	0·3
KCl	0·3	0·1
Polymerisation temperature (°F)	122	41
Reaction time (h)	22	12
% conversion	85	60

Howland and his co-workers have found that, in the case of low-temperature polymerisations, the reaction rate is very sensitive to changes in pH. Their results for the above low-temperature formulation are given in Fig. 7.23, from which it is seen that the optimum rate occurs in the region of pH 6·0. Moving outside of the range from about pH 5·0 to about pH 6·5 causes the reaction rate to fall sharply. Howland and his co-workers state that control of pH is easier if a weak acid, such as acetic, is used to form the soap with the amine, rather than a strong acid, such as hydrochloric. Alternatively, a mixture of weak and strong acids can be used. It should be noted that the system to which the results of Fig. 7.23 refer contained dodecylammonium acetate as the soap.

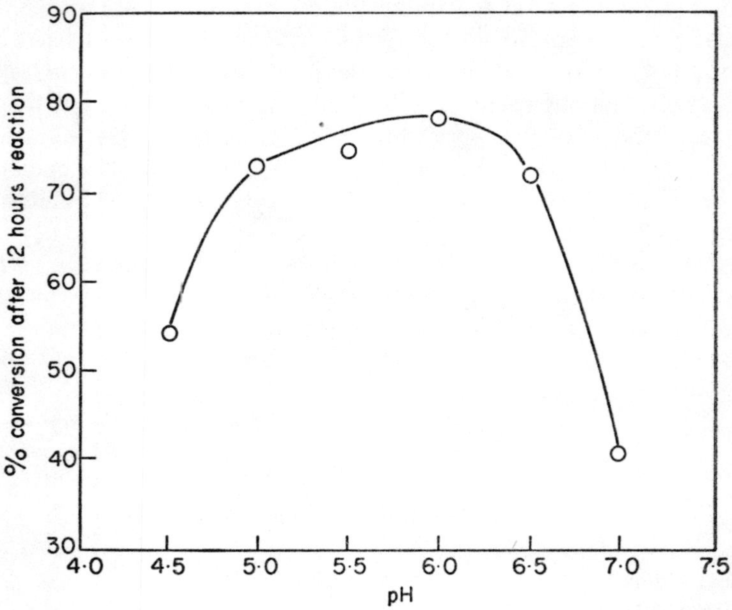

FIG. 7.23. Effect of pH upon conversion after 12 h reaction at 41°F for a styrene–butadiene emulsion copolymerisation using dodecylammonium acetate as micelle generator, according to Howland *et al.*[45] (See text for details of reaction system.)

7.9 NON-IONOGENIC MICELLE GENERATORS

Non-ionogenic micelle generators have found application in emulsion polymerisation systems in those instances where it is desired to produce a latex whose colloidal properties are relatively insensitive to changes of pH over quite wide ranges of acid and alkaline conditions. Most substances in this class which have been used are products made by reacting ethylene oxide with active-hydrogen compounds such as phenols, alcohols, carboxylic acids, amines, amides, etc. The structures (XIII)–(XVII) show the types of compound produced by reaction with each in turn of these classes of substance. In these structures, the group R is hydrophobic in character; usually it is an alkyl group.

The number of different molecular species which can be built up according to this pattern is limitless. In each case, both the hydrophobic group, R, and the number of ethylene oxide units per molecule can be varied. It is further possible to prepare a bewildering number of anionic and cationic

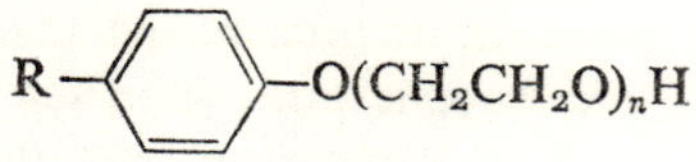

(XIII)

$RO(CH_2CH_2O)_nH$

(XIV)

$RCO.O(CH_2CH_2O)_nH$

(XV)

$RN\begin{cases}(CH_2CH_2O)_mH \\ (CH_2CH_2O)_nH\end{cases}$

(XVI)

$RCO.N\begin{cases}(CH_2CH_2O)_mH \\ (CH_2CH_2O)_nH\end{cases}$

(XVII)

molecules of this type by incorporating suitable functional groups within R or elsewhere in the molecule.

Helin and his co-workers[46] have published the results of a comprehensive survey of a large number of ethylene oxide adducts as potential micelle generators for the emulsion copolymerisation of styrene and butadiene. Polymerisations were attempted at both 122°F and at 41°F. The formulation for the former studies contained potassium persulphate as initiator; for the latter, a cumene hydroperoxide–iron(II)sulphate–potassium pyrophosphate–dextrose initiation system was used. The balance between hydrophilic and hydrophobic character for the ethylene oxide adducts was quantified by means of a so-called 'solubility index', defined as the ratio 100 × (number of ethylene oxide units): (number of carbon atoms in base molecule + constant), the constant being included to allow adjustment for the inherent water-solubility of the base molecule. Lack of space precludes a detailed discussion of the many observations reported by this group of investigators. They summarise their overall conclusions regarding non-ionogenic micelle generators as follows: 'As a group, the nonionic emulsifiers were unsatisfactory because of low conversion rates and the variable stabilities of the resulting latices. The data fail to show a correlation between the solubility index and the conversion rate in the sense that the performance of the emulsifier can be predicted from solubility data alone. It was obvious that at values below a minimum value of the solubility index, which although not very sharply defined is probably in the range of 70 to 120, little or no polymerization occurred. Possibly this was because of poor emulsification of the monomer or because the emulsifier was too insoluble to reach the critical micelle concentration in the water phase. Above this minimum range of values, conversion rates did not depend on the solubility index.'

Whilst there is apparently no general overall correlation between solubility index and performance in a polymerisation system, it appears that, for any one particular chemical type, both polymerisation rate and the stability of the resultant latex increase as the ethylene oxide content of the micelle generator increases. One typical set of results, those for alkyl phenoxypolyethylene ethanols, is given in Table 7.25. It is also stated by Helin and his co-workers that the stability of the latex eventually produced is apparently unrelated to the solubility index of the micelle generator used in its preparation. They further point out that, as a group, ethylene oxide adducts are less soluble in hot water than in cold water because the

TABLE 7.25

Effect of solubility index upon performance of alkyl phenoxyoxyethylene ethanols in low- and high-temperature emulsion copolymerisation of styrene and butadiene (according to Helin *et al.*[46])

(reaction systems: (a) 41°F polymerisation: styrene 28·5, butadiene 71·5, water 180, micelle generator 5, *t*-dodecyl mercaptan 0·18–0·26†, cumene hydroperoxide 0·06–0·125‡, dextrose 1·0, $FeSO_4.7H_2O$ 0·12, $K_4P_2O_7$ 0·17; (b) 122°F polymerisation: styrene 28·5, butadiene 71·5, water 180, micelle generator 5, *n*-dodecyl mercaptan 0·58–0·85†, $K_2S_2O_8$ 0·23)

Micelle generator type: 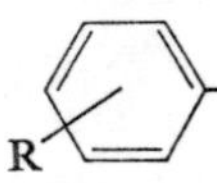 $—O(CH_2CH_2O)_nH$

		Polymerisation at 41°F		*Polymerisation at* 122°F	
Solubility index	*Solubility in water*	*Rate of polymerisation* (% h^{-1})	*Stability and viscosity of latex produced*	*Rate of polymerisation* (% h^{-1})	*Stability and viscosity of latex produced*
44	Insoluble	0	Phase separation	—	—
100	Soluble	1·5	Satisfactory	0·5	Phase separation
167	Soluble	2·9	Very viscous	1·8	Fluid, foamy
212	Soluble	2·6	Slight preflocculation	2·4	Satisfactory, foamy
334	Soluble	2·9	Viscous, foamy	1·8	Slight preflocculation, foamy

† Varied to obtain copolymers with a Mooney viscosity of about 50 ML-4.
‡ Charged on basis of purity.

association which occurs between the ether linkages and water molecules is partially destroyed by heating. To this phenomenon may be due some of the variations between behaviour observed in the 41°F system and that observed in the 122°F system.

Concerning specific types of polyethylene oxide adduct, Helin and his co-workers summarise their experience as follows:

> 'None of the polyoxyethylene ethanols yielded conversion rates comparable to those obtained with the potassium rosin or sodium fatty acid soaps. The best rate at 41°F. was about 58% of that exhibited by the control. The polymerizations were not affected greatly by variations in the pH and the stability of the latices were variable.
>
> 'The ester derivatives provided conversion rates which were comparable to each other but considerably lower than those obtained with the control charges; the best rate was 34% of that exhibited by the potassium rosin acid (KRA) soap emulsified control, and the degree of stability was very inconsistent.
>
> 'Of the nonionics, the alkyl aryl polyoxyethylene amides yielded the best conversion rates. At 41°F., the rates averaged 46% of that for the KRA soap control, and at 122°F. the rates for the charges polymerized without adjustment of the pH averaged 58% of that for the NaFA soap control.
>
> 'Conversion rates obtained with amine derivatives at 41°F. were poor at a low pH (where they presumably act as cationic emulsifiers) and varied from poor to moderately good at a high pH; the best rate was 64% as much as that of the KRA soap control. At 122°F., the rates of polymerization varied from excellent (148% of that of the NaFA soap control) at a low pH, through good (approximately equal to the control) at a pH near the neutral point, to fair (56% of that control) at the high pH levels. However, at 122°F., the latex stability was very poor.'

It was also noted that combinations of micelle generators occasionally give good results. Thus an alkyl phenol–ethylene oxide adduct in combination with an alkylarylsulphonate or an alkylphenoxypolyethylene sulphate gives comparatively good conversion rate at 41°F, together with satisfactory latex stability.

Greth and Wilson[47] have published the results of an extensive survey of the performance of several non-ionogenic micelle generators, and of their blends with anionic substances, in systems for the emulsion polymerisation of both styrene and vinyl acetate. Styrene was polymerised at 50°C using a persulphate–bisulphite redox couple as initiator; vinyl acetate was polymerised at 70°C using unactivated persulphate. The results were interpreted in terms of the so-called 'HLB value' of the micelle generator. The HLB value is a number which is said to quantify the 'hydrophilic–lipophilic balance' of a surface-active substance. It was first introduced by Griffin[48,49] as a basis for predicting the behaviour of surface-active substances, and so assisting with the rational selection of appropriate types for various applications. The scale runs from 1 (indicating perfect lipophilic character) to 20 (indicating perfect hydrophilic character). Griffin[48] stresses that HLB value is not a direct measure of solubility, although there is an overall correlation in the sense that substances having low HLB values tend to be oil-soluble, whilst those having high values tend to be water-soluble. However, two substances can have the same HLB value but exhibit different solubility characteristics. Elsewhere, Griffin[50] has given the following broad correlation between HLB value and behaviour upon addition to water:

HLB range	*Behaviour when added to water*
1–4	No dispersibility in water
3–6	Poor dispersion
6–8	Milky dispersion after vigorous agitation
8–10	Stable milky dispersion
10–13	From translucent to clear dispersion
13+	Clear solution

The determination of HLB value is essentially empirical. It is accomplished, for example, by observing emulsifying behaviour under standardised conditions,[48] from cloud-point measurements,[50] or by calculation from chemical constitution.[49] The latter procedure is especially appropriate for non-ionogenic substances. Thus for most esters of fatty acids and polyhydric alcohols the suggested formula is

$$\text{HLB} = 20(1 - S/A)$$

where S is the saponification number of the ester and A is the acid number of the acid. For simple ethylene oxide addition products in which no polyhydric alcohol, other than the polyoxyethylene chain, is present, the suggested formula is

$$\text{HLB} = E/5$$

where E is the weight percent of ethylene oxide in the adduct. For blends of two or more surface-active substances, the HLB value of the blend is assumed[48] to be the sum of the individual values weighted according to the weight fraction of each in the blend. Thus, when Greth and Wilson present results for various properties as functions of the average HLB value of a given pair of micelle generators, they are virtually presenting data for the effect of blend composition upon those properties. Thus even for those who regard the HLB value itself as an unnecessary fiction, the data themselves are still of great value, because the average HLB value can readily be converted into blend compositions. Indeed, although Greth and Wilson's results have been presented here in terms of HLB values (as in their original paper), the blend composition to which the various results refer are readily accessible from the original paper.

The results of Greth and Wilson for the performance of blends of polyethylene oleyl ether and an alkylarylsulphate in the emulsion polymerisation of styrene are summarised in Figs. 7.24 and 7.25. Fig. 7.24 (a) shows that the rate of polymerisation passes through a maximum in the region of blend composition giving an HLB value of about 13·5. Figs. 7.24 (b) and (c) show that the particle size of the final latex passed through a sharp minimum in this region, and that the viscosity of the latex passes through a sharp maximum. What is immediately obvious from these results is the apparent sensitivity of these properties to HLB value, and therefore to blend composition. The obvious inference from these results is that, for those systems containing blends of micelle generator for which the average HLB value is about 13, the number of loci initiated in the polymerisation is a maximum. It should be noted that HLB values in this region for this pair of substances correspond to blends containing approximately equal parts of the two by weight. It is not, however, clear just why such a blend should give the optimum number of loci for the styrene polymerisation.

Greth and Wilson also give results for the stability of the polystyrene latices produced. It appears that latices of excellent stability can be obtained using blends for which the HLB value is between 13 and 15. Similar variations of performance with HLB value were observed when the micelle generator was a stearic acid–ethylene oxide adduct. In this case, the HLB value was varied by varying the average number of ethylene oxide units per stearic acid molecule. It is perhaps significant that no maximum in polymerisation rate was recorded as the HLB value was increased. Instead, there was observed a sharp increase in rate as the HLB value increased from 13 to 16, followed by immediate entry into a region of colloidal instability as the HLB value was further increased. These observations agree in a general way with those reported by Helin and his co-workers[46] for the emulsion copolymerisation of styrene and butadiene.

The effects of level of micelle generator upon the emulsion polymerisation of styrene at 50°C, using a blend of equal parts of a polyoxyethylene oleyl ether and an alkylarylsulphonate, are shown in Fig. 7.25. As expected,

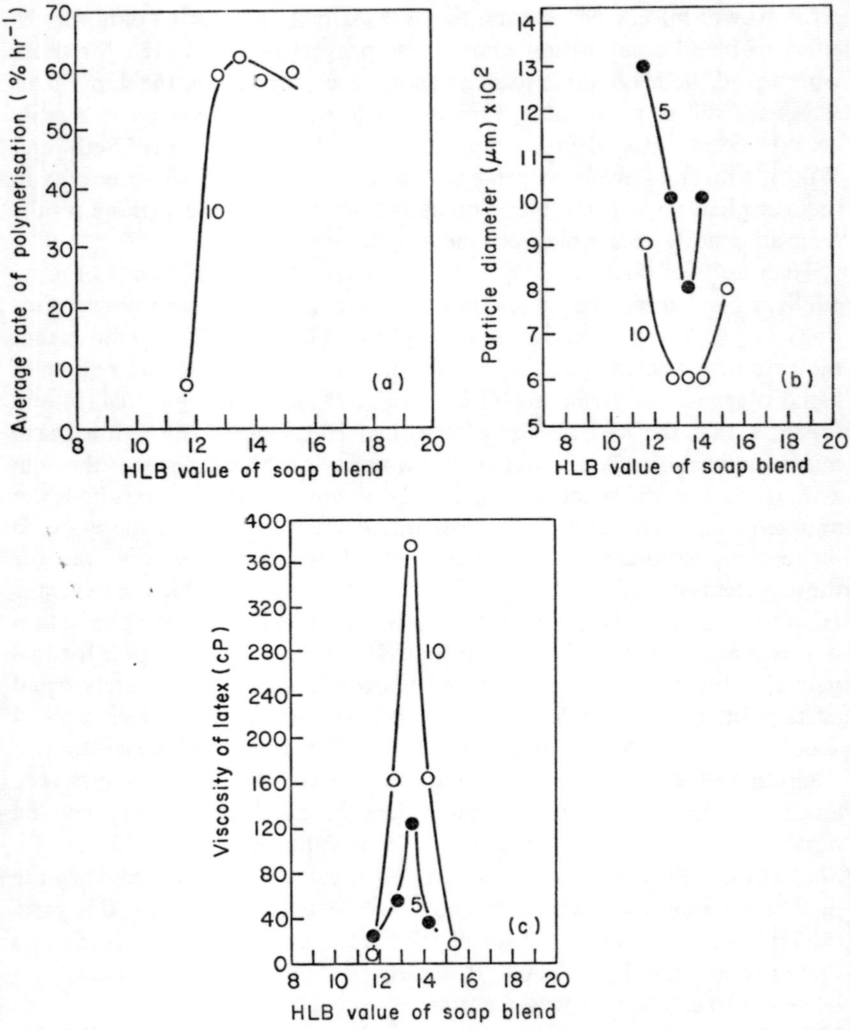

FIG. 7.24. Effect of average HLB value of soap blend (and therefore of blend composition) upon performance of blends of a polyoxyethylene oleyl ether and an alkylarylsulphonate in the emulsion polymerisation of styrene at 50°C, according to Greth and Wilson:[47] (a) effect upon rate of polymerisation; (b) effect upon particle size of latex produced; (c) effect upon latex viscosity. (Reaction systems: styrene 100, water 150, micelle generators—total of 5 or 10,† $K_2S_2O_8$ 0·375, $NaHSO_3$ 0 or 0·15.‡)

† The numbers appended to the curves indicate the total level of micelle generator.

‡ It appears that the sodium bisulphite activator was omitted from the system containing 5 pphm of micelle generator, but was present at the indicated level in the systems containing 10 pphm of micelle generator.

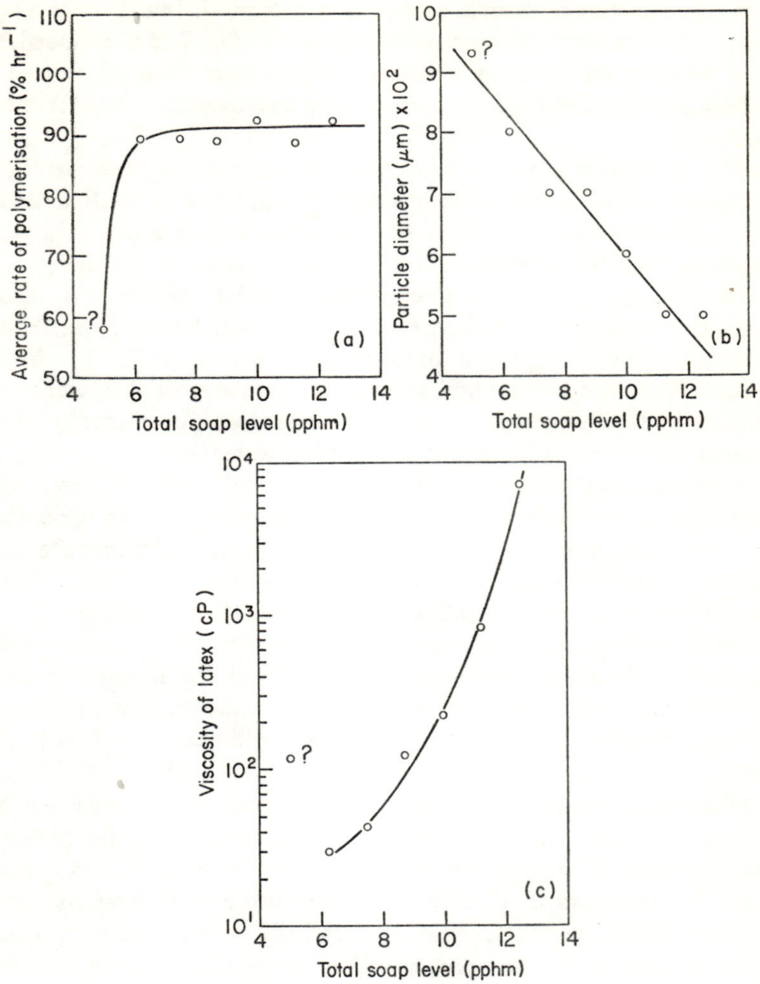

FIG. 7.25. Effect of level of micelle generator upon emulsion polymerisation of styrene at 50°C using as micelle generator a blend of equal parts by weight of a polyoxyethylene oleyl ether and an alkylarylsulphonate, according to Greth and Wilson:[47] (a) effect upon rate of polymerisation; (b) effect upon particle size of latex produced; (c) effect upon latex viscosity. (Reaction system was as given in the caption to Fig. 7.24, except that the level of total micelle generator was variable.) Note that some doubt exists concerning the relevance of the points corresponding to the reaction system which contained 5 pphm of micelle generator, because apparently the sodium bisulphite activator was omitted in this case.

the average particle size of the resultant latex falls, and its viscosity increases, as the level of micelle generator is increased. But if the doubtful point to which reference is made in the caption to Fig. 7.25 is excluded, the rate of polymerisation seems to be essentially independent of the level of micelle generator. Certainly this seems to be true over the range from 6 to 12 pphm.

In the case of the persulphate-initiated emulsion polymerisation of vinyl acetate at 70°C, the variations of performance with HLB value are similar, except that the optimum HLB values tends to be rather higher than for styrene. Thus the most stable latex is obtained at an HLB value of 16·9, and the range for good latex stability was found to be 15 to 18. Using blends of polyoxyethylene oleyl ether and an alkylarylsulphonate, the optimum blend corresponds to an average HLB value of 15. The higher optimum HLB range for vinyl acetate as compared with styrene was attributed by Greth and Wilson to a greater hydrophilic character of the monomer as evidenced by higher solubility in water.

The general conclusion which emerges from this study by Greth and Wilson is that the performance of blends depends critically upon their composition, and this can be translated into an apparent dependence upon HLB value if desired. As a corollary to this sensitivity, variations in blend composition can be used to control such properties as the particle size and viscosity of the final latex, as well as the rate of polymerisation. However, these controls cannot be exercised independently of one another by variations of blend composition alone. But variation of the total level of micelle generator, as well as of its composition, offers additional freedom in this respect.

Brief reference must also be made to a recent study by Medvedev and his co-workers[51] of the 'emulsion' polymerisation of styrene in the presence of non-ionogenic micelle generators. The word 'emulsion' is in quotes, because the initiators used were oil-soluble rather than water-soluble. (They were dibenzoyl peroxide and azobis*iso*butyronitrile.) It appears that, in some cases at least, the polymerisations more closely resembled suspension rather than emulsion reactions. The results of this group of workers for the effect of the composition of a cetyl alcohol–ethylene oxide adduct upon its performance in a system polymerising at 50°C are summarised in Table 7.26. With 8 molecules of ethylene oxide per molecule of cetyl alcohol, the rate is said to be virtually that of a bulk polymerisation carried out under comparable conditions. The particles of the product are large, and the reaction is clearly of the suspension type. As the number of ethylene oxide molecules is increased, so the characteristic features of

TABLE 7.26

Influence of composition of a cetyl alcohol–ethylene oxide adduct upon its performance in 'emulsion' polymerisation of styrene at 50°C (according to Medvedev *et al.*[51])

(reaction system was such that the concentrations of micelle generator and initiator (azobis*iso*butyronitrile) were, respectively, $2{\cdot}0 \times 10^{-4}$ and 1×10^{-4} mole per ml of monomer; the hydrocarbon: aqueous phase ratio was 1:2 by volume)

Number of ethylene oxide units per molecule of cetyl alcohol	*Rate of polymerisation* [mole (ml *of monomer*)$^{-1}$ min^{-1}]	*Average particle diameter of latex* (Å)	*Average molecular weight of contained polymer*†
8	$0{\cdot}11 \times 10^{-4}$	10 000	5×10^{4}
20	$0{\cdot}75 \times 10^{-4}$	1 120	$2{\cdot}66 \times 10^{6}$
30	$1{\cdot}21 \times 10^{-4}$	780	$3{\cdot}90 \times 10^{6}$
40	$1{\cdot}51 \times 10^{-4}$	610	$4{\cdot}70 \times 10^{6}$

† From intrinsic viscosity measurement.

emulsion polymerisation become apparent: the rate of polymerisation increases somewhat, the particle size of the latex product falls, and the molecular weight of the product increases.

On the basis of these results, it appears that rate, degree of subdivision of the product, and polymer molecular weight increase continuously with the ethylene oxide content of the micelle generator. These authors attribute the effect of the length of the hydrophilic part of the adduct upon its performance to an increase in the area of interface which can be stabilised by each molecule of micelle generator as the number of ethylene oxide units is increased. Medvedev and his co-workers have also reported upon the use of mixtures of ionic and non-ionogenic micelle generators. They find that the effects upon polymerisation rate and particle size of the product are not additive. Some of their results resemble those of Greth and Wilson[47] in a general way. The results presented by Medvedev are regarded as indicating the important part which the interfacial layer between particle and aqueous phase plays in determining the course of an emulsion polymerisation reaction.

7.10 OLIGOMERIC AND POLYMERISABLE MICELLE GENERATORS

We now proceed to a brief consideration of two topics upon which papers have recently appeared in the literature, and in which some interest is currently being shown. The first concerns the use of oligomeric micelle generators for emulsion polymerisation systems. These have been described by Roe.[52,53] The oligomers in question are copolymers of acrylonitrile and acrylic acid whose molecular weights have been kept low by polymerising them in the presence of *n*-octyl mercaptan. In fact, it is the alkyl group introduced into most molecules as a result of the transfer-to-mercaptan reaction which provides the hydrophobic part of the molecule. The alkali–metal salts of these copolymers are soluble in water and are able to stabilise polymer latices. Their advantage over conventional colloid stabilisers is their low tendency to foam. The oligomers are produced by free-radical polymerisation in homogeneous phase in a solvent. Recovery of the oligomer is achieved by flashing off the solvent under reduced pressure, either before or after neutralisation with potassium hydroxide. Molecular weights range from a few hundred up to a few thousand. A crude fractionation according to molecular weight can be carried out by dissolving the unneutralised oligomer in acetone and adding increasing amounts of *n*-hexane. A more refined fractionation can be effected by selective adsorption of the potassium salt on to the particles of a polystyrene latex stabilised with a minimal amount of potassium palmitate. The various oligomer fractions all appear to have the same ratio of acrylonitrile to acrylic acid units.

The unusual properties of these oligomers seem to arise from different dependencies upon molecular size of adsorption at the two types of interface presented by a polymer dispersion, *i.e.* the interface between the particles and the aqueous phase, and that between the aqueous phase and air. The reader is referred to the original papers for a very full discussion of these matters. It appears that the lower-molecular-weight oligomers are more strongly adsorbed at the particle interfaces than are the higher oligomers, but are less effective at imparting colloid stability to the particles. On balance, the oligomers of lower molecular weight appear to be more efficient as colloid stabilisers than do the higher oligomers, on a basis of equal weights. Roe also finds that (a) the surface activity at the air–water interface decreases with an increasing oligomer molecular weight, and (b) the maximum adsorption at the air–water interface is nearly independent of molecular weight. Thus the situation is that the particles tend to adsorb preferentially those oligomer molecules which show most activity at the air–water interface. To this characteristic is attributed the high surface free energy and low foaming tendency of systems containing these oligomers.

As regards performance in an emulsion polymerisation system, Roe claims that these oligomers are potent supporters of locus nucleation. Thus he cites a styrene system containing 2·5 pphm of unfractionated oligomer salt, 200 pphm of water, 0·62 pphm of potassium persulphate and 2·6 pphm of dipotassium hydrogen sulphate, which, after polymerising to completion at 50°C, gave a stable latex containing 2×10^{14} particles per millilitre. But he also claims that by the use of these oligomers it is possible to reduce the number of particles (and therefore increase their size) without unduly diminishing the stability of the latex produced. He contrasts the control which is possible using these oligomers with the situation which exists when conventional soaps are used. In the latter case, the cessation of locus nucleation implies virtual exhaustion of soap in the aqueous phase; there is therefore left no reservoir of soap to contribute to the colloidal stability as the particles grow with conversion. Of the latices produced using these oligomers, he says that they are characterised by high surface free energy, good stability to stirring, low foaming tendency, and incomplete adsorption of the oligomer. The unadsorbed oligomer which remains in solution at the end of the conversion process consists of large molecules, and is depleted with respect to the smaller components.

Concerning the mechanism by which such latices are formed, Roe makes the following suggestions: 'Particle nuclei generated during early stages or conversion selectivity adsorb the smallest molecule components of oligomer. Generation of new particles accompanied by selective adsorption continues until a stage is reached such that new nuclei are absorbed by preexisting particles as fast as they are formed. Large molecule emulsifier components are available in the serum at this point but their overall stabilising capacity (adsorption plus intrinsic stabilising capacity) is inadequate to confer permanent identity on new nuclei. As the existing particles develop additional interfacial area, successively larger components of the dissolved emulsifier participate in the adsorption equilibrium, depending on their relative abundance. Despite their limited adsorption, their contribution to stability appears to be crucial and is available without undesired side effects such as development of foaming tendencies.'

The second topic concerns the use of polymerisable micelle generators. These presumably become irreversibly combined with the polymer molecules during the course of the polymerisation reaction. They have been the subject of three papers by Greene and his co-workers.[54–56] The particular susbstance used was the sodium salt of 9- (and 10-) acrylamidostearic acid, and the latex was of a 60/40 styrene–butadiene copolymer. Comparisons were made with latices containing equivalent amounts of the monomeric soap added after polymerisation. It is reported that the *in situ* polymerised soap is more efficient in conferring mechanical stability than is the added monomeric soap, as long as the surface coverage exceeds 20%. In fact, the greater the surface coverage, the more marked is the difference between the two. Greene and his co-workers have also found that the stability of the latex towards added electrolyte is greater (at high surface coverage) with the *in situ* polymerised soap than with the added monomeric soap.

7.11 PREPARATION OF LATICES OF UNIFORM PARTICLE SIZE

Some interest has been shown in the preparation by emulsion polymerisation of latices having a very narrow distribution of particle sizes. It is generally recognised that a prerequisite for the formation of a colloid system of uniform particle size is the presence of a fixed number of nuclei which subsequently grow to a much larger size, there being no further formation of nuclei during the growth period. This generalisation seems to be true whatever may be the nature of the process by which the particles are formed. In the case of emulsion polymerisation, this requirement can be met by the technique of 'seeding' the reaction (see, for example, the paper Willson, Miller and Rowe[57].) The principle is to prepare a 'seed' latex by emulsion polymerisation, and then to use these particles as the loci for further polymerisation, having present just sufficient additional surface-active substances to confer colloidal stability upon the growing particles, but insufficient to allow new loci to develop. Since the number of loci which form is connected very closely with the amount of micelle-generating substances which is present, it is appropriate to consider this subject here.

The process of seeded emulsion polymerisation has been reported upon in some detail in papers by Vanderhoff *et al.*[58] and by Bradford, Vanderhoff and Alfrey.[59] They consider first the theoretical implications of the growth pattern of latex particles during emulsion polymerisation. Let v be the volume of polymer contained in a typical particle, and r the radius of the sphere which this volume of polymer would occupy. (The original argument of Vanderhoff and co-workers is developed in terms of particle diameters, not radii, but this makes no essential difference.) Assume that the growth pattern for the particle depends exponentially upon its radius, *i.e.* that

$$\frac{dv}{dt} = kr^n$$

where n is as yet unknown and k is a constant which may depend upon environment (*e.g.* upon initiator concentration, temperature, etc.) and also possibly upon reaction time. Since $v = 4\pi r^3/3$, we have that $\mathrm{d}v/\mathrm{d}t = 4\pi r^2 \mathrm{d}r/\mathrm{d}t$, *i.e.*

$$4\pi r^2 \frac{\mathrm{d}r}{\mathrm{d}t} = kr^n$$

i.e.

$$\frac{1}{r}\frac{\mathrm{d}r}{\mathrm{d}t} = \frac{kt}{4\pi} r^{n-3}$$

Thus the *relative* rate of growth of particle radius is proportional to r^{n-3}. If $n = 3$, then this rate is independent of particle size and the distribution of sizes would remain constant throughout the growth period. If $n < 3$, then the smaller particles would grow relatively faster than the larger ones and so tend to catch them up. The distribution of sizes is then self-sharpening. If $n < 3$, then the distribution would broaden during the growth stage. The Smith–Ewart theory assumes that $n = 0$ (*i.e.* that $\mathrm{d}v/\mathrm{d}t$ is constant). Theoretically this would lead to a very rapid sharpening of the particle-size distribution as reaction proceeds. Experiments reported by Vanderhoff and his co-workers, using polystyrene seed latex, are consistent with a value of n of about 2·5. This value confirms that seeded emulsion polymerisations should be self-sharpening but not perhaps as rapidly as predicted by the Smith–Ewart theory.

To understand the nature of the experimental evidence upon which the value of $n \cong 2{\cdot}5$ is based, it is necessary to integrate the above expression for $\mathrm{d}r/\mathrm{d}t$. Clearly

$$\int_{r_0}^{r} r^{2-n}\,\mathrm{d}r = \frac{1}{4\pi}\int_0^t k(t)\,\mathrm{d}t$$

where $k(t)$ is written for k in order to emphasise its possible dependence upon time of reaction, and r_0, r are the values of r at $t = 0$ and $t = t$, respectively. Integration gives immediately that

$$r^{3-n} - r_0^{3-n} = \frac{3-n}{4\pi}\int_0^t k(t)\mathrm{d}t$$

Consider now two seed latices whose initial particle sizes are r_0 and R_0, respectively. Allow the emulsion polymerisation of excess monomer to occur in their presence, maintaining the conditions of concentration, temperature, etc., the same for both. Then, whatever may be the actual value of $k(t)$ at any instant in the reaction, its value for both latices should be the same. If, then, the period of growth is the same for both, the value of the integral on the right-hand side of the above equation for r^{3-n} should be the same for both latices. Thus if r and R are the respective particle radii at the end of the growth period, then

$$r^{3-n} - r_0^{3-n} = R^{3-n} - R_0^{3-n}$$

Putting $\alpha = R_0/r_0$ and $\beta = r/r_0$, simple rearrangement gives

$$\frac{R}{r} = \frac{1}{\beta}(\alpha^{3-n} + \beta^{3-n} - 1)^{1/(3-n)}$$

Thus n can be estimated from an experiment in which an emulsion polymerisation reaction is seeded with particles of two different initial sizes, and these particles are then allowed to compete with each other for monomer and free radicals in the same system. α in the above equation is known from the ratio of particle sizes in the initial seed latices. As reaction proceeds, so the size ratio R/r can be determined by electron microscopy, and its variation with the ratio β (which is a measure of the extent of reaction) ascertained It is possible then to compare the observed variation of R/r with β for given α with that predicted on the basis of various values of n. Having carried out an experiment of this type, Vanderhoff and his co-workers found their results to be reasonably consistent with $n \simeq 2{\cdot}5$.

7.12 EMULSION POLYMERISATION IN THE ABSENCE OF ADDED MICELLE GENERATOR

The final topic, of which mention will be made under the general heading of micelle generators, is that of emulsion polymerisation in the absence of substances deliberately added to form micelles or otherwise to assist with the nucleation of new particles. Thus an early paper by Willis[60] describes the production of 'emulsifier-free' styrene–butadiene copolymer latices by emulsion polymerisation, the desire being to produce latices and polymers derived therefrom which are free from deliberately-added surface-active substances. The reaction system comprised monomers, water, potassium persulphate and sodium metaphosphate, the latter being added to function as a mild colloid stabiliser. The particles of the resulting latex were found to be large (a typical value being 5600 Å) relative to those normally found in latices produced using the standard GR-S system. As would be expected, reaction rates were rather slow.

Willis gives some information upon the effect of varying the nature of the reaction system. Increasing the level of potassium persulphate up to 1·6 pphm was found to increase the rate of polymerisation, but higher levels caused retardation. The magnitude of the increase was not, however, large, and the maximum rate obtained was only 0·58% conversion per hour. Varying the level of sodium metaphosphate was found to have little effect upon the rate of conversion, although the presence of this substance was essential for the colloid stability of the resultant latex. The addition of dodecyl mercaptan was observed to have an adverse effect upon the reaction rate. Variation of pH over the range from 3·0 to 9·4 had little effect; outside of these limits, definite retardation was observed. Persulphate activators, such as potassium ferricyanide and sodium cobaltinitrile, tended to cause retardation. However, rapid polymerisation was reported for an emulsifier-free system containing styrene, butadiene and acrylonitrile, and using bisulphite-activated persulphate as initiator.

Kotera, Furusawa and Takeda[61] have recently described the preparation of soap-free polystyrene latices. The method uses initiation by potassium persulphate. The stability of the resulting latex seems to depend upon the presence in the particle–aqueous phase interface of sulphate ions which are chemically combined with the polymer. (The evidence that such combination occurs has been fully discussed in Chapter 6, Section 6.4). A typical procedure involves the use of an aqueous phase containing about 6×10^{-4} mole l^{-1} of potassium persulphate to which is added styrene such that its concentration expressed in terms of the aqueous phase is in the range from 0·1 to 1·0 mole l^{-1}. A speed of agitation of 350 rev/min is suggested. Typical polymerisation conditions are 28·5 h at 70°C. It is claimed that the latices produced are generally stable and contain spherical particles of very narrow size distribution. The particle diameter can be varied over a wide range (*e.g.* 3500–14 000 Å) according to the preparative conditions used. The particle diameter increases with increasing monomer concentration, but the rate of increase is relatively small. Furusawa, Norde and Lyklema[62] have shown that by carrying out the polymerisation at near-neutral pH in potassium bicarbonate, or in a dipotassium hydrogen phosphate buffer, it is possible to attain considerably higher negative surface charge densities than by the procedure of Kotera, Furusawa and Takeda.

Regarding the mechanism by which such latices form, it seems likely that surface-active oligomers are first generated by a persulphate-initiated solution polymerisation, and that these oligomers subsequently aggregate into loci for the subsequent polymerisation reaction. Reaction then proceeds by normal micellar mechanism, except that polar head-groups become attached to polymer molecules at the same rate as those molecules form, and therefore the colloid stability of the developing latex is maintained. The products of this reaction are mainly of interest as model colloids.

In support of the above view of the reaction mechanism, it is interesting to note that Kotera, Furusawa and Takeda[61] failed to obtain satisfactory latices using hydrogen peroxide as initiator in place of potassium persulphate. Unstable latices containing particles having a broad distribution of sizes were obtained. Failure to obtain satisfactory latices was attributed to the absence of Coulombic repulsions between charged species, thereby permitting uncontrolled coagulation of polymer particles to occur during the early stages of the polymerisation.

Goodwin *et al.*[63] have recently reported upon the preparation of monodisperse polystyrene latices in the absence of added surface-active substances by a single-stage process. It is shown that, by suitable choice of ionic strength, monomer concentration, initiator concentration and temperature, it is possible to obtain monodisperse polystyrene latices having volume fractions of contained polymer of the order of 0·1, and particle sizes ranging from about 10^3 to 10^4 Å. Latices of larger particle size can be obtained by using a process comprising several growth stages. It is stressed that the conditions of agitation throughout the reaction are critical. Rapid stirring is required in order to maintain the monomer dispersed in small droplets, and to encourage rapid transfer of monomer from the droplets to the water in order to maintain a monomer-saturated aqueous phase. It is recommended that stirring conditions be standardised as closely as possible from batch to batch if reproducible particle sizes are to be obtained.

REFERENCES

1. K. Shinoda, 'The formation of micelles', in *Colloidal Surfactants: Some Physicochemical Properties* (by K. Shinoda, T. Nakagawa, B. Tamamushi and T. Isemura), Academic Press, New York, 1963, pp. 1–96.
2. W. D. Harkins, *J. Am. chem. Soc.*, 1947, **69**, 1428.
3. M. L. Corrin and W. D. Harkins, *J. Am. chem. Soc.*, 1947, **69**, 683.
4. D. Stigter and K. J. Mysels, *J. phys. Chem., Ithaca*, 1955, **59**, 45.
5. M. E. Hobbs, *J. phys. Chem. Ithaca*, 1951 **55**, 675.
6. K. Shinoda, 'Solvent properties of nonionic surfactants in aqueous solutions', in *Solvent Properties of Surfactant Solutions* (ed. K. Shinoda), Marcel Dekker, New York, 1967, pp. 27–63.

7. H. B. Klevans, *Chem. Rev.*, 1950, **47**, 1.
8. M. E. L. McBain and E. Hutchinson, *Solubilization and Related Phenomena*, Academic Press, New York, 1955.
9. J. W. McBain and P. H. Richards, *Ind. Engng Chem.*, 1946, **38**, 642.
10. R. S. Stearns, H. Oppenheimer, E. Simon and W. D. Harkins, *J. chem. Phys.*, 1947, **15**, 496.
11. W. D. Harkins and H. Oppenheimer, *J. Am. chem. Soc.*, 1949, **71**, 808.
12. E. W. Hughes, W. M. Sawyer and J. R. Vinograd, *J. chem. Phys.*, 1945, **13**, 131.
13. W. D. Harkins, R. W. Mattoon and M. L. Corrin, *J. Am. chem. Soc.*, 1946, **68**, 220.
14. W. D. Harkins, R. W. Mattoon and M. L. Corrin, *J. Colloid Sci.*, 1946, **1**, 105.
15. F. A. Bovey and I. M. Kolthoff, *J. Polym. Sci.*, 1950, **5**, 487.
16. V. J. Frilette and W. P. Hohenstein, *J. Polym. Sci.*, 1948, **3**, 22.
17. R. G. R. Bacon, *Trans. Faraday Soc.*, 1946, **42**, 140.
18. J. H. Baxendale, M. G. Evans and G. S. Park, *Trans. Faraday Soc.*, 1946, **42**, 155.
19. J. H. Baxendale, M. G. Evans and J. K. Kilham, *Trans. Faraday Soc.*, 1946, **42**, 668.
20. L. B. Morgan, *Trans. Faraday Soc.*, 1946, **42**, 169.
21. C. W. Carr, I. M. Kolthoff, E. J. Meehan and D. E. Williams *J. Polym. Sci.*, 1950, **5**, 201.
22. I. M. Kolthoff, E. J. Meehan and C. W. Carr, *J. Polym. Sci.*, 1951, **6**, 73.
23. W. V. Smith, *J. Am. chem. Soc.*, 1948, **70**, 3695.
24. M. Morton, P. P. Salatiello and H. Landfield, *J. Polym. Sci.*, 1952, **8**, 111.
25. M. Morton, P. P. Salatiello and H. Landfield, *J. Polym. Sci.*, 1952, **8**, 215.
26. M. Morton, P. P. Salatiello and H. Landfield, *J. Polym. Sci.*, 1952, **8**, 279.
27. J. W. Wilson and E. S. Pfau, *Ind. Engng Chem.*, 1948, **40**, 530.
28. C. W. Carr, I. M. Kolthoff, E. J. Meehan and R. J. Stenberg, *J. Polym. Sci.*, 1950, **5**, 191.
29. V. L. Kuznetsov and A. V. Lebedev *Soviet Rubb. Technol.*, 1963, **22** (7), 10.
30. T. T. Serafini and E. G. Bobalek, *Off. Dig. Fed. Paint Varn. Prod. Clubs*, 1960, **32**, 1259.
31. E. W. Duck, J. A. Waterman and G. E. La Heij, *J. appl. Chem., Lond.*, 1962, **12**, 469.
32. V. L. Kuznetsov and A. V. Lebedev, *Soviet Rubb. Technol.*, 1964, 23(3), 31.
33. W. D. Stonecipher and R. W. Turner, 'Rosin and rosin derivatives', in *Encyclopedia of Polymer Science and Technology*, John Wiley, New York, 1970, Vol. 12, pp. 139–61.
34. J. T. Hays, A. E. Drake and Y. T. Pratt, *Ind. Engng Chem.*, 1947, **39**, 1129.
35. J. L. Azorlosa, *Ind. Engng Chem.*, 1949, **41**, 1626.
36. M. Feldon, F. L. McKennon, and R. V. Lawrence, *Ind. Engng Chem.*, 1952, **44**, 1662.
37. M. L. Corrin, H. B. Klevens and W. D. Harkins, *J. chem. Phys.*, 1946, **14**, 480.
38. C. F. Fryling and A. E. Follett, *J. Polym. Sci.*, 1951, **6**, 59.
39. H. W. Starkweather, *et al.*, *Ind. Engng Chem.*, 1947, **39**, 210.
40. B. Dogadkin, quoted by A. Talalay and M. Magat, in *Synthetic Rubber from Alcohol: A Survey Based on the Russian Literature*, Interscience, New York, 1945, p. 131.
41. H. Hopff and I. Fákla, *Br. Polym. J.*, 1970, **2**, 40.
42. P. M. Hay, *et al.*, *J. appl. Polym. Sci.*, 1961, **5**, 23.
43. W. A. Schulze, C. M. Tucker and W. W. Crouch, *Ind. Engng Chem.*, 1949, **41**, 1599.
44. D. C. Blackley and A. P. McGilvray, unpublished results; A. P. McGilvray, Ph.D. Thesis, C.N.A.A., 1973.

45. L. H. Howland, V. C. Neklutin, R. W. Brown and H. G. Werner, *Ind. Engng Chem.*, 1952, **44**, 762.
46. A. F. Helin *et al.*, *Ind. Engng Chem.*, 1953, **45**, 1330.
47. G. C. Greth and J. E. Wilson, *J. appl. Polym. Sci.*, 1961, **5**, 135.
48. W. C. Griffin, *J. Soc. Cosmetic Chemists*, 1949, **1**, 311.
49. W. C. Griffin, *J. Soc. Cosmetic Chemists*, 1954, **5**, 249.
50. W. C. Griffin, *Off. Dig. Fed. Paint Varn. Prod. Clubs*, 1956, **28**, 446.
51. S. S. Medvedev, I. A. Gritskova, A. V. Zuikov, L. I. Sedakova and G. D. Berejnoi, *J. Macromol. Sci.-Chem.*, 1973, **A7**, 715.
52. C. P. Roe, *J. Colloid and Interface Sci.*, 1971, **37**, 93.
53. C. P. Roe, 'Emulsion polymerziation with a surface active polyelectrolyte as the emulsifier', in *Polymer Colloids* (ed. R. M. Fitch), Plenum Press, New York, 1971, pp. 139–52.
54. B. W. Greene, D. P. Sheetz and T. D. Filer, *J. Colloid and Interface Sci.*, 1970, **32**, 90.
55. B. W. Greene and D. P. Sheetz, *J. Colloid and Interface Sci.*, 1970, **32**, 96.
56. B. W. Greene and F. L. Saunders, *J. Colloid and Interface Sci.*, 1970, **33**, 393.
57. A. E. Willson, J. R. Miller and E. H. Rowe, *J. phys. Colloid Chem.*, 1949, **53**, 357.
58. J. W. Vanderhoff, J. F. Vitkuske, E. B. Bradford and T. Alfrey, *J. Polym. Sci.*, 1956, **20**, 225
59. E. B. Bradford, J. W. Vanderhoff and T. Alfrey, *J. Colloid Sci.*, 1956, **11**, 135.
60. J. M. Willis, *Ind. Engng Chem.*, 1949, **41**, 2272.
61. A. Kotera, K. Furusawa and Y. Takeda, *Koll. Z. Z. Polym.*, 1970, **239**, 677.
62. K. Furusawa, W. Norde and J. Lyklema, *Koll. Z. Z. Polym.*, 1972, **250**, 908.
63. J. W. Goodwin, J. Hearn, C. C. Ho and R. H. Ottewill, *Br. Polym. J.*, 1973, **5**, 347.

CHAPTER 8

MODIFIERS

8.1 INTRODUCTION

Modifiers are substances which are introduced into free-radical emulsion polymerisation systems primarily in order to reduce the molecular weight of the polymer produced. Usually such substances are extremely active as transfer agents in free-radical reactions, and it is by this means that most are believed to be effective in reducing the average molecular weight of the product. The generalised transfer reaction can be written as follows:

$$\sim\sim\cdot + X—Y \rightarrow \sim\sim X + \cdot Y$$

where $\sim\sim\cdot$ denotes a growing polymer chain, XY is the transfer-active substance, and the radical $\cdot Y$ is capable of propagating a new polymer chain. Clearly, if this is the only reaction in which the substance XY can participate, then it will have no effect upon the kinetic chain length of the reaction, nor upon the rate of reaction (assuming that the reactivity of $\cdot Y$ is comparable with that of a normal propagating centre); its effect will be confined solely to that upon degree of polymerisation. However, modifiers sometimes have other effects upon the individual steps of the polymerisation process. The outstanding instance of this is that of mercaptans functioning as promoters for the persulphate-initiated polymerisation of butadiene and copolymerisation of styrene and butadiene. This matter has already been discussed in some detail in Chapter 6, Section 6.8.

In the present chapter, attention will be confined principally to the molecular-weight-reducing function of modifiers. Much of the information available is concerned with the use of mercaptans, and so an apparent disproportionate emphasis will be placed upon this class of compound. Other substances of which some note will be taken include the xanthogen disulphides and the di-aroyl disulphides.

Two matters of a general nature call for comment by way of introduction to a discussion of the behaviour of modifiers in emulsion polymerisation systems. The first concerns the question of what controls the rate of consumption of modifier relative to the rate of polymerisation of the monomer, and therefore, by implication, the rate at which it exerts its modifying action. The modifier is usually an oil-soluble substance which is initially present mainly in the monomer droplet phase. The reaction by which modifier is consumed, on the other hand, takes place mainly in the reaction loci. The reader may find it helpful to note that at least four distinct processes must occur in order for modifier dissolved in monomer droplets to become consumed at the reaction loci. In principle, any one of these four processes may be rate-determining for modifier consumption. The four steps are as follows: (1) transfer of modifier molecules across the boundary separating the monomer droplet from the aqueous phase; (2) diffusion through the aqueous phase from the vicinity of the droplet to the vicinity of a reaction locus; (3) transfer across the boundary between the aqueous phase and a reaction locus; and (4) reaction within the polymerisation locus.

The second matter concerns the various indices which are used to quantify the ability of a modifier to reduce the molecular weight of the polymer produced. These indices are usually derived from a study of the way in which the free modifier disappears from the reacting system as polymerisation occurs. The not unreasonable twofold assumption is made that (a) in order to be effective as a modifier, a modifier molecule must disappear from the system, and (b) all modifier molecules which have so disappeared have been used in modifying the polymer produced. However, matters are somewhat complicated by the following considerations:

(i) Because the transfer reaction represents a competition between transfer agent and monomer for the propagating polymer radicals, it is natural to consider the rate of disappearance of modifier relative to that of free monomer (*i.e.* relative to the rate of polymerisation) rather than its rate of disappearance *per se*.

(ii) Furthermore, because the propagation and transfer reactions are kinetically of first order with respect to the concentration of monomer and of transfer agent, respectively, it is natural to consider the rates of disappearance of monomer and modifier relative to their initial concentrations rather than the absolute rates of disappearance of either. Since $\delta[M]/[M] \simeq \delta \log [M]$ and $\delta[S]/[S] \simeq \delta \log [S]$ (where M and S denote monomer and modifier respectively), it is found that log $[S]$ is often considered as a function of log $[M]$, rather than $[S]$ as a function of conversion.

(iii) The pattern of disappearance of modifier relative to monomer depends upon both the modifier and the reaction conditions under which it is used.

The fundamental parameter which quantifies the effectiveness of a transfer-active substance in a bulk polymerisation is the so-called *transfer constant*, ϵ_s. This is defined as the ratio of the rate coefficient for the transfer reaction, k_s, to that for propagation, k_p. The two reactions can be written formally as

$$M_i\cdot + M \xrightarrow{k_p} M_{i+1}$$

and

$$M_i\cdot + S \xrightarrow{k_s} M_i + S\cdot$$

where $M_i\cdot$ denotes a growing polymer radical containing i monomer units. Clearly

$$-\frac{\mathrm{d}[M]}{\mathrm{d}t} = k_p[M]\sum[M_i\cdot]$$

if monomer which is consumed during other reactions is neglected. Also

$$-\frac{\mathrm{d}[S]}{\mathrm{d}t} = k_s[S]\sum[M_i\cdot]$$

Thus

$$\frac{\mathrm{d}[S]}{\mathrm{d}[M]} = \frac{\mathrm{d}[S]}{\mathrm{d}t}\bigg/\frac{\mathrm{d}[M]}{\mathrm{d}t} = \frac{k_s[S]}{k_p[M]} = \epsilon_s\frac{[S]}{[M]}$$

i.e.

$$\frac{\mathrm{d}\log[S]}{\mathrm{d}\log[M]} = \epsilon_s$$

i.e.

$$\log\frac{[S]}{[S]_0} = \epsilon_s\log\frac{[M]}{[M]_0}$$

where $[S]_0$ and $[M]_0$ are the initial concentration of $[S]$ and $[M]$, respectively, and thus $[S]/[S]_0$ and $[M]/[M]_0$ are the fractions of modifier and monomer, respectively, which remain unreacted. Thus log $[S]$ or log $([S]/[S]_0)$ plotted against log $[M]$ or log $([M]/[M]_0)$ should give a straight line whose slope is equal to ϵ_s. In fact, such plots for *emulsion polymerisations* are often found to be linear, but the slope is not necessarily equal to ϵ_s. The term *apparent transfer constant* is often used to distinguish this slope from the transfer constant as determined from bulk or solution polymerisation experiments. Apparent transfer constants for emulsion polymerisation systems are denoted here by ϵ_s'.

Usually ϵ_s' is less than, or at most equal to, ϵ_s. This is because, whereas ϵ_s quantifies the effect of only the fourth of the above four processes in controlling the rate of disappearance of modifier relative to that of monomer, ϵ_s' quantifies the possible rate-controlling effects of the other three processes as well. It is reasonable to suppose that ϵ_s' will be equal to ϵ_s if processes (1)–(3) play no part in controlling the rate of disappearance of the modifier, and if the ratio $[S]/[M]$ at the reaction loci is the same as in the polymerising system as a whole. It is further reasonable to suppose that, if the latter condition prevails (and there is no reason to suppose that it will not if processes (1)–(3) are rapid relative to other processes), then always ϵ_s' should approach ϵ_s if the polymerisation is carried out sufficiently slowly. Furthermore, it appears that for all emulsion polymerisation systems it should be possible to make ϵ_s' less than ϵ_s by causing the polymerisation to occur sufficiently rapidly. Thus, whereas ϵ_s is a fundamental parameter for a given modifier–monomer pair at a particular temperature, ϵ_s' is, in principle, dependent upon the rate at which an emulsion polymerisation occurs, and is therefore in the nature of a 'procedural' parameter.

The justification for expecting a linear relationship between $\log [S]$ and $\log [M]$ is thus seen to be the occurrence of simultaneous bimolecular propagation and transfer reactions, one species (the propagating radical) being common to both reactions. However, for some reactions, it is found that a linear relationship exists between $\log [S]$ and some *linear* function of $[M]$, such as $[M]/[M]_0$ or $([M]_0 - [M])/[M]_0$ (the latter being the fractional conversion). In such cases, the slope of the relationship between $\log [S]$ and $([M]_0 - [M])/[M]_0$ is taken as a measure of the effectiveness of the modifier. This slope is sometimes known as the *regulating index* for the modifier in the polymerising system. It is reasonable to enquire what pairs of reactions are kinetically consistent with such a linear relationship. One pair comprises (a) a propagation reaction which is of zero order with respect to monomer concentration and of first order with respect to radical concentration, and (b) a transfer reaction which is first-order in both modifier and radical concentrations. One then has that

$$-\frac{\mathrm{d}[M]}{\mathrm{d}t} = k \sum [M_i\cdot]$$

(k being the rate coefficient for the propagation reaction) and

$$-\frac{d[S]}{dt} = k_s[S]\sum[M_i\cdot]$$

i.e.

$$\frac{d[S]}{d[M]} = \frac{k_s}{k}[S]$$

Integration of this equation gives a linear relationship between log $[S]$ and $[M]$.

It is also worth noting that, for those reactions for which log $[S]$ varies linearly with log $[M]$, values of ϵ_s or ϵ_s' equal to unity imply that modifier and monomer are depleting at the same rates relative to their initial concentrations. The relationship $\log([S]/[S]_0) = \epsilon_s \log([M]/[M]_0)$ then carries the implication that $[S]/[M] = [S]_0/[M]_0$ at all stages in the reaction, *i.e.* that the ratio of modifier to monomer remains constant throughout the reaction.

8.2 MERCAPTANS AS MODIFIERS

It has already been noted that mercaptans are by far the most important class of modifier for emulsion polymerisation reactions. Most of the styrene–butadiene copolymer which is produced (in the Western world at least) by emulsion polymerisation is made using a mercaptan modifier. Usually the mercaptan is an alkyl compound containing from 8 to 14 carbon atoms. As the succeeding paragraphs will endeavour to show, within this restricted class there is a wide range of compounds from which to choose, and the exact behaviour of each compound depends not only upon its structure, but also upon the conditions under which it is used. As a broad generalisation, it appears that *n*-alkyl mercaptans containing about 12 carbon atoms are preferred for styrene–butadiene emulsion copolymerisations which are carried out at about 50°C, whereas *t*-alkyl mercaptans containing about the same number of carbon atoms tend to be preferred for these copolymerisations carried out at low temperature.

The possibility and desirability of 'modifying' the properties of rubbery polymers produced by emulsion polymerisation was first brought to the notice of industrial chemists through experience gained in the preparation of styrene–butadiene copolymers by emulsion copolymerisation. This situation has been well summarised by Snyder *et al.*,[1] as follows: 'When butadiene and styrene are copolymerized in an aqueous emulsion containing only the monomers, an emulsifying agent, and a peroxidic catalyst the product obtained is an insoluble substance which is difficult to process and is of little value as a rubber. Useful copolymers are obtained by adding to the polymerization system a substance known as a 'modifier'. In the preparation of GR-S the modifier used is a mercaptan or a mixture of mercaptans. The copolymer produced in the presence of such a modifier is soluble in benzene, and it can be milled and processed much like natural rubber.'

During roughly the first half of the decade from 1940 to 1950, there existed two theories which purported to account for the modifying action of substances such as mercaptans. The first of these attributed the effect to a re-direction of the butadiene propagation step of the polymerisation, whereby 1,4 addition was favoured relative to 1,2 addition. The modifier was considered either to catalyse the former mode of addition or else to inhibit the latter. In the absence of modifier, the polymerisation was supposed to yield a product in which much 1,2 addition had occurred, with the consequent presence of many vinyl side-chains. The double bonds in the latter were regarded as being more reactive than the internal double bonds resulting from 1,4 addition, and were in consequence supposed to be capable of ready participation in further polymerisation reactions, leading to the formation of branched and crosslinked molecules. According to this view, then, the modifier exerted its action by discouraging the formation of vinyl side-chains, and hence causing a less branched and cross-linked polymer to be formed.

The second theory attributed the effect of the modifier to its ability to participate in transfer reactions, and so reduce the molecular weight of the polymer produced. It is this view which has prevailed, and for various reasons. Perhaps the most convincing of these reasons is that it is precisely those substances which are known from bulk-polymerisation studies to be very effective transfer agents which are also effective as modifiers. This observation extends to a wide variety of chemical types, and is not merely confined to mercaptans.

Evidence of a more direct kind was provided by Snyder *et al.*[1] in respect of the use of mercaptans in the persulphate-initiated emulsion copolymerisation of styrene and butadiene. They showed (i) that *n*-dodecyl mercaptan is an effective modifier for the emulsion polymerisation of styrene alone (modification being interpreted in this case as reduction in molecular weight, and not as improvement in processability); and (ii) that the polymer molecules produced each contain approximately one atom of sulphur. The interpretation of the sulphur contents is obviously complicated to some extent by the fact that combined sulphur would have resulted from the persulphate-initiation mechanism as well as from the mercaptan-transfer reaction. This fact may not have been widely appreciated in 1946 when the work of Snyder and his co-workers was published. As reference to Section 6.4 of Chapter 6 will show, it was during the period 1947–55 that papers appeared in the literature describing work which established that polymers produced by persulphate initiation can contain persulphate-derived sulphur. Notwithstanding these complications, both the above observations are consistent with the transfer hypothesis and inconsistent with the hypothesis of re-direction of propagation. For, if mercaptans are effective in styrene–butadiene copolymerisation because they alter the balance of 1,4- to 1,2-butadiene addition, they can obviously have no effect upon the polymerisation of styrene from which butadiene is entirely absent. Furthermore, it is difficult to reconcile one sulphur atom per eventual polymer molecule with re-direction of propagation. In retrospect, it seems clear that, although some of the combined sulphur was almost certainly derived from the persulphate initiator, the observation that polymers of different molecular weights all contained about one sulphur atom per molecule still strongly supports the transfer hypothesis. The evidence is especially persuasive in that, as the mercaptan level in the reaction system was increased, the molecular weight of the polymer decreased and its sulphur content, expressed as a proportion of the weight of the polymer, increased.

The results of Snyder and co-workers are summarised in Table 8.1. The sulphur contents of purified polymers were determined by a process in which the polymers were first oxidised by the Parr bomb method, and then the sulphur estimated gravimetrically as barium sulphate. The emulsion polymers were isolated and purified by a process which comprised coagulation of the latex with saturated sodium chloride solution containing a little sulphuric acid, steam distillation of the polymer to remove excess styrene, washing with methanol, and then precipitation into excess methanol from a solution in dioxan.

Studies were also carried out on polybutadienes of low molecular weight prepared in the presence of dodecyl mercaptan. These were found to be less satisfactory materials for study than similarly prepared polystyrenes. Thus they were not suitable for oxidation by the Parr bomb method, and, instead, a wet-oxidation procedure was used, for which the oxidising agent was a mixture of fuming nitric acid, bromine and potassium chlorate. The progressive change in physical character of the

TABLE 8.1

Effect of mercaptan level upon molecular weight and sulphur content of polystyrenes produced by persulphate-initiated emulsion polymerisation at 50°C (according to Snyder *et. al.*[1])

Details of reaction system (parts by weight)					Characteristics of polymer produced			
						Molecular weight		
Styrene	2·5% aqueous fatty-acid soap	$K_2S_2O_8$	n-dodecyl mercaptan	Reaction time (h)	Sulphur content (%)	Viscometric	Cryoscopic	Number of S atoms per polymer molecule
100	200	0·3	0·5	18	—	—	—	—
100	200	5·0	10·0	18	1·1	$4{\cdot}40 \times 10^3$	$2{\cdot}80 \times 10^3$	0·96
100	200	6·0	26	18	1·82–1·91	—	$1{\cdot}32–1{\cdot}50 \times 10^3$	0·8–0·9
100	200	6·0	26	22	1·80–1·83	—	$1{\cdot}65 \times 10^3$	0·94
100	200	6·0	26†	20	0·3	$1{\cdot}30 \times 10^4$	$1{\cdot}24 \times 10^4$	1·1–1·2
230	450	8·0	60	18	2·50–2·95	$1{\cdot}20–1{\cdot}50 \times 10^3$	—	1·1–1·35
90	200	1·0	50·7	22	Addition compound $\phi CH_2CH_2SC_{12}H_{25}$ was isolated as the main product			

† Added stepwise as follows: 0·1 parts initially, then 0·2, 0·4, 0·8, 1·6, 3·2, etc., added each hour.

polybutadienes, from rubbery solids through sticky semi-solids to viscous oils, which accompanied an increase in mercaptan level was taken as indicating a decrease in molecular weight with increasing mercaptan.

For polymers prepared by both emulsion polymerisation and by bulk polymerisation (for which the results are not given here), the molecular weight was found to be approximately inversely proportional to the amount of mercaptan in the polymerisation system. Correlation of molecular weight results with sulphur contents indicated about 0·6 sulphur atoms per polymer molecule. In view of the experimental difficulties encountered, this was regarded as agreeing satisfactorily with the theoretical prediction of almost 1·0. Hydrogenolysis of both dodecyl mercaptan-modified polystyrene and styrene–butadiene copolymer using Raney nickel as a catalyst gave dodecane as one of the products. This was taken as evidence that the mercaptan modifier enters the polymer molecule as a consequence of the modifying reaction. In fact, the evidence of Snyder and his co-workers suggests that both the alkyl group and the thiol sulphur of a mercaptan become incorporated in the polymer molecule.

The results of a further study of the average number of sulphur atoms per molecule of styrene–butadiene copolymer was published by Wall, Banes and Sands[2] at the same time as that by Snyder and his co-workers. The work described in this paper was carried out on a range of GR-S rubbers. These were almost certainly prepared by persulphate initiation. It was presumed that, if the mercaptan-transfer reaction had occurred at all, then it would have been of much more frequent occurrence than either initiation or termination. The expected concentration of mercaptan derived combined sulphur atoms would therefore be almost exactly one per polymer molecule.

Wall, Banes and Sands argued that, since an appreciable fraction of the copolymers investigated was insoluble in benzene and similar solvents, it was necessary to remove the gel fraction in order to obtain meaningful molecular weights. Since they used the osmotic pressure method for molecular weight determination, they also considered it desirable to remove low-molecular-weight polymer from the samples because these diffuse through the membrane of the osmometer and so do not contribute to the osmotic pressure. Their procedure was to use that fraction of the copolymer which was soluble in benzene but insoluble in a mixture consisting of equal volumes of benzene and methanol. The number of sulphur atoms per polymer molecule was found to range between 0·7 and 1·1. For various reasons, connected with experimental procedure, the value of 0·7 was considered to be too low. More accurate determinations on a small number of samples indicated that 1·1 was nearer the correct value. In view of the experimental difficulties implicit in the procedures used, the results were considered to be in good agreement with the theoretical prediction on the basis that modification is due to transfer.

Much information has been published concerning the rate of consumption of mercaptans in styrene–butadiene emulsion copolymerisation systems. Thus Smith[3] has reported on the rates of disappearance of mercaptans in both emulsion and bulk polymerisations for *t*-butyl mercaptan in styrene, *n*-amyl mercaptan in styrene, and *n*-amyl mercaptan in methyl methacrylate. In each of these three cases, the rate of consumption of mercaptan relative to the rate of polymerisation was virtually identical in both emulsion and bulk polymerisations. The results are summarised in Fig. 8.1. Residual mercaptan was estimated by amperometric titration using silver nitrate. The method consists essentially of adding a few grams of the latex to 100 ml of ethanol with vigorous stirring. 1 ml of 0·1*N* iron(II) sulphate in 1*N* sulphuric acid is then added. The purpose of adding the iron(II) sulphate is to decompose residual potassium persulphate. After standing for a few minutes, 2 ml of concentrated ammonia solution is added and the amperometric titration with 0·005*N* silver nitrate carried out. A rotating platinum electrode was used by Smith in conjunction with a half-cell prepared by dissolving 4·2 g of potassium iodide and 1·3 g of mercuric iodide in 100 ml of saturated potassium chloride solution, and placing a pool of mercury at the bottom of the solution. In the case of polymers prepared by bulk polymerisation, the mercaptan determination was carried out by adding the sample to ethanol with vigorous stirring, and then titrating with silver nitrate as above. It should be noted that the bulk polymerisations were carried out at different temperatures in order to obtain convenient rates of reaction. Smith states that, notwithstanding this, the rate of mercaptan consumptions in the two types of reaction can be compared because the temperature-dependence of the reaction is small.

The conclusion drawn by Smith from these results is that, using the above-mentioned lower alkyl mercaptans, diffusion of mercaptan from the bulk monomer phase through the aqueous phase to the reaction loci is not a rate-controlling process for mercaptan consumption in emulsion polymerisation. He concludes that (i) both monomer and mercaptan can diffuse through the aqueous phase sufficiently rapidly to make good any losses due to reaction, and (ii) the ratio of mercaptan to monomer at the reaction loci will be the same as that in the droplets, and both will be equal to the overall ratio in the system as a whole.

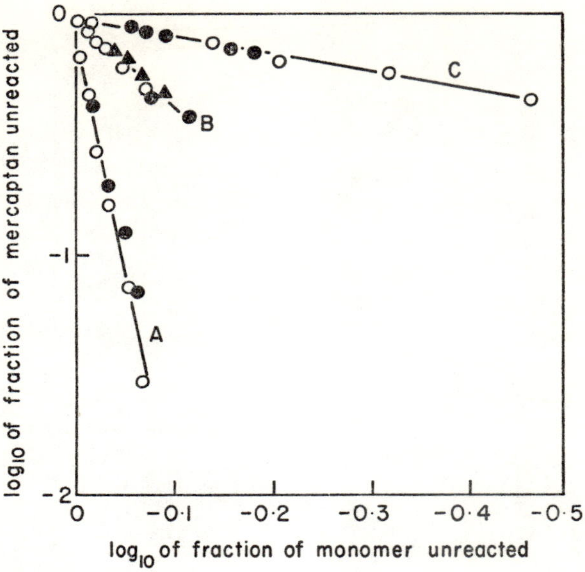

FIG. 8.1. Modifier consumption curves for two low-molecular-weight alkyl mercaptans during bulk and emulsion polymerisation, according to Smith.[3] Curve A: *n*-amyl mercaptan in styrene. Curve B: *t*-butyl mercaptan in styrene. Curve C: *n*-amyl mercaptan in methyl methacrylate. Significance of points: ○ emulsion polymerisation at 62·5°C; ● bulk polymerisation at 62·5°C; ▲ bulk polymerisation at 100°C. (For bulk polymerisations, the charges consisted of monomer and mercaptan. For emulsion polymerisation, the charges consisted of (i) 10 ml of a solution comprising 3·9 g of stearic acid, 0·2 ml of mercaptan, and monomer to 100 ml, and (ii) 15 ml of a solution comprising 0·175 g of $K_2S_2O_8$, 3·51 ml of 12.1% KOH, and oxygen-free water to 100 ml.)

That the rate of consumption of mercaptan relative to polymerisation is not always the same in emulsion and bulk polymerisations is indicated by results published in a second paper by Smith.[4] Some results for the effect of alkyl chain length upon the disappearance of alkyl mercaptans in butadiene emulsion polymerisation are shown in Fig. 8.2. It is seen that, for mercaptans having chain lengths up to C_{10}, the rate of disappearance is virtually independent of molecular weight, but that *n*-alkyl mercaptans having more than 10 carbon atoms are consumed progressively more slowly as the molecular weight increases. There is good reason to believe that the relative rate of consumption in bulk polymerisations does not depend very much upon the molecular size of the mercaptan. (Smith also published information on the rates of butadiene emulsion

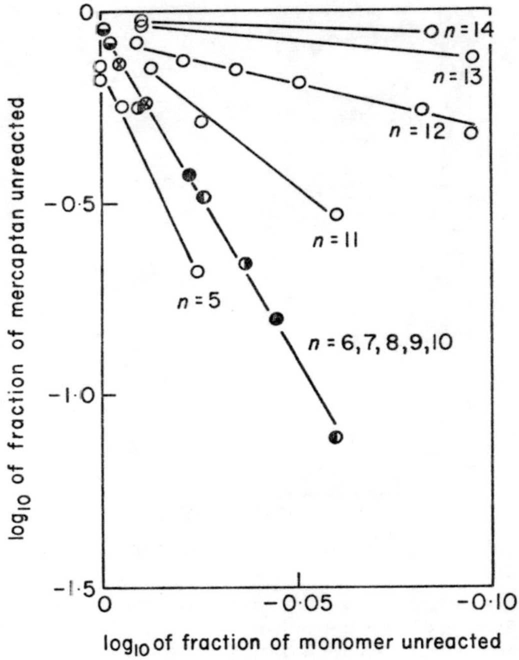

FIG. 8.2. Modifier consumption curves for *n*-alkyl mercaptans of various alkyl chain lengths during persulphate-initiated emulsion polymerisation of butadiene at 50°C, according to Smith.[4] Significance of points: ○ $C_nH_{2n+1}SH$ for value of *n* appended to the appropriate line; ◒ hexyl mercaptan; ⊗ heptyl mercaptan; ◑ octyl mercaptan; ● nonyl mercaptan; ◐ decyl mercaptan. (Typical reaction system: butadiene 70, water 120, fatty-acid soap 3·5, mercaptan 0·1–2·0 ml, $K_2S_2O_8$ 0·21.)

polymerisation in the presence of *n*-alkyl mercaptans from C_5 to C_{14}. He found that the rate of polymerisation is almost independent of chain length for mercaptans having 8 or more carbon atoms, but is significantly less for mercaptans of lower chain length.) Data for the effect of chain length upon the relative rate of disappearance of mercaptans in both styrene and butadiene emulsion polymerisations are summarised in Fig. 8.3. This shows clearly that, in the case of butadiene, the relative rate of disappearance of mercaptan is independent of alkyl chain length up to C_{10}, and that the addition of successive carbon atoms reduces the relative rate progressively. The more limited data for styrene suggest a similar trend. The conclusion which Smith draws from these results is that, when the alkyl chain length exceeds 10 carbon atoms, the rate-controlling process becomes the transfer of mercaptan from monomer

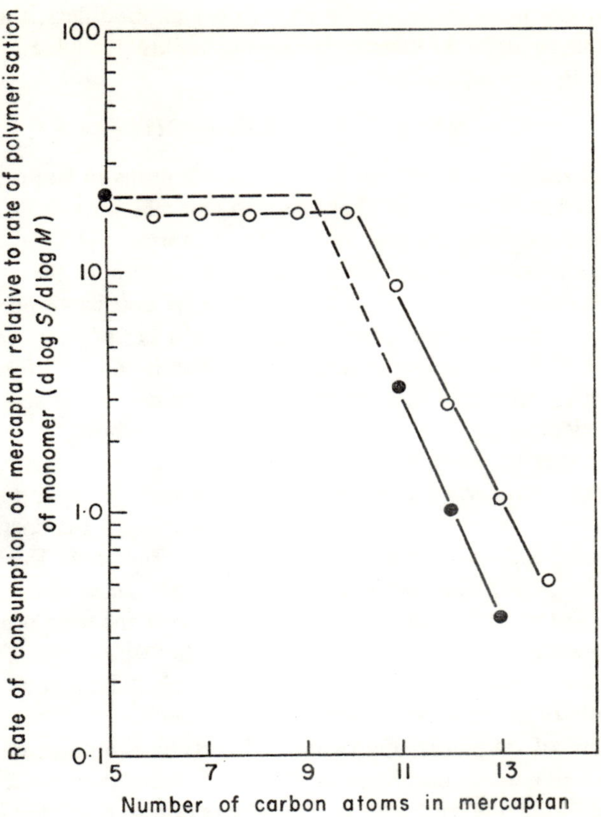

FIG. 8.3. Effect of alkyl chain length upon rate of consumption of *n*-alkyl mercaptan relative to rate of polymerisation of monomer for persulphate-initiated emulsion polymerisation of butadiene at 50°C (points ○) and of styrene at 40°C (points ●), according to Smith.[4] (*S* denotes the level of residual mercaptan and *M* that of residual monomer. See the caption to Fig. 8.3 for details of a typical reaction system.)

droplets through the aqueous phase to the reaction loci. Additional support for this view comes from the observation that the rate of consumption of *n*-alkyl mercaptans having alkyl groups larger than C_{10} can be increased by raising the pH of the aqueous phase. For lower-molecular-weight mercaptans, such as *n*-amyl mercaptan, pH has no effect upon the rate of mercaptan consumption. It appears that, in the case of the mercaptans of higher molecular weight, diffusion of mercaptan through the aqueous phase is supplemented by diffusion of mercaptide ions. Although these ions may not themselves be able to act as modifiers, they are presumably able to make the corresponding mercaptan available by an appropriate shift in the equilibrium:

$$RS^- + H_2O \rightleftharpoons RSH + OH^-$$

The acid dissociation constants for alkyl mercaptans in water are said to be of the order of 10^{-11}, so that the effect of pH upon mercaptide ion concentration should become apparent at about pH 11. The overall conclusion drawn from this investigation by Smith is that, unless the aqueous phase is strongly alkaline, the rate of consumption of *n*-alkyl mercaptans of chain length greater than 10 will be seriously retarded by the presence of the aqueous phase through which it has to diffuse, and that in consequence the ratio of mercaptan to monomer at the reaction locus will be considerably lower than in the system as a whole.

A third paper by Smith[5] is of interest in that it shows that the rate of transfer relative to propagation in an emulsion copolymerisation reaction can be predicted quite accurately from a knowledge of the corresponding quantities for the individual homopolymerisations and the monomer reactivity ratios for the particular pair of monomers which are being polymerised. It must be pointed out, however, that the data presented are for *n*-amyl mercaptan, *i.e.* they are for a mercaptan whose rate of consumption is not controlled by rate of diffusion through the aqueous phase.

The results of an extensive study of the factors which affect the rate of consumption of aliphatic mercaptans in styrene–butadiene emulsion copolymerisation were published by Kolthoff and Harris in 1947.[6] The method used for determining residual mercaptan was that described above in connection with the review of the investigation of Smith.[3] Kolthoff and Harris point out that sampling procedure is very important unless the whole bottle of latex is used for the estimation. They state that extremely large errors in mercaptan analysis can be made if precautions are not taken to obtain a truly representative sample. They give a very full discussion of this problem, and also of that of the co-precipitation of unreacted mercaptan with the polymer at the coagulation stage. They state that when the latex is coagulated by the addition of a mixture containing sulphuric acid and sodium chloride, or by aqueous acid alone, all of the unreacted mercaptan is precipitated with the polymer, and that this mercaptan cannot be extracted quantitatively from the polymer using alcohols as solvent. They further state that it is possible to separate the mercaptan quantitatively from the polymer if alcohol is used to coagulate the latex, and if the conditions of coagulation are such that the polymer is maintained as a small-particle crumb and not allowed to agglomerate into large lumps. The best conditions for coagulation are said to be a large alcohol–latex ratio, vigorous stirring of the alcohol while adding the latex, and addition of the latex to the alcohol dropwise or as a fine spray.

Kolthoff and Harris found that the fraction of unreacted mercaptan remaining in a latex after a given conversion is independent of the initial amount of mercaptan. The evidence upon which this conclusion is based is shown in Fig. 8.4 for (a) a commercial mixture of mercaptans and (b) *t*-dodecyl mercaptan. It appears that the disappearance curves are identical for at least a tenfold variation in the level of mercaptan. According to Kolthoff and Harris, this finding has been confirmed by other workers using different mercaptans.

The results of Kolthoff and Harris for the effect of number of carbon atoms in an alkyl mercaptan upon its consumption during the course of the reaction are shown in Fig. 8.5. At (a) are shown the disappearance curves for primary mercaptans, and at (b) the curves for tertiary mercaptans. The chain length of the alkyl group clearly has a profound effect upon the course of the mercaptan disappearance, especially in the case of the primary mercaptans. For the latter, Kolthoff and Harris conclude

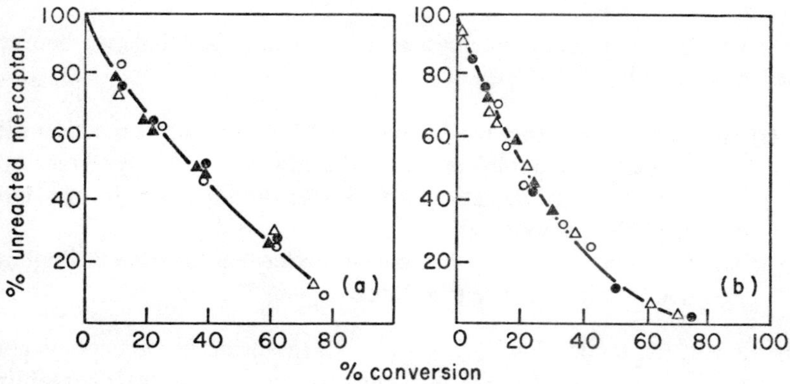

FIG. 8.4. Modifier consumption curves for (a) a commercial mixture of mercaptans and (b) *t*-dodecyl mercaptan, present in various amounts during persulphate-initiated emulsion copolymerisation of styrene and butadiene at 50°C, according to Kolthoff and Harris.[6] Significance of points in relation to mercaptan level in pphm is as follows: (a) ○ 0·10; ● 0·25; △ 0·50; ▲ 0·75; (b) ○ 0·1; ● 0·3; △ 0·5; ▲ 1·0. (Reaction system: styrene 25, butadiene 75, water 180, fatty-acid soap 5, mercaptan—variable, $K_2S_2O_8$ 0·3.) Composition of commercial mixture of mercaptans was as follows: *n*-decyl mercaptan 5%, *n*-dodecyl mercaptan 58%, *n*-tetradecyl mercaptan 25%, impurities 12%.

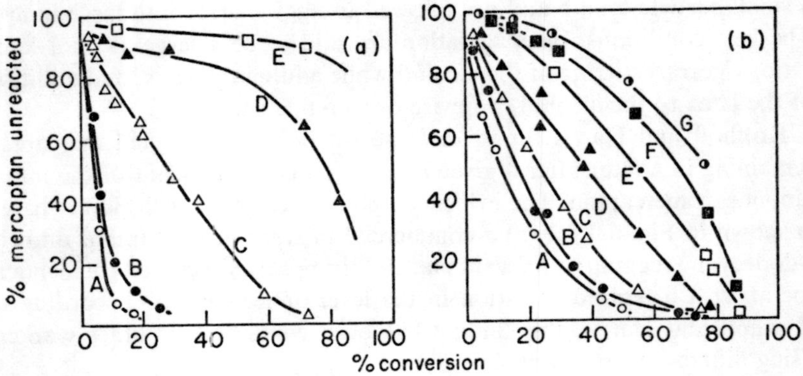

FIG. 8.5. Modifier consumption curves for (a) primary alkyl mercaptans and (b) tertiary alkyl mercaptans of various chain lengths during persulphate-initiated emulsion copolymerisation of styrene and butadiene at 50°C, according to Kolthoff and Harris.[6] Significance of letters appended to curves: (a) A, octyl mercaptan; B, decyl mercaptan; C, dodecyl mercaptan; D, tetradecyl mercaptan; E, hexadecyl mercaptan; (b) A, butyl mercaptan; B, octyl mercaptan; C, decyl mercaptan; D, dodecyl mercaptan; E, tetradecyl mercaptan; F, pentadecyl mercaptan; G, hexadecyl mercaptan. (See caption to Fig. 8.4 for details of reaction system. Mercaptan level believed to be 0·5 pphm.)

that the C_{12} (*i.e.* dodecyl) mercaptan provides the best balance between the two conflicting tendencies of

(i) very rapid disappearance during the early stages of the reaction with consequent over-modification of the low-conversion polymer and rapid mercaptan depletion and under-modification in the later stages of the reaction; and

(ii) slow disappearance with consequent under-modification throughout the whole course of the reaction.

The first of the tendencies is observed with the primary mercaptans containing fewer than 12 carbon atoms, and the second with those containing more than 12. The effects observed with tertiary mercaptans are qualitatively similar, but the rate of disappearance depends less markedly upon alkyl chain length in this case. It is also worth recording in this connection that Starkweather *et al.*[7] concluded that primary mercaptans containing between 11 and 14 carbon atoms appear to be most effective as modifiers for the persulphate-initiated copolymerisation of styrene and butadiene at 40°C using a fatty-acid/rosin-acid mixture as micelle generator and

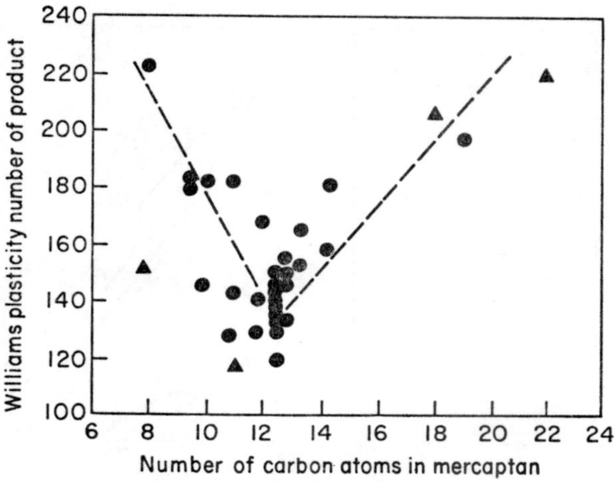

FIG. 8.6. Effect of chain length of primary alkyl mercaptan modifier upon plasticity of styrene–butadiene copolymer produced by persulphate-initiated emulsion copolymerisation for 20 h at 40°C, according to Starkweather *et al.* [7] Significance of points: ● product yield between 85 and 95%; ▲ product yield between 70 and 85%. (Reaction system: styrene 25, butadiene 75, water 150, oleic acid 4, rosin acid 1, sodium naphthalene formaldehyde sulphonate 1, mercaptan 0·75, $K_2S_2O_8$ 1, $K_3Fe(CN)_6$ 0·15, excess NaOH 0·15.)

potassium ferricyanide as activator. Their results are shown in Fig. 8.6, in which the Williams plasticity number of the polymer (which is an inverse measure of plasticity) is plotted against the number of carbon atoms in the modifier. The data are extremely scattered, which was probably partly due to variation in conversion, but they seem to sustain the broad conclusion drawn by the authors. It is interesting that, according to the lines which Starkweather and his co-workers drew through the scatter of points, the maximum product plasticity is obtained when the mercaptan contains 12 carbon atoms. It is also worth noting that Marvel and Lewis[8] have claimed that *n*-octyl mercaptan is a better modifier than *n*-dodecyl mercaptan for the persulphate-initiated emulsion copolymerisation of styrene and butadiene at high temperature (110°C).

The results of Kolthoff and Harris[6] for the comparative rates of consumption of primary, secondary and tertiary mercaptans of a given number of carbon atoms are shown in Fig. 8.7 (a) and (b). The first set of curves is for octyl mercaptans, and the second for dodecyl mercaptans. The differences are not large, especially in the case of the C_{12} mercaptans, where diffusion through the aqueous phase is the rate-controlling process.

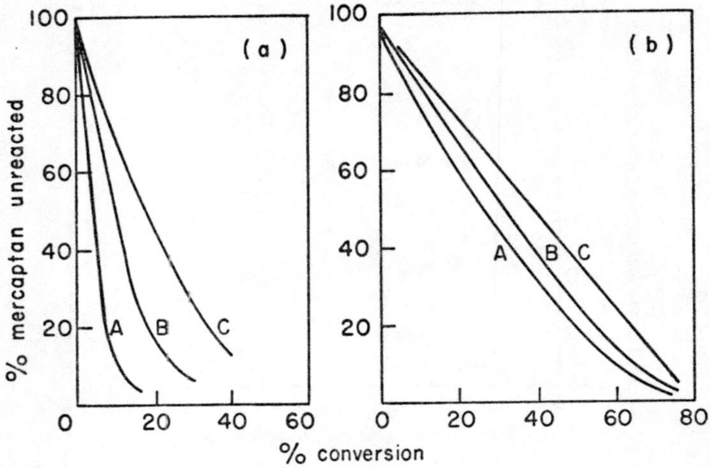

FIG. 8.7. Modifier consumption curves for (a) octyl mercaptans and (b) dodecyl mercaptans during persulphate-initiated emulsion copolymerisation of styrene and butadiene at 50°C, according to Kolthoff and Harris.[6] Curves marked A, B and C refer to primary, secondary and tertiary mercaptans respectively. (See caption to Fig. 8.4 for details of reaction system. Mercaptan level believed to be 0·5 pphm.)

In both cases, the order of increasing rate of consumption is tertiary < secondary < primary.

The effect of mercaptan structure upon rate of diffusion through aqueous media and rate of consumption during emulsion polymerisation has also been the subject of study by Frank *et al.*[9] These workers carefully synthesised five C_{12}-alkyl mercaptans having different isomeric dodecyl groups. The compounds used were

(i) the straight-chain primary C_{12} mercaptan, *i.e.* *n*-dodecyl mercaptan—(structure(I);

(*ii*) a branched-chain primary C_{12} mercaptan, namely 4-*n*-butyl-1-octyl mercaptan—structure (II);

(*iii*) a straight-chain secondary C_{12} mercaptan with the thiol group near the centre of the chain, namely, 6-dodecyl mercaptan—structure (III);

(*iv*) a tertiary C_{12} mercaptan having the maximum amount of branching, namely, di-*n*-butylpropylcarbinthiol—structure (IV); and

(*v*) a tertiary C_{12} mercaptan having the minimum amount of branching, namely, dimethyl-*n*-nonylcarbinthiol—structure (V).

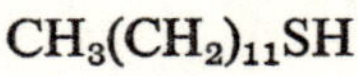

(I)

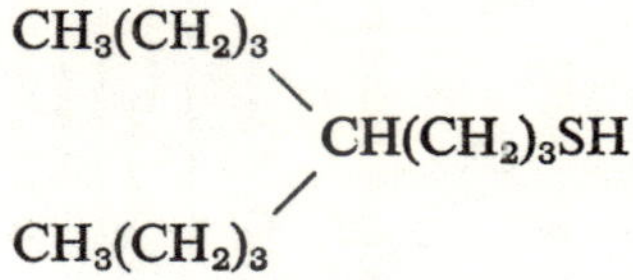

(II)

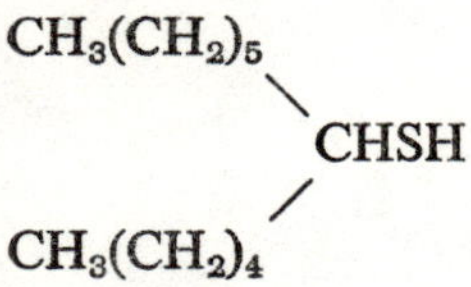

(III)

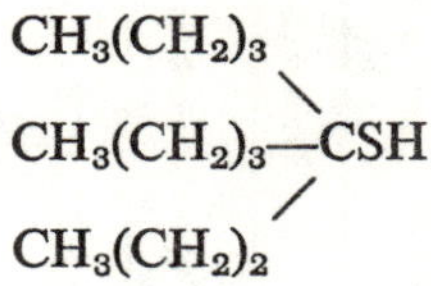

(IV)

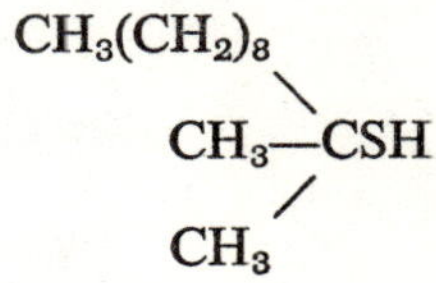

(V)

Rates of diffusion from the oil phase into and through an aqueous phase were determined by measuring the rate of decrease of a droplet of mercaptan suspended in a large volume of soap solution. The rates of mercaptan disappearance during the emulsion copolymerisation of styrene and butadiene were followed by the method of Kolthoff and Harris.[6] The behaviour of the mercaptans in the emulsion copolymerisation was judged by effect upon (i) the rate of polymerisation, as indicated by the percentage conversion after a fixed time, and (ii) the structure of the polymer produced, as indicated by effect upon solubility in benzene and intrinsic viscosity. The results are summarised in Table 8.2. It is clear that the amount of branching in the mercaptan molecule is an important factor in determining diffusion rate, and that the shorter is the longest linear sequence of atoms in the molecule, the faster is the rate of diffusion. The rate of mercaptan consumption correlates well with rate of diffusion if comparisons are restricted to mercaptans of a given type, *i.e.* primary, secondary or tertiary. (The data in Table 8.2 permit comparisons to be made for primary and tertiary mercaptans only.) Effect upon overall conversion is slight, but suggests that, for a given type, the slower-diffusing mercaptans are able to exert a promoting effect over a rather longer time and so increase the overall conversion slightly. The results for intrinsic viscosity also indicate that the slower-diffusing mercaptans are more effective over the reaction as a whole.

Kolthoff and Harris[6] have reported upon the rate of disappearance of mixtures of mercaptans, and found that the curve for the overall disappearance of mercaptan is almost exactly that which would be predicted on a basis of simple additivity from a knowledge of the known rates of disappearance of the individual mercaptans.

TABLE 8.2

Effect of mercaptan structure upon rate of diffusion, rate of consumption, and ability to modify, of C_{12} alkyl mercaptans (according to Frank *et al.*[9])

(reaction system: styrene 25, butadiene 75, water 170, sodium fatty-acid soap 5, *n*-dodecyl mercaptan 0·5, $K_2S_2O_8$ 0·3)

Mercaptan structure (see text for details)	*Longest linear sequence of atoms (including S)*	*Primary, secondary or tertiary mercaptan*	*Rate of diffusion* ($g\ cm^{-2}\ h^{-1} \times 10^4$)	*Mercaptan consumption data: % unreacted mercaptan at conversion indicated*								*Properties of polymer produced*		
				10%	20%	30%	40%	50%	60%	70%	80%	*% conversion‡*	*% soluble in benzene*	*Intrinsic viscosity* ($dl\ g^{-1}$)
(I)	13	Primary	1·32, 1·22†	83	69	54	38	21	7	1	—	77·2, 77·7†	100	2·06, 1·94†
(II)	9	Primary	3·12	76	60	44	29	14	3	—	—	74·2	100	2·37
(III)	12	Secondary	2·09	85	70	55	39	24	12	2	—	75·8	100	2·09
(IV)	9	Tertiary	3·13	85	70	55	40	26	14	6	2	75·8	100	1·89
(V)	11	Tertiary	2·37	87	73	59	44	28	12	3	1	76·5	100	1·89

† These two results refer to samples obtained by different methods.
‡ Reaction time not stated; reaction temperature believed to have been 50°C.

TABLE 8.3

Effect of mercaptan level upon molecular weight of low-conversion styrene–butadiene copolymers produced by persulphate-initiated emulsion copolymerisation at 50°C (according to Kolthoff and Harris[6])

(reaction system: styrene 25, butadiene 75, water 180, fatty-acid soap 5, *n*-dodecyl mercaptan—variable, $K_2S_2O_8$ 0·3)

Mercaptan level (pphm)	*Intrinsic viscosity of polymer at about 5% conversion* (dl g^{-1})	*Viscosity-average molecular weight of polymer*†	$\overline{M}_v$ × *level of mercaptan in reaction system*
0·3	1·14	$1·10 \times 10^5$	$3·3 \times 10^4$
0·24	1·31	$1·35 \times 10^5$	$3·2 \times 10^4$
0·18	1·62	$1·87 \times 10^5$	$3·4 \times 10^4$
0·12	2·16	$2·90 \times 10^5$	$3·5 \times 10^4$
0·06	3·80	$6·80 \times 10^5$	$4·0 \times 10^4$
0·03	4·85‡	$9·80 \times 10^5$‡	$2·9 \times 10^4$‡
0·015	4·95‡	$1·00 \times 10^6$‡	$1·5 \times 10^4$‡

† Calculated from the intrinsic viscosity of polymer in benzene using the equation $\log_{10} M = 4·95 + 1·49 \log_{10} [\eta]$.
‡ Polymers contained gel.

They also point out that, for a given molecular size, the mercaptan structure influences the rate of attainment of solubilisation equilibrium. This is presumably the corollary of the effect of structure upon diffusion rate reported by Frank and his co-workers[9] (see Table 8.2). Thus Kolthoff and Harris state that *t*-dodecyl mercaptan is solubilised in soap solutions about three times as fast as is primary dodecyl mercaptan.

Kolthoff and Harris[6] have also provided information concerning the modifying effects of mercaptans. Their results for the effect of mercaptan level upon the molecular weight of the polymer produced at low-conversion (*i.e. ca.* 5%) are given in Table 8.3. These indicated that the molecular weight is inversely proportional to mercaptan level up to molecular weights of about 1×10^6. The relationship between molecular weight and mercaptan level for polymers at higher conversions is complicated by variation in rate of modification accompanying variation in mercaptan consumption over the course of the reaction. Kolthoff and Harris point out that the disappearance curve for, say, primary decyl mercaptan suggests that low-molecular-weight polymers are formed at low conversion, followed by the formation of greatly under-modified polymer at higher conversions. They provide evidence to show that the intrinsic viscosity of the polymer first formed is very low, but rapidly increases at about 20%

conversion. The polymer soon becomes incompletely soluble in benzene after this rapid rise in intrinsic viscosity has occurred. On the other hand, *t*-hexadecyl mercaptan is consumed so slowly at all conversions that the polymer formed throughout the reaction is of high molecular weight and high intrinsic viscosity. An intermediate situation is presented by primary dodecyl mercaptan. This is consumed at a much slower rate than is primary decyl mercaptan. The intrinsic viscosity at low conversions is therefore much higher than that obtained with an equivalent amount of the latter. Furthermore, the intrinsic viscosity remains at about the same value, though increasing slowly, over much of the conversion range, and begins to increase more rapidly at conversions between 60% and 70%, by which time most of the mercaptan has been consumed. At higher conversions, little modification occurs, and the polymer becomes partially insoluble in benzene.

Kolthoff and Harris also give information on the effect of experimental technique upon the rate of mercaptan consumption and consequent modifying effect. Thus mode and rate of agitation have an important influence, which will be especially evident for those mercaptans for which the process controlling the rate of consumption is transfer from monomer droplets to reaction loci. The initial rate of disappearance of commercial mercaptan mixture in large reactors has been found to be much faster relative to conversion than under laboratory conditions; this has been attributed to a higher rate of agitation in the former case. The rate of consumption of commercial mercaptan mixture is also found to be enhanced by forming the soap *in situ* (*i.e.* adding the fatty acid to the monomer and the alkali to the water as compared to that observed when preformed soap is used. This is attributed to the finer state of subdivision of the droplets obtained in the former case. Again, this is an effect observed for mercaptans for which transfer from droplets to loci is the process controlling the rate of consumption. Kolthoff and Harris have also confirmed the observation of Smith[4] that, where such transfer is the rate-controlling process, the addition of excess alkali enhances the rate of consumption of mercaptan. The effect of temperature upon modification by the commercial mercaptan mixture is such that the amount required to obtain a polymer of given intrinsic viscosity at 70–75% conversion increases as the temperature decreases. This is because the lower the temperature, the faster is the relative rate of consumption of mercaptan, and therefore at lower temperatures less is left for modification at the higher conversions unless more is used. Kolthoff and Harris also point out that stepwise or continuous addition of mercaptan throughout the course of the polymerisation is a logical procedure for producing a polymer of uniform molecular weight throughout the whole course of the reaction. They do not, however, give data to support this view.

Harris and Kolthoff[10] have provided information concerning the problem of predicting the molecular weights and intrinsic viscosities of styrene–butadiene emulsion copolymers from mercaptan consumption data. They point out that a close correlation should exist between the mercaptan consumed during the polymerisation and any property of the polymer which is dependent upon its molecular weight distribution. They also point out that any mercaptan should be as effective as any other mercaptan as a modifier at a given conversion if the same molecular amounts have been consumed at that conversion and at all lower conversions. The latter qualification is important; the nature of the molecular weight distribution in the polymer produced at any conversion depends not only upon the total amount of mercaptan consumed, but also upon its pattern of consumption. In fact, theoretically the number-average molecular weight of the polymer produced should depend only upon total mercaptan consumption and not upon mode of consumption. But other molecular weight averages which depend upon both the location of the distribution and upon its shape should depend upon the pattern of consumption as well as upon the total consumption. The viscosity-average molecular weight is one such average, and the intrinsic viscosity is an example of a property which depends upon the distribution of molecular sizes.

Harris and Kolthoff have found that, for styrene–butadiene copolymers produced by persulphate-initiated emulsion copolymerisation at 50°C using the standard GR-S system, the following empirical relationship enables the intrinsic viscosity in benzene of the polymer produced to be predicted from mercaptan consumption data:

$$[\eta] = \left\{\frac{0{\cdot}22P_\infty}{AS_\infty} + \frac{1{\cdot}12}{A}\int_0^{P_\infty}\frac{\mathrm{d}P}{\mathrm{d}S}\,\mathrm{d}P\right\}^{0{\cdot}66}$$

In this equation, P is the fraction of monomer which is converted to polymer, P_∞ is the final conversion, S is the fraction of the total mercaptan which has reacted at conversion P (so that S_∞ corresponds to P_∞), and A is the initial level of mercaptan in the reaction system. (Actually A is defined as the initial mercaptan level in pphm of mercaptan equivalent to pure dodecyl mercaptan. It can be calculated as the product of the pphm of impure mercaptan and the quotient (% sulphur in mercaptan)/ 15·87.) The first term in the bracket on the right-hand side of the above equation depends only upon the total amount of mercaptan consumed per unit amount of polymer produced. It is independent of the manner in which the mercaptan is consumed. The second term allows for the effect of the shape of the mercaptan disappearance curve upon the overall intrinsic viscosity of the polymer produced. By using an empirical relationship previously established between $[\eta]$ and the viscosity-average molecular weight for these polymers (see Table 8.3), it is possible directly to predict molecular weight from mercaptan consumption data as follows:

$$10^{-5}\overline{M}_v = \frac{0{\cdot}2P}{AS_\infty} + \frac{1}{A}\int_0^{P_\infty} \frac{\mathrm{d}P}{\mathrm{d}S}\,\mathrm{d}P$$

Harris and Kolthoff[11] have taken the matter further by giving some consideration to the question of modifier efficiency. They note that there is wide variation in the amounts of different mercaptans which are required to produce a polymer of given intrinsic viscosity at a given conversion. They then introduce the concept of the theoretical minimum amount of modifier which would be required to produce a polymer of specified intrinsic viscosity at a specified conversion. The problem is to find the minimum value of A for given values of M_v and P_∞, assuming the above relationship between the three quantities to hold. The first step is to rearrange the relationship to read

$$A = \frac{0{\cdot}2P_\infty}{10^{-5}\overline{M}_v S_\infty} + \frac{1}{10^{-5}\overline{M}_v}\int_0^{P_\infty} \frac{\mathrm{d}P}{\mathrm{d}S}\,\mathrm{d}P$$

For given values of $\overline{M}_v$ and P_∞, the first term on the right-hand side has a minimum value when $S_\infty = 1$; a value of S_∞ equal to unity implies that, by the time the conversion has reached its final value P_∞, all the modifier has been consumed.

This seems to be an obvious requirement for maximum modifier efficiency. The second term has a minimum value when $\mathrm{d}P/\mathrm{d}S$ is constant and $S_\infty = 1$ at conversion P_∞. Under these circumstances, the definite integral has the value P_∞^2, and hence the minimum value for A is

$$A_0 = \frac{P_\infty}{10^{-5}\overline{M}_v}(0{\cdot}2 + P_\infty)$$

Thus A_0 can be calculated solely from a knowledge of P_∞ and $\overline{M}_v$. Over the course of a reaction as a whole, the minimum modifier will only be effective in achieving the desired degree of modification if two conditions are fulfilled: (i) modifier must have been added in such a way that $\mathrm{d}P/\mathrm{d}S$ is constant over the course of the polymerisation; and (ii) all of the modifier must have been consumed by the time the end of the reaction has been

TABLE 8.4

Maximum modifier efficiencies and corresponding conversions for various mercaptans in persulphate-initiated styrene–butadiene emulsion copolymerisation at 50°C (according to Harris and Kolthoff[11])

(reaction system: styrene 25, butadiene 75, water 180, fatty acid-soap 5, mercaptan 0·5, $K_2S_2O_8$ 0·3)

Mercaptan	*Maximum efficiency*	*% conversion at which efficiency is a maximum*
Primary decyl mercaptan	0·70	22
Primary dodecyl mercaptan (commercial)	0·65	40–50
n-Dodecyl mercaptan (pure)	0·80	68
t-Dodecyl mercaptan	0·80–1·0	60
t-Octyl mercaptan	0·80	55
t-Hexadecyl	No maximum over range 0–80% conversion	>80
n-Butyl mercaptan-*iso*butylene sulphide adduct	0·90	41
4-*n*-Butyloctyl mercaptan-1	0·85	62
β-di-*n*-Amylaminoethylthiol	0·70	11

reached. In passing, it is of interest to note that, whereas it has been tacitly assumed in the past that a uniform rate of consumption of monomer gives a polymer of constant intrinsic viscosity throughout the reaction, Harris and Kolthoff have shown that, in fact, the intrinsic viscosity may be far from constant if the modifier consumption is linear with respect to conversion.

Harris and Kolthoff then proceed to express the *efficiency of a modifier* in a given reaction system as being the ratio A_0/A, where A_0 is the theoretical minimum modifier requirement calculated from P_∞ and $\overline{M}_v$ using the above formula, and A is the actual amount of modifier required. It is found that, in general, the efficiency so defined is low at low conversions, and gradually increases to a maximum value with increasing conversion. For some modifiers, the maximum occurs at as low as 10% conversion; for others, at higher than 80% conversion. After the maximum, the efficiency decreases with further conversion. Some typical results are given in Table 8.4. The commercial mixture of mercaptans used by Harris and Kolthoff is most efficient at between 40% and 50% conversion. At conversions higher than 50%, the efficiency falls off gradually to a value of about 0·5 at 80% conversion.

Harris and Kolthoff also make suggestions for increasing the efficiency of a given modifier. They point out that, in general, an inefficient modifier can be made more efficient by changing conditions and procedures so as to ensure more nearly linear consumption with conversion, and more complete consumption at the end of the reaction. For mercaptans which are consumed rapidly at relatively low conversions, incremental addition is suggested as an obvious method of increasing efficiency at high conversion.

A further study by Meehan, Kolthoff and Sinha[12] has confirmed that the low apparent consumption of the higher alkyl mercaptans relative to conversion is due to a slow rate of transfer from bulk monomer droplets to reaction loci. Furthermore, evidence is presented to show that the dominant factor which determines the rate of transfer is the tendency for the mercaptan to distribute between aqueous and organic phases. The effect of diffusion coefficient is thought to be relatively small. Particular emphasis is placed upon the fact that the addition of methanol markedly increases the rate of consumption of *n*-tetradecyl mercaptan in the persulphate-initiated emulsion polymerisation of styrene at 50°C. It is argued that the addition of methanol to the aqueous phase can have little effect upon the diffusion coefficient of the mercaptan, but would be expected markedly to affect its solubility in the aqueous phase. It therefore appears from this work that the process which probably limits the rate of transfer of mercaptan from the monomer droplet phase to the reaction loci is mass transfer through the interface between the monomer droplets and the aqueous phase. Consistent with this conclusion is the observation noted previously that increasing the pH of the aqueous phase increases the relative rate of consumption of those mercaptans whose rate of consumption depends upon mass transfer. The formation of mercaptide ions in the layers surrounding the monomer droplets will deplete the concentration of un-ionised mercaptan molecules, and so provide a mechanism whereby transfer across the monomer–water interface can be accelerated.

Booth, Beason and Bailey[13] have published the results of a study of the incremental modification of the styrene–butadiene emulsion copolymerisation using *t*-dodecyl mercaptan. The purpose of such incremental addition is to produce a polymer of more uniform molecular weight throughout the course of the polymerisation, and, in particular, to obviate the formation of material of very high and very low molecular weight. The polymerisations studied in this work were 5°C reactions, for which the initiator was either a hydroperoxide–iron–sulphoxylate system or a hydroperoxide–iron–pyrophosphate system. In the former case, the soap was of the mixed fatty-acid/rosin-acid type; in the latter, it was exclusively rosin acid. A reasonable programme of incremental additions was arrived at from mercaptan consumption data. The general conclusion was reached that, in the system studied at least, incremental addition of modifier produces a polymer whose molecular weight distribution differs very little from that of a polymer produced with the same total quantity of modifier added as an initial charge. This experience accords with that recorded elsewhere for styrene–butadiene copolymers produced by high-temperature emulsion copolymerisation. Furthermore, it was observed that the inherent viscosity of the polymer produced by incremental modification remains almost constant over the course of the reaction (up to 60% conversion), whereas that of the non-incrementally modified polymer increases progressively with conversion. The results are given in Table 8.5. Whilst the general trends are clear, these results present some puzzling features. In particular, it is not entirely clear why the inherent viscosities of the non-incrementally-modified polymers are lower than those of the incrementally-modified polymers when the total amount of modifier added incrementally was about three times that added as a single charge initially. The explanation may be that the initial increment was so small (0·0258 pphm) that a small amount of very high-molecular-weight polymer was produced initially.

Booth, Beason and Bailey present what they regard as a consistent explanation of both these effects in terms of the rate-controlling effect of the process of transfer of mercaptans from monomer droplets to reaction

TABLE 8.5

Variation of inherent viscosity with conversion during low-temperature emulsion copolymerisation of styrene and butadiene with and without incremental addition of modifier (according to Booth, Beason and Bailey[13])

(initiator system: hydroperoxide–iron–sulphoxylate; soap: fatty-acid/rosin-acid blend; polymerisation temperature 5°C)

	Inherent viscosity of polymer†	
% conversion	*Incrementally modified*	*Non-incrementally modified*
10	2·1	1·0
20	2·1	1·1
30	2·1	1·2
40	2·1	1·4
50	2·1	1·6
60	2·0	1·8

† Determined in toluene (*ca.* 0·25% solution) at 30°C; units presumed to be dl g^{-1}.

loci. Their explanation seems to depend upon the observation that monomer in the droplets is depleted much more quickly than is mercaptan because the monomer is relatively rapidly imbibed by the growing polymer particles. Thus the ratio of mercaptan to monomer in the droplets increases as the reaction proceeds, and so also, in consequence, does the rate of transfer out of the droplets and into the water phase. In a sense, a system for which the modification process is controlled by mass transfer of modifier provides its own internal mechanism for making modifier available at an increasing rate as the reaction proceeds. External incremental addition of modifier would do little to supplement the effect of a process which is taking place anyway. This argument has a corollary in that the polymer initially produced in a non-incrementally-modified reaction will tend to be less extensively modified than that produced later in the reaction, although apparently other effects are present to ensure that the overall molecular weight of the polymer produced does rise as conversion proceeds. If the modifier is added incrementally, the balance can be so disturbed in favour of low initial modification as to produce a polymer whose average molecular weight is almost uniform throughout the course of the reaction. The overall conclusion which Booth, Beason and Bailey draw from their study is that incremental addition of modifier should not be attempted with modifiers whose rate of consumption is controlled by the rate of transfer from monomer droplet phase to reaction loci.

Similar observations have been reported by Uraneck and Burleigh[14] for the modification of a styrene–butadiene emulsion polymerisation at 5°C using *t*-dodecyl mercaptan. The initiator was a hydroperoxide–iron–sulphoxylate system. Not only were none of the expected advantages found to accrue from incremental modification, but, on the contrary, the Mooney viscosities and dilute solution viscosities of the polymer produced by incremental addition were in fact considerably higher than those of a control polymer produced by a reaction in which all the mercaptan was added initially. On the other hand, when *t*-nonyl mercaptan was used as modifier, a marked advantage of incremental over non-incremental addition was apparent. The published results show that in order to obtain a polymer of given Mooney viscosity, it is necessary to use rather more *t*-nonyl mercaptan if it is all added initially rather than incrementally. The difference in behaviour between the two mercaptans is attributed to differences in rate of transfer between monomer droplet phase and reaction loci. The molecular size of *t*-nonyl mercaptan is such that this transfer is no longer the rate-controlling process.

Uraneck and Burleigh comment that 'If the attainment of an equilibrium distribution of the increment is greatly delayed because of the size of the mercaptans, the effectiveness of the increment is greatly reduced. Under these circumstances, incremental addition can be less effective than adding all mercaptan originally. This is a possible explanation for the difference in behaviour between *tert*-dodecyl and *tert*-nonyl mercaptans in the systems studied. The overall conclusion reached from this work is that significant differences in modification can be realised by the incremental addition of mercaptans to styrene–butadiene emulsion copolymerisation systems if several conditions are fulfilled. Thus the incremental additions should be made in such a way that no gel is formed during the intervals between successive additions. Furthermore, the molecular weight or solubility of the mercaptan should be such that it distributes itself reasonably quickly between the several phases in the system.

Uraneck and Burleigh[15] have also published the results of an extensive study of the behaviour of various mercaptans in acrylonitrile–butadiene and acrylonitrile–styrene emulsion copolymerisations. In endeavouring to summarise the conclusions from this work, it is important to realise at the outset that generalisations are hazardous because of the several variables (apart from mercaptan type) which are possible to these systems. Not only were variations made in co-monomer type, but also, in the case of the acrylonitrile–butadiene copolymerisation, in the monomer feed composition and in the temperature of polymerisation. Uraneck and Burleigh make a distinction between 'efficient' and 'inefficient' modifiers. The basis for this distinction is the arbitrary, but practical, criterion that an efficient modifier is one which gives a polymer having an inherent viscosity of less than 3 when present in the system at a level of 0·25 pphm. The inherent viscosity is that in chloroform at 25°C. Unfortunately, this criterion can be faulted on two grounds: in the first place, nothing is said regarding the conversion to which the system shall be taken, and in the second place, the mode of addition of the modifier is not specified. It appears that the distinction relates not so much to modifiers *per se*, but rather to modifier added to a particular emulsion polymerisation system in a particular way.

The following is a list of mercaptans classified as inefficient for the emulsion copolymerisation of acrylonitrile and butadiene in 30/70 ratio at 5°C: *n*-tridecyl mercaptan, *n*-octyl mercaptan, mixture of *n*-nonyl and *n*-decyl mercaptans, *n*-dodecyl mercaptan, mixture of *n*-undecyl and tetradecyl mercaptans, cyclododecadienyl mercaptan, 4-cycloöctene-1-mercaptan, dipentene dithiol, 1,10-decanedithiol, 3-chloropropyl mercaptan, ethylene glycol 3-mercaptopropionate, mercaptobenzthiazole, 1,6-hexanedithiol, pentaerythritol mercaptan. Of these, the last five gave polymers containing between 64% and 89% gel, even though used at a level of 0·5 pphm. The remaining nine gave gel-free polymers when used at levels of 0·4–0·5 pphm. The following mercaptans were found to be efficient as modifiers for this copolymerisation: 2-nonyl mercaptan, mixture of 2-nonyl and 2-decyl mercaptans, 2-decyl mercaptan, *sec*-dodecyl mercaptan, *t*-heptyl mercaptan, *t*-nonyl mercaptan, *t*-decyl mercaptan, *t*-dodecyl mercaptan.

Of the secondary mercaptans investigated, the mixture of nonyl and

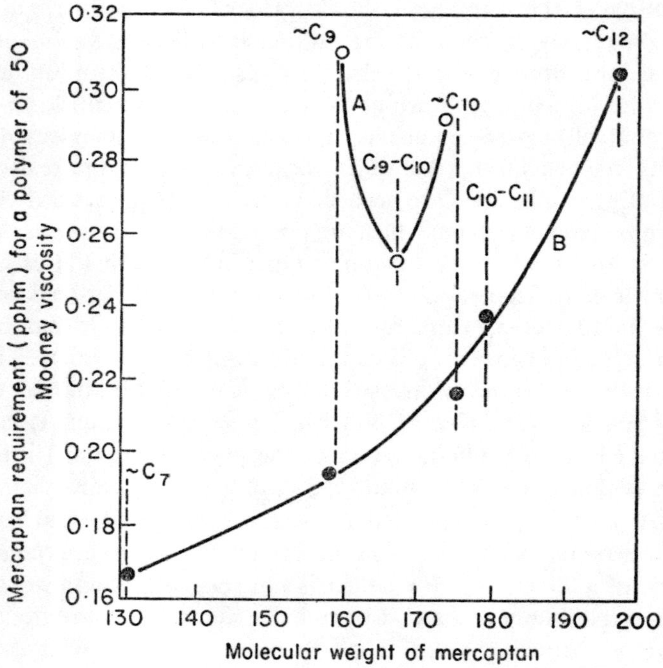

FIG. 8.8. Effect of mercaptan molecular weight upon amount of mercaptan required to give a polymer having a Mooney viscosity of 50 from the low-temperature (5°C) emulsion copolymerisation of acrylonitrile and butadiene, according to Uraneck and Burleigh.[15] Curve A is for secondary alkyl mercaptans; curve B is for tertiary alkyl mercaptans. (Reaction system: acrylonitrile 30, butadiene 70, water 180, potassium fatty-acid soap 4·6, sulphonate stabiliser 0·20, mercaptan—variable, cumene hydroperoxide 0·0268, sodium formaldehyde sulphoxylate dihydrate 0·025, $FeSO_4.7H_2O$ 0·0050, tetrasodium ethylenediamine tetraäcetate 0·020, KCl 0·30, KOH 0·03.)

decyl was considered to be the most efficient, as judged by the amount required to give a polymer having a Mooney viscosity of 50. Of the tertiary mercaptans, the nonyl compound was preferred as being the most efficient of those investigated which did not (unlike *t*-heptyl mercaptan) retard polymerisation. As judged by the criterion of the amount required to give a polymer having a Mooney viscosity of 50, the tertiary mercaptans were found to be considerably more efficient than the corresponding secondary mercaptans. The variation of mercaptan requirement for a Mooney viscosity of 50 with mercaptan molecular weight is shown in Fig. 8.8 for both secondary and tertiary alkyl mercaptans. In the former case, the mercaptan requirement goes through a sharp minimum corresponding to an equimolal mixture of *sec*-nonyl and *sec*-decyl mercaptans. All the polymer obtained using the secondary and tertiary mercaptans listed as efficient were free of gel. The effect of incremental addition upon modification by 2-nonyl and *t*-nonyl mercaptans was reported. Two-thirds of the modifier was added initially and the remaining one-third added during the course of the polymerisation. A greater modifying effect was observed with 2-nonyl mercaptan than with *t*-nonyl mercaptan. In the case of 2-nonyl mercaptan, the Mooney viscosity of the polymer produced by incremental modification was 41% lower than that of the control. The effect upon Mooney viscosity of varying the conversion at which the one-third increment is added is shown in Fig. 8.9 for both 2-nonyl mercaptan and *t*-nonyl mercaptan. Mercaptan consumption measurements showed that some of the inefficient modifiers (*e.g.* *n*-decyl mercaptan and dipentene dithiol) were consumed too quickly for modification to be efficient.

For the low-temperature emulsion copolymerisation of acrylonitrile and butadiene in 20/80 ratio, none of the isomeric series of secondary dodecyl mercaptans with the thiol group in positions 2 to 6 along the chain was rated as an efficient modifier, although the polymers were gel-free. The following mercaptans were found to be efficient modifiers: mixture of 2-nonyl and 2-decyl mercaptans, 2-decyl mercaptans, *t*-nonyl mercaptan, *t*-decyl mercaptan, *t*-dodecyl mercaptan. Tertiary mercaptans were found to be more efficient in the 20/80 copolymerisation than in the 30/70 copolymerisation. The complexities of the situation are evident from the observation that *t*-dodecyl mercaptan is more efficient in systems containing all rosin-acid soap than in those containing all fatty-acid soap. The same trend was reported for *t*-nonyl mercaptan, although the difference was less marked. Mercaptan depletion curves for three isomeric decyl mercaptans in the 20/80 acrylonitrile–butadiene emulsion copolymerisation at 5°C are shown in Fig. 8.10. (The caption to these curves in the original

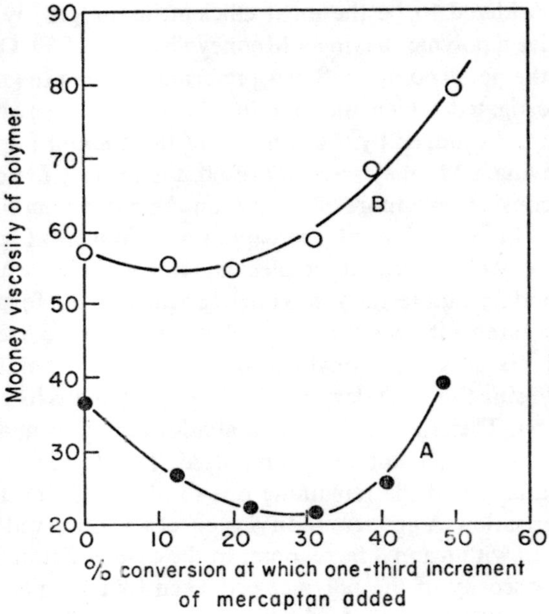

FIG. 8.9. Effect upon polymer Mooney viscosity of conversion at which a one-third increment of mercaptan modifier is added during low-temperature emulsion copolymerisation of acrylonitrile and butadiene, according to Uraneck and Burleigh.[15] Curve A is for 2-nonyl mercaptan; curve B is for *t*-nonyl mercaptan. (See caption to Fig. 8.8 for details of reaction system.)

paper gives the polymerisation temperature as 50°C, but from the context it seems clear that the temperature was in fact 5°C.) It is seen that the order of increasing rate of mercaptan consumption relative to conversion is tertiary < secondary < primary.

Some results which illustrate the performance of various mercaptan modifiers in the persulphate-initiated emulsion copolymerisation of acrylonitrile and butadiene are given in Table 8.6 and in Fig. 8.11. The acrylonitrile–butadiene charge ratio was 20/80. The general conclusion reached by Uraneck and Burleigh is that primary mercaptans are much

† The caption (Fig. 8.10) to these curves in the original paper gives the polymerisation temperature as 50°C, but from the context it seems clear that the temperature was in fact 5°C. The reaction system is believed to have been as follows: acrylonitrile 20, butadiene 80, water 180, potassium fatty-acid soap 4·6, sulphonate stabiliser 0·20, mercaptan 0·5(?), cumene hydroperoxide 0·061, sodium formaldehyde sulphoxylate dihydrate 0·055, $FeSO_4.7H_2O$ 0·011, tetrasodium ethylenediamine tetraäcetate 0·032, KCl 0·30, KOH 0·03.

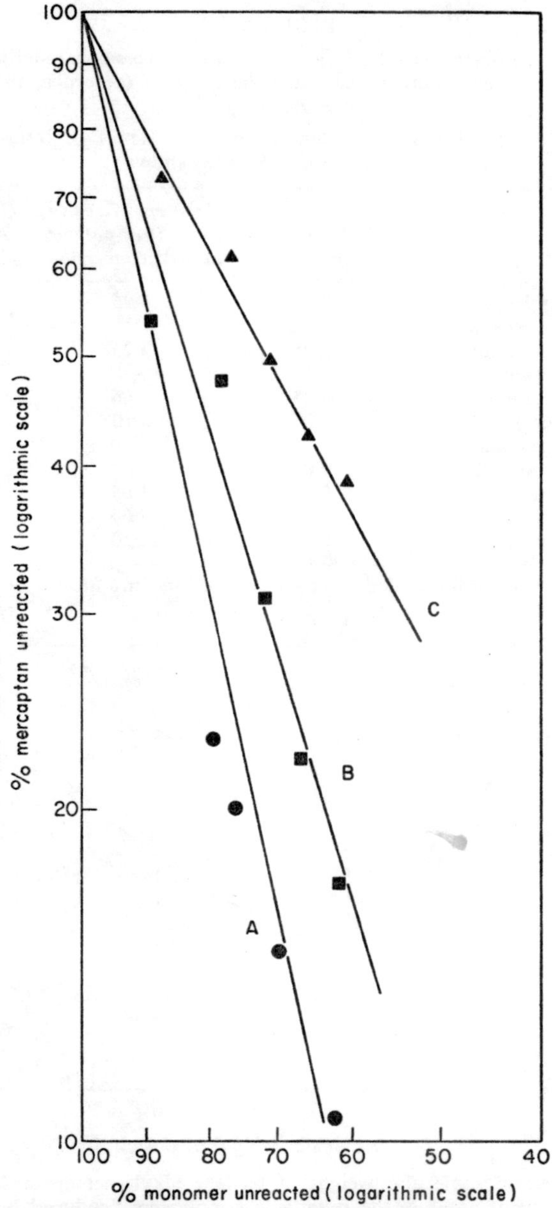

FIG. 8.10. Modifier consumption curves for three isomeric decyl mercaptans during low-temperature (5°C)† (see p. 360) emulsion copolymerisation of acrylonitrile and butadiene, according to Uraneck and Burleigh.[15] Curves A, B and C refer respectively to *n*-decyl mercaptan, *sec*-decyl mercaptan and *t*-decyl mercaptan.

TABLE 8.6

Effect of various primary and tertiary mercaptans in persulphate-initiated emulsion copolymerisation of acrylonitrile and butadiene at 50°C (according to Uraneck and Burleigh[15])

(reaction system: acrylonitrile 20, butadiene 80, water 180, fatty-acid soap 5, mercaptan 0·3, $K_2S_2O_8$ as shown)

Mercaptan	$K_2S_2O_8$ *level* (pphm)	*Inherent viscosity* ($dl\ g^{-1}$) *of polymer at 70% conversion*	*Gel content (%) of polymer at 70% conversion*
Mixture of *n*-undecyl to *n*-tetradecyl mercaptans	0·05	1·95	82
n-Dodecyl mercaptan	0·05	3·25	80
n-Decyl mercaptan†	0·05	—	50
t-Dodecyl mercaptan	0·05	3·66	0
Mixture of *t*-decyl and *t*-undecyl mercaptans	0·05	3·10	0
t-Decyl mercaptan	0·05	2·87	0
t-Nonyl mercaptan	0·05	2·65	0
t-Octyl mercaptan	0·05	2·60	0
t-Heptyl mercaptan	0·10	2·50	0

† Increment of 0·16 pphm added after 4·8 h reaction, in addition to the initial 0·3 pphm.

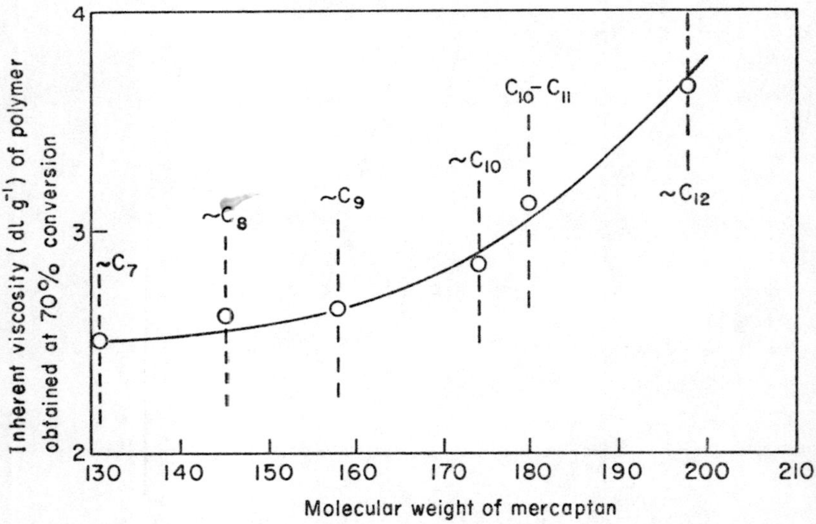

FIG. 8.11. Effect of molecular weight of tertiary alkyl mercaptan modifier upon inherent viscosity of acrylonitrile–butadiene copolymers produced by persulphate-initiated emulsion polymerisation at 50°C, according to Uraneck and Burleigh.[15] (Reaction system: acrylonitrile 20, butadiene 80, water 180, fatty-acid soap 5, mercaptan 0·3, $K_2S_2O_8$ 0·05.)

less efficient as modifiers for this emulsion copolymerisation than are tertiary mercaptans of the same molecular weight. Furthermore, as Fig. 8.11 shows, the molecular weight of the polymer produced using a given level of a tertiary alkyl mercaptan increases as the molecular weight of the mercaptan increases. It is difficult to comment on the effect of mercaptan molecular weight upon the 'molecular efficiency' of tertiary mercaptans (*i.e.* upon the average number of transfer steps per molecule of mercaptan incorporated in the system). This difficulty arises from the lack of any precise information concerning the relationship between inherent viscosity and polymer molecular weight. However, the general appearance of Fig. 8.11 suggests that the molecular efficiency may increase from that for C_7 to that for C_9 mercaptan, pass through a maximum in the range C_9 to C_{11}, and thereafter decrease.

Uraneck and Burleigh[15] also report upon the use of mercaptans as modifiers for the persulphate-initiated emulsion copolymerisation of 25/75 mixtures of acrylonitrile and styrene at 50°C. They found that polymer inherent viscosity did not provide a satisfactory criterion for distinguishing between the behaviour of the various modifiers studied. However, measurements of the melt-flow index of the polymer produced showed that, of the modifiers investigated, *t*-decyl mercaptan was the most efficient. Mercaptan depletion measurements were reported for tertiary alkyl mercaptans having alkyl groups varying from C_8 to C_{13}. Uraneck and Burleigh found that the rate of mercaptan consumption is inversely related to the molecular weight of the mercaptan. For *t*-decyl mercaptan, the relative rate of consumption of the modifier was found to be almost exactly equal to that of the monomer.

Uraneck and Burleigh[16] have also published information concerning the effect of mercaptan modifier upon the molecular weight distribution of polystyrenes and of styrene–butadiene copolymers produced by emulsion copolymerisation. Polymers were prepared at both 5°C (using a hydroperoxide–iron–sulphoxylate redox initiation system) and at 50°C (using a persulphate as initiator). Data are presented for a range of alkyl mercaptans, and comparisons made between observed molecular weight distributions (as determined by gel-permeation chromatography) and those predicted assuming a plausible mechanism for the polymerisation reaction. Styrene was selected as the monomer for which these comparisons were made, because it is less subject to branching and crosslinking during polymerisation than are dienes. Curves published for polystyrenes prepared at 5°C using a range of *t*-alkyl mercaptans as modifiers show a consistent trend as the regulating index of the mercaptan increases. The distribution broadens, the modal molecular weight and the ordinate corresponding to the modal species both fall as the regulating index of the modifier increases. The distribution curve for the polymer produced using the modifier having the highest regulating index (*i.e.* *t*-nonyl mercaptan) has a hump which suggests a bimodal distribution of molecular sizes. For modifiers having low regulating indices, the observed polymer heterogeneities agree well with those predicted theoretically. This confirms earlier observations by Montgomery and Winkler[17] concerning the heterogeneity of emulsion polystyrenes prepared using *t*-dodecyl mercaptan as modifier. The hump which appears on the experimental curve for polymers made using modifiers of higher regulating index is not predicted theoretically. It is attributed by Uraneck and Burleigh to branching.

Similar trends were observed for styrene–butadiene copolymers prepared at 5°C to a constant Mooney viscosity of about 50. Again, the experimental curves for molecular weight distribution resemble those predicted theoretically when the modifier has a relatively low regulating index. But when the regulating index is large, the experimental curves deviate markedly from those predicted theoretically. The curve obtained using cyclododecadienyl mercaptan, a modifier having an unusually high regulating index of 12·7, shows clearly that the distribution of molecular sizes is bimodal. The latter feature is attributed to non-uniform branching which occurs during the later stages of polymerisation in the outer layers of the monomer–polymer particles. Comparison between the polystyrene and styrene–butadiene copolymer distribution curves for polymers made using *t*-nonyl mercaptan confirms that the presence of the humps is probably due to branching; for the curve for the copolymer (in which the branching would be expected to be greater) is rather more distorted than is that for the homopolymer.

Uraneck and Burleigh[16] have also compared the behaviour of *n*- and *t*-dodecyl mercaptans as modifiers for the persulphate-initiated emulsion copolymerisation of styrene and butadiene at 50°C. Marked differences in the molecular-weight distributions of the product were observed. The distribution obtained using *n*-dodecyl mercaptan is definitely bimodal, whereas that obtained using the tertiary mercaptan is probably not, although it does have a very slight hump at the high-molecular-weight end. The *n*-mercaptan certainly gives the broader distribution of molecular sizes, and this correlates with both the easier processing of the polymer made with this mercaptan and also the inferior balance of vulcanisate properties, as compared with those of the polymer made using the tertiary mercaptan.

In a further paper,[18] Uraneck and Burleigh have shown that the reactivity of mercaptans of high molecular weight can be markedly enhanced by pre-emulsifying the modifier before addition to the charge. The modifier can be emulsified either in the neat condition, or as a solution in monomer. The effect of such pre-emulsification is to increase the regulating index as measured by the rate of consumption of the mercaptan. Thus the regulating index of *t*-hexadecyl mercaptan for a low-temperature styrene–butadiene emulsion copolymerisation increases from 0·3 to 3·0 as a result of pre-emulsification, and thereby approaches that of *t*-dodecyl mercaptan in efficiency as judged on a molal basis. The regulating index of *n*-dodecyl mercaptan can be increased from 1·0 to 16, but this latter value is, of course, too high for efficient modification. The technique of modifier pre-emulsification permits the regulating index to be varied continuously over a wide range. It also has the advantage of enabling a given degree of modification to be achieved by more efficient use of a mercaptan of higher molecular weight, and of thereby reducing the level of unpleasant mercaptan odour in the latex produced and in products derived from it.

Uraneck and Burleigh[18] have reported in detail upon certain problems associated with the technique of modifier pre-emulsification. A certain amount of modifier is consumed during the pre-emulsification process and also an abnormally high rate of consumption can occur during the initial stage (up to 10% conversion) of the polymerisation reaction. Thereafter, the mercaptan is consumed at a normal rate, in the sense that plots of the logarithm of the fraction of unreacted mercaptan against the extent of conversion are linear. The extent of these abnormal depletions depends upon the soap which is used in the polymerisation system and for the pre-emulsification. They increase in the order potassium laurate < potassium rosinate < potassium stearate < potassium oleate. Using stearates and oleates, the losses of modifying capacity accompanying these losses of mercaptan outweigh the advantage of increased regulating index gained by pre-emulsification.

The losses of mercaptan during the pre-emulsification stage were found to correlate with the peroxidic oxygen content of the soap, and with its unsaturation, except in the case of stearate. It was found that loss of mercaptan could be prevented by first boiling the soap solution for about 10 minutes with about 2% of sodium dithionite on the dry weight of soap.

This treatment does not, however, overcome the abnormal rate of consumption during the initial stage of the polymerisation. This is associated with the presence in the pre-emulsion of modifier which is solubilised within soap micelles. Subsequent addition of monomer, with solubilisation of monomer into these micelles, give a species of potential reaction locus in which the ratio of mercaptan to monomer is abnormally high. The rate of consumption of mercaptan relative to the polymerisation of monomer is therefore abnormally high due simply to the concentration effect. It is advantageous to pre-emulsify the mercaptan as a solution in part of the styrene if it is desired to reduce the extent of the initial mercaptan consumption. The presence of this monomer is thought to have a diluting effect upon the mercaptan which is solubilised into the soap micelles; for it is reasonable to suppose that the monomer solvent will be solubilised as well as the mercaptan solute. Thus when the main quantity of monomer is added subsequently, a more nearly normal ratio of mercaptan to monomer is achieved in the reaction loci. Pre-emulsification in the presence of monomer creates another problem in that some polymerisation can occur during this process. In fact, pre-treatment of the soap solution with sodium dithionite, as described above, was found to prevent polymerisation as well as loss of mercaptan.

Regarding the effects of the more obvious procedural variables upon the efficiency of a pre-emulsified mercaptan modifier, Uraneck and Burleigh have shown that, for a given intensity of agitation during pre-emulsification, the regulating index of the modifier increases with time of agitation. The pattern of the variation is such that the regulating index increases rapidly during the early stages of agitation, but then levels off to an optimum value. The rate at which the optimum regulating index is attained depends upon the particular mercaptan used; it was found to be especially low for *t*-hexadecyl mercaptan, whereas the time required to develop optimum regulating index in the case of *t*-dodecyl mercaptan was found to be much shorter. The intensity of agitation during pre-emulsification also greatly affects the time required to reach the optimum regulating index for a given mercaptan; as expected, the more intense the agitation, the more rapidly is the optimum attained. The temperature at which the pre-emulsion was formed was only found to be significant if the agitation was mild. Decreasing the water content of the soap solution reduces the regulating index. Pre-emulsifying all the mercaptan with one-half of the amount of soap, and then adding the other half in the normal manner, lessens the abnormal consumption of monomer during the initial stage of the reaction, but not to the same extent as when monomer is present during the pre-emulsification. No particular advantage was observed to accrue from forming the soap by the *in situ* procedure during pre-emulsification. Modification by pre-emulsified *t*-dodecyl mercaptan was responsive to the technique of incremental addition.

Other recently-published papers on the behaviour of mercaptan modifiers in emulsion polymerisation systems include those of Poddubnyĭ and Rabinerzon[19] and of Václavek.[20] The first of these is concerned with a comparison of *t*-dodecyl mercaptan and a non-mercaptan modifier, dis*iso*propylxanthogen disulphide, in both styrene–butadiene and acrylonitrile–butadiene emulsion copolymerisations. The paper by Václavek reports on the modifying behaviour of four aliphatic mercaptans in the emulsion copolymerisation of styrene and butadiene at 5°C, using a redox initiation system comprising di*iso*propylbenzene hydroperoxide, complexed iron(II) and sodium formaldehyde sulphoxylate. The micelle generator used was the sodium soap of disproportionated rosin acid. The apparent transfer constants for *t*-dodecyl, *t*-tetradecyl and *t*-hexadecyl mercaptans were found to decrease in a uniform manner; in fact, the logarithm of the constant was found to decrease linearly with the number of carbon atoms in the alkyl group of the mercaptan. The empirical relationship

$$\log_{10} \epsilon'_s = 1{\cdot}747 - 0{\cdot}138n$$

appears to represent the results adequately, n being the number of carbon atoms in the alkyl group. The apparent transfer constant for *n*-dodecyl mercaptan was found to be considerably smaller than that for *t*-dodecyl mercaptan.

The chemistry of mercaptans, with especial reference to their behaviour in emulsion polymerisation systems, has been discussed in papers by Kharasch, Nudenberg and Mantell[21] and by Kharasch, Nudenberg and Kawahara.[22] The reactions whereby mercaptans effect reduction of polymer weight are usually written as

$$\sim\!\sim\cdot + \mathrm{HSR} \rightarrow \sim\!\sim\mathrm{H} + \cdot\mathrm{SR}$$

$$\mathrm{RS}\cdot + \mathrm{M} \rightarrow \mathrm{RSM}\cdot$$

$$\mathrm{RSM}\cdot + \mathrm{M} \rightarrow \mathrm{RSMM}\cdot, \text{ etc.}$$

where $\sim\!\sim\cdot$ denotes a propagating chain and M denotes a monomer molecule. Kharasch, Nudenberg and Mantell agree that this is the probable mechanism of modification for persulphate-initiated reactions at 50°C, but suggest that, for reactions initiated at −10°C by hydroperoxide–iron redox couples, the effective reaction is

$$\sim\!\sim\cdot + \cdot\mathrm{SR} \rightarrow \sim\!\sim\mathrm{SR}$$

The ·SR radicals are envisaged as being generated by interaction between the mercaptan modifier and components of the redox initiation system, *e.g.*

$$Fe^{3+} + HSR \rightarrow Fe^{2+} + H^{+} + \cdot SR$$

If modification does occur by this mechanism in low-temperature polymerisations, then the mercaptan is behaving as a premature chain terminator, *i.e.* as a retarder, rather than as a transfer agent.

Mercaptans can also add across the double bond of the monomers in a polymerisation system. If peroxides are absent, the addition is in accordance with Markownikoff's rules, *i.e.*

$$R'CH{:}CH_2 + RSH \rightarrow R'\underset{\displaystyle SR}{\underset{|}{C}}H\text{—}CH_3$$

But if peroxides are present, then addition occurs in the manner contrary to that predicted by Markownikoff's rule, *i.e.* the product is $R'CH_2CH_2SR$. The rate of addition to styrene has been found to increase in the order *t*-mercaptans < *sec*-mercaptans < primary mercaptans.

A further reaction can occur if monomers containing electron-attracting groups adjacent to the carbon–carbon double bond are present, and the reaction system is alkaline. This is the Michael condensation. An obvious example of such a monomer is acrylonitrile, for which the reaction mechanism is probably as follows:

$$RSH + OH^{-} \rightleftharpoons RS^{-} + H_2O$$

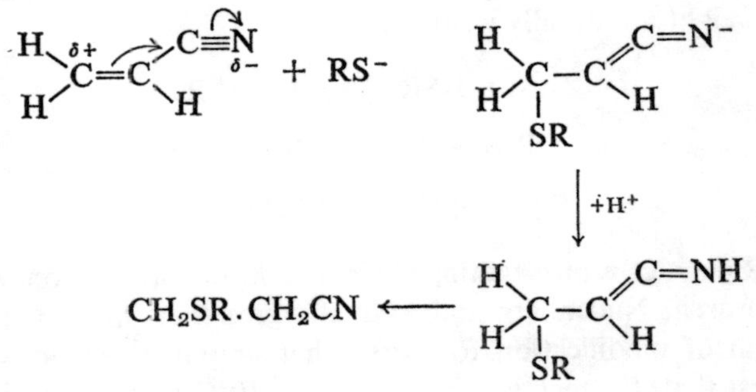

It is convenient to conclude this summary of the behaviour of mercaptans as modifiers for emulsion polymerisation systems by drawing attention to a review which was published by Krause in 1954.[23] Whilst this paper gives little information additional to that noted above, it is of interest in providing a comprehensive list of references on the subject up to the early part of the decade in which it was published.

8.3 MERCAPTAN DERIVATIVES

Frank *et al.*[24] have investigated the possibility of using certain mercaptan derivatives as modifiers for the persulphate-initiated emulsion copolymerisation of styrene and butadiene at 50°C. The motive for the investigation was the desire to find modifiers which would give polymers whose molecular weights did not vary with conversion. The approach was to use mercaptan derivatives of such a nature that they would gradually decompose to liberate free mercaptan as the polymerisation proceeded. Nine derivatives of *n*-dodecyl mercaptan were investigated. They were as follows: *n*-dodecyl thiolacetate (VI), mono-*n*-dodecyl thiolsuccinate (VII), di-*n*-dodecyl thioloxalate (VIII), *n*-dodecyl thiolbenzoate (IX), sodium mono-*n*-dodecyl thiosulphate (X), tri-*n*-dodecyl thiolorthoformate (XI), di-*n*-dodecyl trithiocarbonate (XII), the *n*-dodecyl hemimercaptal of chloral (XIII), and *β*-phenyl-*β*-*n*-dodecylmercaptopropiophenone (XIV).

$CH_3CO.S(CH_2)_{11}CH_3$

(VI)

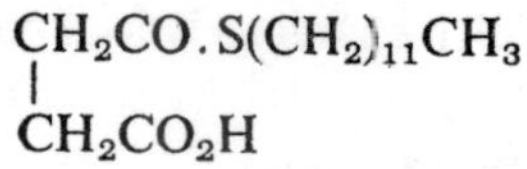

(VII)

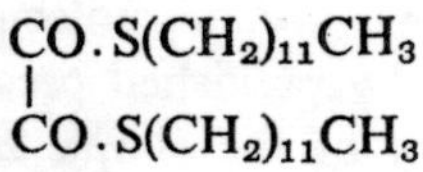

(VIII)

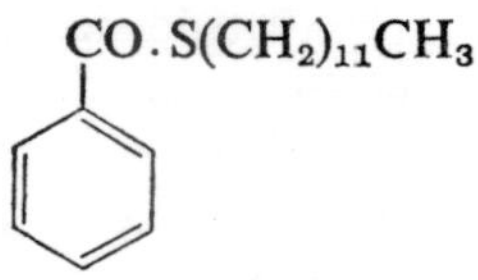

(IX)

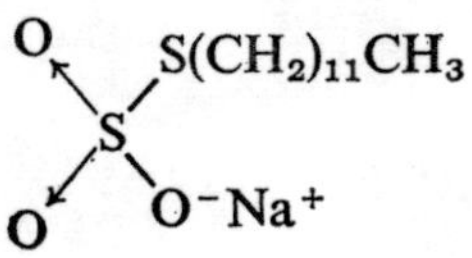

(X)

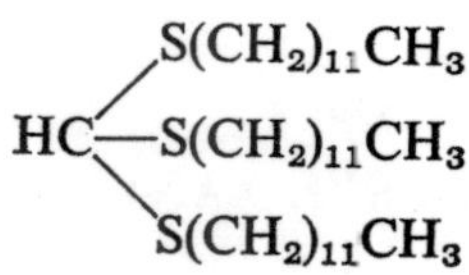

(XI)

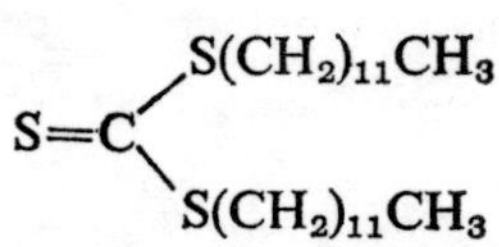

(XII)

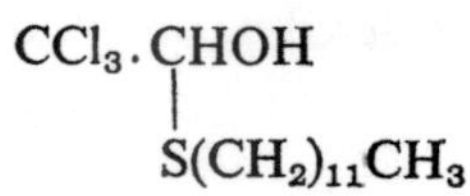

(XIII)

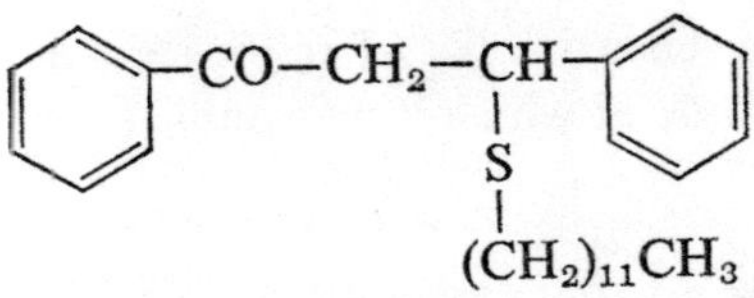

(XIV)

Compounds (VI), (IX), (X), (XII) and (XIV) appeared to be too stable under the conditions of the emulsion polymerisation to act as modifiers, although (VI), (XII) and (XIV) did show some promoting ability at higher levels of addition. But, in all cases, the polymers were of low solubility in benzene. Compounds (VIII), (XI) and (XII) were found to be more active, all being good promoters of polymerisation and (VIII) and (XIII) being modifiers of molecular weight. Compound (VII), added as its sodium salt, appeared to have the effect of eliminating the polymer of low molecular weight which forms in the early stages of polymerisation. Thus when *n*-dodecyl mercaptan was replaced by an equivalent amount of this salt, the polymers obtained between 9 and 15 h conversion were similar to those of the controls modified with *n*-dodecyl mercaptan. But if the reaction was stopped after 3 h, the conversion being below 30%, the over-modification which is characteristic of polymers produced using *n*-dodecyl mercaptan was not observed. The contrast in the appearance of the low-conversion polymers was quite striking: those obtained using *n*-dodecyl mercaptan were almost liquid, whilst those prepared using the succinate salt were rubbery. Further evidence of greater homogeneity of molecular sizes when the succinate salt was used is provided by published precipitation curves. It was also observed that the activity of the simple *n*-dodecyl mercaptan esters could be increased by adding alkali to the reaction system in order to increase their rate of hydrolysis and hence the rate of liberation of mercaptan.

8.4 DIALKYL XANTHOGEN DISULPHIDES

The dialkyl-substituted xanthogen disulphides—structure(XV)—have long been considered as alternatives to mercaptans as modifiers for emulsion

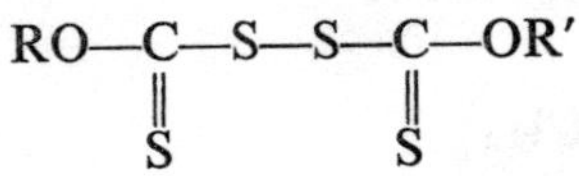

(XV)

polymerisation reactions, in particular, the emulsion copolymerisation of styrene and butadiene. One of the earliest references to their use appears to be contained in a United States Patent granted to Jasco Inc. in 1941.[25] The chemical reactions which underlie the modifying effect of these compounds are obscure. It seems probable that the effect arises from premature termination of propagating polymer chains by RO.CS.S· radicals, rather than by a chain transfer reaction:

$$\sim\!\sim\cdot + \cdot\mathrm{S.CS.OR} \rightarrow \sim\!\sim\mathrm{S.CS.OR}$$

Such radicals might be formed by spontaneous reversible thermal dissociation of the dialkyl xanthogen disulphide molecule, and this would appear to be the most likely mechanism in a polymerisation which takes place at 50°C and does not rely upon a redox couple for initiation. Alternatively, if a redox couple is used for initiation, then the radicals might arise by interaction between the dialkyl xanthogen disulphide and one of the components of the couple. Thus if iron is a component, then radicals might be generated as follows:

$$\mathrm{Fe^{3+} + RO{-}C({=}S){-}S{-}S{-}C({=}S){-}OR \rightarrow Fe^{2+} + RO{-}C({=}S){-}S^{-} + \cdot S{-}C({=}S){-}OR}$$

Snyder *et al.*[1] have given results for the sulphur contents of polystyrenes prepared by persulphate initiation at 50°C using di*iso*propyl xanthogen disulphide as modifier. With the exception of one doubtful result, these sulphur contents correspond to about two sulphur atoms per polymer molecule.

A substantial corpus of information concerning the behaviour of dialkyl xanthogen disulphides in low-temperature styrene–butadiene emulsion copolymerisations is available in two papers by Václavek.[26,27] The polymerisation temperature was 5°C, and a combination of di*iso*propylbenzene hydroperoxide, complexed iron(II) and sodium formaldehyde sulphoxylate was used for initiation. The principal micelle generator used in this study was the sodium soap of disproportionated rosin acid, although, as will appear from what follows, some interesting variations of soap type were investigated.

It was found that the apparent transfer constant for the dialkyl xanthogen disulphides decreases rapidly with increasing size of the alkyl group. The size of the alkyl group also has an effect upon the rate of polymerisation, the lower alkyl derivatives tending to retard the reaction relative to the higher alkyl derivatives, but all xanthogen disulphides tending to retard the reaction to a certain extent relative to that observed when no xanthogen disulphide is present. The situation is that, as the chain length of the alkyl group increases, so the polymerisation rate approaches that obtained in the absence of modifier. The results summarised in Table 8.7 show the effects of size and nature of alkyl groups upon both apparent chain transfer constant and rate of polymerisation. As can be seen from Fig. 8.12, the logarithm of the apparent transfer constant decreases linearly with the number of carbon atoms in the alkyl group, provided that the structural pattern of the alkyl group is unaltered. For those xanthogen disulphides in which the alkyl group is straight-chain, the observed variation of apparent chain transfer constant with number of carbon atoms (n) in the alkyl group is well represented by the relationship

$$\log_{10} \epsilon_s' = 1{\cdot}454 - 0{\cdot}263n$$

TABLE 8.7

Effect of size and nature of alkyl group of a dialkyl xanthogen disulphide upon its behaviour in emulsion copolymerisation of styrene and butadiene at 5°C (according to Václavek[26])

(reaction system: styrene 30, butadiene 70, water 195, sodium soap of disproportionated rosin acid 4·3, dialkyl xanthogen disulphide 0·1, di*iso*propylbenzene hydroperoxide 0·1, sodium formaldehyde sulphoxylate dihydrate 0·1, $FeSO_4.7H_2O$ 0·015, sodium ethylenediamine tetraäcetate 0·025, NaCl 0·387)

R *in* $(RO.CS.S)_2$	*Apparent transfer constant*	*Rate of polymerisation* (% h^{-1})
CH_3—	16·43	2·1
CH_3CH_2—	8·43	3·8
$CH_3(CH_2)_2$—	4·42	5·1
$CH_3(CH_2)_3$—	2·42	5·9
$CH_3(CH_2)_4$—	1·45	6·1
$CH_3(CH_2)_5$—	0·75	6·4
$CH_3(CH_2)_6$—	0·41	6·5
$CH_3(CH_2)_7$—	0·23	6·6
$(CH_3)_2CH$—	2·83	5·9
$(CH_3)_2CHCH_2$—	1·87	6·2
$(CH_3)_2CH(CH_2)_2$—	1·21	6·3
$C_2H_5CH(CH_3)$—	1·65	6·2
$(CH_3)_3CCH_2$—	1·01	4·8
$CH_2{<}\substack{CH_2-CH_2\\ CH_2-CH_2}{>}CH-$	1·78	5·6
No modifier	—	6·7

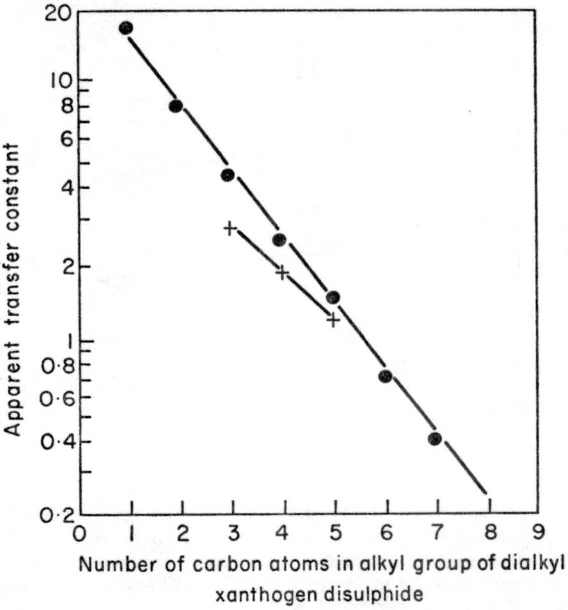

FIG. 8.12. Effect of size and nature of alkyl group upon apparent transfer constant for dialkyl xanthogen disulphides during emulsion copolymerisation of styrene and butadiene at 5°C, according to Václavek.[26] Significance of points: ● xanthogen disulphides for which alkyl group is of form $CH_3(CH_2)_{n-1}$-; + xanthogen disulphides for which alkyl group is of form $(CH_3)_2CH(CH_3)_{n-3}$-. (Reaction system styrene 30, butadiene 70, water 195, sodium soap of disproportionated rosin acid 4·3, dialkyl xanthogen disulphide 0·1 di*iso*propylbenzene hydroperoxide 0·1, sodium formaldehyde sulphoxylate dihydrate 0·1, $FeSO_4.7H_2O$ 0·015, sodium ethylenediamine tetraäcetate 0·025, NaCl 0·387.)

For xanthogen disulphides in which the alkyl group is of the form $(CH_3)_2CH(CH_2)_{n-3}$, the corresponding relationship is

$$\log_{10} \epsilon_s' = 1{\cdot}007 - 0{\cdot}184n$$

It appears that branching within the alkyl group tends to lower the value of the apparent chain transfer constant. Václavek interprets the effect of alkyl chain length upon apparent transfer constant as being due to its effect upon saturation solubility in the aqueous phase. Possible effects upon the reactivity of the xanthogen disulphide or of radicals derived therefrom are regarded as being of secondary importance. On balance, di*iso*propyl xanthogen disulphide was selected as being the most suitable modifier for use under the polymerisation conditions employed in this

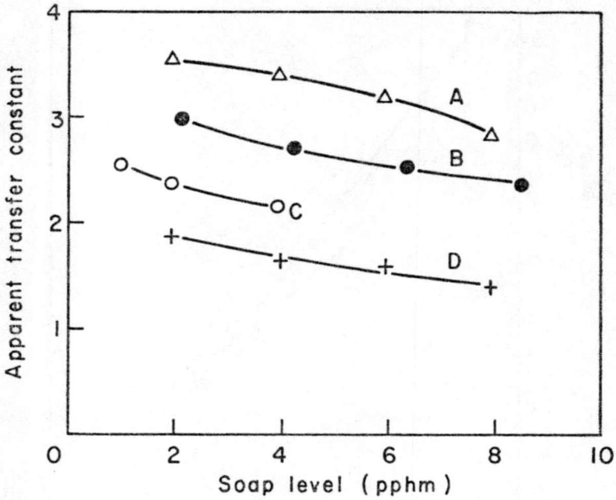

FIG. 8.13. Effect of soap type and level upon apparent transfer constant for di*iso*propyl xanthogen disulphide during emulsion copolymerisation of styrene and butadiene at 5°C, according to Václavek.[27] The letters appended to the curves refer to different micelle generators as follows: A, sodium dibutylnaphthalene sulphonate; B, sodium soap of disproportionated rosin acid; C, sodium oleate; D, sodium C_{14}–C_{20} alkylsulphonate. (Reaction system: as given in caption to Fig. 8.12, with obvious variations in nature and level of micelle generator.)

investigation. Evidence is also presented to show that the modifying efficiency of xanthogen disulphides increases as the polymerisation temperature increases, and that, at any given temperature, the efficiency is affected by variations in the reaction system.

The effect of recipe variations is reported in more detail in a second paper by Václavek.[27] This second paper is again concerned with low-temperature (5°C) polymerisation using the same initiation system as in the previous study. The results summarised in Fig. 8.13 show that the apparent transfer constant of di*iso*propyl xanthogen disulphide depends very much upon the micelle generator used in the reaction system, and also, to a lesser extent, upon the amount used. Václavek suggests that these effects arise because the permeability of the adsorbed soap layer surrounding the monomer–polymer particles to modifier molecules seeking entrance depends upon the nature of the soap molecule and also upon the density of its packing in the interface. The strongly-ionised (and therefore strongly-hydrated) and strongly-adsorbed sodium C_{14}–C_{20} alkylsulphonate apparently provides a formidable barrier to the ingress of modifier molecules,

TABLE 8.8

Effect of soap counterion upon behaviour of di*iso*propyl xanthogen disulphide in emulsion copolymerisation of styrene and butadiene at 5°C (according to Václavek[27])

(reaction system: styrene 30, butadiene 70, water 195, disproportionated rosin-acid soap as shown, di*iso*propyl xanthogen disulphide 0·1, di*iso*propylbenzene hydroperoxide 0·1, sodium formaldehyde sulphoxylate dihydrate 0·1, $FeSO_4.7H_2O$ 0·015, sodium ethylene–diamine tetraäcetate 0·025, electrolyte as shown)

Soap counterion	*Soap level* (pphm)	*Electrolyte type*	*Electrolyte level* (pphm)	*Apparent transfer constant*	*Rate of polymerisation* (% h^{-1})
Li^+	4·08†	LiCl	0·284‡	3·20	5·4
Na^+	4·30	NaCl	0·387	2·74	5·5
K^+	4·50†	KCl	0·500‡	2·39	5·5

† These amounts are equivalent, on a molecular basis, to 4·30 pphm of the sodium soap.
‡ These amounts are equivalent, on a molecular basis, to 0·387 pphm of NaCl.

whereas the equally strongly-ionised but more weakly-adsorbed sodium dibutylnaphthalenesulphonate creates a more permeable layer.

The soap counterion also influences the apparent transfer constant, as is shown by the results summarised in Table 8.8. In the study to which these results relate, the soap was of the disproportionated rosin acid type. The level of soap was adjusted in each case so that equivalent amounts were present on a molal basis. Likewise, the electrolyte was varied so that its cation was the same as the counterion of the soap, and its level such that equimolar quantities were present in each case. The apparent chain transfer constant was found to increase in the order $K^+ < Na^+ < Li^+$ as the counterion was varied. Václavek attributes this effect to the increasing tendency for the soap to hydrolyse as potassium is replaced by sodium, and especially as sodium is replaced by lithium. The adsorbed layers formed by the lithium soap are therefore considered to be more permeable to the modifier molecules because they are less extensively ionised and hydrated.

Another recipe variable which can affect the apparent transfer constant is the pH of the aqueous phase. Václavek's results are reproduced in Fig. 8.14, from which it is seen that the effect of pH change depends upon the soap which is used in the reaction system. If a carboxylate soap is used, then the apparent transfer constant falls markedly with increasing pH; this is attributed to the effect pH upon ionisation, of a high pH giving a

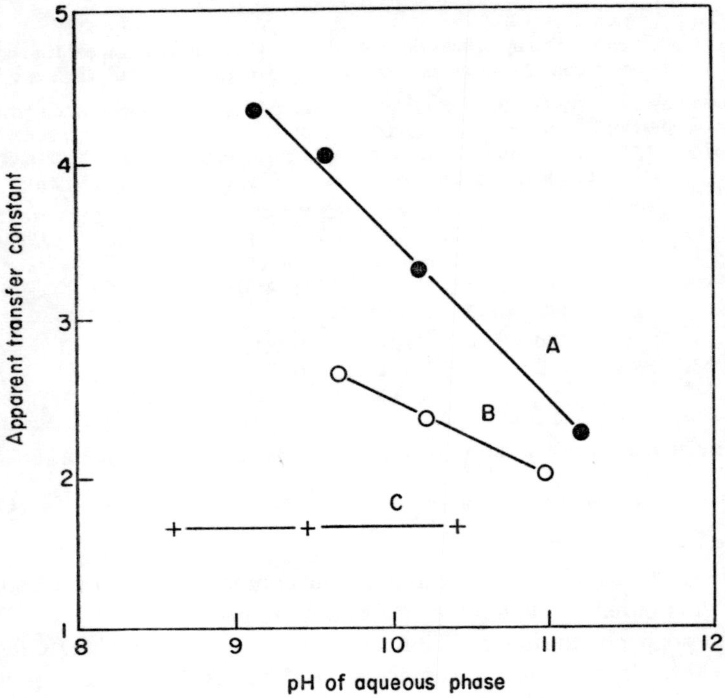

FIG. 8.14. Effect of pH of aqueous phase upon apparent transfer constant for di*iso*-propyl xanthogen disulphide during emulsion copolymerisation of styrene and butadiene at 5°C, according to Václavek.[27] The letters appended to the curves refer to the micelle generators listed in the table below. The reaction system was essentially that given in the caption to Fig. 8.12, with variations in the nature and level of both micelle generator and electrolyte as shown in the table below. Adjustment of pH was effected by replacing a little of the water with an equal amount of 0·1*N* NaOH solution.

Letter appended to curve	*Micelle generator*		*Electrolyte*	
	Type	*Level (pphm)*	*Type*	*Level (pphm)*
A	Sodium soap of disproportionated rosin acid	4·3	Na_2HPO_4	0·939
B	Sodium oleate	2·0	NaCl	0·387
C	Sodium C_{14}–C_{20} alkylsulphonate	4·0	Na_2HPO_4	0·939

soap which is fully ionised and therefore of minimum permeability to the modifier molecules. On the other hand, if a sulphonate soap is used, the apparent transfer constant is lower than that for the carboxylates, and is also virtually independent of pH; this insensitivity to pH change is attributed to the fact that the degree of ionisation of sulphonate soaps is almost unaffected by pH change.

Two further points of interest are apparent from these results for the effect of pH variation. Firstly, it appears that at very high pH (*ca.* 12), the apparent transfer constant given by a carboxylate soap might fall to that given by the long-chain sulphonate soap. And secondly, at any given pH, the apparent transfer constant given by an oleate soap is lower than that given by a rosin-acid soap; this correlates with the fact that the carboxyl group in a long-chain fatty acid is more strongly acidic than is that in a rosin acid. Both these observations support Václavek's view that variations in soap type, level, counterion type and pH affect the apparent chain transfer constant for di*iso*propyl xanthogen disulphide because they affect the permeability of the adsorbed soap layer surrounding the monomer–polymer particles. (It should also be noted that, whilst the apparent chain transfer constants of certain mercaptans also depend upon the pH, both the nature of the variation and its probable explanation differ from that for the xanthogen disulphides. It will be recalled that increasing the pH of the aqueous phase tends to *increase* the rate of consumption of mercaptans for which mass transfer is the rate-controlling process, and that this effect is probably to be explained as due to the formation of mercaptide anions.)

Václavek also shows that adding free rosin acid to a system containing rosin-acid soap as micelle generator has the effect of increasing the apparent transfer constant. This effect is presumably due to a slight reduction in the pH of the aqueous phase, and would disappear if the pH were readjusted to its original value by adding alkali (except that the apparent transfer constant would now tend to be lower than formerly because the total level of soap in the system would have been increased). For general application, Václavek recommends that the pH of the aqueous phase be adjusted to 10·8 for low-temperature polymerisations in which rosin-acid soaps are used as micelle generator.

The final recipe variable upon which Václavek provides information is the composition of the rosin-acid soap which is used in the system. The results are presented in terms of the so-called 'δ-value' of the rosin. This is a spectroscopic parameter which is related to the composition of the rosin; it is a measure of the concentration of conjugated unsaturation.

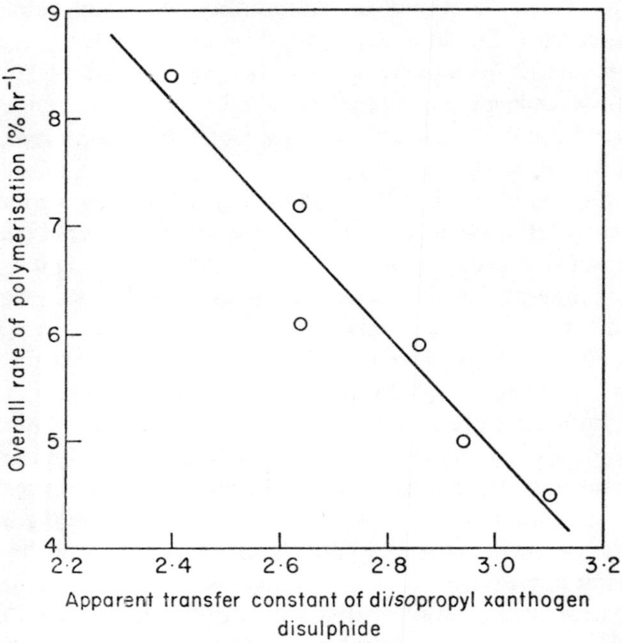

FIG. 8.15. Correlation between apparent transfer constant and overall rate of polymerisation for the emulsion copolymerisation of styrene and butadiene at 5°C in the presence of 0·1 pphm of di*iso*propyl xanthogen disulphide and a fixed level of various types of rosin-acid soap. This correlation has been constructed from data published by Václavek.[27] The reaction system to which it refers is essentially that given in the caption to Fig. 8.12. The pH was 10·8.

What is of particular interest is that Václavek finds an inverse correlation between apparent transfer constant and the rate of polymerisation. His numerical data are plotted in Fig. 8.15 to show this correlation. The inference is that, whilst the presence of conjugated unsaturation tends to slow the polymerisation down, it also tends to enhance the apparent transfer constant of di*iso*propyl xanthogen disulphide. It may be that the latter effect arises from the effect of conjugation upon the acid strength of rosin acids.

The paper by Poddubnyĭ and Rabinerzon,[19] to which reference has been made previously, gives comparisons between the behaviour of di*iso*propyl xanthogen disulphide and *t*-dodecyl mercaptan in styrene–butadiene and acrylonitrile–butadiene emulsion copolymerisations at both 5°C and 30°C. Di*iso*propyl xanthogen disulphide is consumed more rapidly than is *t*-dodecyl mercaptan. Curves published in this paper show that, if present as a single increment added initially, the former modifier has virtually disappeared at between 40% and 50% conversion. In accordance with expectation on the basis of pattern of consumption, the initial polymer formed with the xanthogen disulphide is of rather low molecular weight, but molecular weight increases rapidly as conversion proceeds. Also in accordance with expectation, it was reported that polymers whose molecular-weight distributions are almost independent of conversion can be obtained by adding di*iso*propyl xanthogen disulphide incrementally as the reaction proceeds. By contrast, incremental addition of *t*-dodecyl mercaptan was found to be without effect upon the molecular-weight distribution of the product.

8.5 OTHER MODIFIERS

Frank, Blegen and Deutschman[28] have investigated the possibility of using certain di-aroyl disulphides, general structure (XVI) as modifiers for the persulphate-initiated emulsion copolymerisation of styrene and butadiene.

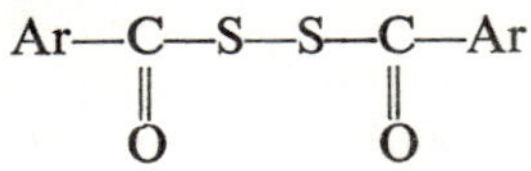

(XVI)

Dibenzoyl, di-*p*-methoxybenzoyl, di-3,4-dimethoxybenzoyl, di-*p*-bromobenzoyl, di-*p*-phenylbenzoyl, di-*p*-benzoylbenzoyl, di-*p*-methylsulphonylbenzoyl, *p*-(*N*,*N*-dimethylsulphonamido)benzoyl and di-α-naphthoyl disulphides were all found to be promoters for the reaction, but to possess little ability to modify the polymer. Their efficiencies in this respect seem to be far less than that of *n*-dodecyl mercaptan. Difuroyl disulphide differed from the other di-aroyl disulphides investigated, in that it showed neither promoting nor modifying ability. In fact, it was found to retard the promoting effect of *n*-dodecyl mercaptan. Being the sulphur analogues of the di-aryl peroxides, it might be thought that the di-aroyl disulphides would themselves initiate polymerisation by thermal decomposition to form free radicals. However, Frank, Blegen and Deutschman provide evidence to indicate that these compounds are not effective as polymerisation initiators.

Many other substances have been proposed from time to time as modifiers for emulsion polymerisation reactions. Considerations of space preclude more than a mere listing of some of these, together with the citing of appropriate references. The more important of these substances include carbon tetrachloride,[29] thiuram disulphides,[30,31] sulphur,[32] selenium,[33] substituted phosphines,[34] hydrazines,[35] amines,[36] Schiff's bases,[37] nitroso compounds,[38] and diazoamino compounds.[39–41]

Seymour and Patel[42] have very recently given a comparison between transfer constants in bulk polymerisation and apparent transfer constants in emulsion polymerisation for a wide range of transfer agents. The monomer used was styrene, and the temperature of polymerisation was 60°C. Initiation of the bulk reaction was by the thermal decomposition of azobis*iso*butyronitrile; initiation of the emulsion reaction was by potassium persulphate. A non-ionogenic micelle generator was used. For three of the substances investigated, namely, two mercaptans (1-propyl mercaptan and 1-dodecyl mercaptan) and carbon tetrachloride, the apparent transfer constant in emulsion is reported as being almost identical with that for the bulk reaction. For all the other substances (a variety of alcohols, acids, amines and esters), the value for the emulsion reaction is less than that for bulk. In some cases, the ratio of apparent transfer constant in emulsion to transfer constant in bulk is almost as low as 0·01. The ratio is especially low for organic substances of high water-solubility. Seymour and Patel suggest a possible correlation between this ratio and the solubility parameter of the transfer agent for substances within a given homologous series.

REFERENCES

1. H. R. Snyder, J. M. Stewart, R. E. Allen and R. J. Dearborn, *J. Am. chem. Soc.*, 1946, **68**, 1422.
2. F. T. Wall, F. W. Banes and G. D. Sands, *J. Am. chem. Soc.*, 1946, **68**, 1429.
3. W. V. Smith, *J. Am. chem. Soc.*, 1946, **68**, 2059.
4. W. V. Smith, *J. Am. chem. Soc.*, 1946, **68**, 2064.
5. W. V. Smith, *J. Am. chem. Soc.*, 1946, **68**, 2069.
6. I. M. Kolthoff and W. E. Harris, *J. Polym. Sci.*, 1947, **2**, 49.
7. H. W. Starkweather *et al. Ind. Engng Chem.*, 1947, **39**, 210.
8. C. S. Marvel and C. D. Lewis, *J. Polym. Sci.*, 1948, **3**, 354.
9. R. L. Frank, P. V. Smith, F. E. Woodward, W. B. Reynolds and P. J. Canterino, *J. Polym. Sci.*, 1948, **3**, 39.
10. W. E. Harris and I. M. Kolthoff, *J. Polym. Sci.*, 1947, **2**, 72.
11. W. E. Harris and I. M. Kolthoff, *J. Polym. Sci.*, 1947, **2**, 82.
12. E. J. Meehan, I. M. Kolthoff and P. R. Sinha, *J. Polym. Sci.*, 1955, **16**, 471.
13. C. Booth, L. R. Beason and J. T. Bailey, *J. appl. Polym. Sci.*, 1961, **5**, 116.
14. C. A. Uraneck and J. E. Burleigh, *J. appl. Polym. Sci.*, 1965, **9**, 1273.
15. C. A. Uraneck and J. E. Burleigh, *J. appl. Polym. Sci.*, 1968, **12**, 1075.
16. C. A. Uraneck and J. E. Burleigh, *J. appl. Polym. Sci.*, 1970, **14**, 267.

17. D. S. Montgomery and C. A. Winkler, *Can. J. Res.*, 1950, **28B**, 416.
18. C. A. Uraneck and J. E. Burleigh, *J. appl. Polym. Sci.*, 1973, **17**, 2667.
19. I. Ya. Poddubnyĭ and M. A. Rabinerzon, *J. appl. Polym. Sci.*, 1965, **9**, 2527.
20. V. Václavek, *J. appl. Polym. Sci.*, 1967, **11**, 1903.
21. M. S. Kharasch, W. Nudenberg and G. J. Mantell, *J. org. Chem.*, 1951, **16**, 524.
22. M. S. Kharasch, W. Nudenberg and F. Kawahara, *J. org. Chem.*, 1955, **20**, 1550.
23. A. H. Krause, *Rubb. Age, N.Y.*, 1954, **75**, 217.
24. R. L. Frank, S. S. Drake, P. V. Smith and C. Stevens, *J. Polym. Sci.*, 1948, **3**, 50.
25. K. Meisenburg, I. Dennstedt and E. Zaucker/Jasco Inc. (mesne), U.S. Pat. 2,248,107(1941).
26. V. Václavek, *J. appl. Polym. Sci.*, 1967, **11**, 1881.
27. V. Václavek, *J. appl. Polym. Sci.*, 1967, **11**, 1893.
28. R. L. Frank, J. R. Blegen and H. Deutschman, *J. Polym. Sci.*, 1948, **3**, 58.
29. A. Carpmael/I.G. Farbenindustrie A.G., B. Pat. 349,499 (1931).
30. C. F. Fryling/The B. F. Goodrich Co., U.S. Pat. 2,376,350 (1945).
31. G. L. Browning/The B. F. Goodrich Co. (mesne), U.S. Pat. 2,385,190 (1945).
32. H. W. Starkweather and M. A. Youker/E. I. du Pont de Nemours & Co., U.S. Pat. 2,234,204 (1941).
33. M. A. Youker/E. I. du Pont de Nemours & Co., U.S. Pat. 2,333,403 (1943).
34. J. L. Parker/E. I. du Pont de Nemours & Co., U.S. Pat. 2,382,812 (1945).
35. O. Bächle/I.G. Farbenindustrie A.G., German Pat. 702,209 (1941).
36. H. Murke/Jasco Inc. (Mesne), U.S. Pat. 2,260,475 (1942).
37. A. Carpmael/I.G. Farbenindustrie A.G., B. Pat. 360,822 (1931).
38. W. D. Wolfe/Wingfoot Corporation, U.S. Pat. 2,217,631 (1940).
39. B. W. Bysow/Resinotrest, B. Pat. 314, 933 (1929).
40. B. A. Dogadkin, K. B. Berezan and M. G. Lapuk, Russian Pat. 46,354 (1936).
41. C. F. Fryling/The B. F. Goodrich Co., U.S. Pat. 2,313,233 (1943).
42. R. B. Seymour and V. Patel, *J. Macromol. Sci-Chem.*, 1973, A7, 961.

CHAPTER 9

OTHER COMPONENTS OF EMULSION POLYMERISATION SYSTEMS (EXCLUDING MONOMERS)

9.1 INTRODUCTION

This chapter is concerned with four types of recipe ingredient concerning which relatively little information seems to be available in the published literature. The absence of information does not necessarily imply relative unimportance of the ingredient; for without one of the ingredients to be considered here, namely, the dispersion medium, no emulsion polymerisation would be possible. In the case of the other three, however, it must be admitted that they are to a lesser or greater degree optional. Indeed, one, the shortstopper, is not strictly a component of an emulsion polymerisation system, but is rather a substance which enables a polymerisation reaction to be brought to an abrupt and premature halt.

9.2 ELECTROLYTES

We consider first the use of simple inorganic electrolytes in emulsion polymerisation systems. It is regular practice to add such electrolytes to systems for the emulsion polymerisation of monomer mixtures whose main component is butadiene. This is the case whether the reaction is carried out at high temperature (*e.g.* 50°C) or at low temperature (*e.g.* 5°C). Doubtless, it is also the practice to add electrolytes to systems for the polymerisation of other monomers. Notwithstanding this, there seems to be remarkably little information in the literature concerning either the effect of electrolytes upon the polymerisation reaction or the interpretation of those effects. Rather more information is available concerning the effects upon the latex end-product. In considering the available information, it should be noted that inorganic electrolytes have also been added to emulsion polymerisation systems as antifreezes, thereby enabling the reaction to be carried out at sub-zero temperatures. This aspect of the use of electrolytes will be discussed in Section 9.3 of this chapter, under the heading of dispersion media. However, it is clearly impossible to distinguish the effects upon polymerisation rate and colloid stability of electrolyte added as an antifreeze from those effects which are observed when the primary objective of the electrolyte addition has been other than depression of the freezing-point of the aqueous phase.

It is perhaps significant that Bovey *et al.*[1] in their important book on emulsion polymerisation, published twenty years ago, dismiss the subject of the effect of electrolytes upon rate of emulsion polymerisation in little over a single page. Basing their remarks on some unpublished results obtained by Williams and Carr at the University of Minnesota in 1944, and referring to the standard persulphate-initiated GR-S emulsion polymerisation system, they state that 'Studies have been made of the effect of different concentrations of potassium chloride and of tripotassium phosphate in the conversion–time curves with different amounts of fatty acid soap ... Salts quite generally increase the initial rate of polymerization with any amount of soap. At higher conversions, after all the soap has been adsorbed by the polymer particles, the salt retards further conversion. At a given electrolyte concentration the retarding effect is more pronounced, the smaller the amount of soap.' The effect of adding potassium chloride in amount sufficient to give a concentration of 0·5 mole l^{-1} on the aqueous phase is said to be as follows. If the standard quantity of soap is present, the conversion after 12 h is reduced from 73% to 67%. If one-quarter of the standard amount of soap is present, then the conversion after 12 h is reduced from 82% to 28%.

Concerning explanations, Bovey and his co-authors continue: 'The increase in the initial rate of conversion is undoubtedly due to increased solubilization of the monomers by the micelles. The retarding effect of the electrolyte in the later stage of polymerization is probably due to agglomeration and coagulation of latex particles. While the number and size of the particles are not of major significance in determining the rate of conversion of these monomers, it appears that a drastic reduction in the number of particles by extensive agglomeration must cause a reduction in the rate of conversion. Indeed, this has been demonstrated directly; when a latex is partially coagulated by careful freezing in dry ice, the polymerization on subsequent warming to 50°C is extremely slow. Since adsorbed soap has a protective effect against agglomeration, the retarding effect of electrolyte becomes less pronounced with increasing amounts of soap.' Elsewhere in the book by Bovey *et al.*, allusion is made to the effect of simple inorganic electrolytes upon the size of the particles produced in the final latex, and upon the fluidity of the final latex.

Some results for the effect of electrolyte upon the relationship between soap level and number of particles produced have been published by Roe.[2] The system to which these results refer is styrene emulsion polymerisation at 50°C, using potassium laurate as soap and potassium persulphate as initiator. The electrolyte was a mixture of potassium carbonate and potassium bicarbonate such that the pH of the aqueous phase was buffered at 10. The results, reproduced in Fig. 9.1, show that increasing the electrolyte level increased the number of particles per unit volume of the final latex. The rate of polymerisation was therefore presumably increased, although no data demonstrating this are presented. The soap being in this case potassium laurate, the critical micelle concentration was rather high (*ca.* 0·5–2·0 × 10^{-2} mole l^{-1}, depending upon the electrolyte concentration),

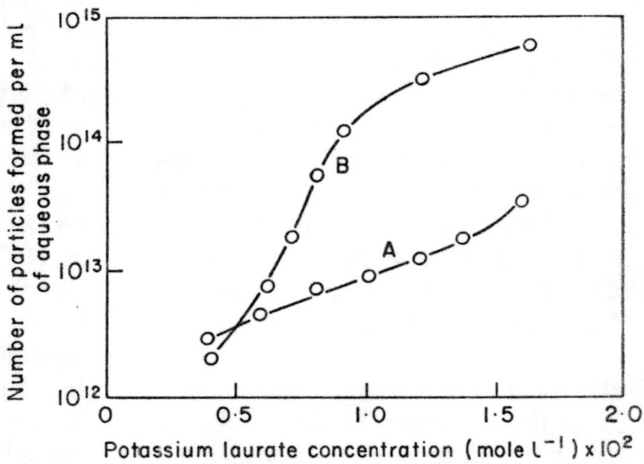

FIG. 9.1. Effect of electrolyte level upon relationship between soap concentration and number of particles formed during persulphate-initiated polymerisation of styrene at 50°C, according to Roe.[2] Curve A refers to a total [K^+] of 0·01 g-ion l^{-1}; curve B to a total [K^+] of 0·1 g-ion l^{-1}. (Reaction system: styrene 20, water *ca.* 250, potassium laurate—variable, $K_2S_2O_8$ 0·3, K_2CO_3–$KHCO_3$ buffer (pH 10) to give requisite total [K^+]. Note: An earlier publication by Roe and Brass[3] seems to imply that the potassium-ion concentration *excludes* any contribution from the potassium laurate. It is not, however, clear whether the contribution from the potassium persulphate is included or excluded. But this contribution is rather small anyway.)

and therefore particularly susceptible to depression by added electrolytes. To this latter effect is attributed in part the increase in the number of particles when the electrolyte level was increased. In effect it can be argued that the addition of electrolyte made available a higher proportion of the soap in micellar form than would otherwise have been the case. However, this can only be a partial explanation, since an enhancement of particle count was observed at soap levels so low that micelles were initially absent at both levels of potassium-ion concentration used. Further reference will be made to these observations, and to their implications for the theory of the mechanism of emulsion polymerisation, in the final chapter of this book.

According to a recent Ph.D. thesis by Kamath,[4] the effect of electrolytes on a 'standard' emulsion polymerisation system using 'regular' amounts of an ionic micelle generator has been studied by Patel.[5] Apparently, the rate of polymerisation increased with the addition of electrolyte. Kamath comments as follows: 'This is obviously due to the greater number of micelles caused by the lowering of the critical micelle concentration, resulting in a higher number of particles. This increase in the number of particles accounts for the observed increase in the rate of polymerization. With further increase in the electrolyte concentration, a maximum is reached in the rate, after which the polymerization rate decreases with increasing electrolyte concentration. Thus, in general, a bell-shaped curve is obtained when the polymerization rate is plotted against the electrolyte concentration. Coalescence of latex particles at high electrolyte concentration has been used to account for the observed decrease in the number of particles and the concomitant decrease in the rate of polymerization.'

Whatever may be the expected effect of electrolyte addition upon the rate of emulsion polymerisation, it seems to be general industrial experience that such additions are accompanied by a decrease in the rate, as evidenced by the overall conversion achieved after a given time of reaction. Thus the results of Schulze, Tucker and Crouch,[6] shown in Fig. 9.2, clearly demonstrate this for the effect of sodium sulphate upon the low-temperature emulsion copolymerisation of styrene and butadiene using a sodium alkyltoluenesulphonate as micelle generator. Schulze, Tucker and Crouch make no comment regarding the explanation for the observed decrease in rate, but point out the obvious practical relevance of these results in that industrial grades of sodium alkylarylsulphonates often contain large amounts of sodium sulphate. It is perhaps worth emphasizing that the decrease in rate accompanying an increase in sodium sulphate

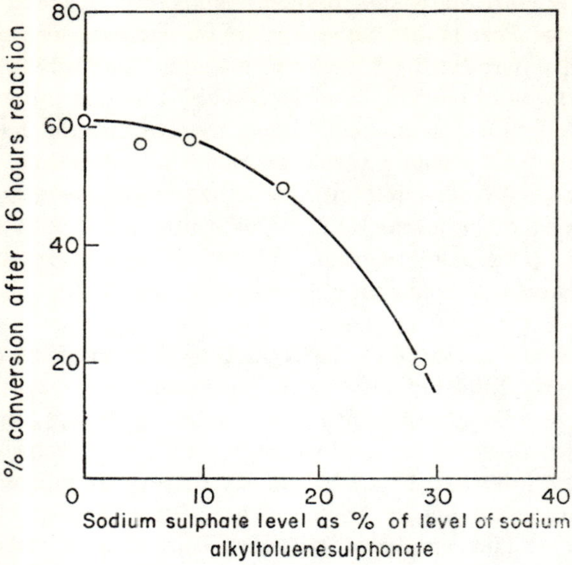

FIG. 9.2. Effect of sodium sulphate impurity upon overall conversion after 16 h reaction in the emulsion copolymerisation of styrene and butadiene at 41°F, according to Schulze, Tucker and Crouch.[6] (Reaction system believed to be: styrene 28, butadiene 72, water 180, sodium alkyltoluenesulphonate 4·7, sodium naphthalene formaldehyde sulphonate 0·10, mixed tertiary mercaptans 0·24, cumene hydroperoxide 0·10, dextrose 1·0, $FeSO_4.7H_2O$ 0·14, $K_4P_2O_7$ 0·177, KCl 0·50, Na_2SO_4—variable, KOH 0·10. In the presentation of the results given here, the Na_2SO_4 level is expressed as a percentage of the level of active sodium alkyltoluenesulphonate; the latter level was maintained constant throughout this series of experiments.)

content shown in Fig. 9.2 is not due merely to dilution of the micelle generator with simple electrolyte. In the work to which these results relate, the precaution was taken of ensuring that each system studied contained the same level of active alkyltoluenesulphonate soap.

Results for the effect of potassium chloride level upon both conversion–time relationship and appearance of the resultant latex have been published by Fryling and Follett.[7] The system to which these observations refer is the emulsion copolymerisation of styrene and butadiene at −10°C, using a mixed rosin-acid/fatty-acid soap and a hydroperoxide–iron redox couple as initiator. From the results given in Table 9.1, it appears that some electrolyte is necessary in order to sustain a reasonable rate of reaction, but that higher concentrations cause the polymerisation to be seriously retarded. Fryling and Follett suggest 0·10 pphm as the optimum level in

TABLE 9.1

Effect of level of potassium chloride upon conversion–time relationship and appearance of final latex for emulsion copolymerisation of styrene and butadiene at −10°C (according to Fryling and Follett[7])

(reaction system: styrene 30, butadiene 70, water 192, methanol 48, potassium soap of disproportionated rosin acid 3·5, potassium fatty-acid soap 1·5, mixed tertiary mercaptans 0·25, di*iso*propylbenzene hydroperoxide 0·3 mmole, $FeSO_4.7H_2O$ 0·0834, $Na_4P_2O_7.10H_2O$ 0·134, KCl—variable)

KCl *level* (pphm)	*% conversion after time shown* 4 h	7 h	24 h	*Appearance of latex after* 24 h *reaction*
0·00	27	33	36	Viscous
0·05	27	45	76	Viscous
0·10	22	40	76	Slightly viscous
0·15	19	30	71	Slightly viscous
0·20	22	35	73	Very slightly viscous
0·25	18	32	71	Very slightly viscous
0·30	18	31	71	Fluid
0·50	14	22	58	Fluid
0·75	12	17	40	Fluid
1·00	10	15	40	Fluid

this case. Their judgement is based not only on the effect upon reaction rate, but also on the effect upon the viscosity of the latex produced. Further comment will be made concerning this latter aspect of electrolyte addition subsequently.

Kamath[4] has investigated the effect of a range of simple inorganic electrolytes upon the rate of emulsion polymerisation of styrene at 50°C. The initiator was potassium persulphate, and the micelle generator a mixture of tridecyloxypolyethyleneoxyethanol (presumably an ethylene oxide adduct of tridecyl alcohol, *i.e.* of $C_{13}H_{21}OH$) and a small amount of sodium lauryl sulphate. The electrolytes used were sodium chloride, potassium chloride, ammonium chloride and potassium bromide. Kamath gives complete sets of conversion–time curves for each electrolyte at several levels of addition. His results are summarised in Table 9.2 and in Fig. 9.3. In the table, the electrolyte levels are given as pphm. In the figure, which shows rate of polymerisation as a function of electrolyte concentration, the latter is expressed as moles per litre of aqueous phase. Fig. 9.3 illustrates the significant observation that, for these electrolytes at least, the relationship between polymerisation rate and molal concentration for the several electrolytes can be represented by a single curve. It also

TABLE 9.2

Effect of additions of various inorganic electrolytes upon persulphate-initiated emulsion polymerisation of styrene at 50°C (according to Kamath[4])

(reaction system: styrene 100, water 180, sodium lauryl sulphate 0·05, tridecyloxypolyethyleneoxyethanol 5, $K_2S_2O_8$ 0·3, inorganic electrolyte—variable)

Electrolyte level (pphm)	*Rate of polymerisation (% per hour) in presence of each electrolyte at level indicated*			
	NaCl	KCl	NH_4Cl	KBr
0·01	11·5	11·0	11·0	11·0
0·02	—	11·0	—	11·0
0·03	11·5	11·0	11·0	11·0
0·05	9·0	10·0	9·0	10·0
0·10	7·5	8·0	6·5	9·0
0·15	6·0	7·0	6·0	—
0·20	—	6·5	—	—

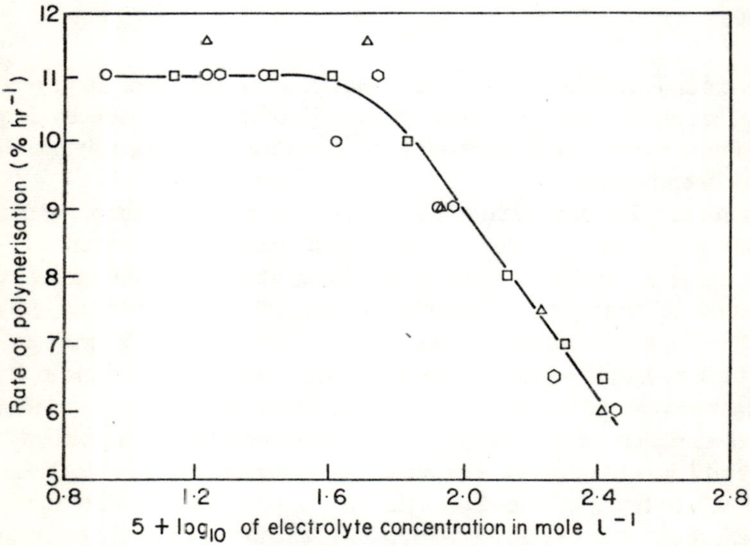

FIG. 9.3. Effect of various inorganic electrolytes upon rate of presulphate-initiated emulsion polymerisation of styrene at 50°C, according to Kamath.[4] Significance of points: ○ KBr; □ KCl; △ NaCl; ⬡ NH_4Cl. (Reaction system: styrene 100, water 180, sodium lauryl sulphate 0·05, tridecyloxypolyethyleneoxyethanol 5, $K_2S_2O_8$ 0·3, inorganic electrolyte—variable.)

shows that the added electrolyte has no effect upon polymerisation rate up to a concentration of about 0·25 mole l^{-1} on the aqueous phase. Thereafter, a decrease in rate is observed. It should be emphasised that, in the system used for these studies, the principal micelle generator was a non-ionogenic polyethylene oxide adduct. A minor amount only of the anionic micelle generator was present; in fact, the level was considerably below that corresponding to the critical micelle concentration. Kamath interprets the initial insensitivity of polymerisation rate to electrolyte level as being due to the presence of non-ionic micelles, whose number and size are presumably little affected by electrolyte concentration. The subsequent fall in polymerisation rate is presumably to be interpreted in terms of the effect of higher electrolyte levels in promoting particle coalescence, perhaps by a mechanism which involves partial dehydration of the interfacial regions between particles and aqueous phase.

Kamath[4] also gives results for the effect of added potassium hydroxide upon the persulphate-initiated emulsion polymerisation of styrene. His observations are summarised in Fig. 9.4. In accordance with expectation from the previous experiments with inorganic electrolytes, the rate

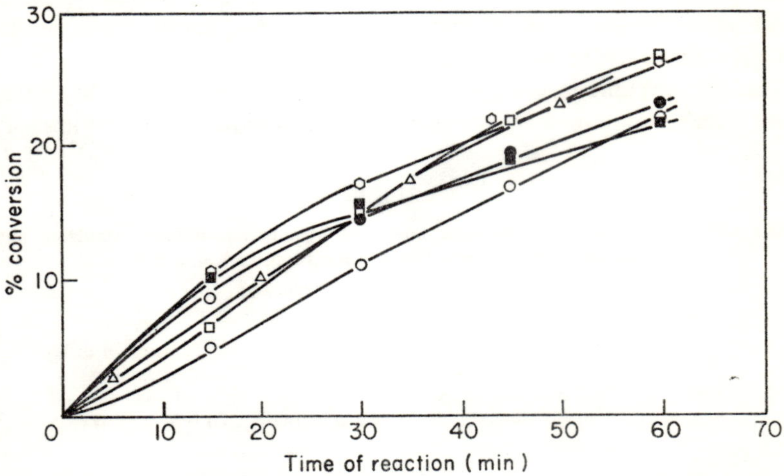

FIG. 9.4. Effect of added potassium hydroxide upon conversion–time relationship for persulphate-initiated emulsion polymerisation of styrene at 50°C, according to Kamath.[4] Significance of points in relation to level of added KOH in pphm: ○ 0; □ 0·01; △ 0·02; ⬡ 0·03; ● 0·05; ■ 0·07. (Reaction system: styrene 100, water 180, sodium lauryl sulphate 0·05, tridecyloxypolyethyleneoxyethanol 5, $K_2S_2O_8$ 0·3, KOH—variable.)

of polymerisation was found to depend little upon level of potassium hydroxide up to about 0·07 pphm. Thereafter, the steady polymerisation rate tends to fall. Kamath draws attention to the fact that the initial rate of polymerisation tends to increase slightly with increasing concentration of potassium hydroxide. This effect is also illustrated by the results given in Table 9.3, which shows the conversions after 10 and 20 minutes reaction and also the induction period and the initial pH of the aqueous phase. It appears that potassium hydroxide is able to accelerate the reaction during that stage which is conventionally associated with locus initiation. Kamath explains these results in terms of the effect of pH upon the decomposition of the persulphate initiator. He argues that, whereas under acid conditions hydroxyl radicals can be formed by two reactions, namely,

$$SO_4^{\bar{\cdot}} + H_2O \rightarrow HSO_4^- + \cdot OH$$

and

$$SO_4^{\bar{\cdot}} + H_3O^+ \rightarrow H_2SO_4 + \cdot OH$$

under alkaline conditions only the first of these reactions can occur. Therefore fewer hydroxyl radicals will be formed under alkaline conditions. Furthermore, wastage of radicals by re-combination in the aqueous phase will be less under alkaline conditions because the mutually-repelling sulphate radical-ions are expected to interact less readily than the electrically-neutral hydroxyl radicals.

Added inorganic electrolytes have an effect upon the stability of the latex which is obtained by emulsion polymerisation, as well as upon the

TABLE 9.3

Effect of added potassium hydroxide upon the persulphate-initiated emulsion polymerisation of styrene at 50°C (according to Kamath[4])

(reaction system: see Table 9.2)

KOH *level* (pphm)	pH *of aqueous phase*	*Induction period* (min)	*% conversion after times shown* 10 min	20 min
0	6·0	70	2·5	7·0
0·01	6·0	10	4·0	9·5
0·02	8·0	0	5·0	10·0
0·03	8·5	0	7·5	13·5
0·05	9·0	0	6·5	11·5
0·07	11·0	0	7·5	12·5

rate of polymerisation. The results of Howland *et al.*,[8] shown in Fig. 9.5, are for the effect of added electrolyte (potassium sulphate) upon the amount of prefloc formed during the low-temperature emulsion copolymerisation of styrene and butadiene in the absence of mercaptan modifier. These workers found that the stability of the latex during polymerisation and processing greatly decreased as the Mooney viscosity of the contained polymer was increased by reducing the level of modifier in the reaction system. Thus systems which gave a latex of satisfactory stability if normal amounts of modifier were present, developed massive amounts of coagulum when the modifier was reduced in amount or eliminated altogether. It was further found that the stability of the unmodified latex could be improved by increasing the level of electrolyte. Fig. 9.5 gives curves showing amount of prefloc as a function of electrolyte level for four levels of water in the polymerisation system. It is apparent that some electrolyte is necessary in order to prevent the formation of prefloc, and that, if no potassium sulphate is present in the system, as much as

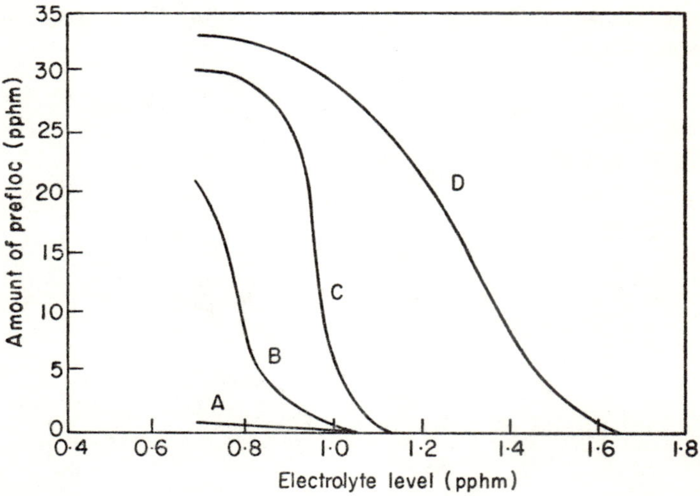

FIG. 9.5. Effect of level of electrolyte and level of water upon stability of latex obtained by emulsion copolymerisation of styrene and butadiene at 50°F, according to Howland *et al.*[8] The four curves refer to different amounts of water in the reaction system as follows: A, 60 pphm; B, 70 pphm; C, 85 pphm; D, 100 pphm. (Reaction system: styrene 30, butadiene 70, water—variable, potassium tall-oil soap 1·25, potassium salt of condensed alkylnaphthalenesulphonic acid 1·5, di*iso*propylbenzene hydroperoxide 0·22, sodium formaldehyde sulphoxylate dihydrate 0·052, $FeSO_4.7H_2O$ 0·10, $K_4P_2O_7$ 0·25, K_2SO_4—variable, KOH 0·02.)

30% of the polymer which is formed may flocculate. There is evidently at each water content a minimum concentration of electrolyte necessary if the formation of prefloc is to be prevented. Fig. 9.5 shows that this level increases with the water content of the system, and so evidently it is the concentration of electrolyte in the aqueous phase which is of importance. Furthermore, except for the system with the highest and lowest water contents studied, the transition between unstable and stable systems as the electrolyte level is increased is remarkably sharp.

Fig. 9.6, which is also from the paper by Howland *et al.*,[8] shows the effect of electrolyte level upon the average particle size of the final latex as determined by light-scattering. The particle size increases in an approximately linear fashion. This presumably reflects the tendency of the electrolyte to cause particle agglomeration during the later stages of the reaction. However, it may well be enquired just why it is that addition of an electrolyte in sufficient quantity to prevent the formation of prefloc also causes a latex of increased particle size to form. It may be that the presence of the electrolyte allows a self-terminating partial agglomeration of the particles to occur, which in turn enhances the colloid stability of the latex because of the more favourable ratio of soap to interfacial area which is thereby established.

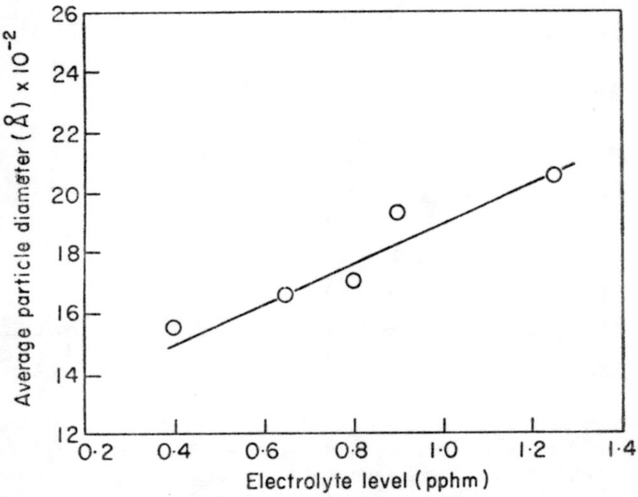

FIG. 9.6. Effect of level of electrolyte upon average diameter (by light-scattering) of particles in latex obtained by emulsion copolymerisation of styrene and butadiene at 50°F, according to Howland *et al.*[8] (See caption to Fig. 9.5 for details of reaction system. The level of water is not specified, but was probably about 70 pphm.)

A similar explanation was considered and rejected by Howland and co-workers in order to account for the drastic reduction in latex stability which was found to accompany the omission of the mercaptan from the polymerisation system. The ground for rejecting this explanation was that, within experimental error, latices made with and without modifier had the same average particle size at a given electrolyte level. The true explanation of the effect of modifier in relation to latex stability is presumably to be found in the change in viscoelastic properties of the particles concomitant with the omission of the modifier. In particular, it seems reasonable to suppose that the omission of the modifier will cause the formation of harder, more elastic particles which are less able to dissipate by viscoelastic deformation the kinetic energy associated with their relative velocities.

The results of Fryling and Follett[7] for the effect of potassium chloride upon the low-temperature emulsion copolymerisation of styrene and butadiene, given earlier in Table 9.1, show clearly the effect of electrolyte in making the latex less viscous. It is apparent that some electrolyte is necessary in order to ensure a free-forming latex. Indeed, it is well known[9,10] that if a synthetic latex is made entirely free of extraneous simple electrolytes, then it tends to take the form of a reversible gel. According to Fryling and Gindler,[10] in order to achieve such gels, it is necessary to use non-electrolyte initiators such as 1,3-diphenyltriazene, azobis*iso*butyronitrile, or cyclohexanone peroxide. The micelle generator should be ionic, either cationic or anionic, and it must be carefully neutralised or slightly under-neutralised. There is no evidence of gel formation if a small amount of electrolyte is included in the polymerisation system. Addition of a small amount of electrolyte to the latex gel causes it to transform into a free-flowing liquid. Such electrolyte additions are accompanied by a fall in pH; sometimes a slight initial increase in pH occurs which precedes the fall.

It is inappropriate to discuss this phenomenon in detail here. It is sufficient to point out that Fryling and Gindler[10] have postulated the following mechanism to account for the formation of the reversible gels and their peptisation by added electrolytes. The latex particle is regarded as being, in effect, a single large, spherical, polyelectrolyte molecule. For it consists of a sphere of hydrocarbon polymer on to the surface of which are strongly adsorbed the anions of a surface-active molecule such as a long-chain carboxylate. The hydrocarbon chains are firmly embedded in the polymer and the carboxyl groups are orientated towards the aqueous phase. Being, in effect, a polyelectrolyte, its ionic groups should be almost entirely undissociated, although not necessarily un-ionised. There is therefore at best a low concentration of positive counterions in the aqueous phase. Despite their escaping tendency, most must remain firmly bound to the polymer particle in ion-pair association with the adsorbed carboxylate ions. These ion pairs will in turn tend to have water molecules firmly associated with them. Thus each particle is surrounded by an adsorbed layer of water molecules which effectively prevents coalescence of the particles and enables them to retain their identities. Such particles have almost no overall electric charge. They can therefore approach one another closely. Hydrogen bonding between the layers of adsorbed water binds them into a three-dimensional network which gives the gel its rigidity; but coalescence of the particles is prevented by the layers of water. Furthermore, the adsorbed layer of water will tend to increase the effective hydrodynamic size of the particle. Addition of electrolyte will, by primary salt effect, increase the activity of the surface electrolyte and also provides an aqueous phase which encourages the release of counterions from the particle surface. The result is to cause the particle to acquire a negative charge, and also to disrupt the hydrogen bonds which hold together the particles in the gel. The structure breaks down, and the gel becomes a free-flowing liquid. At higher levels of electrolyte addition, a secondary salt effect becomes evident, namely suppression of ionisation by the common ion effect. As the electrolyte concentration is increased, so the electrical double layer surrounding the particle becomes compressed and the electrokinetic potential at the particle interface falls again. However, Fryling and Gindler regard the situation as being quite different from that which pertains in the absence of electrolytes. Because of the strong attractive forces of ions towards the highly-polar water molecules, the surface ion pairs are unable to compete for water with the highly-solvated ions dissolved in the aqueous phase. The particles are therefore no longer protected by a tightly adsorbed mantle of water molecules, and thus coalescence of the particles can occur as they collide. In this way, Fryling and Gindler explain the formation of stable, electrolyte-free gels, their peptisation by small amounts of electrolyte, and their destabilisation by the addition of larger amounts.

The above explanation accounts for the observation that the electrokinetic potential passes through a maximum with increasing electrolyte addition. The slight rise in pH which can accompany the initial addition of electrolyte may be due to carboxylate–anion hydrolysis which is made possible by the dissociation of the ion pairs; the subsequent fall in pH may be due to release of hydrogen ions into the aqueous phase as hydrogen bonding between the particles is disrupted and peptisation occurs.

TABLE 9.4

Effect of various inorganic electrolytes upon polymerisation rate and resulting latex in emulsion copolymerisation of styrene and butadiene at −10°C (according to Fryling and St. John[9])

(reaction system: styrene 30, butadiene 70, water 180, methanol 40, potassium laurate 5, mixed *t*-mercaptans 0·12, cumene hydroperoxide 0·177, $FeSO_4.7H_2O$ 0·31, $Na_4P_2O_7.10H_2O$ 0·56, electrolyte 0·14)

Electrolyte	*% conversion after* 17·5 h	*Character of latex produced*†
NaF	86	f
$NaPO_3$	81	g
$(.CHOH.CO_2K)_2.\frac{1}{2}H_2O$ (potassium tartrate)	81	f
$Na_3Co(NO_2)_6$ (sodium cobaltinitrite)	80	v
Na_2CO_3	80	v
$HO.C(CO_2K)(CH_2CO_2K)_2.H_2O$ (potassium citrate)	78	v
$K_2HPO_4.H_2O$	77	v
$HOC_6H_4CO_2K$ (potassium‡ salicylate)	76	v
$NaH_2PO_2.H_2O$	74	v
KSCN	72	v
$C_6H_5CO_2Na$	70	v
HCO_2Na	70	f
$KClCO_3$	69	v
Control (no added electrolyte)§	64	v
$NaHCO_3$	60	v
$Na_2B_4O_7$	58	pc, v
$NaHSO_4$	55	f, pc
$K_4Fe(CN)_6$	49	f
KCN	38	v
$K_2S_2O_8$	8	—
KIO_3	1	—
$NaBO_3.4H_2O$ (sodium perborate)	0	—
$NaBrO_2$	0	—

† v = viscous, f = fluid, pc = precoagulum, g = gelled.
‡ Attention is drawn to the fact that the original reference cites *potassium* salicylate as the salicylic acid salt used in these experiments, whereas the *sodium* salt is cited for the experiments to which Table 9.5 refers.
§ Control contained 0·7 pphm of $Na_4P_2O_7.10H_2O$.

Fryling and St. John[9] have published information concerning the effect of a range of different inorganic electrolytes upon both rate of emulsion polymerisation and the character of the latex which is produced as a result of the reaction. This information, which relates to the emulsion copolymerisation of styrene and butadiene at $-10°C$, is summarised in Table 9.4. In many instances, the rate of polymerisation was found to be increased. In some cases, fluid latices were also obtained. Retardation was also observed, sometimes to the point of complete suppression of the reaction. The performance of sodium fluoride is particularly interesting, in that it apparently allows the formation of a fluid latex at a fast rate of reaction. No explanation is given for these observations.

Fryling and St. John also give information, summarised in Table 9.5, for the effect on conversion of varying the level of some of the more effective electrolytes. They also make the interesting observation that the effect of potassium chloride on the conversion–time relationship depends to some extent upon the way in which it is added. They state that 'The experiments in which potassium chloride was added to the activator solution showed no tendency to die out at low conversions whereas potassium chloride added to the soap solution has no perceptible effect on the rate of polymerization or on the tendency of the reaction to die out.' These workers conclude that inorganic electrolytes exhibit four distinct effects in emulsion polymerisation reactions; these they describe as peptisation, micellisation, activation and promoter action. The distinction between the last two of these is not entirely clear.

In a further paper,[11] Fryling has reported upon the rheological behaviour of electrolyte- and soap-deficient latex gels. In this paper, it is also pointed out that the peptisation of electrolyte-free gels by electrolytes is reversible, since gelation can be re-induced by dialysing the latex.

The effect of simple inorganic electrolytes upon the flow behaviour of synthetic latices is also the subject of two papers by Brodnyan and Kelley.[12,13] They regard the high viscosity of electrolyte-free latices as being due essentially to the large effective hydrodynamic volume of the particles, which in turn is attributed to the highly expanded character of the electrical double-layer surrounding the particles. The addition of electrolyte causes the layer to reduce in thickness, and hence the effective size of the particles is also reduced.

TABLE 9.5

Effect of varying levels of certain inorganic electrolytes upon the extent of polymerisation in the emulsion copolymerisation of styrene and butadiene at −10°C (according to Fryling and St. John[9])

(reaction system: styrene 30, butadiene 70, water 192, methanol 48, potassium laurate 5, mixed *t*-mercaptans 0·25, cumene hydroperoxide 0·084, $FeSO_4.7H_2O$ 0·139, $Na_4P_2O_7.10H_2O$ 0·223, electrolyte—variable).

Electrolyte level (pphm)	% conversion after 7·2 h reaction using electrolyte indicated											
	NaH_2PO_3	$HO.C(CO_2K)(CH_2CO_2K)_2.H_2O$ (*potassium citrate*)	KSCN	$NaPO_3$	$Na_2HPO_4.12H_2O$	$Na_2CO_3.H_2O$	$C_6H_5CO_2Na$	$(.CHOH.CO_2K)_2.\frac{1}{2}H_2O$ (*potassium tartrate*)	$HOC_6H_4CO_2Na$ (*sodium*† *salicylate*)	HCO_2Na	Na_2SO_4	NaF
0·00	24	26	26	28	28	25	24	24	25	25	21	21
0·05	33	31	34	29	31	32	29	34	23	26	26	27
0·10	35	30	36	27	32	30	30	33	26	—	30	33
0·15	35	34	36	28	34	8	33	34	31	33	34	36
0·20	38	31	36	33	33	0	33	33	33	37	34	40
0·25	39	32	37	32	34	0	34	32	29	39	38	42

† Attention is drawn to the fact that the original reference cites *sodium* salicylate as the salicylic acid salt used in these experiments, whereas the *potassium* salt is cited for the experiments to which Table 9.4 refers.

9.3 DISPERSION MEDIA

Water is the normal dispersion medium for emulsion polymerisation. Indeed, part of the industrial attractiveness of the process is associated with the ready availability and cheapness of the dispersion medium, and with the lack of any hazards attending its use. In retrospect, it is probably true to say that the industrial viability of the process would have been in serious doubt had it not been possible to use water as the dispersion medium.

Until comparatively recently, it might also have been said that the use of water as the dispersion medium does not incur any pollution problems. However, whilst this claim is still true in the sense that pollution *by* water is obviously not a problem, the world is being increasingly made aware of the problem of pollution *of* water. In common with many other chemical processes, emulsion polymerisation can be accompanied by the serious problem of the safe disposal of effluent water, some of which may be derived from the water which was used as the dispersion medium.

If one accepts that an emulsion polymerisation system must contain micelles if initiation is to take place, then the principal technical requirements for the dispersion medium appear to be as follows:

(1) It should be a solvent for the micelle generator and for at least one component of the initiation system.
(2) It should induce the molecules of dissolved micelle generator to aggregate together to form micelles.
(3) It should preferably induce micelle formation at a rather low critical concentration of micelle generator, so that most of the added micelle generator is in fact present as micelles.
(4) It should preferably be such that the monomers are relatively insoluble in it.
(5) Its molecules should either not be efficient scavengers for free radicals, or, if they are, the products of the scavenging reaction should be such that at least one is able itself to initiate polymerisation.
(6) It should be a liquid of low viscosity and able to permit transfer of heat between reaction loci and the walls of the reactor in which the polymerisation is taking place.
(7) The physical properties of the dispersion medium should be such that it permits the carrying-out of the polymerisation reaction over a wide range of temperature and pressure.

For the majority of the common monomers of interest, water admirably satisfies most of the above criteria, in addition to possessing the advantages of cheapness, availability and freedom from hazard. Its principal shortcoming is that its physical properties are not such as to permit polymerisation to be carried out over as wide a range of temperature as is desirable. In particular, the freezing-point of 0°C places a lower limit to the reaction temperature, whereas, as will become apparent from Chapter 10, there has been considerable interest in carrying out butadiene polymerisations and copolymerisations at sub-zero temperatures.

It is proposed at this juncture to confine attention to what may fairly be described as 'conventional' emulsion polymerisations. The technique of emulsion polymerisation using non-aqueous dispersion media is relatively novel and, as yet, little developed or exploited. It represents such a radical departure from conventional systems that it will be dealt with separately in Chapter 11. In this present section, note will be taken of methods which have been proposed for the modification of aqueous dispersion media such that they can be used at sub-zero temperatures. All these modifications take the form of the addition of antifreezes.

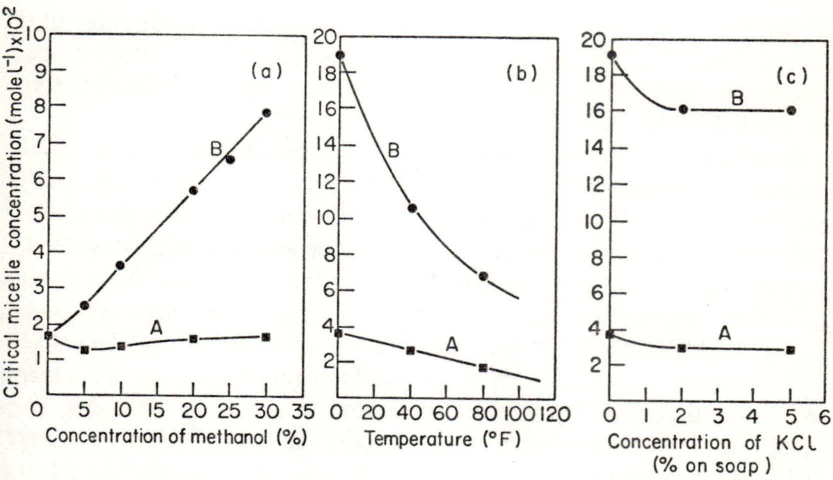

FIG. 9.7. Effect of certain variables upon critical micelle concentration of a fatty-acid soap (A) and a disproportionated rosin-acid soap (B) in methanol–water mixtures, according to Howland *et al.*[14] (a) Effect of methanol concentration upon critical micelle concentration at 80°F. (b) Effect of temperature upon critical micelle concentration in 25% aqueous methanol. (c) Effect of an inorganic electrolyte (KCl) upon critical micelle concentration in aqueous methanol at 0°F.

Proposed additives fall into two classes: (a) non-electrolytes such as alcohols and glycols; and (b) electrolytes such as simple inorganic salts.

We consider first the use of non-electrolyte antifreezes. Howland *et al.*[14] have described the use of methanol as an antifreeze for the emulsion copolymerisation of styrene and butadiene at 0°F. They replaced one-quarter of the water of the aqueous phase by methanol. It is stated that, although lower concentrations of methanol could be successfully used in small reactors, this was not possible in large vessels because the necessarily lower coolant temperature caused freezing of the latex on the walls of the reactor.

The effects of methanol concentration, temperature and electrolyte concentration on the critical micelle concentration of both a fatty-acid soap and a disproportionated rosin-acid soap are shown in Fig. 9.7 (a), (b) and (c), respectively. The effects of increasing methanol concentration in the aqueous phase and of reducing the temperature are much more pronounced in the case of the rosin-acid soap than in that of the fatty-acid soap. Both increasing the methanol concentration and reducing the temperature result in an increased critical micelle concentration. The effect of the methanol is entirely as expected: it reduces the polarity of the aqueous phase and also the ability of the water molecules to form ordered structures; there is therefore less tendency for the hydrophobic part of the soap molecule to be expelled from the aqueous phase. Electrolyte concentration was found to have little effect upon the critical micelle concentration at 0°F in the presence of 25% methanol.

Howland and his co-workers point out that although, under similar conditions, the critical micelle concentration of the rosin-acid soap is many times higher than that of the fatty-acid soap, it is still considerably below the concentration which is present in the aqueous phase of a normal emulsion polymerisation system. They also noted that, as the methanol concentration is increased, and particularly as the temperature is decreased, so the definition of the critical micelle concentration becomes less sharp. This they take to imply a more gradual transition between monomeric and micellar soap. They suggest that, even at relatively high soap concentrations, the number of micelles which form may be relatively small, and that therefore the rate of polymerisation is controlled by the number of micelles available for initiation, rather than by the activity of the initiation system.

Howard *et al.*[14] have also reported that the rate of polymerisation in sub-zero systems containing methanol as an antifreeze depends very much on the ratio of rosin-acid to fatty-acid soap used as micelle generator.

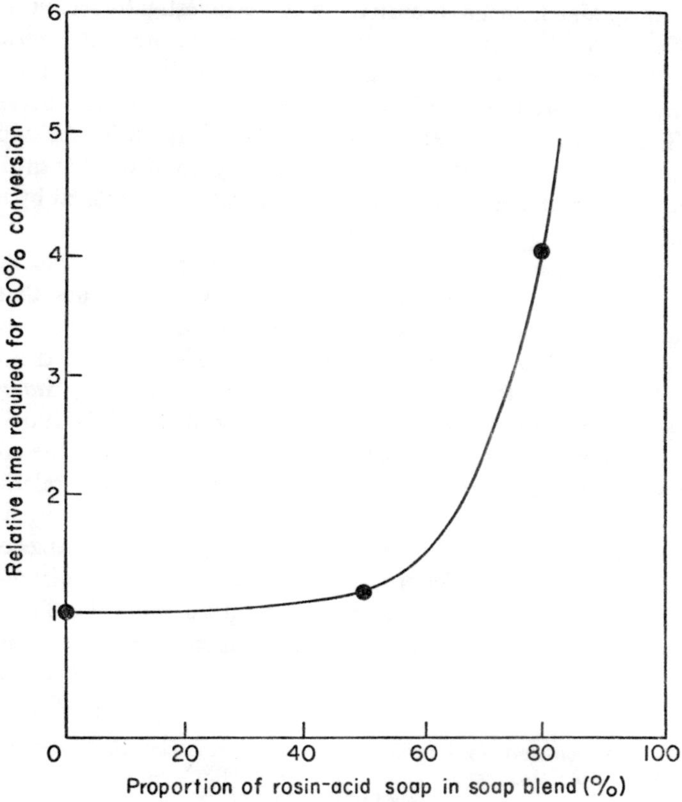

FIG. 9.8. Effect of balance of fatty-acid and rosin-acid soap upon relative rate of emulsion copolymerisation of styrene and butadiene at 0°F using blends of the two soaps as micelle generator, according to Howland *et al.*[14] (Reaction system believed to be as follows: styrene 29, butadiene 71, water 180, methanol 60, total soap 4·7, sodium naphthalene formaldehyde sulphonate 0·05, *t*-dodecyl mercaptan 0·20, di*iso*propylbenzene hydroperoxide 0·16, triethylenetetramine 0·29, K_2SO_3 0·50, $FeSO_4.7H_2O$ 0·002, ethylenediamine tetraäcetic acid 0·025, KOH 0·02.)

Their results are shown in Fig. 9.8. It appears that replacement of up to 50% of the fatty-acid soap with rosin-acid soap has but a slight retarding effect upon the reaction, but replacement of more than 50% of the fatty-acid soap is accompanied by serious retardation. Thus the reaction time to 60% conversion in the case of a micelle generator comprising 80% of rosin-acid soap and 20% of fatty-acid soap was found to be four times that required when exclusively fatty-acid soap was used. Another curious, and largely unexplained, result reported by these investigators concerns the effect of soap level upon reaction rate. Apparently, the reaction rate can go through a maximum as the amount of soap is increased, a situation being reached where the addition of further soap causes the reaction rate to fall. The comment is made: 'The data serve to emphasise the unpredictable effect of employing methanol as a freezing point depressant but do not necessarily controvert the hypothesis that rate is controlled by micelle concentration.'

An attempt was also made to improve reaction rates by replacing part of the methanol antifreeze with ethylene glycol. It appears that such replacement is accompanied by an increase in rate of polymerisation. Other non-electrolyte antifreezes which have been suggested for this application include ethanol, glycerol, acetone, formamide, mono-alkyl ethers of ethylene glycol, and ethylene oxide adducts of the mono-alkyl ethers of ethylene glycol. Fryling *et al.*[15] have reported upon the effect of various non-electrolyte antifreezes on the conversion–time relationship for a styrene–butadiene emulsion copolymerisation initiated by a diazo thioether. The particular initiator used was 2-(4-methoxybenzenediazomercapto)naphthalene (which is prepared by coupling *p*-anisidine and thio-β-naphthol). The polymerisation temperature was 5°C. The results are summarised in Table 9.6. Fryling and his co-workers comment as

TABLE 9.6

Effect of various non-electrolyte antifreezes upon the conversion–time relationship for a diazo thioether-initiated emulsion copolymerisation of styrene and butadiene at 5°C (according to Fryling *et. al.*[15])

(reaction system: styrene 30, butadiene 70, water 250 − x, antifreeze x, sodium soap of disproportionated rosin acid 5, mixed tertiary mercaptans 0·4, 2-(4-methoxybenzenediazomercapto) naphthalene 0·3, $K_3Fe(CN)_6$ 0·3, $Na_3PO_4.12H_2O$ 0·5)

Antifreeze	*Ratio* pphm *antifreeze* (x):pphm *water* (250 − x)	% *conversion after time shown*		
		4 h	7·5 h	24 h
— (control)	0:250	68		
*iso*Propanol	50:200	1		3
Ethanol	50:200	0		4
Methanol	50:200	16	30	62
Acetone	50:200	4		25
Acetonitrile	50:200	24	36	38
Dioxan	50:200	13	30	60
Ethylene glycol	50:200	47	71	88
Glycerol	75:175	61	70	72

follows: 'Unfortunately, all of the freezing point depressants tested resulted in some retardation of reaction... Glycol and glycerol retarded only slightly. Methanol, although somewhat of a retardant, was selected as most practical on the basis of ease of recovery. However, at lower temperatures of polymerization (−10° and −18°C), closely comparable rates of copolymerization were subsequently obtained with methanol and ethylene glycol.'

Carr and Johnson[16] have urged the advantages of formamide as an antifreeze, on the basis of its performance in a 25/75 styrene–butadiene system using an alkylarylsulphonate micelle generator and a dibenzoyl peroxide–sugar redox initiator. The polymerisation temperature was −10°C. Comparing the behaviour of formamide with that of glycerol, Carr and Johnson state that 'A great viscosity advantage in low temperature recipes is given by the use of formamide as antifreeze as compared with glycerol. For example, an aqueous solution containing sufficient glycerol to prevent freezing at −40°C. had a viscosity of 1600 centipoises at −35°C (Brookfield viscometer), whereas an equivalent solution utilizing formamide as antifreeze had a viscosity of only 26 centipoises. Moreover, formamide would have an advantage over methanol, the commonly used antifreeze: at extremely low temperatures, where high concentrations are required, formamide is largely immiscible with monomers, whereas methanol tends to blend monomers with the continuous medium.' Little is said about the effect of formamide upon polymerisation rate, except that the latter is low. The comment is made that 'It is expected that improvement in the rate of polymerization in low temperature aqueous-formamide recipes can be accomplished by further changes in the recipe.'

The use of simple inorganic electrolytes as antifreezes for the emulsion copolymerisation of styrene and butadiene has been investigated by St. John, Uraneck and Fryling.[17] The advantages claimed over organic antifreezes include cheapness and the tendency to encourage, rather than discourage, micelle formation. It is suggested that it would thereby be possible to use more soluble micelle generators of lower molecular weight than are conventionally employed. In their preliminary experiments, St. John, Uraneck and Fryling made the significant observation that the rate of polymerisation can be affected by the choice of electrolyte antifreeze. Thus the effect of sodium, potassium and calcium chlorides was investigated in a styrene–butadiene emulsion copolymerisation at −10°C using a hydroperoxide–iron–pyrophosphate initiation system and an alkylaryl polyether alcohol as micelle generator. The conversions after 72 h were, respectively, 15%, 23% and 42% for the three electrolytes.

A simple screening test was then applied to ascertain the solubility of some 100 possible micelle-forming substances in strong salt solutions at −10°C. In general, it was found that rosin-acid soaps, tall-oil soaps, high-molecular-weight fatty-acid soaps, high-molecular-weight fatty-amine soaps, alkylarylsulphonates, fatty-acid esters, branched alkyl phosphates and alkylamide phosphates showed suitable solubility behaviour.

The substances most compatible with the electrolytes were low-molecular-weight alkyl sulphates and phosphates, low-molecular-weight fatty-acid soaps and fatty-amine soaps, alkylarylsulphonates, alkylaryl polyether alcohols, and certain other ethylene oxide addition products. Several emulsion polymerisation recipes containing these micelle generators and inorganic-electrolyte antifreezes were developed. In addition to the electrolytes mentioned above, the following were also investigated: sodium acetate, ammonium chloride and magnesium chloride. Each recipe was found to exhibit certain individual characteristics. Thus using potassium caprylate, the copolymer was found to precipitate as it formed. The latex could be completely coagulated by further addition of water. The use of sodium hexyl sulphate gave a lower rate of polymerisation but the resultant latex was more fluid. By suitable control, the copolymer could be precipitated as it was formed. The use of an alkylaryl polyether alcohol resulted in a latex which was stable at normal temperatures, but tended to coagulate at higher temperatures; this is presumably due to the inverse solubility characteristics of substances of this type.

As regards rates of polymerisation, the following are typical. With sufficient electrolyte to reduce the freezing point of water to −10°C, a combination of sodium chloride and potassium caprylate gave 60% conversion after 4·6 h of reactions, whereas a combination of calcium chloride and sodium hexyl sulphate gave 41% conversion in 24 h, and a combination of calcium chloride and an alkylaryl polyether alcohol gave 58% conversion in 18·5 h. St. John, Uraneck and Fryling reported the surprising observation that, although the low-molecular-weight fatty-acid soaps form some curd with certain salts (presumably those of bivalent cations especially), the rate of polymerisation is not greatly affected by such curd formation. Presumably, the precipitated soap is still sufficiently soluble to function as an efficient micelle generator.

Another interesting observation recorded is that the amount of mercaptan required for satisfactory modification of the polymer is greater than in the case of conventional recipes. Furthermore, mercaptans of lower molecular weight could be used; *t*-butyl mercaptan was found to be especially effective.

9.4 SEQUESTERING AGENTS

It is common industrial practice to add very small amounts of metal ion sequestering compounds to emulsion polymerisation systems, quite apart from any which may be added to complex the iron component of a redox initiation couple. The substance most commonly used as a sequesterant is ethylenediamine tetraäcetic acid, or one of its alkali–metal salts. The advantages conferred by its addition are: (i) faster rates of polymerisation, (ii) less variable rates of polymerisation, and (iii) an enhancement of the colloid stability of the latex produced. The usual explanation given for this effect is de-activation of traces of calcium and magnesium ions which may be present in the water used, or, indeed, in any of the other components of the reaction system. Contamination with traces of heavy-metal ions may also occur through contact with defective reactor linings.

According to Pryor, Harrington and Druesedow,[18] the addition of a small amount of a salt of ethylenediamine tetraäacetic acid to the water of a low-temperature GR-S polymerisation system can reduce the reaction

TABLE 9.7

Effect of calcium and magnesium ions, and of sequestering agent, upon low-temperature emulsion copolymerisation of styrene and butadiene (according to Pryor, Harrington and Druesedow[18])

(reaction system is merely described as the 'standard cold GR-S recipe', without details being given; it was presumably initiated by a hydroperoxide–iron–pyrophosphate–sugar system, with a rosin-acid soap as the micelle generator)

Concentration of added heavy-metal ions (parts per million on water)		pphm *of tetrasodium ethylenediamine tetraäcetate*	*Reaction time to 60% conversion* (h)
Ca^{2+}	Mg^{2+}		
4	0	0	12·0
4	0	0·025	10·4
8	0	0	13·0
8	0	0·025	10·4
16	0	0	13·7
16	0	0·025	11·7
0	4	0	15·5
0	4	0·025	9·9
0	8	0	†
0	8	0·025	12·2
0	16	0	†
0	16	0·025	†

† Reaction ceased before 60% conversion.

time by as much as 4 h. These authors cite the results given in Table 9.7 as illustrating the facts that (i) trace amounts of calcium and magnesium ions can have a severe retarding effect upon the polymerisation, and (ii) the addition of a sequestering agent can have a corrective effect. It is not clear what exactly was the formulation to which these observations refer; it is merely described as the 'standard cold GR-S recipe'. Presumably, these effects are observable in most systems. Certainly, sequestering agents are widely used in high-temperature persulphate-initiated systems, as well as in low-temperature systems.

9.5 SHORTSTOPPERS

Included among the papers in which reference is made to shortstoppers for the emulsion copolymerisation of styrene and butadiene are those by Smith *et al.*,[19] by Kluchesky and Wakefield,[20] by Wakefield and Bebb,[21] by Smith *et al.*,[22] by Antlfinger and Lufter,[23] and by Howland *et al.*[14] Those by Wakefield and Bebb and by Antlfinger and Lufter describe the results of extensive surveys of various chemical types.

The requirements which the ideal shortstopper should fulfil are several, and include the following:

(i) It should swiftly bring the polymerisation reaction to a halt when added to the reaction system in small quantity.
(ii) It should discourage further chemical modification of the polymer (*e.g.* by degradation or by crosslinking) once polymerisation has ceased. Such chemical modification might tend to occur, for instance, during the monomer stripping stage.
(iii) It should not affect the colloid stability of the latex.
(iv) It should not adversely affect the physical or chemical properties of the polymer obtained from the latex.
(v) It should not tend to be left behind in reactors after the short-stopped latex has been removed, otherwise the subsequent polymerisation batch may be severely retarded or inhibited.
(vi) It should not cause discoloration.
(vii) It should be cheap, readily available, and have no hazards associated with its use.
(viii) For convenience of handling, it should be readily soluble in water, and should be capable of being stored as an aqueous solution over long periods of time.
(ix) The ideal shortstopper would in addition be of very wide applicability, and would certainly display all the above attributes when

employed with a given monomer type initiated by several different methods.

The primary effect of adding a shortstopper is to bring to an abrupt halt the propagation stage of the polymerisation reaction. This will normally be brought about either by a transfer reaction with the production of a relatively unreactive free radical, or by a copolymerisation reaction, again with the formation of a relatively unreactive free radical. In either case, the free radical which is formed may itself be effective in terminating a second propagating radical by mutual combination.

Although curtailment of propagation is the primary effect of the shortstopper, it should ideally have a secondary effect, in that it should react with and destroy the radical-initiation system. In this way can be met the requirement that subsequent chemical modification of the polymer be prevented. A consequence of this consideration is that the effectiveness of a given shortstopper may depend to some extent upon the initiation system with which it is used. It appears that low-temperature polymerisation reactions are more prone to post-conversion polymer changes than are those carried out at higher temperatures. This is attributed to the longer average lifetime of free radicals at low temperatures, and also to a retardation of reactions which would result in the destruction of the initiation system. Thus it is that the requirements for shortstoppers for low-temperature reactions tend to be rather more stringent than are those for high-temperature reactions.

A wide range of substances has been used for the shortstopping of styrene–butadiene emulsion polymerisations. These include compounds containing, or capable of forming, quinoid structures, nitro and nitroso compounds, oxygen, aromatic polyhydroxy compounds such as catechol and pyrogallol (which could be classified among the compounds which are capable of forming quinonoid structures), water-souble dithiocarbamates, sulphur, and 2,4-dinitrochlorobenzene and certain of its derivatives.

The results of a survey of many possible substances as shortstoppers for the persulphate-initiated emulsion copolymerisation of styrene and butadiene have been published by Kluchesky and Wakefield.[20] The temperature of polymerisation was in the range between 30° and 50°C. Hydroquinone has been widely used for the shortstopping of this reaction and their results confirm its effectiveness if present in excess of about 0·05 pphm. Commenting upon the correlation between the structure of phenols and their performance as shortstoppers, Kluchesky and Wakefield state that 'An interesting difference exists between *p*-phenylphenol and *p*-cresol:

the phenylphenol is an excellent stopping agent, whereas the *p*-cresol is ineffective. On the other hand, *o*-phenylphenol is devoid of stopping power. Both α- and β-naphthol are good stopping agents but certain alkylated β-naphthols lack effectiveness; the alkyl groups are believed to be in the alpha position, so that here, as with *o*-phenylphenol, steric hindrance probably is preventing the oxidation of the compound to a quinone.' Other compounds found to be effective include *p,p'*-dihydroxybiphenyl sulphide, tetramethylthiuram disulphide, sodium sulphide and dinitrobenzene.

Comment is also made about the use of oxygen as a shortstopper. The presence of a certain amount of oxygen is desirable for effective initiation, but too much causes complete inhibition of the reaction. But because of the risk of explosion, it is considered too hazardous to use oxygen as a shortstopper. Another interesting matter to which Kluchesky and Wakefield refer is the fact that it is much easier to prevent an emulsion poly-

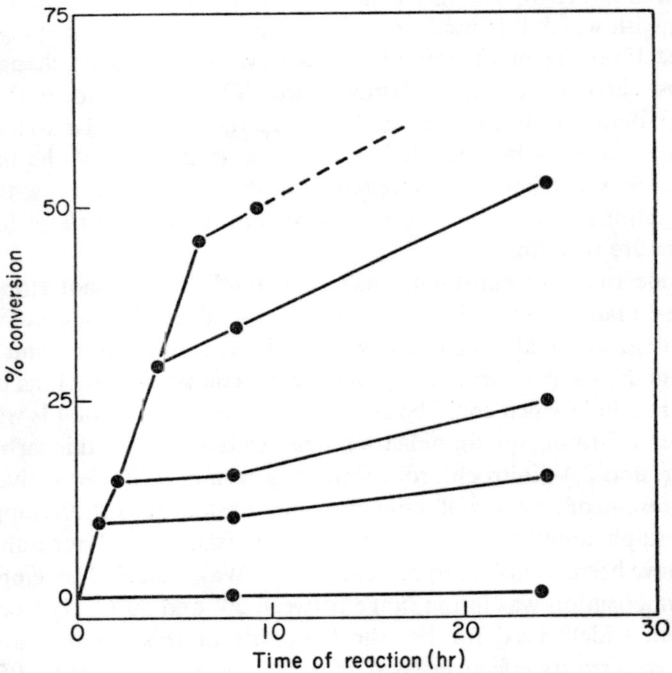

FIG. 9.9. Effect of *p*-cresol added to a persulphate-initiated styrene–butadiene emulsion copolymerisation at various stages of conversion, according to Kluchesky and Wakefield.[20] (Reaction system is said to have been a 'GR-S recipe', but exact details are not given, nor is the level of added *p*-cresol.)

merisation from commencing than to stop it once it has begun. This is illustrated in Fig. 9.9 for *p*-cresol (a relatively ineffective shortstopper) added to a standard GR-S system at various conversions. Added initially, it inhibits polymerisation completely; added after the commencement of the reaction, it seems to exert a progressively smaller retarding effect the longer its addition is delayed. Apart from the inherent interest of this effect, it has the practical consequence that it is unwise to try to infer shortstopping performance from ability to inhibit polymerisation when present in the system before polymerisation has started.

As has been pointed out above, hydroquinone has been widely used as a shortstopper for persulphate-initiated reactions carried out at about 50°C. The results summarised in Table 9.8 illustrate the considerable effectiveness of hydroquinone for such polymerisations, as compared with a mixture of *p*-phenylphenol and hydroxylamine. The latter combination would be

TABLE 9.8

Comparison of hydroquinone and mixture of *p*-phenyl phenol and hydroxylamine as shortstoppers for persulphate-initiated emulsion copolymerisation of styrene and butadiene at 50°C (according to Wakefield and Bebb[21])

(reaction system: styrene 25, butadiene 75, water 180, fatty-acid soap 5, dodecyl mercaptan 0·45, $K_2S_2O_8$ 0·3)

	Shortstopper			
	0·25 pphm *of hydroquinone*		0·25 pphm *of p-phenylphenol* + 0·1 pphm *of hydroxylamine*	
Property	Initial†	After ageing‡	Initial†	After ageing‡
Conversion (%)	72	72	74	86
Williams plasticity of contained polymer (mm)	2·89	2·95	3·29	3·73
Recovery of contained polymer (mm)	0·42	0·52	0·56	0·99
Gel content of contained polymer (%)	1·1	0·9	0·4	49
Intrinsic viscosity of contained polymer (dl g^{-1})§	2·05	1·95	2·30	1·50

† These values refer to the latex and its contained polymer immediately after the addition of the shortstopper.
‡ These values refer to the latex and its contained polymer after being heated for a further 20–24 h following the addition of the shortstopper.
§ These values presumably refer to only the sol fraction of the contained polymer.

expected to be especially effective in view of the fact that Kluchesky and Wakefield[20] reported *p*-phenylphenol by itself to be quite effective. The results given in Table 9.8 are taken from the paper by Wakefield and Bebb.[21] They show that not only does hydroquinone suppress further polymerisation; it also suppresses any tendency for the polymer to become chemically modified in any significant way as revealed by the properties listed in Table 9.8. Wakefield and Bebb also give information concerning the effect of shortstopper upon the properties of an EPC-black-filled vulcanisate prepared from the copolymer. Readers who require information on this matter are referred to the original paper.

It would be expected that the effectiveness of a shortstopper depends to some extent upon the amount which is used. The results summarised in Table 9.9 show that this is indeed the case. They are taken from the paper by Kluchesky and Wakefield,[20] and refer to the persulphate-initiated emulsion copolymerisation of styrene and butadiene. The results primarily of interest are those for hydroquinone, but results for *p*-aminophenol are also included to show that the effect of level of addition is apparently quite general. It appears that, for any given shortstopper, there is a certain minimum level of addition, below which it is relatively ineffective.

TABLE 9.9

Effect of level of shortstopper upon its effect in persulphate-initiated emulsion copolymerisation of styrene and butadiene (according to Kluchesky and Wakefield[20])

(reaction system is said to be 'GR-S recipe', but exact details are not given)

Additive	*Level of addition* (pphm *of monomer initially present*)	*% conversion at which added*	*% conversion after further* 15 h *reaction*	*Increase in % conversion*
Hydroquinone	0·01	60·0	79·5	19·5
	0·025	49·5	62·0	12·5
	0·05	48·0	49·0	1·0
	0·10	47·5	48·0	0·5
p-Aminophenol	0·01	57·5	90·5	33·0
	0·025	51·0	52·5	1·5
	0·05	51·5	52·5	1·0
	0·20	52·5	53·0	0·5
Control: no additive	—	51·5†	99·0	47·5

† Presumably this figure refers to roughly the conversion at which a shortstopper would have been added to the reaction system if one was to be added.

Hydroquinone is presumed to exert its shortstopping action through first being oxidised by the persulphate to *p*-benzoquinone which then inhibits polymerisation by a copolymerisation mechanism. Some of the persulphate initiator is also destroyed in the oxidation process. Hydroquinone is not, however, effective for low-temperature reactions which use hydroperoxide–iron–pyrophosphate–sugar redox initiation systems. It has been suggested[14] that the sugar suppresses the oxidation of hydroquinone to *p*-benzoquinone. Howland *et al.*[14] state that hydroquinone can be used as a shortstopper for sugar-free low-temperature systems, provided that either the latex is heated immediately after shortstopping, or a small amount of hydrogen peroxide is used in conjunction with the hydroquinone. It appears that, even in sugar-free systems, the hydroquinone is not oxidised sufficiently rapidly by the residual peroxide, so that either the temperature must be raised or a secondary oxidising agent must be employed. These authors further point out that the oxidation of hydroquinone only occurs rapidly in an alkaline environment such as pertains in a typical styrene–butadiene emulsion copolymerisation system. Thus because of the mildly acidic nature of both hydroquinone and hydrogen peroxide, it is possible to premix and store them at room temperature before use.

According to Howland *et al.*,[14] di-*t*-butylhydroquinone was the shortstopper used when styrene–butadiene rubbers were first produced by low-temperature polymerisation. The isomer believed to have been used is di-2,5-*t*-butylhydroquinone. This compound is not water-soluble, and so had to be added as a ball-milled dispersion. This led to handling difficulties, and to somewhat erratic shortstopping performance because of incompatibility with the latex. Thus it appears that it could cause flocculation when added to the latex, unless special precautions were taken. A further disadvantage was that it could not be added to latex in reactors, because traces tended to be left behind when the latex was subsequently removed, and these traces caused inhibition and retardation of the next polymerisation batch. Consequently, it was necessary to transfer the latex to the stripping vessel before adding the shortstop. Smith *et al.*[19] comment that: 'In actual plant operation this requires very careful scheduling and, at the same time, may prevent operation at full capacity, because an empty stripping vessel must always be available whenever a batch is to be stopped.' However, in itself, di-2,5-butylhydroquinone is said to be an excellent shortstopper for low-temperature reactions under ideal conditions of usage.

2,4-Dinitrochlorobenzene has also been investigated as a shortstopper for low-temperature styrene–butadiene emulsion copolymerisations.

It was less expensive than di-*t*-butylhydroquinone and more compatible with the latex. It could conveniently be added as a solution in styrene. Its great effectiveness as a shortstopper for low-temperature reactions has not been questioned. It does, however, suffer from a number of serious disadvantages. It is toxic, being said to cause, *inter alia*, violent and prolonged skin irritation. Furthermore, it causes a pronounced yellow coloration in both the latex and the polymer obtained from it. This coloration, which is certainly most objectionable if the latex is to be used as such, is said to arise from hydrolysis to 2,4-dinitrophenol.

Wakefield and Bebb[21] have investigated the possibility of using derivatives of 2,4-dinitrochlorobenzene, which, although effective as shortstoppers, would be less toxic in nature. An early observation that 2,4-dinitrophenylbenzthiazyl sulphide (I) is also an effective shortstopper led to the belief that the chlorine atom in 2,4-dinitrochlorobenzene was not

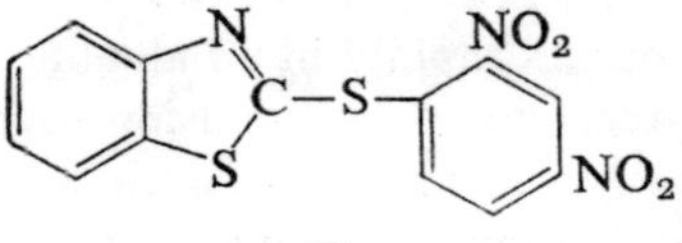

(I)

essential to its shortstopping ability. (The compound (I) is prepared by reacting 2,4-dinitrochlorobenzene with the sodium salt of mercaptobenzthiazole.) Since the compound (I) is virtually non-toxic, the idea was conceived that other derivatives might be found which would combine effective shortstopping ability with freedom from toxicity and tendency to discolour. Wakefield and Bebb confirmed the effectiveness of compound (I), not only as a terminator of polymerisation, but also as a preventer of polymer modification. In fact, of the derivatives examined, compound (I) was found to be the best. As close second came the derivatives 2,4-dinitrobenzenethiol (II) and 2,4-dinitrophenylpyridinium chloride (III). The latter compound was, however, found to cause some preflocculation

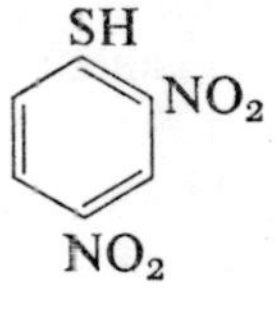

(II)

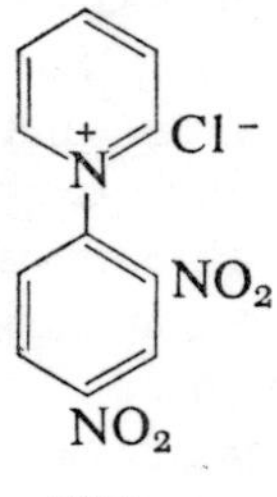

(III)

of the latex. This is hardly surprising in so far as the cation which it contains is presumably mildly surface-active, and therefore tends to be adsorbed at the surfaces of negatively charged particles with partial neutralisation of charge and loss of colloid stability. Wakefield and Bebb consider that the compounds (I), (II) and (III) are probably effective by different mechanisms. They state that: 'The benzothiazyl sulfide may well owe its action to an activation of a point on the ring by the nitro groups, whereas the pyridinium compound is known to react in the presence of base to give a form of glutaconaldehyde, leaving behind dinitroaniline. The different forms of these materials are undoubtedly highly active toward free radicals, and the resulting adducts would be greatly stabilized by resonance. The thiol could react in the same manner as an overactive mercaptan modifier, acting as a chain terminator instead of a transfer agent.

Wakefield and Bebb regard the presence of nitro groups as necessary but not sufficient to ensure shortstopping ability. They envisage the nitro groups as probably activating a position or another group within the molecule, thereby encouraging reaction at that point. They also comment that the solubility of the compound appears to have no direct bearing upon its shortstopping ability. Thus compound (II) above is very soluble in water, whereas compound (I) is insoluble. Thus, in particular, it should be noted that oil-solubility is not essential, and therefore neither, it would appear, is a favourable partitioning of the shortstopper between particle and aqueous phases. On the other hand, as has been pointed out above, it would be expected that compound (III) would be at least weakly adsorbed from the aqueous phase on to the surface of the particles.

Wakefield and Bebb[21] also investigated the use of sodium nitrite as an auxiliary shortstopper. They found that generally this compound is without effect. They did, however, report the interesting observation that, when used in conjunction with 2,5-di-*t*-butylhydroquinone, the polymer underwent a marked softening during the ageing period to which the latex was subjected. In fact, it was found that a high initial gel content was reduced to zero by this treatment. The effect of sodium nitrite upon post-shortstopping conversion was found to be small, because 2,5-di-*t*-butylhydroquinone is very effective in this respect without an auxiliary shortstopper.

The use of water-soluble dialkyldithiocarbamates as shortstoppers has been mention by Smith *et al.*,[22] and described in more detail by Howland *et al.*[14] Water-soulbe dialkyldithiocarbamates are not only very convenient to use, relatively cheap, and very effective as shortstoppers for low-temperature polymerisations; they are also non-toxic, non-discolouring and non-staining. Dimethylammonium dimethyldithiocarbamate is said to be particularly suitable.[22] Although the sodium salt is cheaper, it tends to cause some coagulum formation when added to the latex. However, this tendency can be eliminated by incorporating a small amount of soap in the shortstopper solution. Howland *et al.*[14] have reported that potassium dimethyldithiocarbamate can also be used without any tendency to cause colloidal instability. Aqueous solution of higher concentration can be used, and this is important when it is desired to make high-solids latices by emulsion polymerisation.

It is, of course, well known that dithiocarbamates are powerful accelerators of the sulphur-vulcanisation of diene rubbers. It is therefore of interest to note the following comment by Howland and his co-workers: 'Because of the fact that dithiocarbamates are used as vulcanization accelerators, there was considerable hesitancy on the part of many compounders in accepting these materials as components of the polymer. However, careful investigation showed that as long as excesses of dithiocarbamate were avoided, the rate of cure of the solid polymer was not affected significantly. It is reasonable to assume that whatever excess shortstop is present is either washed out or decomposed during the coagulation.'

As far as the present writer is aware, little is known concerning the mechanism by which dialkyldithiocarbamates act as shortstoppers. It is certainly known that the dithiocarbamate ligand of certain heavy-metal dialkyldithiocarbamates is able to participate in free-radical reactions, and, furthermore, in radical reactions which suppress the oxidation of hydrocarbon rubbers by a chain mechanism. It is likely that, when added to a low-temperature emulsion polymerisation system, oxidation by the hydroperoxide to the corresponding thiuram disulphide would occur by some such reaction as the following:

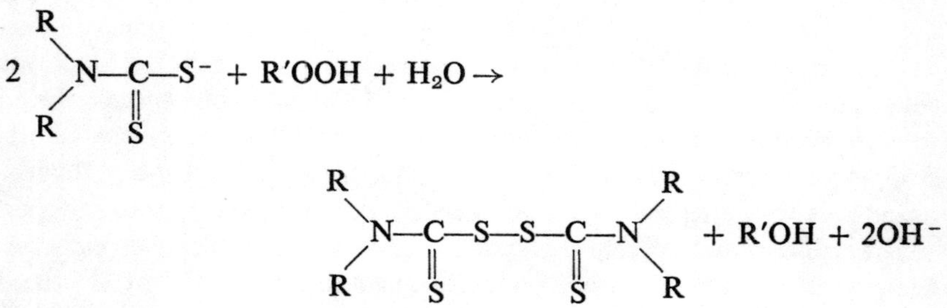

The tetralkylthiuram disulphides are themselves known to be effective as shortstoppers; possible they dissociate thermally into two $R_2NCSS\cdot$ radicals, which are then able to terminate propagating chains. The above oxidation reaction would account for the destruction of the hydroperoxide component of the redox couple. It is also possible that the dithiocarbamate anion could be directly effective as a shortstopper, by some such reactions as:

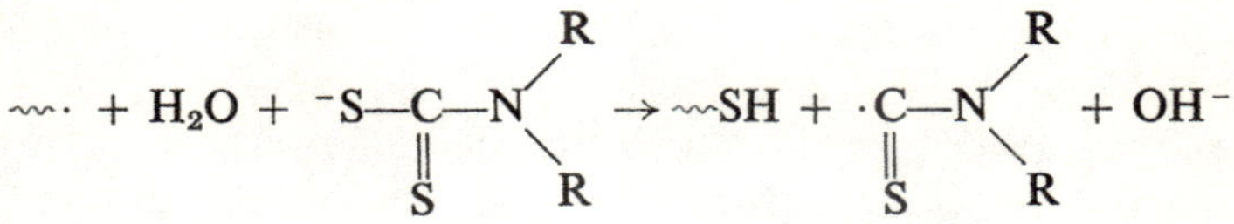

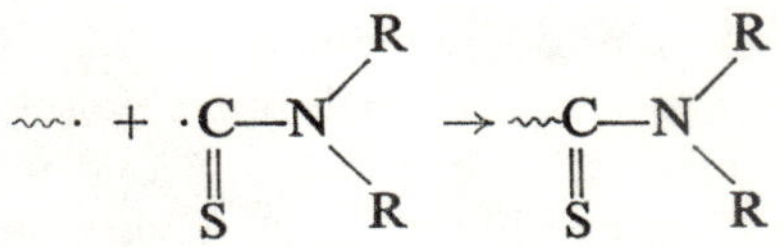

These reactions, as given, would seem to require the simultaneous presence of both water and dithiocarbamate anions in the reaction loci. It could be argued that neither of these requirements is likely to be fulfilled in practice.

An alternative suggestion is that a two-stage mechanism is operative. The first stage occurs in the aqueous phase and results in the generation of dialkylamino radicals:

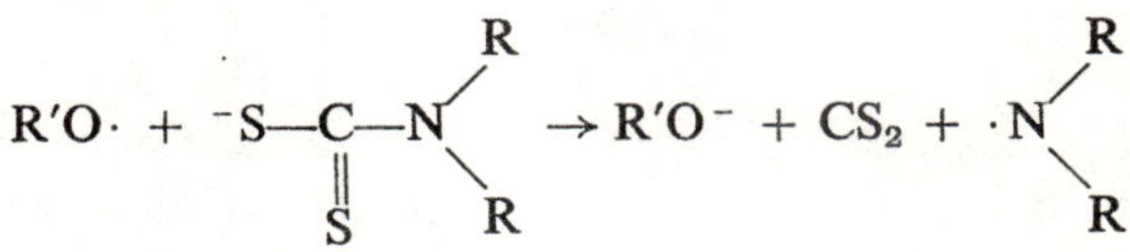

the R′O· radicals being derived from the hydroperoxide. The second stage, which is interaction between the propagating chains and the dialkylamino radicals, occurs after the latter have diffused into the reaction loci:

$$\sim\!\sim\cdot + \cdot NR_2 \rightarrow \sim\!\sim NR_2$$

It is also possible that the dithiocarbamate ions deactivate the iron component of the redox system by precipitating the iron as an insoluble dithiocarbamate.

Antlfinger and Lufter[23] have reported the results of an extensive survey of possible substances as shortstoppers for the emulsion copolymerisation of styrene and butadiene at 5°C using a hydroperoxide–iron–pyrophosphate–sugar initiation system. The following were among the compounds found to be effective at a level of about 0·2 pphm: *p*-*t*-butylcatechol, tetramethylthiuram disulphide, phenylhydrazine, di-*t*-amylhydroquinone, sodium dimethyldithiocarbamate, *p*-nitrosodimethylaniline and α-nitroso β-naphthol. Sodium hydrosulphite was found to be effective at a level of 0·4 pphm.

Sulphur at a level of 0.2 pphm was also found to be a fairly good shortstop. It was added as a solution in styrene. Since sulphur is known to be a chain terminator, the idea was conceived that a combination of sulphur with a substance which would destroy the initiation system would provide an inexpensive but efficient shortstopper. An obvious way of destroying the initiation system would be to eliminate the oxidative component of the redox system. Thus many of the reducing agents already found from the first survey to be effective as shortstoppers were then investigated in combination with elemental sulphur. It was found that a combination of sulphur and sodium dimethyldithiocarbamate was the best of the pairs investigated. The optimum combination of these two substances which would function as an adequate shortstopper was found to be 0·5 pphm of sulphur and 0·03 pphm of the dithiocarbamate. This combination allowed the use of much less sulphur than would be required if it alone were used. The amount of dithiocarbamate is also less than that required when it is used alone. In fact, a synergistic effect appears to be operative. The quality of rubber produced using this shortstopper combination was judged to be equal to that shortstopped with 2,4-dinitrochlorobenzene, although the rate of cure is said to be somewhat enhanced.

A further development has been the use of sodium polysulphide as a water-soluble equivalent for elemental sulphur. Antlfinger and Lufter[23] report that the combination of sodium polysulphide and sodium dimethyldithiocarbamate is as effective as sulphur and the dithiocarbamate, but has the advantage of complete water-solubility. They also state that, in addition to furnishing sulphur, the polysulphide also acts as a reducing agent. They might also have added that it could be a source of sulphide ions as well, and that these could be effective in de-activating the iron component of the redox system by precipitation—even though iron(II) sulphide has itself been suggested as a source of iron(II) ions for redox couples. A combination of 0·04 pphm of each of the components is suggested. It is also stated that the persulphate-initiated emulsion copolymerisation of styrene and butadiene can be adequately shortstopped either by 0·15 pphm of sodium dimethyldithiocarbamate, or by 0·10 pphm of a combination with sodium polysulphide.

REFERENCES

1. F. A. Bovey, I. M. Kolthoff, A. I. Medalia and E. J. Meehan, *Emulsion Polymerization*, Interscience, New York, 1955, pp. 346–7.
2. C. P. Roe, *Ind. Engng Chem.* 1968, **60**(9), 20.
3. C. P. Roe and P. D. Brass, *J. Polym. Sci.*, 1957, **24**, 401.
4. V. R. Kamath, 'Emulsion polymerization of styrene: kinetics', Ph.D. thesis, University of Akron, 1973.
5. C. Patel, M.S. thesis, University of Akron, 1965.
6. W. A. Schulze, C. M. Tucker and W. W. Crouch, *Ind. Engng Chem.*, 1949, **41**, 1599.
7. C. F. Fryling and A. E. Follett, *J. Polym. Sci.*, 1951, **6**, 59.
8. L. H. Howland, V. C. Neklutin, R. W. Brown and H. G. Werner, *Ind. Engng Chem.*, 1952, **44**, 762.
9. C. F. Fryling and W. M. St. John, *Ind. Engng Chem.*, 1950, **42**, 2164.
10. C. F. Fryling and E. M. Gindler, *Proceedings of International Rubber Conference*, Washington, 1959, p. 456.
11. C. F. Fryling, *J. Colloid Sci.*, 1963, **18**, 713.
12. J. G. Brodnyan and E. L. Kelley, *J. Colloid Sci.*, 1964, **19**, 488.
13. J. G. Brodnyan and E. L. Kelley, *J. Colloid Sci.*, 1965, **20**, 7.
14. L. H. Howland, V. C. Neklutin, R. L. Provost and F. A. Mauger, *Ind. Engng Chem.*, 1953, **45**, 1304.
15. C. F. Fryling, S. H. Landes, W. M. St. John and C. A. Uraneck, *Ind. Engng Chem.*, 1949, **41**, 986.
16. E. L. Carr and P. H. Johnson, *Ind. Engng Chem.*, 1949, **41**, 1588.
17. W. M. St. John, C. A. Uraneck and C. F. Fryling, *J. Polym. Sci.*, 1951, **7**, 159.
18. B. C. Pryor, E. W. Harrington and D. Druesedow, *Ind. Engng Chem.*, 1953, **45**, 1311.
19. H. S. Smith, H. G. Werner, J. C. Madigan and L. H. Howland, *Ind. Engng Chem.*, 1949, **41**, 1584.
20. E. F. Kluchesky and L. B. Wakefield, *Ind. Engng Chem.*, 1949, **41**, 1768.
21. L. B. Wakefield and R. L. Bebb, *Ind. Engng Chem.*, 1950, **42**, 838.
22. H. S. Smith, H. G. Werner, C. B. Westerhoff and L. H. Howland, *Ind. Engng Chem.*, 1951, **43**, 212.
23. G. J. Antlfinger and C. H. Lufter, *Ind. Engng Chem.*, 1953, **45**, 182.

CHAPTER 10

MONOMERS

10.1 INTRODUCTION

In one sense, the monomer is clearly the most important component of an emulsion polymerisation system. It is in order to effect the polymerisation of the monomer that the system is brought into existence. Yet emulsion polymerisation is only one mode among several for effecting free-radical polymerisation. For this reason, more emphasis has often been placed upon the mode of polymerisation than upon the details of the monomer–polymer interconversion itself. Indeed, this has largely been the point of view from which this book has been written; emulsion polymerisation has, in the main, been presented as a generalised chemical reaction having certain distinguishing characteristics, to which all monomers which are sparingly soluble in water may be subjected. It would therefore be inappropriate in this chapter to give detailed discussions of all the contributions which have appeared in the literature concerning the emulsion polymerisation of individual monomers. Not only would such a course be inappropriate; it would also swell the size of this book to an unreasonable extent. What has therefore been done is to note more or less briefly some of the papers which have appeared dealing with the emulsion polymerisation of the more common monomers, and to give summaries of the more important conclusions to which those papers point. It is appropriate to draw attention here to a recent review of emulsion polymerisation by Vanderhoff[1] which summarises, *inter alia*, the behaviour of a wide range of monomer types. This review is especially valuable in that it contains an extensive compilation of references up until about 1966.

The subject-matter of this chapter has been arranged under the following headings: styrene and butadiene; ethylene; vinyl esters of carboxylic acids; esters of acrylic acid and methacrylic acid; and vinyl chloride and vinylidene chloride. Some comment is necessary regarding the first of these groups of monomers. This group compromises those monomers which have been widely used for the production of various synthetic rubbers and resins. In terms of tonnage of polymer produced, this is undoubtedly the most important group of monomers. For this reason, many of the fundamental characteristics of emulsion polymerisation have been worked out using systems containing these monomers. Especially is this true of styrene and of mixtures of styrene and butadiene. Reference to Chapters 6, 7, 8 and 9 will reveal how frequently the properties of initiators, micelle generators, modifiers and other recipe ingredients have been reported in terms of their behaviour in styrene–butadiene emulsion copolymerisation systems. It must therefore be borne in mind that many of the matters concerning the first group of monomers which might have found a place in this chapter have already been dealt with in detail in earlier chapters. To a lesser extent, these comments apply to other monomers also.

By way of a final introductory comment, it is necessary to remind the reader that no attempt has been made to deal systematically with the subject of emulsion copolymerisation. The subject of free-radical copolymerisation is extensive in its own right. Without giving detailed consideration to the phenomenon as it is observed in bulk and solution polymerisations, it would be impossible to discuss the perturbations which are encountered through carrying out the reaction in emulsion.

10.2 STYRENE AND BUTADIENE

Styrene has frequently been regarded as an ideal monomer with which to carry out emulsion polymerisation studies. This is due not so much to the industrial importance of polystyrene, but rather to a fortunate conjunction of at least four circumstances: (a) styrene is a monomer which is convenient to handle in the laboratory and which polymerises at reasonable rates; (b) it is only slightly soluble in water; (c) a large amount of information is available in the published literature concerning its free-radical polymerisation in both bulk and solution; and (d) its free-radical polymerisation behaviour, as observed in bulk and in solution, has usually been found to be consistent with a simple sequence of reactions for the overall conversion of monomer to polymer.

Van der Hoff[2] has reported the results of an investigation into various aspects of the kinetics of the emulsion polymerisation of styrene. The objective was to test the validity of the Smith–Ewart theory for systems containing more than 10^{14} particles per millilitre of aqueous phase. Polymerisations were carried out using cumene hydroperoxide as initiator and various potassium fatty-acid soaps as micelle generator. The conversion–time curves were found to consist mainly of two essentially straight lines. At almost 55% conversion, there was observed a transition to a lower rate of polymerisation. The extent of the swelling of the latex particles, measured during polymerisation, revealed that the change in reaction rate occurs at the point where the monomer droplets disappear.

From measurements of rate of polymerisation and from particle counts by electron microscopy, it was found that the rate of polymerisation per particle varies as the particle number to the power $-0{\cdot}83$, the range of particle concentration being from 5×10^{14} to 5×10^{15} per millilitre of aqueous phase. The proportionality constant was found to depend upon the nature of the soap and upon its concentration. Data for the molecular weight of the polymer produced were interpreted as indicating that the micelle generator plays a role in the polymerisation process even after the micelles have disappeared.

In a second paper,[3] van der Hoff reported the effect of varying the initial soap level from well below to well above that corresponding to the critical aqueous-phase concentration for micelle formation. Above the critical concentration, the number of particles formed was found to agree with the number predicted by the Smith–Ewart theory. As the soap concentration was reduced below the critical level, so the number of particles decreased sharply. In this concentration range, the number of particles was found to be independent of the amount of initiator, but strongly dependent upon the soap concentration. The kinetic features suggest a similarity between particle formation in both these polymerisations and in the homogeneous precipitation of insoluble inorganic salts. Systems containing particles whose diameters were smaller than about $0{\cdot}12\ \mu m$ were found to obey the Smith–Ewart kinetics. In the case of larger particles, it appeared that more than one radical can coexist in a single particle for significant lengths of time. Rate coefficients for termination and for transfer-to-monomer were calculated from data for polymerisation rate and product molecular weight. The termination coefficient was found to be two or three orders of magnitude smaller than values obtained in solution polymerisation. The discrepancy was attributed to the operation of the Trommsdorff 'gel' effect.

Burnett and Lehrle[4] have reported the results of a critical examination of the emulsion polymerisation of styrene using a very simple system comprising monomer, water, potassium stearate and potassium persulphate. Particular attention was paid to the measurement of reaction rate, the number of latex particles and the concentration of monomer at the reaction loci. From these measurements, an estimate was made of the rate constant for propagation for styrene, using the Smith–Ewart theory. The value obtained was found to differ little from those reported for the bulk polymerisation of styrene.

Bartholomé *et al.*[5,6] have reported the results of an extensive study of the validity of the Smith–Ewart 'Case 2' relationships for the emulsion polymerisation of styrene initiated by potassium persulphate in the presence of an alkali-metal salt of a sulphonated C_{18} olefin as the micelle generator. The principal anion in the latter is believed to have been $CH_3(CH_2)_{15}CHOHCH_3SO_3^-$. The polymerisation temperature was 45°C. The results reported in these papers are consistent with the predictions of the Smith–Ewart theory (although this does not in itself thereby necessarily prove the validity of the concepts underlying the Harkins–Smith–Ewart theory). The rate of polymerisation was observed to vary linearly with the number of particles, and both rate and number were proportional to the 0·6th power of the concentration of micelle generator and the 0·4th power of the initiator concentration. The rate of polymerisation, number of particles, monomer concentration within the particles, and degree of polymerisation of the product were found to be independent of the ratio of aqueous phase to styrene over the range from about 6 to about 20. The estimate for the concentration of monomer in the particles when monomer droplets are still present, obtained from the conversion–time curve, is in good agreement with that reported by other workers using different techniques.

However, marked deviations from the predictions of the Smith–Ewart theory were observed by Gerrens[7] in the emulsion polymerisation of styrene to high conversions. The observations were interpreted in terms of a Trommsdorff 'gel' effect, the retardation of the termination reaction permitting the number of active radicals present within a single locus to increase.

Robb[8] has studied the variation with conversion of the number of particles per unit volume of latex for the emulsion polymerisation of styrene. His results show that, contrary to the prediction of the Smith–Ewart theory, the number of particles increases during the stage of constant rate. This conclusion was confirmed by studying the variation with conversion of the particle-size distribution. The conclusion was reached that new particles can be nucleated even though no soap micelles are present.

Further evidence of this possibility was provided by the observation that latex particles can form in systems for which the concentration of micelle generator is, in fact, below that required for micelle formation. Robb also showed that the chain length of the soap used has a marked effect upon the rate of polymerisation. This effect was shown not to be due to the effect of chain length upon critical micelle concentration, but possibly to the stronger adsorption of the soaps of longer chain length at the polymer–aqueous phase interface.

Further evidence that new particles can be nucleated during the emulsion polymerisation of styrene, even though the concentration of micelle generators is below the critical value, comes from the work of Schulz and Romatowski.[9] They allowed the polymerisation to proceed to 11·5% conversion, then diluted the latex by a factor of 5·7, added monomer and restored the concentration and therefore the rate of decomposition of the initiator to its original level. The rate of polymerisation decreased by about 30%, and two types of particle were observed in the reaction product: a large type, which was attributed to the initial seed latex, and a smaller type which was claimed to have been produced after dilution. Since upon dilution the soap concentration fell to below the critical value for micelle formation, it seems that the smaller particles must have been produced by a non-micellar initiation mechanism.

Harada and his co-workers[10] state that, for the emulsion polymerisation of styrene, the Smith–Ewart theory over-estimates the number of polymer particles by a factor of two or three. They propose a reaction model which, they claim, can give an adequate explanation of the whole course of an emulsion polymerisation of a monomer which is very sparingly soluble in water. Among other matters, the process whereby polymer particles are generated is examined in detail. A new parameter is introduced which allows for the degree of difficulty of monomer initiation within micelles. The view is taken that monomer initiation occurs more easily in existing particles than it does in micelles.

Grancio and Williams[11] have reported upon the changes which occur in product molecular weight during the region of constant rate of polymerisation using styrene as monomer. They found a conflict between observation and the requirements of the conventional Smith–Ewart theory, in that substantial (*i.e.* six- to ninefold) increases in various molecular weight averages were observed as reaction progressed. Concerning the interpretation of these results, Grancio and Williams state that 'While we have no firm explanation for the behaviour just described, we believe that the most straightforward way to interpret these findings is that, although the radicals may be generated with 100% efficiency, they are not utilized with 100% efficiency.' They appear to reject the idea that the increase in molecular size is due to branching, such as might occur through the reaction of transfer to polymer.

Most workers have assumed that styrene, being a solvent for its own polymer, is able uniformly to swell the growing polymer particles during emulsion polymerisation. In a second paper, Grancio and Williams[12] have proposed a model in which the growing particle consists of an expanding polymer-rich core surrounded by a monomer-rich shell which serves as the major locus of polymerisation. This so-called 'core–shell' model was suggested by kinetic studies which indicate that the actual swelling of the particles during the reaction is non-equilibrium, and that the monomer concentrations within the particle are not uniform. Supporting evidence for the suggested morphology was obtained by electron microscope observation of ultra-thin sections of latex particles. Further comment upon the 'core–shell' model, and upon the controversy which it has aroused recently, will be made in the final chapter of this book.

Papers dealing with the emulsion polymerisation of styrene derivatives include those by Gerrens and Köhnlein[13] and by Paoletti and Billmeyer.[14] Gerrens and Köhnlein studied the emulsion polymerisations of (a) vinyltoluene and (b) 2,4-dimethylstyrene, comparing both with that of styrene. The reaction conditions were essentially those used by Bartholomé *et al.*[5,6] for the emulsion polymerisation of styrene, namely, potassium persulphate as initiator, an alkali-metal salt of a sulphonated C_{18} olefin as micelle generator, and a polymerisation temperature of 45°C. All three monomers were observed to obey the Smith–Ewart prediction in that the rate was proportional to the 0·6th power of the micelle-generator concentration and the 0·4th power of the initiator concentration. The reactivity of the three monomers, as judged by the rate of polymerisation under similar conditions of soap and initiator concentration, was observed to decrease in the order styrene > vinyltoluene > dimethylstyrene. But whereas for styrene the variations of particle number with soap and initiator concentrations were found to fit the Smith–Ewart requirement, the variations for the other two monomers did not. For vinyltoluene, the respective exponents for micelle-generator and initiator concentrations were reported as 0·72 and 0·18; for dimethylstyrene, the corresponding exponents are 1·0 and 0·09. These results were interpreted in terms of steric control of the propagation and termination reactions by the presence of the methyl substituents.

The investigation of Paoletti and Billmeyer[14] was concerned with the emulsion polymerisations of styrene, *o*-methylstyrene, *p*-methylstyrene, and a mixture of *m*- and *p*-methylstyrene, using potassium persulphate as initiator and sodium lauryl sulphate as micelle generator. The temperature range was 30–70°C. Good conformity to the predictions of the Smith–Ewart theory was found at temperatures below 70°C. The observed differences between the various substituted styrenes were attributed to an increased radical stability arising from the methyl group contributions to the resonance structures and to steric hindrance caused by a methyl group in the *ortho* position relative to the vinyl group.

Turning now to butadiene and styrene–butadiene mixtures, there is a diverse group of matters which call for some comment. These include the effect of conversion upon butadiene vapour pressure, the effect of polymerisation temperature upon the microstructure of the butadiene units which become incorporated in the polymer chains, the effect of conversion upon the gel content, molecular weight, and molecular-weight distribution of the product, and the problem of the formation of 'popcorn' polymer.

The effect of conversion upon butadiene vapour pressure has been of some interest for various reasons. Not only does it throw some light upon the mechanism of the reaction—in particular, the point at which monomer droplets as a separate phase disappear; it also provides a rough but convenient method of monitoring the progress of a butadiene emulsion polymerisation without having to take samples. Furthermore, in copolymerisation systems, the change of pressure with conversion provides information concerning the relative rates of entry of the two monomers into the copolymer. The data summarised in Fig. 10.1(a), taken from a paper by Meehan,[15] show the nature of the variation of vapour pressure with conversion for the emulsion homopolymerisation of butadiene at 50°C. It is seen that the vapour pressure–conversion relationship does not depend upon the amount of soap in the system, and that, for approximately the first half of the reaction, the vapour pressure remains constant at that for pure butadiene. This is consistent with the view that a separate pure butadiene phase is present. The disappearance of this phase is evidenced by the sharp drop in vapour pressure which then follows. At 50°C this transition occurs at about 52% conversion, and corresponds to a solubility of 0·92 grams of butadiene per millilitre of polybutadiene. The solubility is approximately the same at 30°C, and is unaffected by large variations in the soap level. After the butadiene droplets have disappeared, the curve is essentially that which would be expected for a diminishing

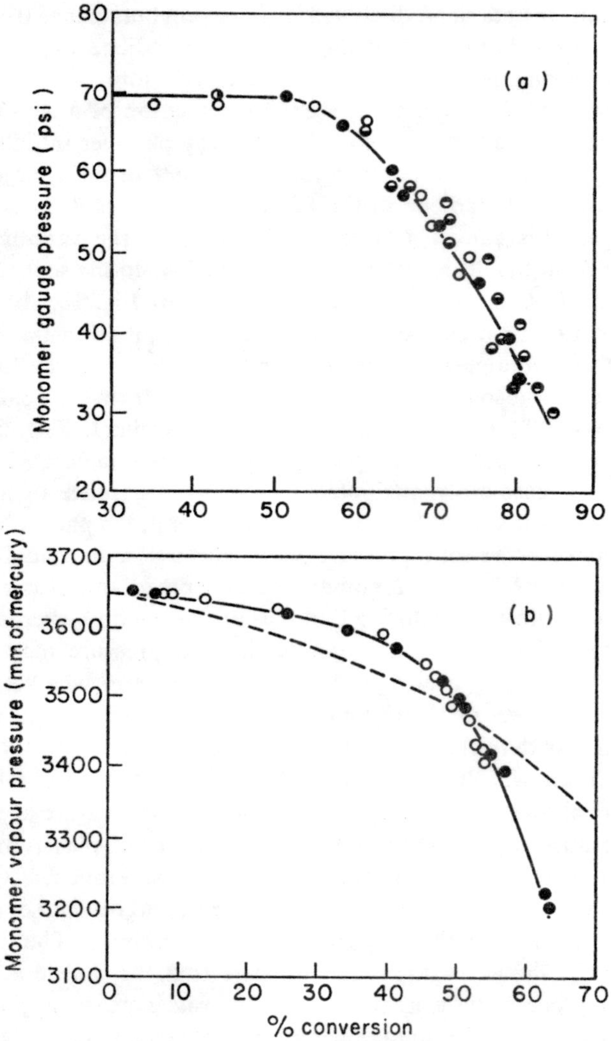

FIG. 10.1. (a) Variation of butadiene vapour pressure with conversion during emulsion homopolymerisation at 50°C, according to Meehan.[15] (Reaction system: butadiene 100, water 180, fatty-acid soap—variable, *n*-dodecyl mercaptan 0·5, $K_2S_2O_8$ 0·3. Data for systems containing different amounts of soap are distinguished as follows: ● 10 pphm; ◒ 5 pphm; ◓ 5/4 pphm; ○ 5/8 pphm.) (b) Variation of total monomer vapour pressure with conversion during emulsion copolymerisation of styrene and butadiene at 50°C, according to Meehan.[15] (Reaction system: styrene 25, butadiene 75, water 180, fatty-acid soap 5, *n*-dodecyl mercaptan 0·5, $K_2S_2O_8$ 0·3. Different types of point refer to two separate experiments.)

concentration of butadiene dissolved in pure polybutadiene. It should be noted that, towards the end of the reaction, butadiene vapour pressure provides a reasonably sensitive measure of conversion.

A similar curve for the emulsion copolymerisation of a 25/75 styrene–butadiene mixture is shown in Fig. 10.1(b). The polymer initially formed contains relatively more butadiene than is present in the charge, so that the proportion of butadiene in the residual monomers decreases during the reaction. This causes a continual decrease in the vapour pressure quite apart from the decrease which is observed when the separate monomer droplet phase disappears. The broken line of Fig. 10.1(b) gives the variation of vapour pressure predicted on the assumption that the composition of the monomer droplet phase is that of the total residual monomer mixture. It is also assumed that the monomers form an ideal solution. Up to about 50% conversion, agreement is excellent. Thereafter, the observed pressure becomes markedly less than the calculated pressure. Again, this is reasonably attributed to the disappearance of a separate monomer phase. It should be noted that, in general, the pattern of change of total monomer vapour pressure with conversion depends upon the relative rates at which the monomers are consumed and also upon their relative vapour pressures. During that part of the reaction where monomer droplets are still present, it is possible that the pressure may decrease, remain unchanged, or even increase, depending upon whether the proportion of the more volatile monomer in the copolymer is greater than, equal to, or less than that in the feed mixture.

It is well known that the physical properties of styrene–butadiene rubber vulcanisates depend very much upon the temperature at which the copolymerisation of the styrene and butadiene took place. In general, the lower the polymerisation temperature, the less branched and cross-linked is the rubber, the lower is its gel content, and the higher are properties such as the tensile strength of the vulcanisate. The results of Smith *et al.*,[16] shown in Fig. 10.2, indicate that the tensile strength of vulcanised styrene–butadiene rubber latex deposits increases approximately linearly as the polymerisation temperature is reduced. Other papers on the subject of the effect of polymerisation temperature upon the properties of these rubbers include those of Howland *et al.*,[17] of Johnson and Bebb,[18] and of Brown, Messer and Howland.[19]

In part, the effect of polymerisation temperature upon properties is explained by the effect upon the relative numbers of 1,2(vinyl), *cis*-1,4 and *trans*-1,4 butadiene units which become incorporated in the polymer chain. Papers giving information on this matter include those of Beu

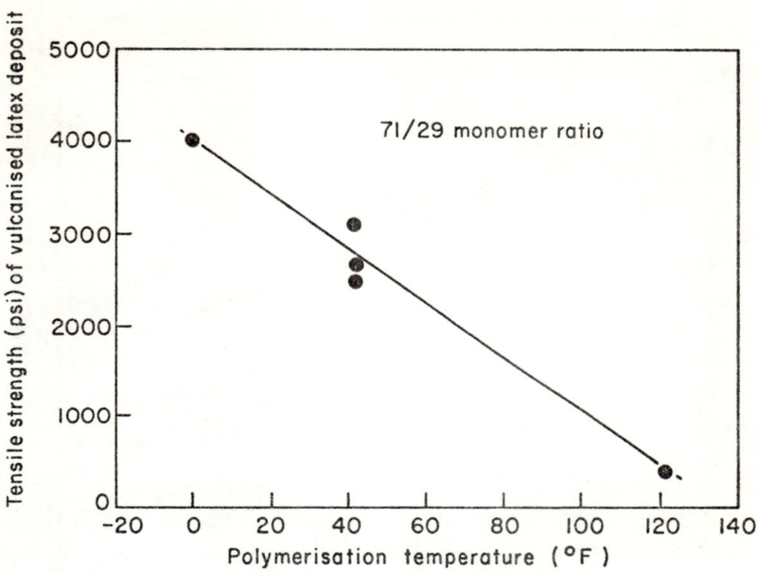

FIG. 10.2. Effect of emulsion polymerisation temperature upon tensile strength of vulcanised deposits from styrene–butadiene rubber latex, according to Smith *et al.*[16] (Vulcanisation recipe: rubber 100, ZnO 3, zinc dibutyldithiocarbamate 0·5, zinc mercaptobenzthiazolate 1·5, sulphur 2.)

et al.,[20] of Hart and Meyer,[21] of Meyer,[22] and of Hampton.[23] Typical of the published results are those of Hampton, which are reproduced in Fig. 10.3. It is seen that the proportion of vinyl groups present in the polymer varies little over the range of polymerisation temperatures from −20°C to 100°C, being about 20%. By contrast, the balance between *cis*-1,4 and *trans*-1,4 structures is markedly affected by polymerisation temperature, the ratio of *cis* to *trans* units falling as the polymerisation temperature is decreased. Whereas polybutadienes produced at low temperatures contain approximately 70% of *trans*-1,4 units, the remainder being roughly 20% of vinyl and 10% of *cis*-1,4 units, those produced at higher temperatures contain about 50% of *trans* and 30% of *cis*, in addition to the 20% of vinyl. The improved physical properties of rubbers prepared at low temperatures have been attributed, in part at least, to the greater regularity of structure occasioned by the increase in the content of *trans*-1,4 units, as well as to the presence of less branching and crosslinking. (It is also worth noting—see, for instance, the paper by Fryling *et al.*[24]—that in the past the improved physical properties of vulcanisates from

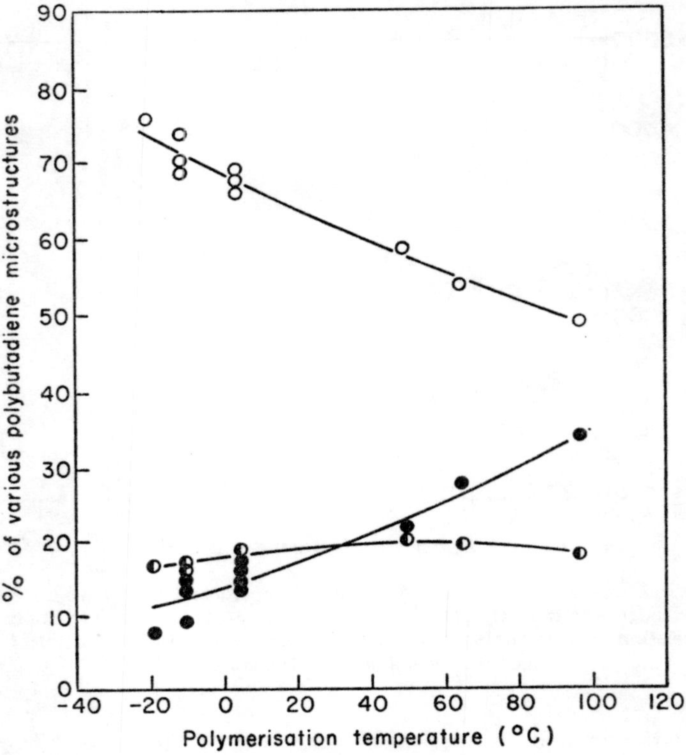

FIG. 10.3. Effect of polymerisation temperature upon proportions of ***trans*-1,4** (points ○), *cis*-1,4 (points ●) and 1,2 (points ◐) microstructures in polybutadienes prepared by emulsion polymerisation, according to Hampton.[23]

rubbers prepared at low temperatures has also been attributed to the fact that the polymerisation occurred more slowly at the lower reaction temperatures. This view is now largely discredited.)

A comprehensive assemblage and review of published data for the effect of polymerisation temperature upon polybutadiene microstructure has been given by Condon.[25] His conclusions are summarised by the following quotation from his paper:

'(1) The percentage of vinyl, *trans*, and *cis* double bonds in emulsion polymers of butadiene are determined primarily by the temperature of polymerization in the fundamental way that is characteristic of systems of simultaneous competing reactions.

'(2) Other environmental factors, such as type of recipe, and including the presence of a comonomer, *e.g.* up to 29% of styrene, have comparatively little effect on the relative amounts of vinyl, *trans*, and *cis* double bonds.

'(3) Differences in the physical properties of emulsion polymers of butadiene *prepared at the same temperature* probably must be attributed to differences in chemical structure other than differences in the relative proportions of vinyl, *trans*, and *cis* double bonds, such as the average molecular weight, the molecular weight distribution, and the presence of a comonomer.' (The italics are due to Condon.)

An important factor which determines the amount of gel in a copolymer containing butadiene is the conversion to which the system is taken. Initially, the gel content is rather low, and may, in fact, be zero. As the reaction proceeds, so the amount of gel increases, as also does the tightness of the gel. The results summarised in Fig. 10.4 show the effect of conversion upon the gel content of a styrene–butadiene copolymer produced by emulsion copolymerisation at 5°C using a hydroperoxide redox initiation system. The system used did not contain an added modifier. At conversions below about 20%, the polymer was free from gel. Above 20% conversion, the gel content of the polymer increased approximately linearly with conversion. Poddubnyĭ and Rabinerzon,[26] from whose paper these results are taken, found that the point at which gel first appeared depends to some extent upon the composition of the polymerising system, and, in particular, upon the nature of the micelle generator. They also found that gel formation can be suppressed by adding incrementally, and at high conversion as well as at low, small amounts of a modifier such as di*iso*propylxanthogen disulphide.

Data for the effect of conversion upon the intrinsic viscosity of the sol content of styrene–butadiene copolymer produced by emulsion at 5°C, have been published by Václavek.[27] The results summarised in Fig. 10.5 show the nature of the relationship for systems containing various mercaptan modifiers, and also for systems containing several levels of one particular mercaptan, *t*-dodecyl mercaptan. It can be seen that, up to 60% conversion at least, conversion appears to have little effect upon the intrinsic viscosity of the sol fraction when an efficient modifier (see Chapter 8, Section 8.2) such as *t*-dodecyl mercaptan is used. On the other hand, when an inefficient modifier, such as *t*-hexadecyl mercaptan, is used, the intrinsic viscosity of the sol polymer, which is high initially, falls as the reaction proceeds.

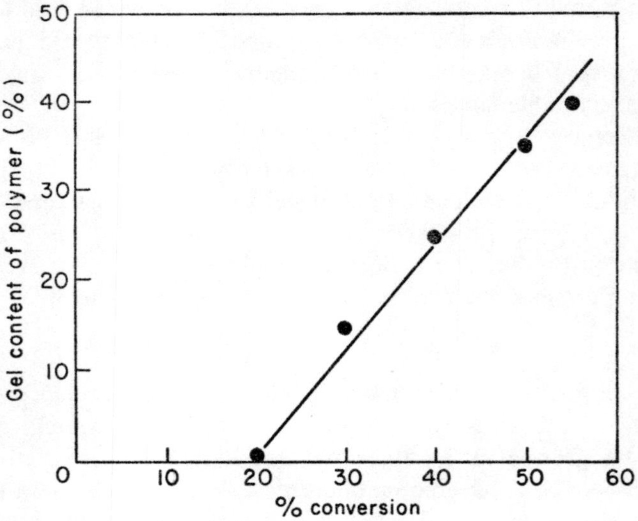

FIG. 10.4. Effect of conversion upon gel content of a styrene–butadiene copolymer produced by emulsion polymerisation at 5°C in the absence of added modifier, according to Poddubnyĭ and Rabinerzon.[26] (The styrene/butadiene ratio was 30/70, the micelle generator sodium dibutylnaphthalene sulphate (sulphonate?), and the initiator a redox system containing cumene hydroperoxide.)

Uraneck and Burleigh[28] have published the results given in Table 10.1 for the inherent viscosities and Mooney viscosities of styrene–butadiene rubbers produced by a similar polymerisation reaction using *t*-dodecyl mercaptan as modifier. These results indicate a tendency for both types of viscosity to increase with conversion.

Uraneck and Burleigh also give information on the effect of conversion upon molecular size distribution. Also included in Table 10.1 are some of their results for the weight- and number-average molecular weights of the polymers, and for the ratio of the two. It appears that molecular weight tends to increase with conversion, whereas it is doubtful whether the heterogeneity of the polymer, as judged by the ratio between the two averages, undergoes any significant change over the region covered by these results. Concerning the molecular size distributions themselves, Uraneck and Burleigh found that conversion has very little effect, as revealed by the technique (gel permeation chromatography) which they used for their studies.

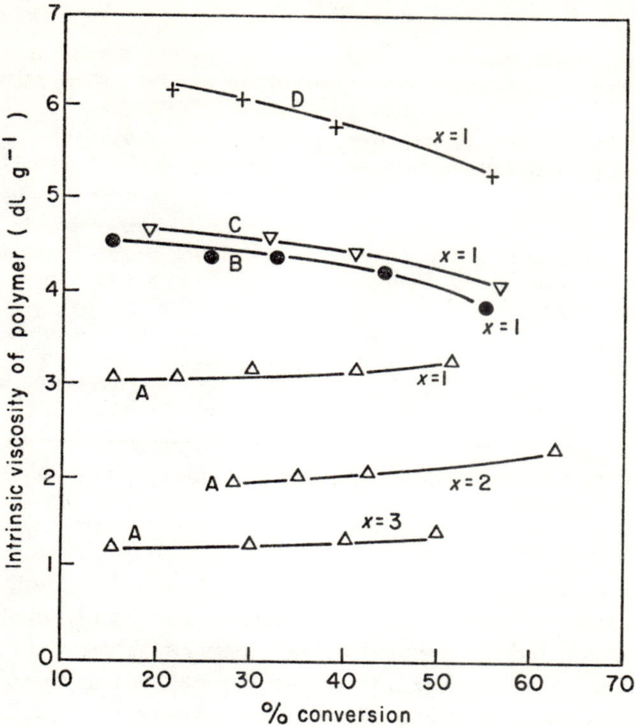

FIG. 10.5. Effect of conversion upon intrinsic viscosity of styrene–butadiene copolymers produced by emulsion polymerisation at 5°C in the presence of various mercaptan modifiers in varying amount, according to Václavek.[27] The letters appended to the curves distinguish the mercaptans as follows: A, *t*-dodecyl mercaptan; B, *n*-dodecyl mercaptan; C, *t*-tetradecyl mercaptan; D, *t*-hexadecyl mercaptan. The value of x appended to each curve refers to the level of mercaptan in the reaction system. (Reaction system: styrene 30, butadiene 70, water 194, sodium soap of disproportionated rosin acid 4·3, mercaptan 0·0748x, di*iso*propylbenzene hydroperoxide 0·1, sodium formaldehyde sulphoxylate dihydrate 0·1, $FeSO_4.7H_2O$ 0·020, disodium ethylenediamine tetraäcetate 0·0333, $Na_2HPO_4.12H_2O$ 1·15.)

A problem which is peculiar to butadiene handling is that of 'popcorn' polymer formation. 'Popcorn' polymer is the name given to a hard, benzene-insoluble material, which derives its name from its appearance. It has been encountered in plants for both the manufacture and the recovery of butadiene. An unusual and troublesome feature of this material is its ability to grow rapidly when exposed to fresh monomer. Thus Welch *et al.*[29] reported that the weight of the molecule increases exponentially

TABLE 10.1

Effect of conversion upon the inherent viscosity, Mooney viscosity and molecular weight averages for styrene–butadiene copolymer produced by emulsion polymerisation at 5°C (according to Uraneck and Burleigh[28])

(reaction system: full details not given; initiation system comprised sodium formaldehyde sulphoxylate and a hydroperoxide; the modifier was 0·185 pphm of *t*-dodecyl mercaptan)

Conversion	*Inherent viscosity* (dl g^{-1})	*Mooney viscosity*	*Molecular weight averages*† $\overline{M}_w$	$\overline{M}_n$	$\overline{M}_w/\overline{M}_n$
50	1·78	44	3·20 × 10^5	1·07 × 10^5	2·98
61	1·90	53	3·39 × 10^5	1·14 × 10^5	2·95
70	2·12	64	3·72 × 10^5	1·12 × 10^5	3·08

† Obtained from gel permeation chromatographs using a linear calibration and applying the Tung correction.

with time, and this has been confirmed by Miller, Alumbaugh and Brotherton.[30] Popcorn formation has caused serious fouling of butadiene distillation plants, and has even led to the bursting of pipes.

According to Welch *et al.*,[29] popcorn formation can be initiated in various ways, but the presence of both iron and oxygen play an important role. The growth of popcorn can be retarded but not inhibited by certain vulcanisation accelerators; conventional antioxidants have little effect. A mechanism for popcorn formation has been proposed by Kharasch and his co-workers.[31] They state that the formation of the initial popcorn 'seed' is a slow process which is relatively easy to inhibit, whereas the growth of the seed is a rapid process which is much more difficult to control. They list the following as the experimentally-observed facts which any satisfactory theory must explain:

'(1) Active popcorn seed grows rapidly, not only in styrene or butadiene or mixtures of the two, but also in other unsaturated monomers such as isoprene or methyl methacrylate. The only pertinent distinction among these various monomers is that a given sample of popcorn seed grows in them at different rates.

'(2) Active popcorn seeds have been collected from numerous industrial plants, and all give strong tests for organic peroxides. Moreover, old specimens of elastomers (both polybutadiene and butadiene–

styrene copolymer), which have not been treated with an antioxidant and which therefore contain peroxides, yield benzene-insoluble polymers when incubated with styrene. On the other hand, elastomers which have been treated with an antioxidant at the time of coagulation show no such activity.

'(3) Soluble organic hydroperoxides (*e.g. tert*-butyl hydroperoxide) and soluble polymers containing peroxides do not cause the formation of popcorn polymer when incubated with styrene. For example, in the course of the work here reported, an active sample of popcorn seed was fractionated into its benzene-soluble and benzene-insoluble portions. Although both fractions gave strong tests for peroxides, nevertheless, when they were incubated with styrene, only the insoluble portion proved active as popcorn seed.

'(4) Certain inactive materials insoluble in styrene (*e.g.* peroxide-free polybutadiene) are transformed into active popcorn seed by exposure to oxygen and light.'

Kharasch and his co-workers give two prerequisites for the growth of popcorn seeds: (a) the seed must be insoluble in, but swollen by, the monomer to which it is exposed; and (b) the seed must contain a relatively large number of groups which can give rise to free radicals. They consider that, in practice, the growth of popcorn seed is due to the presence of peroxide groups, and cite as evidence the observation that there is a strong positive correlation between the peroxide content of the seed and its activity. On the other hand, Miller, Alumbaugh and Brotherton[30] state that 'It is certain, however, that even though peroxides are necessary to initiate popcorn formation, peroxides are not necessary for the spontaneous growth in the presence of monomer.' They believe that the variation of seed activity with mode of generation is independent of peroxide content. As far as the present writer is aware, the problem of exactly how the popcorn seed is initiated remains unsolved. Initiation seems to require the presence of oxygen, of iron in more than one valency state, and possibly of water. The rapid growth of the seed is, however, in all probability an extreme example of the Trommsdorff 'gel' effect, in that mutual termination of radicals is suppressed by immobilisation through being combined in an insoluble polymer network. According to Kharasch *et al.*,[31] two distinct methods may be employed for the prevention of popcorn growth in a plant already contaminated with active seed. The first entails shutting down the plant and introducing into the system (which contains active seeds but no monomers) a substance which greatly reduces or destroys altogether the activity of the seed. The second entails maintaining constantly in the system a concentration of an inhibitor which is sufficient to reduce the activity of the seed to an acceptable level. These workers report that, of the many substances investigated, a dilute aqueous solution of sodium nitrite was found to be by far the most effective agent for inhibiting the growth of popcorn seeds. The phenomenon of popcorn seed formation and growth may have an immediate relevance for butadiene emulsion polymerisations, quite apart from its relevance to the problems of the handling of butadiene monomer. Emulsion polymerisation systems containing butadiene sometimes develop over-size particles of a rather hard nature, known as 'grain'. The presence of grain can be troublesome in certain latex applications. It is possible that the phenomena of grain formation and popcorn seed formation have similar causes.

The effects of various impurities upon the emulsion copolymerisation of styrene and butadiene is the subject of a paper by Frank *et al.*[32] Nineteen possible impurities in butadiene and styrene were added in varying amounts to mixtures of butadiene and styrene, and their effects upon the polymerisation rates of the mixtures determined. In amounts of 1% or less of the monomer used, the following appear to have no effect: ethylbenzene, *o*-xylene, phenylacetylene, methylphenyl carbinol, acetophenone, acetaldehyde, propylene, allene, isoprene and ethylacetylene. The butenes and straight-chain pentenes were found to show some retardation at the 1% level. At higher levels, these substances cause a marked decrease in rate of polymerisation. Polymer conversions are lowered sharply by 1,4-pentadiene and less markedly by 1-vinylcyclohex-3-ene. *p*-divinylbenzene and vinylacetylene at the 1% level have little or no effect upon polymerisation rate, although they do, of course, cause crosslinking in the polymer produced.

The preparation of acrylonitrile–styrene copolymer latices by emulsion polymerisation has been the subject of a recent paper by Seymour and his co-workers.[33]

10.3 ETHYLENE

The emulsion polymerisation of ethylene has been described in a series of papers by Helin, Stryker, Mantell and others.[34,39] Stryker, Mantell and Helin[40] have also published a brief review of this subject which summarises most of the information available up until about 1968. These latter authors give the following formulation for a typical ethylene emulsion polymerisation reaction system, all levels being in parts by weight:

(1) *Autoclave charge*

Water	85
t-Butyl alcohol	15
Myristic acid	1·42
Potassium hydroxide	0·41
Potassium phosphate	0·59
Ethylene at 4500 psi	25

(2) *Initiator charge*

Water	5
Potassium persulphate	0·3

They also give the following description of the process:

> 'To conduct a polymerization, an autoclave is charged with water, *t*-butyl alcohol, emulsifier, and buffer; is purged by evacuating and flushing with ethylene; and is pressured to 1500 psig with ethylene. After the agitator is started and the charge is heated to an operating temperature of 70–90°C, a solution of potassium persulphate initiator is pumped into the reactor and more ethylene is admitted to raise the pressure to 4500 psig, which is maintained until the solids content of the latex reaches about 20–30%. After the autoclave is cooled and the residual ethylene vented, the latex is removed and concentrated at a reduced pressure to remove the alcohol and to increase the solids to about 40%. Because ethylene is supplied at constant pressure during the polymerization, the monomer can never be completely consumed; hence the extent of polymerization is measured by the solids content of the latex rather than by percent of monomer consumed.
>
> When the ingredients are put into an autoclave and agitated at conditions specified for polymerization, they form two layers; one of foam containing about 15–20% ethylene and another of gases consisting largely of ethylene. Actually, the foam is like an emulsion; but the ethylene, being above its critical temperature, forms small bubbles instead of droplets. The foam is not stable, however, and quickly breaks down to a liquid phase when agitation is stopped. If no polymerization has occurred, the liquid phase contains 1–2% dissolved ethylene. If polymer is present, however, the amount of dissolved ethylene increases owing to its solubility in the polymer.'

Stryker, Mantell and Helin[40] state that the solubility of ethylene in the reaction system does not depend greatly upon the composition of the aqueous phase. At 4500 psi pressure and a temperature of 80°C, the solubility is given as between 1% and 2%, *i.e.* 0·3–0·6 mole l^{-1}. This is considerably greater than the solubility of styrene in a typical emulsion polymerisation aqueous phase. In terms of molecular concentration, the difference is especially marked because of the low molecular weight of ethylene; the above figures suggest a molecular solubility in the range $2{\cdot}4 \times 10^{20}$ molecules per millilitre. Thus, although ethylene is normally considered to show very limited solubility in water, being a hydrocarbon, under the conditions normally used for its emulsion polymerisation it must be regarded as being a relatively water-soluble monomer like, say, vinyl acetate. The enhanced solubility of ethylene in the aqueous phase is due primarily to the high pressure under which the reaction is carried out. the presence of the butyl alcohol also no doubt promotes solubility, although a comment by Stryker, Mantell and Helin suggests that this is a factor of minor significance. These authors state that the presence of the micelle-generator in the reaction system approximately doubles the apparent solubility of ethylene in the aqueous phase, implying that approximately one-half of the monomer is present as solubilisate within the micelles. They calculate that, if one assumes approximately 10^{18} micelles per millilitre of aqueous phase, then each micelle must contain about 180 ethylene molecules.

The first of the six papers by Helin and his co-workers noted above, which appeared in 1965, contains references to, and a brief review of, earlier work on the subject. Stable polyethylene latices can be obtained using a variety of anionic and non-ionogenic micelle generators, but much of the work described in these six papers was carried out using fatty-acid soaps or polyethyoxylated alkylphenols. Potassium persulphate is said to be a satisfactory initiator at concentrations within the range 0·05 to 0·50 pphm. Substitution of up to 15% of the water with *t*-butyl alcohol increases both the size of the particles in the latex, and the molecular weight of the contained polymer. Typical results are summarised in Table 10.2. Practical polymerisation rates, and stable latices which can be concentrated by evaporation to 40–50% solids content, can be obtained using pressures in the range $3\text{–}5 \times 10^{3}$ psi and temperatures in the range 80–100°C. The distribution of particle diameters in the final latex appears to be logarithmic Gaussian. The average particle diameter is unaffected by the initiator level over the range normally used, but decreases as the amount of micelle generator is increased. During the course of the polymerisation, the number of particles slowly increases if *t*-butyl alcohol is present, but decreases if *t*-butyl alcohol is absent. In the absence of the

TABLE 10.2

Effect of partial replacement of water by *t*-butyl alcohol in persulphate-initiated emulsion polymerisation of ethylene at *ca.* 90°C (according to Helin, Stryker and Mantell[34])

(reaction system: ethylene as required, water $100-x$, *t*-butyl alcohol x, polyethoxylated alkylphenol 3, $K_2S_2O_8$ 0·25 initially + increments of 0·06 pphm as required to sustain the reaction)

Level of t-butyl *alcohol*(x)	pH *of final latex*	*Average particle diameter* (μm)	*Standard deviation of particle diameter* (μm)	*Inherent viscosity of contained polymer* (dl g^{-1})
0	2·2	0·153	0·068	0·225
7·5	1·9	0·465	0·372	0·440
15	3·1	0·129	0·029	0·878
30	2·1	0·219	0·027	0·850

alcohol, the number of particles is proportional to the 1·1th power of the concentration of micelle generator.

In the third paper of the series, Stryker, Helin and Mantell[36] consider some of the factors which affect the colloid stability of the resultant polyethylene latex. The method used for the assessment of stability was in part subjective and in part objective. The subjective part consisted of a series of arbitrary stability ratings according to the visual appearance of any solids which separated under various conditions. The objective part consisted in determining the amount of material retained by a 50 μm screen. As judged by these procedures, the stability of the latex was found to be related to three factors: (a) the type and amount of post-reaction stabiliser added to the latex, (b) the average size of the latex particles, and (c) the solids content. Generally, the appearance was better when the latices had low (*ca.* 30%) solids content, a large (800 Å) average particle diameter, and enough post-reaction stabiliser to cover most of the polymer surface. Some assessment of resistance to destabilisation by mechanical shear was also attempted. Good stability under these conditions is only obtained if the polymer surface is completely covered with adsorbed emulsifier. When polyethyoxylated alkylphenols are used as post-reaction stabilisers, an inverse relationship appears to exist between the stability of the latex and the average number of ethylene oxide units per molecule. Furthermore, the area occupied by each adsorbed molecule appears to increase linearly with the number of ethylene oxide units. This effect is illustrated in Fig. 10.6; the slope of the relationship between molecular adsorption area and

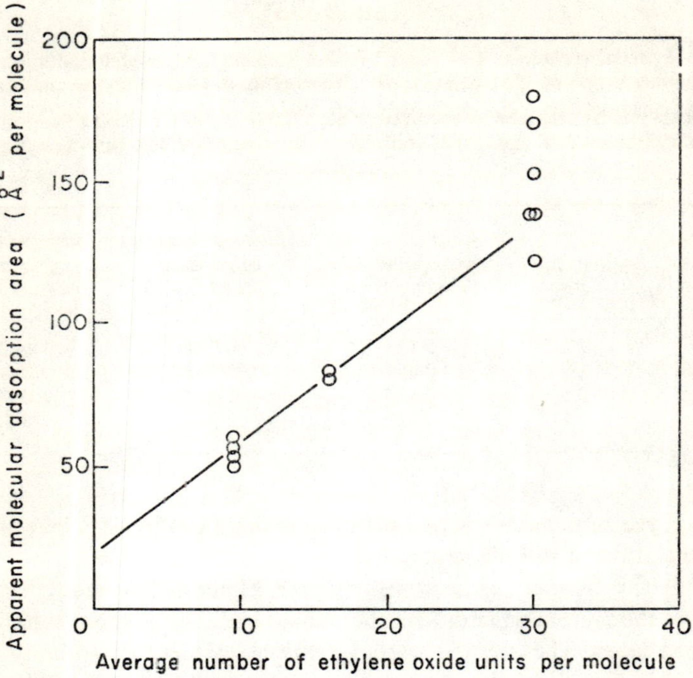

FIG. 10.6. Dependence of apparent molecular adsorption area of polyethoxylated alkylphenols upon number of ethylene oxide units in molecule for polyethylene latices, according to Stryker, Helin and Mantell.[36]

TABLE 10·3

Correlation between the inherent viscosity of the contained polymer and the micelle generator which becomes chemically combined in polyethylenes produced by emulsion polymerisation (according to Mantell *et al.*[37])

(reaction system similar to that given in Table 10.3 with $x = 0$; the micelle generator was an ethylene oxide adduct)

Inherent viscosity of polymer (dl g^{-1})	*Chemically-combined micelle generator* (%)
0·87	1·65
0·83	3·90
0·44	5·28
0·40	5·62
0·34	6·02
0·19	14·51

number of ethylene oxide units per molecule is such as to correspond to about 4 Å^2 per ethylene oxide unit.

An unusual feature of ethylene emulsion polymerisation using either fatty-acid soaps or polyethyoxylated alkylphenols is that, in addition to fragments derived from the initiator, the polymer formed also contains chemically-combined micelle generator. Furthermore, as shown by the results summarised in Table 10.3, the amount of chemically-combined micelle generator, expressed as a proportion of the whole, increases as the molecular weight of the polymer decreases. The amounts of combined soap were determined by infrared spectrophotometry. It has also been found that the inherent viscosity of the contained polymer falls as the level of micelle generator is increased, and that, in consequence, the molecular weight of the contained polymer appears to increase with the particle size of the resultant latex. These observations suggest that a transfer reaction occurs between the growing polymer chains and soap molecules adsorbed on the surface of the polymer particles. A further peculiarity of the ethylene emulsion polymerisation is that the rate of polymerisation

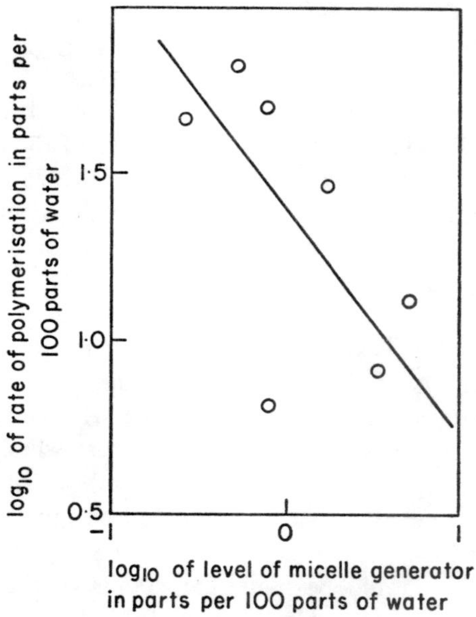

FIG. 10.7. Effect of level of micelle generator upon rate of persulphate-initiated emulsion polymerisation of ethylene at 85°C and 4·5 × 10^3 psi pressure, according to Stryker, Mantell and Helin.[38] (Reaction system: ethylene as required, water 100, polyethoxylated alkylphenol—variable, $K_2S_2O_8$?)

decreases as the level of micelle generator is increased. Although the data summarised in Fig. 10.7 show considerable scatter, they do suggest that the logarithm of the polymerisation rate falls approximately linearly with increasing logarithm of the concentration of micelle generator.

The mechanism which Stryker, Mantell and Helin[39] have proposed to explain the main features of the ethylene emulsion polymerisation is entirely conventional, except that they invoke the reaction of transfer to soap to account for several of the unusual effects noted above. Since the soap is adsorbed at the surface of the particle, such transfer reactions will be expected to be proportionately more important for small-particle systems than for those containing large particles. The fall of polymerisation rate with increasing level of soap is accounted for by either or both of two hypotheses: the radicals produced by the reaction of transfer to soap may be relatively inactive in initiating new polymer chains; or they may tend to be desorbed into the aqueous phase and so lost from the reaction loci. The order of reaction with respect to initiator concentration is approximately 0·5, and so calls for no special comment.

10.4 VINYL ESTERS OF CARBOXYLIC ACIDS

An extensive literature exists on the subject of the emulsion polymerisation of vinyl acetate. This is hardly surprising in view of the industrial importance of the polymer, especially in the form of a latex as a binder or vehicle for surface coatings. Whereas monomers such as styrene and butadiene (although not ethylene under normal emulsion polymerisation conditions) can be fairly regarded as almost insoluble in water, vinyl acetate is an important example of a monomer which, although showing limited miscibility with water, is sufficiently soluble for this to have a perturbing effect upon the emulsion polymerisation reaction.

It is appropriate at this point to note the wide differences in water-solubility among a range of common monomers which are readily susceptible to polymerisation by the emulsion method. A list of solubilities expressed both as percentage by weight and as molecules per millilitre is given in Table 10.4. This table is taken from a review by Lindemann,[41] to which reference should be made for details of the primary sources from which these data come. Also given in Table 10.4 are the weight ratios of monomer to polymer in the latex particle for the various monomers. It is seen that again there are marked differences between the various monomers. For the relatively water-insoluble monomers such as styrene

TABLE 10.4

Water solubilities and equilibrium monomer:polymer ratios in latex particles for various common monomers (according to Lindemann[41])

Monomer	*Temperature* (°C)	*Water solubility* Weight %	*Water solubility* Molecules per ml^{-1}	*Weight ratio of monomer to polymer in latex particle*
Vinylxylene	45	$0{\cdot}6 \times 10^{-2}$	$2{\cdot}7 \times 10^{17}$	0·9–1·7
Vinyltoluene	45	$1{\cdot}2 \times 10^{-2}$	$6{\cdot}1 \times 10^{17}$	0·6–0·9
Styrene	45	$3{\cdot}6 \times 10^{-2}$	$2{\cdot}1 \times 10^{18}$	1·1–1·7
Butadiene	25	$8{\cdot}2 \times 10^{-2}$	$9{\cdot}1 \times 10^{18}$	0·8
Chloroprene	25	$1{\cdot}1 \times 10^{-1}$	$7{\cdot}5 \times 10^{18}$	1·7
Isoprene				0·85
Vinyl chloride	50	1·06	$1{\cdot}0 \times 10^{20}$	0·84
Methyl methacrylate	45	1·50	$9{\cdot}0 \times 10^{19}$	2·5
Vinyl acetate	28	2·5	$1{\cdot}75 \times 10^{20}$	6·4
Methyl acrylate	45	5·6	$3{\cdot}9 \times 10^{20}$	6–7·5
Acrylonitrile	50	8·5	$9{\cdot}6 \times 10^{20}$	

and butadiene, the ratio is of the order of unity, whereas for the more water-soluble monomers it rises to approximately seven. Such differences can have important implications for the nature of the immediate environment in which the polymerisation reaction is occurring.

It is also convenient to note at this juncture the complicating influence which a high monomer water-solubility has upon the kinetics of emulsion polymerisation. Gershberg[42] has reviewed the emulsion polymerisation kinetics of 'relatively' water-soluble monomers and has found that the order of reaction with respect to the concentration of micelle generator falls significantly as the solubility of the monomer in water increases. As shown in Fig. 10.8, a broad correlation appears to exist between the order of reaction and the solubility of the monomer in water. In general terms, increased water-solubility of the monomer can be expected to have at least two effects. In the first place, it will favour solution polymerisation in the aqueous phase relative to polymerisation in the particles. But, secondly, it will increase the tendency for propagating radicals to become desorbed from the particles. This follows because, in general, the more soluble is a monomer in water, the more hydrophilic does its polymer tend to be. It has already been noted above in Section 10.3 that, under the conditions which prevail in the systems normally used for its emulsion polymerisation, ethylene should be regarded as a relatively water-soluble monomer, notwithstanding it being a hydrocarbon. However, the unusual

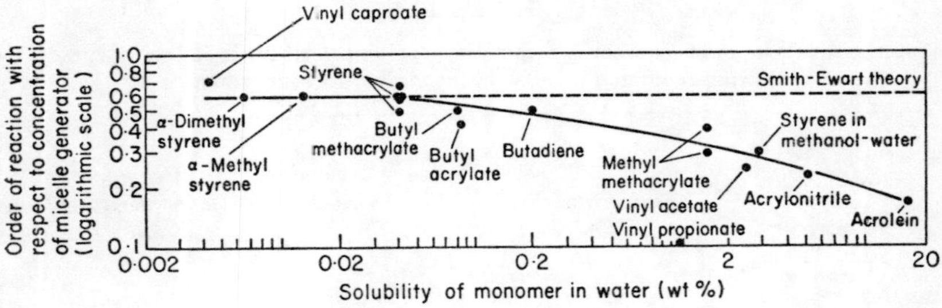

FIG. 10.8. Order of reaction with respect to concentration of micelle generator as a function of solubility of monomer in water, according to Gershberg.[42] Reference should be made to Gershberg's paper for the sources of the data represented here.

conditions under which it is polymerised, together with its untypical polymerisation behaviour and the relative scarcity of published information concerning the kinetics of its emulsion polymerisation, make comparisons with other relatively water-soluble monomers inappropriate.

Turning now specifically to the emulsion polymerisation of vinyl acetate, it is desirable to note at the outset the presence of certain complicating influences which do not usually have to be taken into account when considering the emulsion polymerisation of, say, styrene. Enhanced water-solubility is only one such factor. Further complications arise from the greater reactivity of the polyvinyl acetate free radical as compared with the polystyrene free radical. Transfer reactions to other components of the polymerisation system therefore become correspondingly more important. Still more complication arises from the fact that, in industrial practice, it has become customary to use a 'protective colloid', notably polyvinyl alcohol, as the principal surface-active substance in the system, rather than the soap-like micelle generators of the conventional styrene–butadiene emulsion copolymerisation system. More conventional micelle generators may also be added to the system, but their role seems to be secondary to that of the protective colloid. It must also be noted that the particular protective colloid which is usually used, namely polyvinyl alcohol, is a singularly ill-defined substance. It is made by the partial hydrolysis of polyvinyl acetate, and, as such, is subject to considerable structural variation and uncertainty.

Important papers dealing with the emulsion polymerisation of vinyl acetate and with related topics include those of Priest,[43] of O'Donnell, Mesrobian and Woodward,[44] of French,[45] of Patsiga, Litt and Stannett,[46] of Okamura and Motoyama,[47] of Napper and Parts,[48] of Napper and

Alexander,[49] of Elgood, Gulbekian and Kinsler,[50] of Dunn and Taylor,[51] of Netschey, Napper and Alexander,[52] of Litt, Patsiga and Stannett,[53] of Breitenbach *et al.*,[54] of Dunn and Chong,[55] of Noro,[56] of Shiraishi,[57] of Litt, Stannett and Vanzo,[58] of Klein, Kuist and Stannett,[59] of Friis and Nyhagen,[60] and of Elgood and Gulbekian.[61] Recent reviews of vinyl acetate emulsion polymerisation include those by Lindemann[62] and by Gulbekian and Reynolds.[63] The latter review is concerned principally with the role of polyvinyl acetate as a protective colloid in the reaction system.

The investigation reported by Priest[43] was concerned with the emulsion polymerisation of vinyl acetate at 50° and 70°C in the presence of partially-neutralised polymethylacrylic acid as protective colloid and using potassium persulphate as initiator. The product of this reaction contains particles having a relatively narrow size distribution and of a size much larger than those normally obtained as a result of reactions which use soap-like micelle generators. The number of particles formed falls sharply as the ratio of vinyl acetate to aqueous phase increases, both in the presence and in the absence of the polymethacrylic acid protective colloid. Varying the initiator concentration over a wide range has little effect upon the average particle size. Increasing the reaction temperature from 50° to 70°C reduces the average volume of the particles by a factor of about three. Addition of sodium chloride to the reaction system results in the formation of larger particles, whether polymethacrylic acid is present or not. The addition of small amounts of polymerisation inhibitors to the system increases the induction period for the reaction as expected, but also reduces the particle size of the latex produced. Data are presented to show that the particles obtained when sodium laurylsulphonate is used as the 'micelle generator' are several orders of magnitude smaller in volume than are those obtained using polymethacrylic acid, notwithstanding the fact that, apparently, no micelles were present in at least two of the systems investigated.

In interpreting these results, Priest regards vinyl acetate as being sufficiently soluble in water for solution polymerisation in a saturated aqueous solution to occur at a fairly high rate. The view is taken that, if a water-soluble initiator is present, then the reaction will commence as a solution process, and that, although the aqueous phase may not be an important locus for polymerisation in terms of the relative amounts of monomer converted to polymer, the species formed in the aqueous phase are of importance in their influence on the nature of the latices formed at high conversion. In fact, it is estimated that almost all of the polymerisation (*ca.* 99·9%) occurs within swollen polymer droplets, and a maximum of

0·1% accompanies the formation of primary particles. But the 0·1% conversion to primary particles is of crucial importance in providing the reaction loci for the remainder of the reaction. Loci are regarded as being formed from molecules which are growing by solution polymerisation and in consequence become of such a size that they tend to precipitate and agglomerate together. This is considered to be a process which continues for most of the reaction. Thus Priest states that 'Since solution polymerization continues to occur throughout the whole range of conversion to monomer (polymer ?), new particles should form at an approximately constant rate so long as monomer concentration is maintained at saturation. This is in contrast to the micellar systems, in which no further primary particles are produced when micelles disappear from solution, owing to depletion of surfactants by adsorption on polymer particles. The maximum number of primary particles to be developed in a solution polymerization leading to a latex is equal to the number of polymer chains long enough to give water-insolubility formed throughout the course of the polymerization. The number actually produced may be reduced by supersaturation effects and also by combination of incompletely developed chains with polymer droplets before the former are adequately stabilized.' The tendency of particles to coalesce is related to the stabilising efficiency of surface-active substances which are present in the system; sodium laurylsulphonate is apparently much more effective than is partially-neutralised polymethacrylic acid, and this has apparently nothing to do with its ability to form micelles. The effect of inhibitors in reducing the particle size of the latex eventually formed is attributed to the formation of surface-active species by interaction with the persulphate initiator. These species may include a considerable amount of low-molecular-weight polymer having sulphate end-groups. Protective colloids such as polyvinyl acetate are seen as acting by enveloping the particles in a 'hull' of water-soluble colloid, which discourages the contact of the interiors of the droplets when they collide. The fact that the number of particles formed falls as the ratio of monomer to aqueous phase (of constant composition) increases is taken as evidence that interparticle coalescence occurs and represents an important process by which particles grow in size.

O'Donnell, Mesrobian and Woodward[44] have reported the results of an investigation into the kinetics of the emulsion polymerisation of vinyl acetate at 70°C using potassium persulphate as initiator and a partially-hydrolysed polyvinyl acetate as protective colloid. The initial quantity of monomer used was always considerably in excess of that required to saturate the aqueous phase at 70°C. Conversion–time curves are given

TABLE 10.5

Effect of concentration of polyvinyl alcohol upon rate of emulsion polymerisation of vinyl acetate at 70°C (according to O'Donnell, Mesrobian and Woodward[44])

(reaction system: vinyl acetate 20, polyvinyl alcohol solution 80, $K_2S_2O_8$ 0·02)

Concentration of polyvinyl alcohol in aqueous phase (g ml^{-1})	*Rate of polymerisation* (*% per minute per* ml *of aqueous phase*)	*Rate*/[PVAlc]$^{3/5}$† (% min^{-1} ml$^{-2/5}$ g$^{-3/5}$)
2·04 × 10^{-2}	3·0 × 10^{-2}	0·31
1·52 × 10^{-2}	2·35 × 10^{-2}	0·29
1·01 × 10^{-2}	1·7 × 10^{-2}	0·27
0·755 × 10^{-2}	1·2 × 10^{-2}	0·23

† In the original paper, the heading of this column is given as 'Rate [PVA-A]$^{3/5}$, g^{-1} (cc. H_2O)$^{-2/5}$ min.$^{-1}$'. Both the text and arithmetical considerations indicate that the column should have been headed either 'Rate [PVA-A]$^{-3/5}$' or 'Rate/[PVA-A]$^{3/5}$', in which case the units should presumably have been similar to those given above. The units as given above are equivalent to % min^{-1} ml^{-1} (g ml^{-1})$^{-3/5}$, and are certainly consistent with the arithmetical relationship between the entries in the three columns of the table.

for systems containing four different colloid concentrations but constant initiator concentration and initial amount of monomer. It was observed that, after an induction period of approximately 30 minutes, the conversion–time relationship is essentially linear up to conversions of at least 60 to 70%, indicating that the reaction rate is of zero order with respect to the total monomer concentration. The observed dependence of rate of polymerisation upon the concentration of colloid expressed in terms of the aqueous phase is summarised in Table 10.5. O'Donnell, Mesrobian and Woodward regard the near-constancy of the ratio of the rate of polymerisation to the 3/5th power of the colloid concentration as establishing an order of reaction of 3/5 with respect to the colloid concentration. However, reference to Table 10.5 shows that the ratio, as given by these authors, cannot really be regarded as constant. There is also some confusion as to just what quantity has been tabulated in the final column, as the footnote to Table 10.5 shows.

Conversion–time curves are also given to demonstrate the effect of varying the initiator over a tenfold range, keeping the levels of initial monomer and colloid constant. After an induction period which increased as the amount of initiator decreased, it was found that the polymerisation rate was constant up to at least 60% conversion. As shown in Fig. 10.9, the relationship between the logarithm of the steady rate and that of the

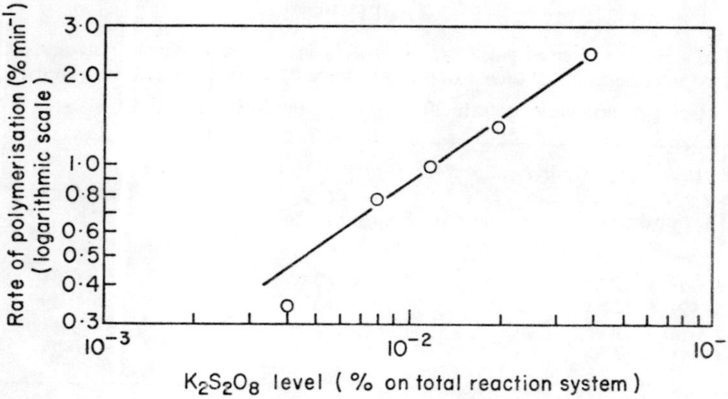

FIG. 10.9. Effect of initiator concentration upon rate of persulphate-initiated emulsion polymerisation of vinyl acetate at 70°C, according to O'Donnell, Mesrobian and Woodward.[44] (Reaction system: vinyl acetate 20, 1·01% polyvinyl alcohol solution 80, $K_2S_2O_8$—variable.)

initiator concentration was found to be linear, apart from the result for the system containing the lowest level of initiator. The slope of this relationship indicates an order of reaction of 0·7, which is almost double the order of 2/5 predicted by the Smith–Ewart theory. O'Donnell, Mesrobian and Woodward found partially-hydrolysed polyvinyl acetate to be mildly surface-active and to have an apparent critical micelle concentration of about 0·25%. But light-scattering measurements failed to provide evidence of micelle formation. They therefore conclude that each molecule of polyvinyl alcohol may be a possible locus for the reaction.

French[45] has reported on the emulsion polymerisation of vinyl acetate at 65–70°C using potassium persulphate as initiator and the adducts of ethylene oxide and polypropylene oxide as micelle generator. These adducts have the general structure (I); the central polypropylene oxide block gives the molecule its hydrophobic character, whereas the polyethylene

$$H(OCH_2CH_2)_x(OCH(CH_3)CH_2)_y(OCH_2CH_2)_zOH$$

(I)

oxide ends impart hydrophilic character. The number of particles in the reaction system was observed to decrease rapidly over the first 20% of conversion, but thereafter remained constant within experimental error over the range from 20% to 100% conversion. This is an interesting result, since the reaction system contained 55% by weight of monomer, and therefore the opportunities for particle agglomeration as reaction proceeded from 20% to 100% conversion were plentiful. In accordance with general expectation, the number of particles present after 20% conversion was observed to increase as the amount of micelle generator in the reaction system was increased. What is surprising is that the number of particles increases much more sharply with concentration of micelle generator than is the case for styrene, where an approximate proportionality to the 3/5th power of the soap concentration is observed, in accordance with the prediction of the Smith–Ewart theory. An immediate consequence of this

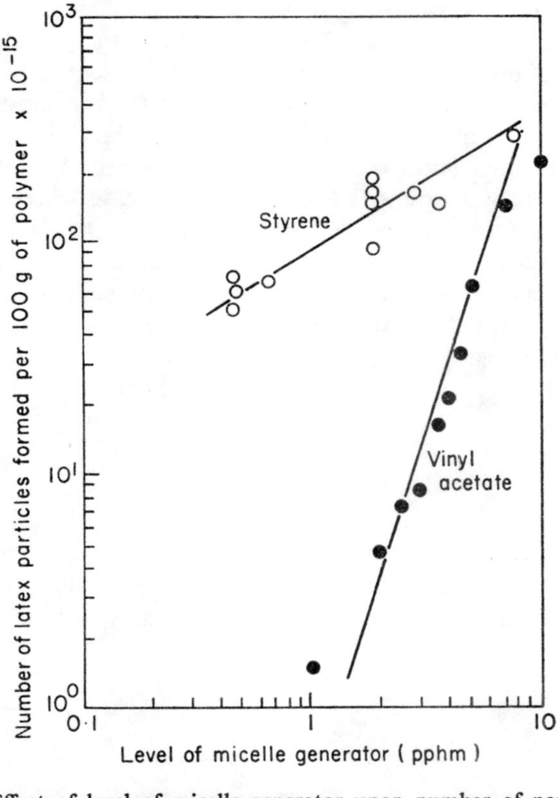

FIG. 10.10. Effect of level of micelle generator upon number of particles formed during persulphate-initiated emulsion polymerisations of vinyl acetate (points ●) and of styrene (points ○), according to French.[45] The data for styrene are due to Smith.[64] (Reaction systems: (1) vinyl acetate 55, water 45, ethylene oxide–polypropylene oxide adduct—variable, $K_2S_2O_8$ 0·041, temperature 65–70°C; (2) the styrene polymerisations were carried out at 50°C using a fatty-acid soap, 0·175% aqueous $K_2S_2O_8$, and an amount of styrene sufficient to give a 35% latex.)

observation is that, weight for weight, at the lower levels of micelle generator the ethylene oxide–polypropylene oxide adducts used by French are very much less efficient in generating particles in vinyl acetate emulsion polymerisation than are conventional fatty-acid soaps in generating particles in the styrene system. The contrasting behaviours of styrene and vinyl acetate are shown in Fig. 10.10, in which the number of particles is shown as a function of level of micelle generator, using logarithmic scales. The data for styrene are taken from a paper by Smith,[64] whereas those for vinyl acetate are by French. The former points are well represented by a line of slope 3/5, the latter by a line of slope 3. The latter line shows that, for vinyl acetate polymerising in the presence of ethylene oxide–polypropylene oxide adducts, the number of particles formed is proportional to the cube of the amount of micelle generator present.

French has given a very simple explanation for this dependence. The assumption is that all the monomer has been imbibed within the polymer particles by the time micellar soap has been consumed. A similar argument has been given by Gerrens.[65] For the simplest case where the particles are spheres of uniform size, the argument can be presented as follows. Consider a system which contains S grams of surface-active material. Let the adsorption area per gram be ϖ. The maximum number, N, of particles which will form will be related to their radius, r, at the point of micelle disappearance, by the equation

$$\varpi S = 4\pi N r^2 \tag{10.1}$$

i.e.

$$r^2 = \frac{\varpi S}{4\pi N} \tag{10.2}$$

If the system also contained, initially, M grams of monomer of density ρ, then the volume of monomer must have been M/ρ. If this is all imbibed within the particles by the time the micelles have disappeared, then we must have that

$$\frac{M}{\rho} = \frac{4}{3}\pi N r^3 \tag{10.3}$$

i.e.

$$r^3 = \frac{3M}{4\pi\rho N} \tag{10.4}$$

Dividing eqn (10.4) by eqn (10.3) gives

$$r = \frac{3M}{\rho\varpi S} \tag{10.5}$$

But eqn (10.1) gives $N = \varpi S/4\pi r^2$, and hence, substituting for r from eqn (10.5), there is obtained the prediction that

$$N = \frac{\rho^2 \varpi^3 S^2}{36\pi M^2} \tag{10.6}$$

This is of the form observed by French.

Three points should be especially noted concerning this derivation. In the first place, it assumes that, whatever is the process which causes the nucleation of new particles, that process ceases as soon as the aggregate area of all the particles is equal to the total adsorption area of the surface-active substances present in the system. In the second place, the derivation becomes inexact if a distribution of particle sizes is present. For eqn (10.2) then gives the square of the surface-area average particle radius, whereas eqn (10.4) gives the cube of the volume–average particle radius. In forming the ratio given on the right-hand side of eqn (10.5), one then obtains the volume/surface area average particle radius, and this is *not* the appropriate average to then substitute in eqn (10.1) rearranged to give N. However, the errors are unlikely to be large unless the distribution of particle sizes is very broad. Thirdly, the derivation assumes that monomer has disappeared as a separate phase by the time that particle nucleation is complete. French provides convincing evidence to show that the monomer is completely imbibed into the particles at a much earlier stage in the reaction than is the case with styrene. He shows that, whereas at about 7% conversion vinyl acetate is largely present as separate droplets, by the time the reaction has proceeded to about 14% reaction those droplets have virtually disappeared. It is not clear, however, whether it is legitimate to infer from these observations that a separate monomer droplet phase is absent by the time particle nucleation ceases. An alternative point of view is that the disappearance of the monomer droplet phase precedes the cessation of particle nucleation, and is indeed a necessary prerequisite for the cessation of particle nucleation. The simple mathematical analysis given by French certainly seems to be consistent with such a model.

French also reported that, for his reaction system, the aggregate surface area of particle–aqueous phase interface is directly proportional to the amount of micelle generator in the system. This observation has an obvious explanation in terms of eqn (10.1) above, *i.e.* the reservoir of micellar surface-active material disappears at the point where the aggregate surface are of monomer-swollen polymer particles is just equal to the total adsorption area of all the surface-active substance present. From data pertaining to this observation, French was able to calculate molecular adsorption areas for his micelle generators; although the molecular weights were much higher than those of conventional soaps (by a factor of 20 or more), the molecular adsorption areas fell on the same approximately linear relationship between adsorption area and molecular volume as is obeyed by a range of lower-molecular-weight soaps. French also records the interesting observation that the logarithm of the viscosity of the final latex increases linearly with the amount of micelle generator in the system, and therefore, apparently, with the aggregate area of interface between particles and aqueous phase.

The paper by Patsiga, Litt and Stannett[46] compares the emulsion polymerisation of vinyl acetate at 60°C with that of styrene at 60°C using, in both cases, seeded reactions and potassium persulphate as initiator. The seed latices were prepared using sodium lauryl sulphate as micelle generator. The seeded reactions were carried out under conditions such that no new particles were formed during the reaction. For styrene, direct proportionality was observed between the rate of polymerisation and the number of seed particles. From the observed relationship between rate and number of particles, and assuming that the reaction conforms to 'Case 2' of the Smith–Ewart theory, it was possible to estimate a value for the rate coefficient for the propagation step of the styrene polymerisation reaction. Good agreement with values obtained by other methods was obtained.

For vinyl acetate, Patsiga, Litt and Stannett state that the observed results for the variation of rate of polymerisation with number of particles can be represented by a relationship of direct proportionality between the rate and the 0·2th power of the number of particles. However, they also show that the results can be adequately represented by a linear relationship which predicts a positive polymerisation rate when no particles are present. Such a relationship is reasonably explained by the simultaneous occurrence of polymerisation reactions in both aqueous phase and particles. Furthermore, the relative magnitudes of the two rates are consistent with the view that most of the polymerisation occurs in the aqueous phase. However, Patsiga, Litt and Stannett succeeded in detecting the presence of polyvinyl acetate in aqueous phases separated from some of their latices. They also reported that the rate of polymerisation does not depend upon the concentration of soap over a wide range, that the rate of polymerisation is first-order with respect to the concentration of potassium persulphate, and that, after an initial period, the rate of polymerisation remains constant until 80–85% conversion, notwithstanding that the monomer droplets have disappeared as a separate phase at or before 35% conversion. Evidence substantiating these claims is given in a comprehensive second paper from these authors which appeared some ten years after the first.[53] Furthermore, the later paper also reports the interesting observation that the introduction of additional potassium persulphate into a non-seeded vinyl acetate emulsion polymerisation brings about an increase in the rate of polymerisation which is proportional to the amount of additional initiator added. Again, there is a marked contrast to the behaviour observed with styrene.

Patsiga, Litt and Stannett[46] envisage the emulsion polymerisation of vinyl acetate to occur by the following mechanism:

> 'During the initial phase of polymerization, soluble polymer molecules are produced which are stabilized in that form by the soap; these probably function as nuclei for further polymerization and thus become monomer swollen polymer particles. When the surface area of the particles exceeds the covering power of the soap, the remaining polymer molecules and micro-particles are partly destabilized and can be swept up by larger particles.
>
> 'During the main portion of the reaction, after the separate monomer phase has vanished, from about 20 to 80% conversion, the number and size of the particles have become constant and the course of polymerization is envisioned as follows: Initiation and much, or most of the polymerisation is in the water phase. The growing polymer in the water phase is partly stabilized by soap molecules which delays its absorption by a large particle. There probably is chain transfer to monomer and polymer in the water phase, but this does not appear in the kinetic picture. Eventually, the polymer molecule is absorbed by a particle, and polymerization then proceeds in the particle until chain transfer to monomer occurs. The monomer free radical diffuses out of the particle and polymerization continues in the water phase until the new polymer molecule is again swept up. It can be shown that the low molecular weight free radical produced by chain transfer will diffuse without reaction from a particle of 0·1 μ diameter at least 99 times out of 100.'

It is also concluded that 'the radical will spend only 4 to 20% of its time in the polymer phase. The exact proportion will depend on the monomer concentration in the particles since the rate of chain transfer to monomer depends on this. It will be 4% initially and rise slowly as the monomer is depleted until it amounts to 20% of the total time at 80% conversion.' The constancy of the polymerisation rate over the range from 20% to 85% conversion is attributed mainly to the monomer-swollen latex particles acting as reservoirs of monomer, by contact with which the aqueous phase remains saturated. At first sight, it might be thought that the equilibrium concentration of monomer in the aqueous phase would fall more or less in proportion to the weight fraction of monomer in the particles. This should not, in fact, be the case since, for mixtures of polymers and low-molecular-weight liquids, the thermodynamic activity of the solvent varies little with the weight fraction of the solute until quite high values of the latter. Thus the activity of the vinyl acetate in the monomer swollen polymer particles would be expected to be virtually unity throughout most of the reaction.

In their second paper, Litt, Patsiga and Stannett[53] develop these mechanistic ideas much more fully, as well as presenting additional data for both non-seeded as well as seeded polymerisations. Electrolytes such as potassium sulphate were observed to increase the rate of polymerisation in non-seeded systems and also to reduce significantly the order of reaction with respect to persulphate initiator. Experiments were reported to show that the high order of reaction with respect to initiator in the absence of added electrolyte is not due to the decrease in pH which accompanies the reaction when persulphate is used as initiator. In seeded systems, an order of reaction of 0·8 was observed for initiator compared with about unity for the non-seeded systems. A twofold variation in the monomer:aqueous phase ratio produced no significant change in rate of polymerisation (expressed in terms of unit volume of aqueous phase) in the case of non-seeded systems. By contrast, for seeded systems, the rate was found to be proportional to the 0·35th power of the volume of the particles, as their volume was varied by adding different amounts of monomer.

The mechanistic model used for the discussion of the results is essentially that of the earlier paper. The polymer chain propagating within the aqueous phase is pictured as being swept up into a particle at a degree of polymerisation between 50 and 200. Chain-transfer to monomer can occur in either phase, and is postulated as resulting in the formation of a butyrolactonyl radical by a cyclisation reaction of the following type:

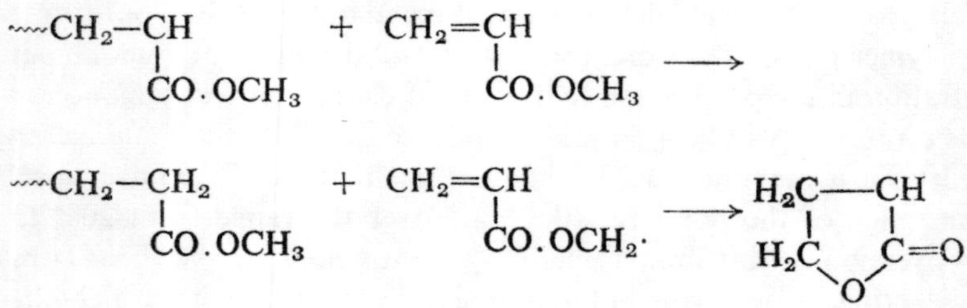

The butyrolactonyl radical, and low-molecular-weight vinyl acetate oligomers derived from it, would be expected to partition strongly in favour of the aqueous phase. The main termination reaction in media of low ionic strength is considered to be the reaction of the butyrolactonyl radical with radicals propagating in the aqueous phase. A secondary termination reaction postulated for such systems is the acquisition of a radical by a particle which already contains a radical. For systems containing aqueous phases of high ionic strength, the latter is regarded as the dominant termination process. The zero-order dependence of rate of polymerisation upon monomer concentration over most of the reaction is explained in terms of propagation occurring mainly in the particles, and termination occurring mainly in the aqueous phase.

A recent paper by Klein, Kuist and Stannett[59] is concerned with the effect of ionic strength of the aqueous phase and of temperature upon the solubility of vinyl acetate monomer in polyvinyl acetate particles which are stabilised by the adsorption of ionic species. The solubility of the monomer increases sharply with increasing ionic strength, and this is explained in terms of decrease in interfacial surface free energy associated with partial collapse of the electrical double layers at the interface. Temperature affects solubility through its effect on both the interfacial tension and the polymer–solvent interaction parameter. The latter was found to vary linearly with reciprocal absolute temperature. The temperature coefficient of interfacial tension was found to be negative at low ionic strength, and to become increasingly more positive as the ionic strength increases.

Okamura and Motoyama[47] have compared the emulsion polymerisation behaviours of vinyl acetate, styrene and vinyl caprate in order to ascertain whether the observed differences between styrene and vinyl acetate are due mainly to differences in water-solubility or to the high reactivity of vinyl acetate radicals making the polymerisation susceptible to the presence of other substances. Vinyl caproate was chosen as a monomer which resembles vinyl acetate in chemical nature but styrene in water-solubility. The emulsion polymerisations were carried out at 60°C using ammonium persulphate as initiator and either sodium dodecyl sulphate or an adduct of ethylene oxide and dodecyl alcohol. They found that, whereas the polymerisation of styrene in pure water proceeds very slowly and increases sharply as the concentration of added soap exceeds that required for micelle formation, vinyl acetate polymerises even when no added surface-active substances are present, and the rate is relatively insensitive to the concentration of added soaps. In particular, no sharp increase in rate was observed as the critical micelle concentration was exceeded. Using a non-ionogenic micelle generator, the rate of polymerisation was found to decrease with increasing concentration. By contrast, vinyl caproate resembled styrene rather than vinyl acetate in obeying essentially the Smith–Ewart prediction of rate dependence upon concentration of micelle generator, thereby providing convincing evidence that the 'anomalous' behaviour of vinyl acetate is due to its high water-solubility, and not to its chemical nature.

Further confirmation of the essential correctness of this conclusion comes from the observation that styrene polymerising in aqueous methanol (in which its solubility is enhanced) using azobis*iso*butyronitrile as initiator resembles vinyl acetate polymerising in water. That vinyl caproate resembles styrene in its emulsion polymerisation behaviour has been confirmed by Litt, Stannett and Vanzo.[58]

Napper and Parts[48] have reported upon the solution polymerisation of vinyl acetate in aqueous medium at pH 3·80 (maintained by an acetic acid–sodium acetate buffer) and 40°C in the absence of added stabilising agents, using potassium persulphate as initiator. Polymer was formed as an extremely stable suspension. Stirring of the polymerising system produced no visible coagulation, nor did it affect the polymerisation rate. The typical conversion–time curve is such as to suggest an initial region of accelerating rate, followed by a region of decelerating rate, there being no significant region of constant rate. The relationship between the square root of the conversion and the time of reaction was found to be approximately linear during the period of acceleration. The slope of the relationship was proportional to the square root of the initiator concentration but independent of the monomer concentration. The number of particles present was independent of conversion beyond about 16%, and was also independent of the monomer concentration. Polymerisation was assumed to occur mainly in the particles formed during the initial stages of the reaction by the precipitation of polymers formed in aqueous solution.

Napper and Alexander[49] have reported upon the effect of low concentrations of added surface-active substances upon this reaction. In all cases, the concentration of added substance was less than the critical level for micelle formation. Anionic soaps accelerate the reaction, the effect increasing with the length of the hydrophobic part of the soap molecule. Non-ionogenic substances have little effect. Cationic soaps have a retarding effect which does not seem to depend upon the chain length of the hydrophobic part of the molecule. Reaction systems containing cationic soaps and taken to 70% conversion were found to contain exclusively negatively-charged particles. Furthermore, the anionic soaps cause a substantial increase in particle concentration, whereas the cationic soaps cause a decrease in particle concentration and give a very broad distribution of sizes. These effects are explained in terms of interaction between the soaps and the negatively-charged oligomeric radical species propagating in the aqueous phase. The anionic soaps discourage aggregation by adsorption of the hydrophobic part of the soap molecule on to the polymer; increasing the chain length of the soap increases the tendency of the molecule to be adsorbed, Coulombic forces in this case tending to discourage adsorption because both soap and polymer are negatively charged. The cationic soaps encourage aggregation by a charge-neutralisation effect which is largely independent of the chain length of the hydrophobic part of the soap molecule.

Dunn and Taylor[51] have also studied the persulphate-initiated polymerisation of vinyl acetate in aqueous solution at 60°C. They confirm that stable suspensions can be formed by the reaction even if no added surface-active substances are added. Using an iron(III)-hydrogen peroxide initiation system, however, rates were much slower for the same rate of initiation, and the reaction tended to cease before completion. The reaction in the absence of added soaps or colloids occurred in three stages: an initial acceleratory stage, a stage of maximum and constant rate between 20% and 45% conversion, and then a final period of deceleration. Stirring had no effect upon either the rate or the particle size of the product. The maximum rate was found to be of order 0·64 with respect to initiator concentration, and to increase with initial monomer concentration at fixed initiator concentration. The particle concentration increased with increasing amount of initiator. Increasing monomer concentration produced fewer but larger particles. Addition of a neutral electrolyte (potassium sulphate) retarded the solution. Addition of sodium dodecyl sulphate had little effect upon the maximum rate, but markedly reduced the particle size. Polyvinyl alcohol reduced the maximum rate and the particle size. Knowing the particle concentration and the monomer concentration in the particles, and assuming a value for the rate coefficient for the propagation step of the vinyl acetate polymerisation, Dunn and Taylor were able to estimate the average number of active radicals per particle necessary to support the observed rate of polymerisation. The values varied greatly with reaction condition and conversion, ranging from 0·17 to 61·5.

Elgood, Gulbekian and Kinsler[50] have reported upon the emulsion polymerisation of vinyl acetate under conditions such that monomer, surface-active substances and initiator were all added to the polymerisation system continuously. The interesting observation is made that, of the various conceivable 'particle-related' factors presented by the reacting system, the aggregate surface area of polymer–aqueous phase interface seems to be the one which correlates most closely with the rate of polymerisation.

Netschey, Napper and Alexander[52] have found that the rate of polymerisation of vinyl acetate in seed latices is affected by the nature of the stabilising layer surrounding the seed particle. Polymerisation occurred much more slowly in seed latices protected by an adsorbed non-ionogenic stabiliser (a block copolymer of vinyl acetate and ethylene oxide) than in those stabilised electrostatically (using sodium hexadecyl sulphate). Polymerisation in seed latices stabilised by a combination of non-ionogenic and anionic surface-active molecules occurred even more slowly. In all cases, polymerisation in the seed latices was initiated by means of potassium persulphate. Although electrostatic stabilisation by adsorbed anions should discourage the ingress of negatively-charged species into the particle, it evidently provides a less formidable barrier than does the electrically-neutral but rather viscous layer which surrounds the particles when a non-ionogenic stabiliser is used. The use of a combined stabiliser system gives an even slower rate of polymerisation; the presence of both viscous and electrostatic barriers is evidently more formidable than is either separately.

A recent paper by Friis and Nyhagen[60] is concerned with the kinetics of vinyl acetate emulsion polymerisation at 60°C in the presence of sodium lauryl sulphate and using potassium persulphate as initiator. It is reported that the rate of polymerisation is proportional to the 0·5th power of the initiator concentration and the 0·25th power of the number of particles. The number of particles was in turn found to be proportional to the 0·5th power of the soap concentration. The limiting viscosity number of the polymer contained in the particles is independent of the initiator concentration and of the number of particles. The conversion–time curve for the system studied is apparently linear for much of the course of the reaction. The linearity is explained as being due partly to a decrease in the rate of desorption of propagating radicals from polymer particles as polymerisation proceeds, and partly to a concomitant decrease in the rate coefficient for the termination reaction. Taken together, these factors compensate for the fall in rate of polymerisation which would otherwise be expected to result from a fall in monomer concentration at the reaction loci as the polymerisation proceeds.

The final paper on the subject of vinyl acetate emulsion polymerisation of which note will be taken here is that of Elgood and Gulbekian.[61] This is concerned with redox-initiated polymerisation at 50°C, the initiation system comprising a combination of ammonium persulphate, a reducing agent and a salt of a metal of variable valency. It is pointed out that the use of conventional redox initiation systems with vinyl acetate is not straightforward, often leading to low conversions only. Several systems are available for polymerising at the relatively high temperatures of 70° and 80°C, but none of these systems is adequate for industrial application at lower temperatures. These authors investigated the use of three reducing agents—sodium formaldehyde sulphoxylate, sodium hydrophosphite and hydrazine hydrate—in combination with the chlorides of titanium(III), iron(III), cobalt(II) and copper(II). The principal surface-active substances used for most of the work described in this paper were adducts of ethylene oxide and nonylphenol; small amounts of sodium dodecylbenzenesulphonate were also present in the reaction systems. Of the three reducing agents investigated only sodium formaldehyde sulphoxylate was found to work with all four metal ions; the other two worked only with copper(II) ions.

Elgood and Gulbekian also report the interesting phenomenon of a 'minimum polymerisation temperature' (*i.e.* a temperature below which polymerisation occurs only very slowly) which varies with the metal ion concentration in such a way as (apparently) to rise very sharply as the metal ion concentration is reduced below a certain critical level. For maximum conversion of monomer to polymer, it is necessary intermittently to adjust the pH of the system. The initial pH was approximately 3·5 and dropped, as the reaction proceeded, to about 2·2 at completion unless adjusted by the addition of small amounts of an alkali such as potassium hydroxide. The indications are that for each metal ion there is an optimum final pH for maximum conversion. The optimum is especially pronounced for systems containing copper(II), and occurs at about pH 4. The order of decreasing effectiveness of the various metal ions in promoting conversion in systems optimised in respect of final pH was found to be Ti(III) > Fe(III) > Co(II) > Cu(II).

The need to increase the pH as reaction proceeds is associated by these authors with an increase in the ratio of polymer to water as the reaction proceeds. They state that 'As this ratio increases, a greater proportion of the available monomer will partition into the particle, and in order to polymerise it the locus of polymerisation must transfer progressively from the water phase to the polymer particle. It will thus be necessary to reduce the solubility of the metal compound in the aqueous phase and increase its solubility in the polymer particles. This is likely to be achieved by increasing the pH of the system, hence favouring the formation of the less water-soluble metal soap.' Confirmation of the importance of the solubility of metal soap formed from the anionic surface-active substance in the system is provided by the results of solubility measurements on the dodecylbenzenesulphonates of the four metals studied. The order of increasing solubility in vinyl acetate is also the order of increasing effectiveness in promoting conversion noted above. The order of increasing effectiveness is also approximately the reverse of the order of increasing solubility of the metal soap in water.

Turning now to the emulsion polymerisation of other vinyl esters of carboxylic acids, note has already been taken of the general similarity in behaviour of vinyl caproate and styrene. Some information is available on the behaviour of the vinyl esters of long-chain carboxylic acids. Thus Port *et al.*[66] have reported upon the free-radical polymerisation of the vinyl esters of caprylic, capric, lauric, myristic, palmitic and stearic acids by various methods, including, *inter alia*, emulsion polymerisation. They state that 'The development of a suitable recipe for emulsion polymerization of the long-chain vinyl esters presented some difficulties, the major one being the difficulty of achieving stable emulsification both before and throughout the polymerization. The usual emulsifiers, such as soaps and alkylaryl sulfonates, were unsatisfactory. After investigation of a large number of commercially available emulsifiers, it was found that a mixture of non-ionic emulsifiers, such as fatty acid esters of anhydrosorbitol or anhydromannitol, and ionic emulsifiers, such as branched-chain alcohol sulfates, gave the best emulsions' A persulphate-initiated system gave conversions of 80–99% after 7 h at 55–75°C. The ratio of total emulsifier to monomer was found to be not critical; as much as 16% and as little as 1% were used successfully. A special procedure was employed to obtain stable initial emulsions. Monomer and emulsifiers were gently heated in an atmosphere of nitrogen until a uniform mixture was obtained. The mixture was then agitated with oxygen-free distilled water in an atmosphere of nitrogen.

Brief descriptions of the persulphate-initiated emulsion polymerisation of vinyl stearate have been given by Parts and Moore,[67,68] and a more detailed account by Moore.[69] This monomer was chosen for study because of its extremely low solubility in water. Thus, whereas 1 ml of a saturated solution of styrene in water contains about 2×10^{18} molecules of solute, the corresponding figure for vinyl stearate is about 4×10^{14}. It was found possible to effect polymerisation using a non-ionogenic substance as the sole surface-active compound added to the system. The particular substance used was an ethylene oxide adduct of a branched octylphenol. The variation of rate of polymerisation with time of reaction was found to resemble that observed for styrene, except that no Trommsdorff 'gel' effect is apparent in the later stages of the reaction. After an initial period of acceleration, the rate passes through a maximum and then falls. The maximum rate of polymerisation is of order 0·8 in the initiator concentration, but the number of particles produced is independent of initiator concentration. The effect of increasing the level of ethylene oxide adduct upon the conversion–time relationship is unusual in that, as the level of the adduct is increased, so the rate of polymerisation passes through a maximum. As expected, the number of particles produced increases with increasing concentration of ethylene oxide adduct, but the variation is not of such a nature as to be expressible in simple exponential form.

In discussing the mechanism of the reaction, Moore[69] concludes that there is no reason to believe that its peculiarities are due to restricted transfer of monomer from droplets to reaction loci by diffusion through the aqueous phase owing to its very low solubility in water. Moore believes that the initiation of the polymerisation commences in the aqueous phase, notwithstanding the low solubility of the monomer; he estimates the stationary concentration of radicals in the aqueous phase to correspond to about one per 35 dissolved monomer molecules. Transfer of monomer into reaction loci to replenish that consumed by polymerisation is suggested as occurring principally by collision of particles with droplets. Although both droplet and particle retain their identities after collision, they are imagined as remaining in contact for sufficiently long for monomer to diffuse from droplet into particle. The maximum in rate of polymerisation as a function of soap concentration can then be explained in terms of collisions between micelles and particles and between small droplets and particles as being of shorter duration than collisions between larger droplets and particles.

The emulsion copolymerisation of the vinyl esters of certain synthetic branched carboxylic acids with vinyl acetate has been described by Vegter and Grommers,[70] and by Vegter.[71] Potassium persulphate can be used as initiator, and a combination of a sodium alkylbenzenesulphonate and an ethylene oxide adduct of nonylphenol as surface-active substance. The polymerisation process which has been developed for those monomers utilises the principle of so-called 'delayed monomer emulsion addition.' An initial reactor charge, consisting of part of the water, part of the anionic

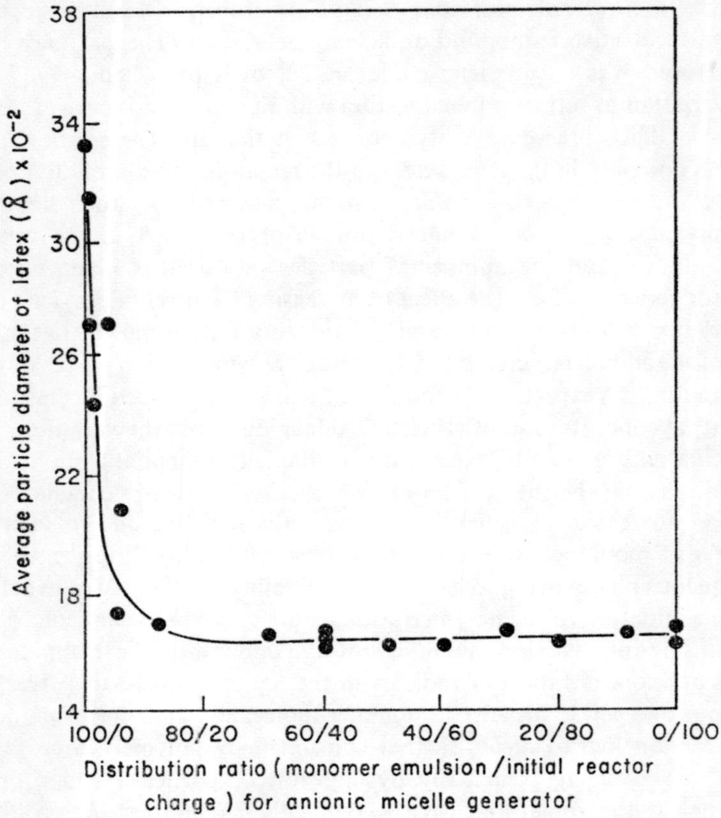

FIG. 10.11. Effect of distribution of anionic micelle generator between monomer emulsion and initial reactor charge upon particle size of latex produced by persulphate-initiated emulsion copolymerisation of vinyl acetate and vinyl esters of synthetic branched carboxylic acid at 80°C, according to Vegter and Grommers[70] and Vegter.[71] (Total reaction system: vinyl acetate 52·5, vinyl esters of synthetic branched carboxylic acids 46·5, acrylic acid 1·0, water 85, sodium alkylbenzenesulphonate 1·0, 20:1 ethylene oxide–nonylphenol adduct 2·5, $K_2S_2O_8$ 0·5, $Na_2B_4O_7.10H_2O$ 0·5.)

soap, part of the initiator and part of a neutralising agent (borax), is heated to the reaction temperature. A monomer feed emulsion, containing all the monomers and non-ionogenic soap, together with the balance of the other ingredients, is then gradually added. A period is then allowed for the completion of the reaction. It has been found that the particle size of the resultant latex is greatly affected by the way in which the anionic

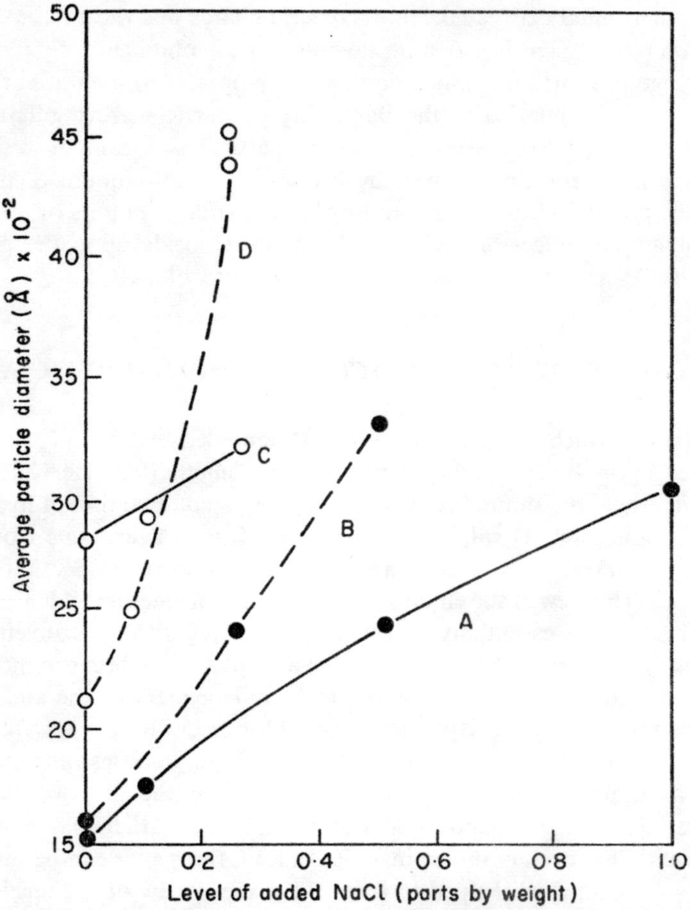

FIG. 10.12. Effect of a simple inorganic electrolyte (NaCl) upon particle size of latex produced by various persulphate-initiated emulsion copolymerisations of vinyl acetate and vinyl esters of synthetic branched carboxylic acids at 80°C, according to Vegter and Grommers.[70] The letters appended to the curves refer to different total reaction systems believed to be as follows:

	A	B	C	D
Vinyl acetate	52·5	52·5	79	79
Vinyl esters of synthetic branched carboxylic acids	46·5	46·5	20	20
Acrylic acid	1	1	1	1
Water	85	98	85	98
Sodium alkylbenzenesulphonate	1·0	0·25	1·0	0·25
20:1 ethylene oxide–nonylphenol adduct	2·5	2·0	2·5	2·0
$K_2S_2O_8$	0·5	0·5	0·5	0·5
$Na_2B_4O_7.10H_2O$	0·5	0·5	0·5	0·5

soap is distributed between the initial reactor charge and the monomer feed emulsion (see Fig. 10.11). A reproducible latex is obtained only if a relatively large proportion of the anionic soap is present in the initial reactor charge, but this eliminates the possibility of particle-size regulation by variation of the distribution ratio. The particle size can, however, be regulated in a reproducible way by varying the total amount of anionic soap, the type of anionic soap, and by adding small amounts of a neutral electrolyte. The effect of addition of sodium chloride upon the particle size given by several reaction systems is shown in Fig. 10.12.

10.5 ESTERS OF ACRYLIC ACID AND METHACRYLIC ACID

In a paper which appeared in 1946, Baxendale, Evans and Kilham[72] reported upon the emulsion polymerisation of methyl methacrylate using cetyltrimethylammonium bromide as soap and a combination of hydrogen peroxide and iron(II) sulphate as initiator. The contributions from this group of workers have been considered in some detail in Section 3.5 of Chapter 3. The view of the emulsion polymerisation reaction which emerges from this work is essentially that which is now regarded as conventional. Initiation was considered to occur in the aqueous phase because the initial rate of polymerisation was observed to be independent of the amount of monomer present as a separate phase. Monomer disappears from the separate monomer phase more rapidly than it polymerises, and this was taken as indicating imbibition of monomer into the growing polymer particles. These ideas are now commonplace, but, as earlier chapters of this book have shown, were not so in 1946. Indeed, this paper can be regarded as one of those which helped to form the present view of the mechanism of emulsion polymerisation, insofar as a single view can be said to exist now.

The technical aspects of acrylic ester emulsion polymerisations have been described in papers by Mast, Smith and Fisher,[73] and by Mast and Fisher.[74] The micelle generator used is said to have a profound effect upon the properties of the latex which is formed. As compared with diene emulsion polymerisations, systems for acrylates require higher reaction temperatures, less reaction time, less initiator, a pH of 7 or lower, and the use of synthetic micelle generators rather than of soaps from natural sources. Modifiers are not required, nor are initiator activators. Control of pH is important in order to prevent hydrolysis of the monomer. Emulsion polymerisation carried out in media of high pH is accompanied by some saponification of monomer, whereas hydrolysis is negligible if the pH is 7 or lower. Information concerning the emulsion polymerisation of esters of methacrylic acid is also available in an extensive paper by Trommsdorff, Köhle and Legally,[75] although it should be noted that much of this paper is concerned with the free-radical polymerisation of these monomers by non-emulsion procedures.

The emulsion polymerisation of methyl methacrylate has been studied in some detail by Zimmt[76] using potassium persulphate as initiator and an anionic micelle generator. The polymerisation temperature was 60°C. A comparison was made with a styrene emulsion polymerisation under comparable conditions. The most significant difference observed between the two monomers was the stronger autoacceleration in the case of methyl methacrylate. This was attributed to the correspondingly stronger Trommsdorff 'gel' effect observed in the bulk polymerisation of methyl methacrylate as compared with styrene. Zimmt concluded that the emulsion polymerisation of methyl methacrylate can be regarded as proceeding in accordance with the Harkins theory if due allowance is made for the early retardation of mutual termination due to the Trommsdorff 'gel' effect. Thus Zimmt states that 'Due to the gel effect the termination constant is low from the start of the polymerization, thus allowing the rate of polymerization to accelerate at smaller particle volumes or lower conversions than for styrene. Once the free monomer has been used up, constant conditions no longer prevail within the monomer–polymer particle, and the rate of polymerization becomes extremely rapid due to a continuous decrease of the rate of termination. However, up to the point at which free monomer disappears ... existing theories can be used to predict the rate of reaction with considerable success.'

Brodnyan *et al.*[77] have reported on the emulsion polymerisation of methyl methacrylate and of *n*-butyl methacrylate. They have concluded that the solubility of the monomer in water is relatively unimportant in determining the dependence of particle number and rate upon the concentrations of initiator, micelle generator and monomer. The orders of reaction with respect to initiator and soap concentrations are both given as approximately 0·5. The number of particles formed is independent of the initiator concentration and the monomer concentration, but proportional to the cube of the soap concentration. (It will be recalled—see Section 10.4—that French[45] observed a similar variation of particle number with soap concentration for the persulphate-initiated emulsion polymerisation of vinyl acetate in the presence of surface-active adducts of ethylene oxide and polypropylene oxide, and that he gave a very simple explanation for this variation.) The rate of polymerisation per unit surface area of particle–aqueous phase interface is roughly constant, although the rate per article changes significantly as the soap concentration is varied. These observations are regarded as being inconsistent with the Harkins theory, and more consistent with a view (to which further reference is made in Chapter 12) that the principal locus of the reaction is the interfacial region of the particle rather than its bulk. The conclusions drawn by Brodnyan and his co-workers therefore seem to be at variance with those drawn by Zimmt.[76]

Two papers by McCurdy and Laidler[78,79] report results for (a) the heats of polymerisation and (b) the rates of polymerisation of various acrylates and methacrylates polymerised in emulsion, using a microcalorimetric technique. The reaction systems contained cetyltrimethylammonium bromide as micelle generator and an iron(II)–hydrogen peroxide redox couple as initiator. The polymerisation temperature was 25°C. The heats of polymerisation of both acrylates and methacrylates increase regularly as the hydrogen of the acid is successively replaced by methyl, ethyl and *n*-butyl groups. This effect is attributed to a progressive increase in the proportion of head-to-head and tail-to-tail additions, relative to head-to-tail and tail-to-head, as the chain length of the alkyl substituent is increased, thereby leading to a decrease in steric hindrance. Acrylates have considerably higher heats of polymerisation than do the corresponding methacrylates, again indicating the effect of reduced steric hindrance, The abnormally low heats of polymerisation observed with hydroxyethyl methacrylate and with 2-hydroxypropyl methacrylate are attributed to a greater degree of hydration of the monomer being possible as compared with the corresponding polymer. The rates of emulsion polymerisation observed with the acrylates were consistently higher than those observed with the corresponding methacrylates. This is attributed to less steric hindrance. In both the acrylates and methacrylates, the rates go through a maximum as one ascends the series methyl, ethyl, butyl, hexyl; the observed changes are interpreted in terms of inductive and steric effects.

The mechanism of particle nucleation in the polymerisation of methyl methacrylate in aqueous solutions has been discussed in a recent paper by Fitch, Prenosil and Sprick.[80] Particle formation, as evidenced by the onset of Tyndall scattering, was found to occur when the polymer concentration reached approximately 0·03 g per 100 ml, regardless of the rate of polymerisation. This was taken as evidence that, in this system at least, the formation of particles is controlled by equilibrium factors such as the solubility in water of the oligomeric species which are formed by solution polymerisation. Two possible mechanisms for particle formation are postulated: (a) coagulation of colloidal macromolecules prior to termination; and (b) mutual termination of macroradicals whilst still in solution, followed by phase separation and coagulation of the inactive polymer species. The evidence presented in this paper suggests that the second mechanism, rather than the first, is operative in the polymerisation of methyl methacrylate in aqueous solution in the absence of added surface-active substances. Further reference to the Fitch theory of particle formation will be found in Section 12.2 of Chapter 12.

10.6 VINYL CHLORIDE AND VINYLIDENE CHLORIDE

The emulsion polymerisation (and, indeed, the bulk polymerisation) of both these monomers is complicated by the fact that the polymer shows only limited miscibility with its own monomer, and therefore tends to precipitate within the reaction locus. Papers dealing with the emulsion polymerisation of vinyl chloride include those by Lazor,[81] by Peggion, Testa and Talamini,[82] by Hopff and Fákla,[83,90] by Giskehaug,[84] by Testa and Vianello,[85] by Mørk,[86] by Mørk and Ugelstad,[87] by Ugelstad *et al.*,[88] and by Ugelstad and Mørk.[89] Ley and Fowler[91] have described the emulsion copolymerisation of vinyl chloride with various alkyl acrylates. The emulsion polymerisation of vinylidene chloride is described in a paper by Wiener,[92] and in three papers by Sweeting and his co-workers.[93,95] Marker, Sweeting and Wepsic[96] have reported upon the emulsion copolymerisation of vinylidene chloride and acrylonitrile.

The reaction system used by Lazor[81] for the emulsion polymerisation of vinyl chloride at 52°C contained sodium lauryl sulphate as soap and a combination of potassium persulphate and sodium bisulphite as initiator. The primary objective of the work was to develop regression equations to represent the effects of varying the levels of the several recipe ingredients upon the concentration of particles produced in the final latex. It was reported that, at sub-critical soap concentrations, the number of particles produced by this reaction system is dependent mainly upon the levels of sodium lauryl sulphate and sodium bisulphite, and, to a lesser extent, upon the level of potassium persulphate. As the critical micelle concentration is approached, the sodium bisulphite level ceases to have an effect. This was interpreted as indicating a change in the particle initiation mechanism as the soap concentration approaches the critical level for micelle formation.

Peggion, Testa and Talamini[82] also used the persulphate–bisulphite redox couple for their study. Several surface-active substances were used. The order of the reaction with respect to potassium persulphate concentration was found to be between 0·6 and 0·7. As long as the concentration of the surface-active substance exceeded that required for micelle formation, it had little effect upon the rate of polymerisation. The number of particles in the latices was observed to remain constant throughout the reaction after about 7% conversion. The particle concentration is independent of initiator concentration, but increases with increasing soap level. The main features of the reaction are explained by these authors in terms of simultaneous aqueous solution and micellar polymerisations. Particle nucleation is by way of solution polymerisation in the aqueous phase followed by precipitation. When the number of particles has become large enough to adsorb the emulsifier completely, the polymer radicals produced continuously in the aqueous phase precipitate in the preformed polymer particles, as they no longer have sufficient soap to stabilise them. Thus the number of particles remains constant thereafter. The autocatalytic character of the conversion–time relationship is explained in terms of the immobilisation of propagating polymer radicals within the polymer which precipitates as it forms. Termination is therefore suppressed relative to propagation.

Similar conclusions have been reached by Giskehaug.[84] In this work, the polymerisation was carried out at 50°C in the presence of a polyvinyl chloride seed latex in order to exclude the particle-forming reaction. The polymerisations were stopped at low conversion. Potassium persulphate was the initiator and sodium dodecylbenzensulphonate the soap. For reaction systems with particle diameters greater than 1000 Å, the rate of polymerisation was found to be of order 0·5 with respect to initiator concentration and nearly independent of the particle diameter, the number of particles, and the weight of the seed. With smaller particles (150–300 Å), the order of reaction with respect to initiator rose to 0·6. The rate in the presence of the smallest particles was considerably greater than in the presence of those larger than 1000 Å. The results were taken as indicating that, with the larger particles, the reaction occurs mainly in the aqueous phase, whereas, in the case of the smaller particles, reaction within, or on the surface of, the particles may make a significant contribution.

Testa and Vianello[85] have considered the applicability of the so-called 'HLB system' (see Chapter 7, Section 7.9) for selecting surface-active substances for the emulsion polymerisation of vinyl chloride. Although useful for styrene and for vinyl acetate (see Chapter 7), the concept of hydrophilic–lypophilic balance seems to be of little use for vinyl chloride. This is attributed by Testa and Vianello to the perturbation caused by the insolubility of polyvinyl chloride in vinyl chloride monomer.

The inhibition of the persulphate-initiated emulsion polymerisation of vinyl chloride by oxygen has been studied by Mørk.[86] The length of the inhibition period is inversely proportional to the square root of the initiator concentration, directly proportional to the square root of the initial amount of oxygen, and nearly independent of the concentration of micelle generator. A dependence upon the volumes of both aqueous and gaseous phases was also observed. A cross-termination reaction between radicals of the type RM· and $RO_2\cdot$ is suggested as the main reason for the observed inhibition. A mechanism is proposed whose kinetic consequences are consistent with the observed features of the inhibition. In a second paper, Mørk and Ugelstad[87] report upon the effect of bases on the inhibition time and polymerisation rate for the oxygen-inhibited emulsion polymerisation of vinyl chloride. The addition of bases reduces the inhibition time and increases the polymerisation rate. This is ascribed to an increase in the rate of radical production, brought about by a base-catalysed decomposition of polyvinyl chloride peroxides which are formed by copolymerisation of vinyl chloride with oxygen. The rates of production of radicals from the polymer peroxides is proportional to the concentration of base.

The kinetics of the persulphate-initiated emulsion polymerisation of vinyl chloride at 50°C have been investigated by Ugelstad *et al.*[88] The order of reaction with respect to initiator concentration is 0·5. The rate of polymerisation increases with conversion. The number of latex particles is constant during the polymerisation from 10% to 90% conversion, and is independent of the initiator concentration. The order of reaction with respect to number of particles increases from 0·05 to 0·15 over the range from 10^{13} to 10^{16} particles per millilitre of aqueous phase. The order decreases slightly with conversion. A mechanism which involves a rapid desorption and re-absorption of radicals to and from the aqueous phase is proposed. A marked increase in rate occurs at between 70% and 80% conversion. A closer investigation of this phenomenon was made by the technique of continuously adding vinyl chloride at a pressure below its saturation vapour pressure at the reaction temperature. It appears that, under these conditions, polymerisation rates of about twice those in ordinary runs can be obtained. No change in the number of particles was observed, either during the polymerisation or as compared with an ordinary reaction carried out using the same concentrations of initiator and micelle generator. The effect of reducing the vinyl chloride pressure progressively was also reported. The polymerisation rate passes through a maximum at about 87% of the saturation vapour pressure. Further consideration of the kinetics of the reaction is given in the paper by Ugelstad and Mørk.[89]

The general features of the free-radical polymerisation of vinylidene chloride in both aqueous soap solution and in emulsion have been reported by Wiener.[92] The initiator in this work was potassium persulphate and the micelle generator potassium laurate. The polymerisation temperature was 25°C. Conversions in aqueous solution were determined by following the change of vinylidene chloride vapour pressure over the solution; conversions in emulsion were determined dilatometrically and gravimetrically. In solution, the rate is directly proportional to the square root of the initiator concentration, and also to the monomer concentration.

The rate of disappearance of monomer as a function of monomer and initiator concentration is well represented by the equation

$$-\frac{\mathrm{d}[M]}{\mathrm{d}t} = \mathrm{k}[M][I]^{1/2} \tag{10.7}$$

where $[M]$ and $[I]$ denote monomer and initiator concentrations, respectively, and k is a constant whose value depends upon the concentration of soap in the system. k rises rapidly with increasing soap concentration over the range from 0 to 0·05 mole l^{-1}, but shows little further change at higher soap concentrations. In emulsion the rate of polymerisation per unit volume of aqueous phase is proportional to the square root of the initiator concentrations and the 3/5th power of the soap concentration, but independent of the amount of monomer emulsified. The rate remains constant up to high conversions. Assuming that the 'true' order (*i.e.* the order at the reaction loci) of reaction with respect to monomer concentration is unity, Wiener was able to calculate rate coefficients corresponding to the k of eqn (10.7) above for the reaction in aqueous soap solution. Values similar to those for the solution reaction were obtained. Wiener points out that, whilst he has interpreted his results mainly in terms of the kinetics of bulk free-radical polymerisation, his data are also consistent with the predictions of the Smith–Ewart theory inasmuch as he finds a dependence of rate upon the 3/5th power of the soap concentration and it is experimentally difficult to distinguish between orders of 1/2 and 2/5 with respect to the initiator concentration. The solubility measurements reported in this paper provide evidence that the monomer in a vinylidene chloride emulsion polymerisation system is adsorbed on to the surfaces of the particles. The solubility of the monomer in polyvinylidene latices is up to ten times greater than its solubility in water, and depends upon both the polymer content and soap concentration. The usual explanation of imbibition of monomer into the particles cannot be invoked in this instance, since the polymer is not normally swollen by the monomer. Therefore the monomer is probably adsorbed on to the particle surfaces, the obvious inference then being drawn that the surface could be the principal locus of reaction.

Reference has been made in earlier chapters to the investigation of Hay and his co-workers[93] into the emulsion polymerisation of vinylidene chloride using sodium lauryl sulphate as micelle generator and a redox couple comprising ammonium persulphate and sodium metabisulphite as initiator. The effects of varying the levels of initiator and micelle generator are shown in detail in Figs. 6.16 and 7.18, respectively. The general shape of the curve remains constant throughout these variations, and is illustrated in Fig. 10.13. It suggests that the reaction falls into three fairly well-defined stages: a fast initial stage, a slow intermediate stage and a

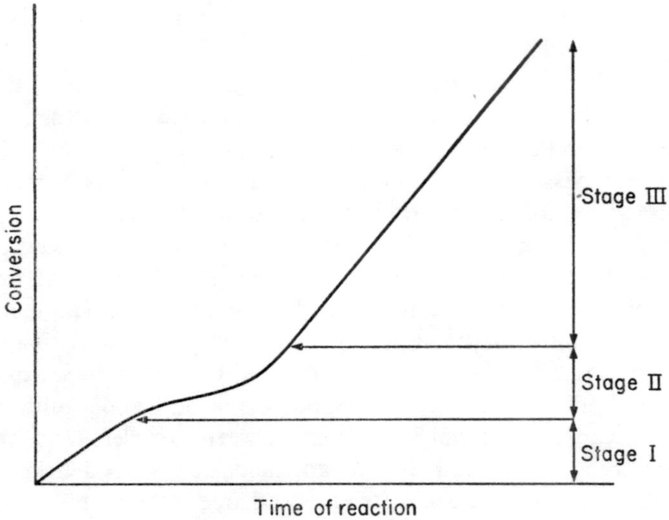

FIG. 10.13. General shape of conversion–time curve for activated persulphate-initiated emulsion polymerisation of vinylidene chloride at 32°C, with dilineation of three stages of reaction, according to Hay *et al.*[93]

faster final stage. The order of reaction with respect to the concentration of initiator is 0·60 for the first stage and 1·0 for the third. For micelle generator, the order of reaction is 0·60. Two explanations are offered for the unusual features of this reaction. One hypothesis is that aggregation of particles may occur after the first stage, thereby disturbing the steady state which has been established for the transport of monomer from emulsified droplets through the aqueous phase to the reaction loci. Hay and his co-authors consider that, whilst particle agglomeration plays some part in determining the kinetics of the reaction, it is not by itself sufficient fully to explain the observed effects. The second hypothesis is that there is a change in the transfer mechanics for the monomer following the first stage. It is suggested that collisions between polymer particles and emulsified droplets may become important, and that each reaction locus may subsequently contain more than one propagating free radical.

The two subsequent papers from this group of workers[94,95] develop these ideas further. One considers the effect of the coalescence of polymer particles upon the kinetics, and the other the effect of stirring. Particle coalescence is assumed to commence as a second-order process as soon as micellar soap is consumed, and the rate of polymerisation is assumed to increase in proportion to the volume of the particle (since it is assumed that each particle can contain more than one propagating radical). The equations derived on the basis of these assumptions predict a minimum in the polymerisation rate at some time after the commencement of coalescence. The effect of increased stirring speed differs for each of the three stages. In the first, the rate of polymerisation decreases; in the second, it increases; and in the third, it has no effect. The fall in polymerisation rate as the system passes from the first stage into the second is interpreted as being due to monomer depletion at the reaction loci caused by a fall in rate of transfer from the monomer droplets due to the coalescence of the latter. The effect of increased agitation in the second stage is to break up the droplets and so increase the rate of polymerisation. The increase in polymerisation rate as the system passes from the second to the third stage is interpreted as being due to enhanced contact between monomer droplets and polymer particles, such contacts being less discouraged than formerly because the repulsive forces between droplets and particles have now become reduced due to an increasing soap deficiency. The inverse relationship between stirring speed and rate of polymerisation observed in the first stage of the reaction is presumably due to an enhancement of the rate of particle agglomeration brought about by increased agitation.

REFERENCES

1. J. W. Vanderhoff, 'The mechanism of emulsion polymerization,' in *Vinyl Polymerization*, (ed. G. E. Ham), Dekker, New York, 1969, Vol. 1, Pt II, pp. 1–138.
2. B. M. E. van der Hoff, *J. phys. Chem., Ithaca*, 1956, **60**, 1250.
3. B. M. E. van der Hoff, *J. Polym. Sci.*, 1960, **44**, 241.
4. G. M. Burnett and R. S. Lehrle, *Proc. R. Soc.*, 1959, **A253**, 331.
5. E. Bartholomé, H. Gerrens, R. Herbeck and H. M. Weitz, *Z. Elektrochem.*, 1956, **60**, 334.
6. E. Bartholomé, H. Gerrens, R. Herbeck and H. M. Weitz, *Z. Elektrochem.*, 1957, **61**, 522.
7. H. Gerrens, *Z. Elektrochem.*, 1956, **60**, 400.
8. I. D. Robb, *J. Polym. Sci.*, Pt A-1, 1969, **7**, 417.
9. G. V. Schulz and J. Romatowski, *Makromolek. Chem.*, 1965, **85**, 195.
10. M. Harada *et al.*, *J. appl. Polym. Sci.*, 1972, **16**, 811.
11. M. R. Grancio and D. J. Williams, *J. Polym. Sci.*, Pt A-1, 1970, **8**, 2733.
12. M. R. Grancio and D. J. Williams, *J. Polym. Sci.*, Pt A-1, 1970, **8**, 2617.
13. H. Gerrens and E. Köhnlein, *Z. Elektrochem.*, 1960, **64**, 1199.
14. K. P. Paoletti and F. W. Billmeyer, *J. Polym. Sci.*, Pt A, 1964, **2**, 2049.
15. E. J. Meehan, *J. Am. chem. Soc.*, 1949, **71**, 628.
16. H. S. Smith, H. G. Werner, J. C. Madigan and L. H. Howland, *Ind. Engng Chem.*, 1949, **41**, 1584.
17. L. H. Howland, W. E. Messer, V. C. Neklutin and V. S. Chambers, *Rubb. Age N.Y.*, 1949, **64**, 459.
18. P. H. Johnson and R. L. Bebb, *Ind. Engng Chem.*, 1949, **41**, 1577.
19. R. W. Brown, W. E. Messer and L. H. Howland, *Ind. Engng Chem.*, 1953, **45**, 1322.
20. K. E. Beu, W. B. Reynolds, C. F. Fryling and H. L. McMurry, *J. Polym. Sci.*, 1948, **3**, 465.
21. E. J. Hart and A. W. Meyer, *J. Am. chem. Soc.*, 1949, **71**, 1980.
22. A. W. Meyer, *Ind. Engng. Chem.*, 1949, **41**, 1570.
23. R. R. Hampton, *Anal. Chem.*, 1949, **21**, 923.
24. C. F. Fryling, S. H. Landes, W. M. St. John and C. A. Uraneck, *Ind. Engng Chem.*, 1949, **41**, 986.
25. F. E. Condon, *J. Polym. Sci.*, 1953, **11**, 139.
26. I. Ya Poddubnyĭ and M. A. Rabinerzon, *J. appl. Polym. Sci.*, 1965, **9**, 2527.
27. V. Václavek, *J. appl. Polym. Sci.*, 1967, **11**, 1903.
28. C. A. Uraneck and J. E. Burleigh, *J. appl. Polym. Sci.*, 1970, **14**, 267.
29. L. M. Welch *et al.*, *Ind. Engng Chem.*, 1947, **39**, 826.
30. G. H. Miller, R. L. Alumbaugh and R. J. Brotherton, *J. Polym. Sci.*, 1952, **9**, 453.
31. M. S. Kharasch, *et, al. Ind. Engng Chem.*, 1947, **39**, 830.
32. R. L. Frank, J. R. Blegen, G. E. Inskeep and P. V. Smith, *Ind. Engng Chem.*, 1947, **39**, 893.
33. R. B. Seymour, *et al.*, *Ind. Engng Chem. Prod. Res. Develop.*, 1973, **12**, 198.
34. A. F. Helin, H. K. Stryker and G. J. Mantell, *J. appl. Polym. Sci.*, 1965, **9**, 1797.
35. H. K. Stryker, A. F. Helin and G. J. Mantell, *J. appl. Polym. Sci.*, **9**, 1965, 1807.
36. H. K. Stryker, A. F. Helin and G. J. Mantell, *J. appl. Polym. Sci.*, 1966, **10**, 81.
37. G. J. Mantell, H. K. Stryker, A. F. Helin, D. R. Jamieson and C. H. Wright, *J. appl. Polym. Sci.*, 1966, **10**, 1845.
38. H. K. Stryker, G. J. Mantell and A. F. Helin, *J. appl. Polym. Sci.*, 1967, **11**, 1.
39. H. K. Stryker, G. J. Mantell and A. F. Helin, *J. Polym. Sci.*, Pt C, 1969, No. 27, 35.

40. H. K. Stryker, G. J. Mantell and A. F. Helin, 'Mechanism of the emulsion polymerization of ethylene', in *Vinyl Polymerization* (ed. G. E. Ham), Dekker, New York, 1969, Vol. 1, Pt II, pp. 175–86.
41. M. K. Lindemann, *Vinyl Polymerization* (ed. G. E. Ham), Dekker, New York, 1967, Vol. 1, Pt I, p. 260.
42. D. Gershberg, A.I.Ch.E.-I.Chem.E. Symposium Series No. 3, 1965 (London: Instn chem. Engrs).
43. W. J. Priest, *J. phys. Chem., Ithaca*, 1952, **56**, 1077.
44. J. T. O'Donnell, R. B. Mesrobian and A. E. Woodward, *J. Polym. Sci.*, 1958, **28**, 171.
45. D. M. French, *J. Polym. Sci.*, 1958, **32**, 395.
46. R. Patsiga, M. Litt and V. Stannett, *J. phys. Chem., Ithaca*, 1960, **64**, 801.
47. S. Okamura and T. Motoyama, *J. Polym. Sci.*, 1962, **58**, 221.
48. D. H. Napper and A. G. Parts, *J. Polym. Sci.*, 1962, **61**, 113.
49. D. H. Napper and A. E. Alexander, *J. Polym. Sci.*, 1962, **61**, 127.
50. B. G. Elgood, E. V. Gulbekian and D. Kinsler, *Polymer Letters*, 1964, **2**, 257.
51. A. S. Dunn and P. A. Taylor, *Makromolek. Chem.*, 1965, **83**, 207.
52. A. Netschey, D. H. Napper and A. E. Alexander, *Polymer Letters*, 1969, **7**, 829.
53. M. Litt, R. Patsiga and V. Stannett, *J. Polym. Sci.*, Pt A-1, 1970, **8**, 3607.
54. J. W. Breitenbach, K. Kuchner, H. Fritze and H. Tarnowiecki, *Br. Polym. J.*, 1970, **2**, 13.
55. A. S. Dunn and L. C.-H Chong, *Br. Polym. J.*, 1970, **2**, 49.
56. K. Noro, *Br. Polym. J.*, 1970, **2**, 128.
57. M. Shiraishi, *Br. Polym. J.*, 1970, **2**, 135.
58. M. Litt, V. T. Stannett and E. Vanzo, *Polymer Preprints*, 1972, **13**(1), 434.
59. A. Klein, C. H. Kuist and V. T. Stannett, *J. Polym. Sci.*, Polymer Chemistry Edn, 1973, **11**, 2111.
60. N. Friis and L. Nyhagen, *J. appl. Polym. Sci.*, 1973, **17**, 2311.
61. B. G. Elgood and E. V. Gulbekian, *Br. Polym. J.*, 1973, **5**, 249.
62. M. K. Lindemann, 'The mechanism of vinyl acetate polymerization', in *Vinyl Polymerization* (ed. G. E. Ham), Dekker, New York, 1967, Vol. 1, Pt I, pp. 207–329.
63. E. V. Gulbekian and G. E. J. Reynolds, 'Polyvinyl alcohol in emulsion polymerization', in *Polyvinyl Alcohol: Properties and Applications*, (ed. C. A. Finch), John Wiley, London, 1973, pp. 427–60.
64. W. V. Smith, *J. Am. chem. Soc.*, 1948, **70**, 3695.
65. H. Gerrens, *Fortschr. HochpolymForsch.*, 1959, **1**, 234.
66. W. S. Port, J. E. Hansen, E. F. Jordan, T. J. Dietz and D. Swern, *J. Polym. Sci.*, 1951, **7**, 207.
67. D. E. Moore and A. G. Parts, *Makromolek. Chem.*, 1960, **37**, 108.
68. A. G. Parts and D. E. Moore, *J. Oil Colour Chem. Ass.*, 1962, **45**, 648.
69. D. E. Moore, *J. Polym. Sc.*, Pt A-1, 1967, **5**, 2665.
70. G. C. Vegter and E. P. Grommers, *J. Oil Colour Chem. Ass.*, 1967, **50**, 72.
71. G. C. Vegter, 'Emulsion copolymerization of vinyl esters of branched carboxylic acids with vinyl acetate', in *Addition and Condensation Polymerization Processes*, Advances in Chemistry Series, American Chemical Society, U.S.A., 1969, No. 91, pp. 178–87.
72. J. H. Baxendale, M. G. Evans and J. K. Kilham, *J. Polym. Sci.*, 1946, **1**, 466.
73. W. C. Mast, L. T. Smith and C. H. Fisher, *Ind. Engng Chem.*, 1945, **37**, 365.
74. W. C. Mast and C. H. Fisher, *Ind. Engng. Chem.*, 1949, **41**, 790.
75. E. Trommsdorff, H. Köhle and P. Legally, *Makromolek. Chem.*, 1947, **1**, 169.
76. W. S. Zimmt, *J. appl. Polym. Sci.*, 1959, **1**, 323.
77. J. G. Brodnyan, J. A. Cala, T. Konen and E. L. Kelley, *J. Colloid Sci.*, 1963, **18**, 73.

78. K. G. McCurdy and K. J. Laidler, *Can. J. Chem.*, 1964, **42**, 818.
79. K. G. McCurdy and K. J. Laidler, *Can. J. Chem.*, 1964, **42**, 825.
80. R. M. Fitch, M. B. Prenosil and K. J. Sprick, *J. Polym. Sci.*, Pt C, 1969, No. 27, 95.
81. J. T. Lazor, *J. appl. Polym. Sci.*, 1959, **1**, 11.
82. E. Peggion, F. Testa and G. Talamini, *Makromolek. Chem.*, 1964, **71**, 173.
83. H. Hopff and I. Fákla, *Makromolek. Chem.*, 1965, **88**, 54.
84. K. Giskehaug, 'Kinetics of the emulsion polymerisation of vinyl chloride', in *The Chemistry of Polymerisation Processes*, S.C.I. Monograph No. 20, S.C.I., London, 1966, pp. 235–48.
85. F. Testa and G. Vianello, *Polymer Preprints*, 1966, **7**, 641.
86. P. C. Mørk, *European Polymer J.*, 1969, **5**, 261.
87. P. C. Mørk and J. Ugelstad, *European Polymer J.*, 1969, **5**, 483.
88. J. Ugelstad, P. C. Mørk, P. Dahl and P. Rangnes, *J. Polym. Sci.*, Pt C, 1969, No. 27, 49.
89. J. Ugelstad and P. C. Mørk, *Br. Polym. J.*, 1970, **2**, 31.
90. H. Hopff and I. Fákla, *Br. Polym. J.*, 1970, **2**, 40.
91. D. E. Ley and W. F. Fowler, *J. Polym. Sci.*, Pt A, 1964, **2**, 1863.
92. H. Wiener, *J. Polym. Sci.*, 1951, **7**, 1.
93. P. M. Hay, *et al.*, *J. appl. Polym. Sci.*, 1961, **5**, 23.
94. J. C. Light, L. Marker, A. T. Santonicola and O. L. Sweeting, *J. appl. Polym. Sci.*, 1961, **5**, 31.
95. C. P. Evans, *et al.*, *J. appl. Polym. Sci.*, 1961, **5**, 39.
96. L. Marker, O. Sweeting and J. G. Wepsic, *J. Polym. Sci.*, 1962, **57**, 855.

CHAPTER 11

NON-AQUEOUS EMULSION POLYMERISATION

11.1 INTRODUCTION

The main concern of this book is with emulsion polymerisation systems of the conventional type, in which the dispersion medium is essentially water, conventional surface-active substances are used as micelle generators, and the monomers have strictly limited miscibility with water. The purpose of this chapter is to review very briefly some of the attempts which have been made, and are currently being made, to carry out so-called emulsion polymerisation reactions in non-aqueous dispersion media. No attempt has been made to give an exhaustive discussion of this topic, but it is considered desirable to take some note of it. The small number of references given at the end of this chapter will provide access to a wider range of references for those who wish to inform themselves more fully concerning this subject.

It has been found convenient to classify the subject-matter under three headings:

(1) attempts to carry out essentially conventional emulsion polymerisations in non-aqueous dispersion media;
(2) so-called 'inverse' emulsion polymerisations; and
(3) so-called 'dispersion' polymerisations in organic media.

In the first of these approaches, the non-aqueous dispersion medium is used, as far as possible, as a straightforward replacement for the water of a conventional system. The second approach is concerned with the emulsion polymerisation of monomers which are readily soluble in water, but show limited miscibility with non-polar liquids. The third is concerned with the solution polymerisation of monomers in organic liquids in which the polymer is insoluble. Thus, as polymerisation proceeds, so the polymer tends to precipitate from the solvent and form discrete particles in which further reaction can occur. It appears that, if certain conditions are fulfilled, it is possible by this means to prepare stable dispersions of polymers in organic media.

It is an open question which, if any, of these processes fall strictly within the scope of the formal definition of an emulsion polymerisation reaction given in Chapter 1. Concerning this matter, the reader must draw his own conclusions. The justification for making brief allusion to each of these topics is that, if omitted, the reader would be left unaware of an aspect of the subject to which some attention has been directed in past and recent years.

11.2 CONVENTIONAL EMULSION POLYMERISATIONS USING NON-AQUEOUS DISPERSION MEDIA

The feasibility of carrying out conventional styrene–butadiene emulsion copolymerisations in non-aqueous media was considered by Carr and Johnson[1] in a paper which appeared in 1949. These authors presumed the first requirement to be lack of miscibility between the liquid medium and the monomer. A secondary requirement is lack of interference in the progress of the polymerisation reaction. Of the possible polar liquids which might be used, Carr and Johnson selected formamide, formic acid and liquid ammonia. They regarded the very low melting and boiling points of ammonia as being especially advantageous in low-temperature polymerisations, and its high volability as facilitating easy separation of the polymer and recovery of the medium. The low viscosity of liquid ammonia was regarded as a further advantage in that it would facilitate agitation.

Preliminary tests were carried out on the solubility of certain vinyl monomers in both formamide and formic acid. Emulsification tests using selected surface-active substances were carried out using styrene. The procedure was to attempt to dissolve 0·1 g of surface-active substance in 5 ml of formamide or formic acid, and then to shake with 2 ml of styrene. The rate of separation of the subsequent emulsion was compared with that for a similar, but emulsifier-free, system. In these tests, it was found that the surface-active substance dissolved to give either a clear solution or a cloudy solution similar to an aqueous soap solution. The following comments are recorded regarding the various substances investigated. Dodecyl*iso*thiouronium hydrobromide dissolved in formamide to give a clear solution from which white crystals soon separated, presumably formed by chemical interaction. This substance was soluble in formic acid without reaction, but its emulsifying capacity was negligible. Fatty-acid soap dissolved readily in formamide if the mixture was warmed to 40°C or above, but on cooling to room temperature, the translucent solution set to a firm gel, similar to the gel formed in water. The emulsifying capacity was small. Sodium oleate was moderately-easily soluble in formamide, but exhibited only slightly better emulsifying capacity than the fatty-acid soap. Cetyldimethylbenzylammonium chloride was a very effective emulsifier in both formamide and formic acid. Part of its effectiveness was attributed to the fact that some water was present, it being added as a paste. A substance described as a 'higher alkyl sulphonic acid salt' was found to be a good emulsifier in both formamide and formic acid. Sodium lauryl sulphate was reported to be an effective emulsifier in formamide, but not in formic acid.

Preliminary experiments to assess suitability as dispersion media for the persulphate-initiated copolymerisation of styrene and butadiene showed that, of the surface-active substances referred to in the preceding paragraph, only sodium lauryl sulphate in formamide was effective. Using formic acid as the dispersion medium, no polymerisation was observed with any of the substances. Several recipe variations were investigated, without success in all cases. For details, the original paper should be consulted. A possible explanation for the failure to effect polymerisation is that the formic acid scavenges the radicals produced by the thermal decomposition of the initiator. In the case of initiation by persulphate, reactions such as the following might occur:

$$SO_4^{\cdot -} + H.CO_2H \rightarrow HSO_4^- + \cdot CO_2H$$

$$2 \cdot CO_2H \rightarrow (CO_2H)_2$$

Carr and Johnson do not mention whether any oxalic acid was formed in the course of the attempted polymerisations in formic acid. They do, however, suggest the possibility that the formic acid interferes with the decomposition of the initiating substances, for example by causing it to decompose by a different mechanism. Although at first sight this might not seem to be a reasonable postulate because failure to polymerise was observed with several initiators, in fact all these initiators were peroxidic in nature. It is therefore at least possible that they are all directly reduced by the formic acid with the consequent oxidation of the latter to carbon dioxide and water. It is also possible to envisage oxidation to water and carbon dioxide as the overall result of interaction between the radicals produced by normal thermal decomposition of the initiator and the formic acid. The following reactions seem plausible for the case where the initiator is persulphate:

$$SO_4^{\cdot -} + H\text{—}\underset{\underset{O}{\|}}{C}\text{—}OH \rightarrow HSO_4^- + \cdot\underset{\underset{O}{\|}}{C}\text{—}OH$$

$$SO_4^{\cdot -} + H_2O \rightarrow HSO_4^- + HO\cdot$$

$$HO\cdot + \cdot\underset{\underset{O}{\|}}{C}\text{—}OH \rightarrow \left[HO\text{—}\underset{\underset{O}{\|}}{C}\text{—}OH\right] \rightarrow CO_2 + H_2O$$

This sequence of reactions would require the initial presence of a small amount of water. This requirement would probably be fulfilled in practice even though nominally anhydrous dispersion media were used.

Using formamide as the dispersion medium, polymerisation was observed in one case only, namely when sodium lauryl sulphate was used as the surface-active substance. It should be pointed out that, if any of the above explanations is correct for the absence of polymerisation when formic acid is used as the dispersion medium, then it is difficult to see why polymerisation is possible at all using formamide. For analogous series of reactions producing such substances as oxamide or carbon dioxide and ammonia would be expected to occur.

Carr and Johnson then report the results of a more extensive investigation of the polymerisation of a 25/75 styrene/butadiene monomer mixture in formamide using sodium lauryl sulphate and potassium persulphate as micelle generator and initiator, respectively. The reaction temperature was 50°C. Table 11.1 gives the conversion–time data and polymer properties for one particular reaction system. Carr and Johnson comment that 'The sodium lauryl sulfate and potassium persulfate dissolved readily to a clear solution. As polymerization set in free monomer was largely taken up at a low conversion level to form a bluish translucent latex with a high degree of transparency. As the conversion increased beyond 50%, the transparency was gradually replaced by the milkiness of typical water latices'. Concerning the end-product of the reaction, it is said that 'The latices resulting from polymerization and copolymerization in the formamide system were mobile and stable, and possessed the faintly bluish cast typical of aqueous latices. The stability was such that no tendency towards separation of flocculation was detectable in periods as long as

TABLE 11.1

Conversion–time data and polymer properties for persulphate-initiated emulsion copolymerisation of styrene and butadiene at 50°C in formamide as dispersion medium (according to Carr and Johnson[1])

(reaction system: styrene 25, butadiene 75, formamide 180, sodium lauryl sulphate 5, dodecyl mercaptan 0·7, $K_2S_2O_8$ 0·5)

	Time of reaction (h)	*Conversion* (%)
	21·5	57·5
	24·0	63·0
	26·5	72·0
	27·6	77·5
Gel content of raw polymer (%)		64·0
Gel content after 10 passes through mill (%)		45·5
Intrinsic viscosity of polymer after 10 passes through mill (dl g^{-1})		1·6

1·5 years'. The data of Table 11.1, and other results given by Carr and Johnson, indicate that the rates of polymerisation in formamide systems are considerably lower than in corresponding aqueous systems. It also appears that mercaptan modifiers are much less efficient in formamide systems. Hydroquinone, which is readily soluble in formamide, can be used as a shortstopper.

Information is also given for the performance of tolyldiazothio-β-naphthyl ether as an initiator in formamide/sodium lauryl sulphate systems. Results for four formulations are given. The first contained monomers, formamide, soap and 0·2 pphm of initiator. The second contained increased initiator (1·0 pphm) and also trisodium phosphate in order to produce the more alkaline environment which has been found to be advantageous with this type of initiator in aqueous systems. The third contained, in addition to trisodium phosphate, potassium ferricyanide as an activator. In the fourth system, sufficient metallic sodium was added to raise the pH of the formamide from its initial value of 6·94 up to 10·38. Presumably hydrogen was evolved and the salt $H.CONH^-Na^+$ formed. The consequent high concentration of $H.CONH^-$ ions then undoubtedly suppressed the tendency for the remaining formamide to ionise into $H.CONH^-$ and H^+ ions; in consequence, the pH would rise. In none of the four cases was polymerisation observed over a period of 16 h at 40°C. A formulation using sodium oleate as micelle generator and a cumene hydroperoxide–iron(II) sulphate–sodium pyrophosphate–dextrose system as initiator was also found not to polymerise.

Concerning the use of liquid ammonia as a dispersion medium, Carr and Johnson report as follows: 'In the case of liquid ammonia the establishing of mechanical systems simulating aqueous emulsion recipes was more difficult because of the high degree of miscibility of monomers with the liquid medium. At 0°C. isoprene, butadiene, and butadiene–styrene (75/25) mixtures are miscible with ammonia. At −24°C. the butadiene–styrene combination is still miscible with ammonia. Isoprene and butadiene have limited solubilities in ammonia at −24°C. while isoprene (but not butadiene) has limited solubility at −18°C.' On the other hand, 'Other ingredients of typical aqueous recipes—*e.g.* sodium lauryl sulfate and other emulsifiers—and potassium persulfate had very low solubility in liquid ammonia.' Notwithstanding these difficulties, attempts were made to effect polymerisations at −28°C using liquid ammonia as the dispersion medium, but without success.

Attention is drawn to the existence of a patent,[2] published in 1937, which suggests the use of formamide, ethylene glycol and glycerol as dispersion media for the emulsion polymerisation of monomers such as chloroprene.

11.3 'INVERSE' EMULSION POLYMERISATION

As pointed out in the introduction to this chapter, the term 'inverse' emulsion polymerisation is used to imply a heterogeneous polymerisation system in which the monomer is readily soluble in water, but only sparingly soluble in non-polar liquids. Thus, in the typical inverse emulsion polymerisation, an aqueous solution of a hydrophilic monomer is dispersed in a continuous hydrophobic oil medium using a surface-active substance which promotes the formation of water-in-oil emulsions. The polymerisation is then initiated with either oil-soluble or water-soluble initiators. If an oil-soluble initiator is used, then formally at least, the system is an almost exact 'mirror-image' inverse of a conventional emulsion polymerisation system. (It is necessary to add the word 'almost', because conventional systems rarely contain an additive analogous to the water of the inverse system. This difference arises because hydrophilic monomers are usually solid at reaction temperature, whereas hydrophobic monomers are liquid, if not gaseous, in their normal state.) The radicals which initiate polymerisation are formed in the external phase, and, presumably, become subsequently entrapped in the polymerisation loci. However, to what extent micelles are present in such a system, and, if present, to what extent they solubilise monomer and so become potential polymerisation loci, are very open questions. As far as the end-product of the reaction is concerned, it seems clear that inverse latices are less stable than are their aqueous counterparts, and so tend to stratify or flocculate more readily. Left to stand without agitation, their particles settle out over a period of hours or days. In some cases, the separated particles can be re-dispersed by gentle agitation. Continuous, gentle agitation is said to maintain these latices as colloidal dispersion indefinitely.

An extensive review of the phenomenon of inverse emulsion polymerisation was published by Vanderhoff *et al.*[3] They list the following as typical hydrophilic monomers which are suitable as subjects for the reaction: sodium *p*-vinylbenzenesulphonate, sodium vinylbenzylsulphonate, 2-sulphoethyl acrylate, acrylic acid, acrylamide, vinylbenzenetrimethylammonium chloride and 2-aminoethyl methacrylate hydrochloride. As typical oil-soluble initiators, dibenzoyl and dilauroyl peroxides are cited. Sorbitol monostearate is said to be a water-in-oil emulsifier which is suitable for this application. The following is the description given of a preparative procedure: 'The emulsions were formed by dissolving the emulsifier in xylene and adding the aqueous monomer solution with stirring. Frequently the crude emulsions were homogenized (Cenco hand homogenizer) to decrease the average droplet size and increase the emulsion stability. The emulsions were heated with stirring at 40°C to 70°C to effect polymerisation. The time required for essentially complete conversion varied from a few minutes to several hours. The viscosities of the final latices were high—*e.g. ca.* 400 centipoises, as compared to *ca.* 10 centipoises for a conventional polystyrene latex of the same concentration.' By the term 'concentration' as used in this context is presumably meant the amount of polymer contained in unit volume of dispersion as a whole. It must be borne in mind that, unless steps are taken to remove the water, the dispersion will contain, in addition to polymer and dispersion medium, the water which was originally used to dissolve the monomer. In fact, the typical latex produced by this method consists of a dispersion in, say, xylene of submicroscopic spheres of hydrophilic polymer which are swollen with the water in which the monomer was initially dissolved. The particles can be as small as 300 Å. Vanderhoff and his co-authors comment that 'the two criteria for an emulsion polymerization system—*i.e.* a segregated free radical system and a number of loci for polymerization within a few orders of magnitude of the number of free radicals existent at a given time, appear to be met by these systems.'

Vanderhoff and his co-workers give the results of a comprehensive investigation into the inverse emulsion polymerisation of sodium *p*-vinylbenzenesulphonate using dibenzoyl peroxide as the initiator, *o*-xylene as the dispersion medium and sorbitol monostearate as the surface-active substance. Emulsions of aqueous monomer solution in xylene were prepared without initiator. They were homogenised, degassed by heating, and then charged into dilatometers to enable subsequent conversion–time relationships to be established by following the contraction of the system as polymerisation occurred. Dibenzoyl peroxide was added as a solution in xylene at the polymerisation temperature. Polymerisations were carried out at 50°C, 60°C and 70°C. After the polymerisation had reached a high conversion, the latex was cooled, shortstopped, and analysed for residual monomer by coulometric bromination. The polymer was recovered by coagulating the latices with absolute alcohol, washing with absolute alcohol and drying under vacuum at 35°C. Polymer molecular weights were determined viscometrically in 0·5 mole l^{-1} aqueous sodium chloride as solvent. The number of particles per unit volume of xylene were determined by electron microscopy of dried deposits obtained from latices

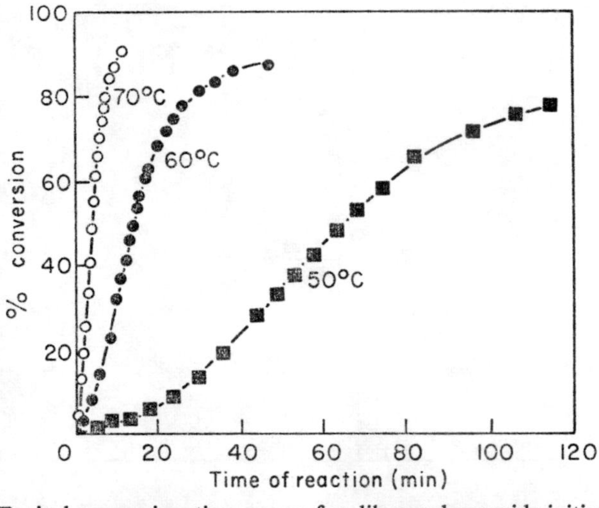

FIG. 11.1. Typical conversion–time curves for dibenzoyl peroxide-initiated 'inverse' emulsion polymerisation of sodium *p*-vinylbenzenesulphonate in *o*-xylene at 50°C, 60°C and 70°C, according to Vanderhoff *et al.*[3] (Reaction system: sodium *p*-vinylbenzenesulphonate 6, water 24, *o*-xylene 70, sorbitan monostearate 0·875, dibenzoyl peroxide 1·4.) Note that these curves have all been shifted to a common origin, Induction periods of up to 40 minutes were observed at 50°C; induction periods at 70°C were very short.

extensively diluted with xylene. After dilution, it was necessary to remove the water by distillation.

Fig. 11.1 shows typical conversion–time curves for the reaction at the above three temperatures. The resemblance between the shape of these curves and those for conventional aqueous systems is quite striking. The great rapidity of the reaction is also notable; at 70°C, conversion is almost complete in 10 min. The effect of emulsifier concentration upon overall rate of polymerisation is shown in Fig. 11.2. At 50°C, the overall rate appears to be independent of emulsifier level, whereas at 60° and 70°C the rate increases with emulsifier level. The effect of emulsifier level upon the number of particles present in the final latices is shown in Fig. 11.3. At low emulsifier levels, the number of particles increases

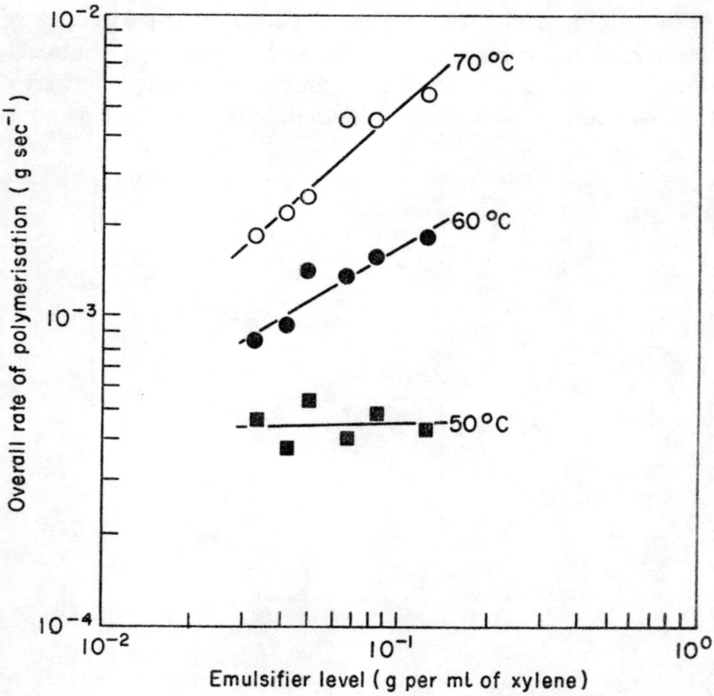

FIG. 11.2. Effect of emulsifier level upon overall rate of dibenzoyl peroxide-initiated 'inverse' emulsion polymerisation of sodium *p*-vinylbenzenesulphonate in *o*-xylene at 50°C, 60°C and 70°C, according to Vanderhoff *et al.*[3] (The reaction system was as given in the caption to Fig. 11.1, except that the level of sorbitan monostearate was varied.)

sharply with increasing emulsifier level, the initial slope of the double logarithmic plots suggesting that particle number is proportional to a power of the emulsifier concentration between 2·5 and 4·5. At higher emulsifier levels, the particle number levels off and appears to become essentially independent of emulsifier concentration. These results differ markedly from the prediction of the Smith–Ewart theory; on the basis of the latter, the double logarithmic plots of Fig. 11.3 should, of course, yield straight lines of slope 0·6. On the other hand, these observations do conform to a certain extent with the prediction given by the French theory (see Chapter 10, Section 10.4), that the number of particles formed should vary as the cube of the emulsifier concentration.

Fig. 11.3 also provides information concerning the effect of temperature upon number of particles. At high emulsifier levels, the particle number is independent of temperature. As the emulsifier level is reduced, so the particle number becomes increasingly sensitive to polymerisation temperature: as the temperature is increased, so the particle number decreases. Fig. 11.4 shows that a linear relationship exists between the logarithm of the rate of polymerisation per particle and the average volume of the particle. As expected, the rate increases with increasing volume and, for a given volume, with increasing temperature. But the dependence of rate upon particle volume tends to diminish as the temperature is increased. Data are given for the degrees of polymerisation of the reaction product obtained under various conditions of temperature and emulsifier level. The trend is for the degree of polymerisation to increase with increasing

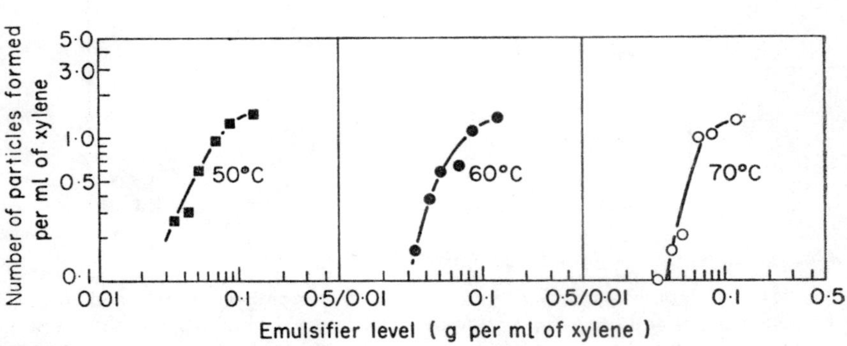

FIG. 11.3. Effect of emulsifier level upon the number of particles per unit volume of dispersion medium formed during the dibenzoyl peroxide-initiated 'inverse' emulsion polymerisation of sodium *p*-vinylbenzenesulphonate in *o*-xylene at 50°C, 60°C and 70°C, according to Vanderhoff *et al.*[3] (The reaction system was as given in the caption to Fig. 11.1, except that the level of sorbitan monostearate was varied.)

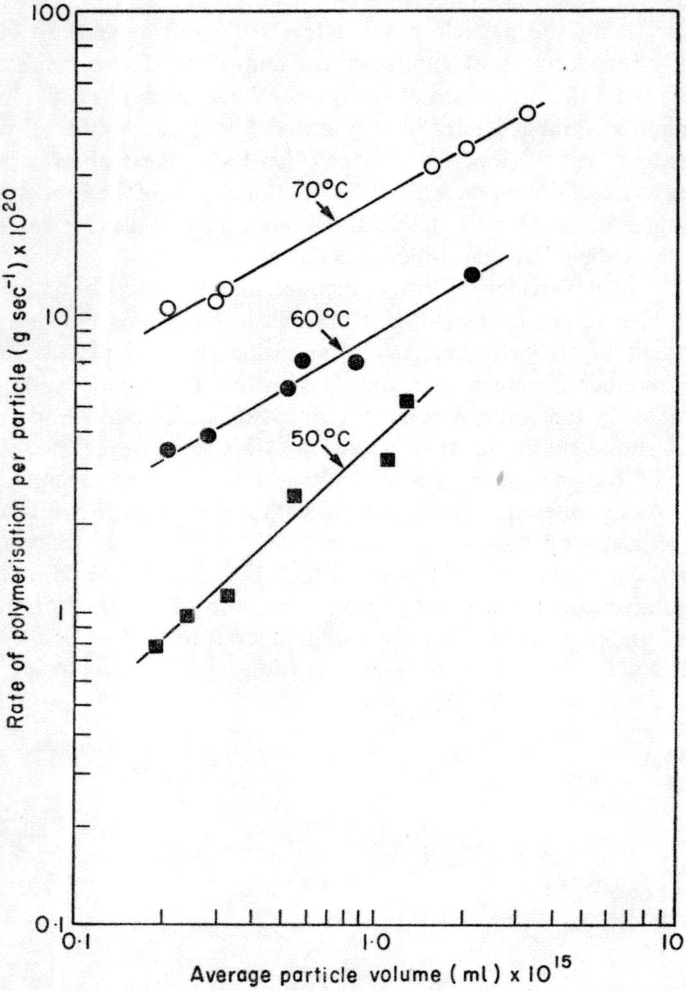

FIG. 11.4. Relationship between rate of polymerisation per particle and average particle volume for dibenzoyl peroxide-initiated 'inverse' emulsion polymerisation of sodium *p*-vinylbenzenesulphonate in *o*-xylene at 50°C, 60°C and 70°C, according to Vanderhoff *et al.*[3] (The reaction system was as given in the caption to Fig. 11.1, except that the level of sorbitan monostearate was varied.)

emulsifier level and to decrease with increasing temperature; but the published data show much scatter. Degrees of polymerisation obtained were of the order of 3×10^3.

Vanderhoff and his co-authors conclude that the unusual variations of particle number with both emulsifier level and temperature may indicate that the emulsified aqueous monomer droplets provide loci for the initiation of polymerisation in addition to those provided by the micelles derived from the emulsifier. That such micelles are undoubtedly present is indicated by the fact that, in all the systems investigated, the emulsifier was present in amounts exceeding that corresponding to the critical concentration for micelle formation. Experiments on the emulsification of aqueous sodium poly-*p*-vinylbenzenesulphonate in xylene, using sorbitol monostearate as emulsifier, indicated that emulsion droplets of a size comparable to that of the smallest particles found in the latices could be formed. Therefore the conclusion was reached that, although micelles containing solubilised monomer can provide loci for particle initiation, many of the emulsified droplets of monomer solution are small enough also to provide loci. Thus particle initiation probably occurs in both loci. No explanation is given for the fall in particle number with increasing temperature at low emulsifier concentration.

Vanderhoff and his co-authors have also analysed their kinetic data for this reaction in the light of the theory of the kinetics of conventional emulsion polymerisation as expounded in Chapter 4 of this book. They conclude that their results are consistent with the theory only if polymerisation is initiated by radicals which are generated in, or enter, the reaction loci in pairs. If an oil-soluble initiator, such as dibenzoyl peroxide, is used, this implies that the only radicals which are effective in initiating polymerisation are those which are formed by the thermal decomposition of peroxide molecules which have entered the loci intact and decompose there. The corollary is that most of the radicals formed by decomposition of the peroxide dissolved in the continuous hydrocarbon phase are ineffective in initiating polymerisation. Although Vanderhoff and his co-authors do not discuss this problem further, it is perhaps not unreasonable to suppose that most of the radicals of the type $\phi.COO\cdot$ which initially form in the hydrocarbon phase may have further decomposed to phenyl radicals and carbon dioxide before they have diffused into the reaction loci, and that phenyl radicals will tend to partition far less favourably in the aqueous reaction loci than either dibenzoyl peroxide or the benzoyloxy radicals produced by the initial scission of the peroxide link.

Results are also given for the inverse emulsion polymerisation of sodium

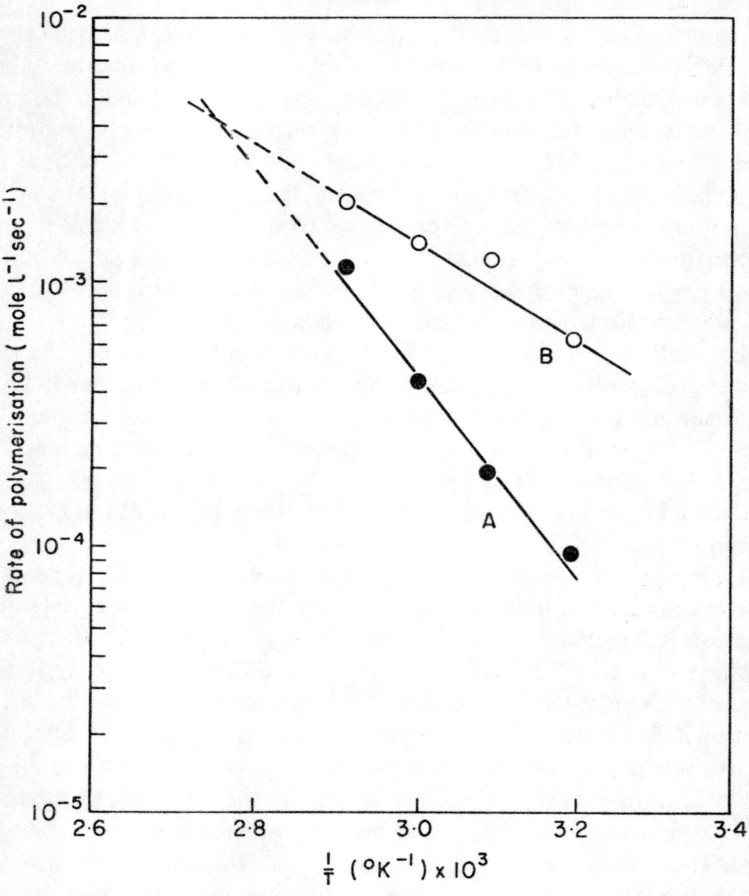

FIG. 11.5. Effect of temperature upon rate of persulphate-initiated aqueous solution polymerisation of sodium *p*-vinylbenzenesulphonate (curve A), and upon rate of persulphate-initiated 'inverse' emulsion polymerisation of sodium *p*-vinylbenzenesulphonate in *o*-xylene (curve B), according to Vanderhoff *et al.*[3] (The reaction system to which curve A refers comprised an aqueous solution containing 20% of sodium *p*-vinylbenzenesulphonate and 0·2% of $K_2S_2O_8$; that to which curve B refers was similar to that for the dibenzoyl peroxide-initiated 'inverse' emulsion polymerisation (see caption to Fig. 11.1), except that the dibenzoyl peroxide was replaced by 0·06 parts of $K_2S_2O_8$.)

p-vinylbenzenesulphonate in *o*-xylene initiated by potassium persulphate. At 50°C, the polymerisation in the inverse emulsion occurs considerably more rapidly than does the corresponding polymerisation in solution, although the conversion–time curves are similar in shape. However, the difference between rates in emulsion and in solution decreases with increasing temperature, as is shown in Fig. 11.5. Extrapolation indicated that the rates should become equal at about 90°C. The slopes of the lines in Fig. 11.5 indicate that the energy of activation for the solution reaction is greater than that for the emulsion reaction. This presumably indicates that it is not sufficient merely to regard the latter as a 'micro-solution' reaction.

11.4 DISPERSION POLYMERISATION IN ORGANIC MEDIA

The term 'dispersion polymerisation' is currently applied to processes in which a monomer is dissolved and polymerised in an organic liquid which is a solvent for the monomer but not a solvent for the polymer, an additional substance being present which enables the precipitated polymer to form a stable dispersion. It is convenient to refer to the liquid in which the monomer but not the polymer is soluble as the *diluent*. The term *dispersion medium* is then used to denote the mixtures of diluent and unpolymerised monomer not imbibed in reaction loci, in which also may be dissolved minor amounts of initiating and stabilising substances. It should be noted that the composition of the dispersion medium will change continuously throughout the reaction. In the initial stages, it comprises virtually the whole reaction system, whereas in the later stages of the reaction it will be almost entirely diluent.

The phenomenon of polymer precipitation during free-radical polymerisation has been known for many years. It inevitably occurs in the course of the bulk polymerisation of monomers such as vinyl chloride and acrylonitrile, for which the corresponding polymer is insoluble in the monomer. In such cases, homogeneous free-radical polymerisation is only possible if it can be carried out in the presence of a liquid which is a solvent for both monomer and polymer. In the case of monomers which are solvents for their own polymers, precipitation during polymerisation can be induced if the monomer is first dissolved in a liquid which is not a solvent for the polymer. An example which has been known for many years is that of styrene polymerising in the presence of sufficient methanol to cause precipitation of the polystyrene as it forms. It is a characteristic of such reactions that they tend to proceed considerably faster than do analogous homogeneous reactions in which the same rate of initiation is established. This effect is conventionally attributed to retardation of the termination reaction relative to propagation. Retardation of the termination reaction is in turn attributed to partial immobilisation of the growing polymer radicals as they precipitate from the reaction medium.

In effect, dispersion polymerisation is a special case of such precipitation polymerisation in which gross flocculation is prevented and the particle size of the resultant dispersion is controlled. As a consequence, not only is a useful polymer dispersion obtained, but also less variability in reaction rate is achieved. The diluents for such systems are typically aliphatic hydrocarbons. Ionic stabilisation associated with the presence of electric charges on the particles is not therefore available as a means of preventing flocculation. Instead, it is necessary to make use of the phenomenon of steric stabilisation.[4] In order for such a stabilisation mechanism to be effective, it is necessary for two conditions to be fulfilled. In the first place, the surface of the particle must have firmly anchored in it macromolecular substances such that one part of the adsorbed molecule has a strong tendency to mix with the molecules of the dispersion medium. But, secondly, the coverage of the particles must be reasonably complete. If these two conditions are fulfilled, the adsorbed stabiliser molecules can be envisaged as forming a mechanical barrier surrounding the particles and thus preventing them from approaching one another sufficiently closely for van der Waals-type attractive forces to become effective in promoting coalescence.

The concept of the mechanical barrier, however, gives a rather crude representation of the true situation. A more accurate explanation of the mechanism of stabilisation sees the effect as arising from an increase of free energy which accompanies the close approach of two particles. As two particles approach one another, so a stage is reached where the peripheral layers of the one tend to overlap those of the other. The effective concentration of adsorbed solute in the region between the particles therefore increases; a concomitant reduction in the chemical potential of the dispersion medium in the region between the particles inevitably occurs. Additional solvent then tends to flow into this space to reduce the solute concentration. The particles will therefore tend to be forced apart until the adsorbed layers no longer overlap. An alternative, but equivalent, way of regarding the situation is to see the overlapping of the adsorbed layers as analogous to the squeezing-out of solvent from a elastomer network which has been swollen to equilibrium with the solvent. If contact with excess solvent is maintained, there will be set up a strong osmotic tendency for the solvent to diffuse back into the network and so oppose the force which is tending to squeeze the liquid out.

Whereas homopolymers and random copolymers are often effective in providing steric stabilisation for inorganic particles dispersed in organic media, they are usually too weakly adsorbed on polymer surfaces to be effective in the application under consideration here. In a recent review of dispersion polymerisation in organic media, Barrett[5] states that the most effective stabilisers for this purpose are block or graft copolymers of amphipathic character. One component should tend readily to mix with both the initial dispersion medium and the diluent itself, and the other to be insoluble in both. The presence of the insoluble component ensures that the stabiliser is virtually irreversibly adsorbed on to the surface of the polymer particles. The attached soluble component then becomes a permanent lyophilic layer of polymer surrounding the particles and preventing flocculation. Barrett has discussed in some detail the structural requirements for block or graft copolymer to be used as stabiliser for this application. It appears that, whilst effective grafts can be formed during the polymerisation reaction itself if a suitable backbone precursor macromolecule is added to the system initially, it is preferable that the copolymer be formed separately from the main reaction. It further appears that the most effective stabilisers are 'comb-like' graft copolymers in which the 'spine' of the 'comb' is a polymer which is miscible with the precipitating polymer, and the 'teeth' are sequences of units which are miscible with the diluent liquid. This structure provides for maximum anchorage into the particle surface together with the presence of a high surface density of lyophilic chains penetrating the external phase. Barrett pictures such molecules as orientating themselves at the particle–dispersion medium interface as shown in Fig. 11.6.

A further disclosure by Waite[6] has revealed that it is also necessary for the 'anchor' moiety of the stabiliser (*e.g.* the 'spine' of the 'comb') to be flexible rather than rigid at the temperature at which the reaction is carried out. It appears that the reaction should be carried out at a temperature above the environmental glass-transition temperature of the anchor moiety if successful stabilisation is to be assured. Thus apparently anomalous situations have been encountered in which polymerisations can be carried out successfully at a higher temperature but fail at a lower temperature. The addition of a solvent plasticiser to the dispersion medium can also be helpful in promoting stabilisation. It appears that strong

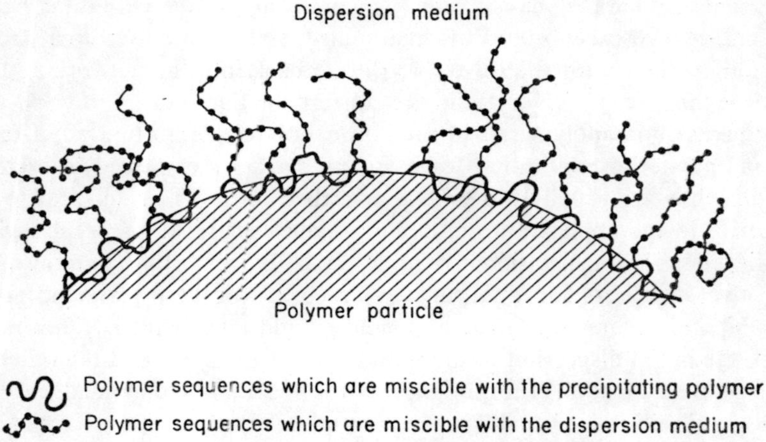

FIG. 11.6. Orientation of 'comb-like' graft copolymer at interface between particle and dispersion medium, as envisaged by Barrett.[5]

adsorption on to the growing particles is encouraged by segmental mobility of the units present in the anchor moiety. A possible reason is that, if the moiety is in a glassy condition, it will tend to be irreversibly entangled with other moieties, and so unable to disperse, be adsorbed, and effect stabilisation.

Thus dispersion polymerisation in organic media is usually effected by polymerising a monomer in an organic liquid in which it is soluble but the polymer is not, there being present a suitable preformed block or graft copolymer of amphipathic character which acts as a stabiliser. By such a process, it is possible to produce dispersions which contain at least 50% of polymer and yet are stable, free-flowing substances of the low viscosity. Control of particle size is possible, and dispersions of uniform particle size can be produced. Particle sizes are typically within the range 0·1 to 1 μm.

Regarding the actual process of dispersion polymerisation, Barrett[5] gives the following general description: 'In the simplest form of dispersion polymerisation, a monomer such as methyl methacrylate, together with a dispersant and an initiator, is dissolved in petrol and heated to reflux with stirring. Initially, the reaction mixture is clear and transparent but after a few seconds, a faint opalescence appears which rapidly whitens to an opaque latex. Polymerisation continues without much further visible change.... A typical dispersion polymerisation begins slowly, gradually accelerates to a maximum rate followed by a diminution in rate as monomer is consumed. Normally, conversions of up to 99% are attained in an hour or so. Polymers of high molecular weight are readily obtained and, for example, poly(methyl methacrylate) dispersions have been prepared with molecular weights of up to several million. When required, polymers of lower molecular weights can be prepared by the use of chain transfer agents, such as mercaptans.' The level of stabiliser to be used is determined by the particle size which is required in the final dispersion. Stabiliser levels are typically a few per cent on the amount of monomer in the system.

Barrett further states that dispersion polymerisation is usually a rapid process, and that the low viscosity and small particle size of the product permit rapid heat transfer, especially under reflux conditions. For many monomers, dispersion polymerisation is an inherently much faster process than the corresponding solution polymerisation under comparable conditions. The reason is essentially the same as for the increase in rate of polymerisation which is usually observed whenever a polymer is precipitated as it forms, namely, retardation of termination relative to propagation.

Whilst apparently analogous in many respects to a conventional emulsion polymerisation reaction, dispersion polymerisations undoubtedly display several important differences. The principal features of similarity are that the propagation step occurs mainly in isolated loci under conditions such that the ratio of polymer to monomer is high and the viscosity is high. Furthermore, the initiating radicals are produced externally to the main reaction loci. The principal differences are that the bulk monomer droplet phase is replaced by monomer dissolved in the dispersion medium, and that each reaction locus can simultaneously contain several propagating radicals. This latter feature arises because of the restricted rate of mutual termination caused by the high viscosity of the medium in which reaction is occurring, together with the fact that the radicals which enter the loci from the external phase are also probably oligomeric in nature. And, furthermore, it is precisely because several macromolecular radicals can coexist in each locus that the suppression of termination relative to propagation becomes kinetically significant. The situation is clearly one in which the phenomenon of restricted termination leads to a condition where the suppression of termination can have increasing significance in determining the rate of polymerisation. This is, of course, the generally accepted explanation for the very marked increases in rate of polymerisation which are observed to accompany the classical Trommsdorff 'gel' effect; it also seems to give a plausible explanation for the very high rates of polymerisation which are observed in dispersion systems.

A further interesting feature of dispersion polymerisation is that the stage of initial polymer particle formation is usually completed very rapidly, often within a few minutes. The conversion at the completion of this stage is low, being usually well under 1%. The stabilised particle nuclei produced are probably less than 0·01 μm in diameter. Thereafter, the number of particles remains almost constant unless the solvent characteristics of the dispersion medium are drastically altered or substantial additional quantities of the polymeric stabiliser are added as the reaction proceeds. After a brief initial stage of particle formation, further polymerisation occurs essentially by the continued growth of existing particles. According to Barrett,[5] 'The size and number of the polymer particles formed initially are dependent on two main factors. The greater the solvency of the diluent phase towards the polymer being produced, the fewer and larger are the particles produced since the polymer nuclei will reach a larger size before precipitation. The initial concentration of the graft dispersant also controls nucleation in that the number of particles produced increases (and their size decreases) with the use of increased amounts of dispersant. In general, the total surface area of polymer particles produced is directly proportional to the amount of dispersant present.'

The reason for the reluctance to form new particles once the brief initial stage of particle nucleation is completed seems to be connected with the tendency for oligomeric radicals to precipitate from the dispersion medium once they have acquired a few monomer units. The tendency to form new particles is suppressed because such oligomeric radicals will tend to precipitate on to existing particles rather than aggregating to form new ones. A further consequence is a considerable decrease in the effective radical concentration in the dispersion medium. Indeed, from the kinetic point of view, so rapid is the ingress of radical centres into the particles that one can apparently regard initiation as occurring almost exclusively within the particle phase, even though the typical initiators used are partitioned between the dispersion medium and the particles.

An extensive study of the kinetics of the free-radical-initiated polymerisation of methyl methacrylate in *n*-dodecane to produce stable dispersions has been published by Barrett and Thomas.[7] Azobis*iso*butyronitrile was used as initiator. The amphipathic copolymer used as a dispersion stabiliser for the polymer particles was a 'comb-like' graft copolymer of 12-hydroxystearic acid and methyl methacrylate. 12-hydroxystearic acid was condensed with itself by heating in the presence of stearyl alcohol and *p*-toluenesulphonic acid in toluene at 180–190°C until the acid value was less than 1 mg of KOH per gram. The stearyl alcohol functioned as a monofunctional chain terminator, its amount being that required to produce the desired average degree of polymerisation of the product. The overall condensation reaction can be represented as follows:

$$nC_6H_{13}CH(OH)(CH_2)_{10}CO_2H + C_{18}H_{37}OH \longrightarrow HO\left[\begin{array}{c} C_6H_{13} \\ | \\ CH(CH_2)_{10}CO\cdot O \end{array}\right]_n C_{18}H_{37} + nH_2O$$

For this application, the average value of n was approximately 5. This oligomer was then reacted with methacrylic anhydride to give the 'macromonomer' of structure (I) below, which was then copolymerised with an equal weight of methyl methacrylate to give the polymer of structure (II).

$$\begin{array}{c} CH_3 \\ | \\ CH_2{:}C.CO.O \end{array}\left[\begin{array}{c} C_6H_{13} \\ | \\ CH(CH_2)_{10}CO.O \end{array}\right]_n C_{18}H_{37}$$

(I)

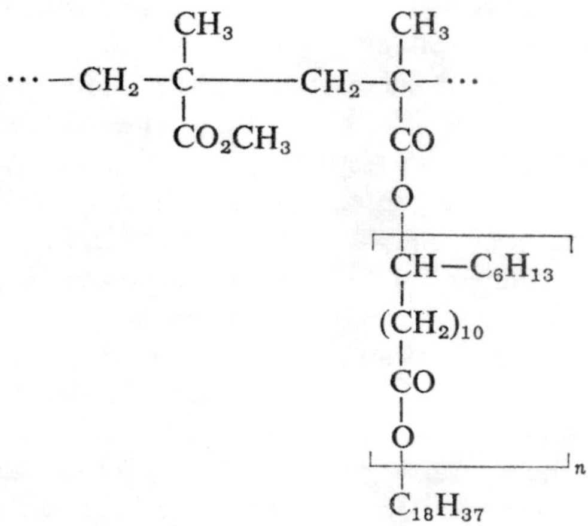

(II)

This polymerisation was carried out in butyl acetate as solvent, using azobis*iso*butyronitrile as initiator. The resulting polymer was used as a 50% solution in n-butyl acetate. The polymethyl methacrylate 'backbone' is insoluble in the dispersion medium for the same reason that the polymethyl methacrylate homopolymer produced by the subsequent dispersion polymerisation is insoluble. Since the polymethyl methacrylate 'backbone' shows a strong tendency to mix with the polymethyl methacrylate produced in the reaction, and the oligomeric poly-12-hydroxystearic acid grafts tend to mix with the dispersion medium, the conditions for steric stabilisation are adequately fulfilled. It will be observed that, if (II), above truly represents the structure of the product the reactions used to form the graft copolymer, then the requisite 'comb-like' structure will have been achieved, with the correct relationship between the 'spine' and the 'teeth'.

Rates of polymerisation were determined by following the evolution of the heat of polymerisation as reaction proceeded. The technique of differential scanning calorimetry was used. Its application to the problem of following polymerisation reactions has been described in another paper by Barrett.[8] Measurements were made under isothermal conditions. The polymerising system was contained in a sealed sample holder in order to prevent loss of methyl methacrylate by volatilisation. The published conversion–time curves show a strong resemblance to those for, say, a conventional persulphate-initiated styrene–butadiene emulsion copolymerisation using an aqueous dispersion medium. The curves are sigmoidal in shape. After an initial period of acceleration, a region of fast reaction is reached, which in turn is followed by strong retardation after about 85% conversion. The kinetics of the reaction in the fast stage are described as 'pseudo-first-order', by which is presumably meant that the reaction appears to be first-order with respect to remaining unreacted monomer. Accordingly, many of the subsequent observations are summarised in terms of calculated 'pseudo-first-order' rate coefficients.

Fig. 11.7 gives comparisons between polymerisations carried out in solution, in dispersion in the absence of an added stabiliser, and in dispersion in the presence of an added stabiliser. In all these cases, the reaction conditions were similar. It is apparent that both the dispersion polymerisations proceed much more rapidly than did the solution polymerisation, and the conversion–time curves differ in shape from that for the solution polymerisation. The curves for the two dispersion polymerisations are of the same general shape, but the reaction in the absence of stabiliser proceeds markedly more slowly. Barrett and Thomas conclude that the same basic mechanism probably operates in both cases, but the diffusion of monomer into the polymer aggregates can exert a rate-controlling effect, especially if the latter are large, as they are in the absence of stabiliser. Barrett and Thomas also give molecular weight data which suggest that the degree of polymerisation of the product of

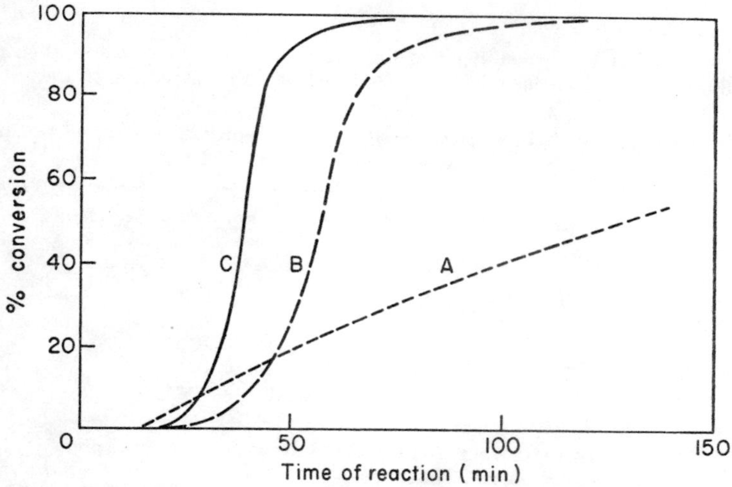

FIG. 11.7. Conversion–time curves for free-radical polymerisation of methyl methacrylate at 80°C in solution in benzene (curve A), in *n*-dodecane without added stabiliser (curve B), and in *n*-dodecane with added stabiliser (curve C), according to Barrett and Thomas.[7] (In each case, the monomer and solvent were present in equal amounts, and the initiator was 0·1% of azobis*iso*butyronitrile. It is not entirely clear whether the 0·1% refers to the monomer only or to monomer + solvent; it is presumed that the latter is the case. The amount of stabiliser in the reaction system to which curve C refers is not specified; chemically, it was the 'comb-like' graft copolymer of 12-hydroxystearic acid and methyl methacrylate described in the text.)

dispersion polymerisation is approximately three times that of a polymer produced by solution polymerisation under comparable conditions. That the former product should have a considerably higher degree of polymerisation than the latter is entirely in accordance with expectation, if the high rate of dispersion polymerisation is due to the operation of a Trommsdorff 'gel' effect.

The effect of stabiliser level upon the apparent first-order rate coefficient for the fastst age of the reaction is shown in Table 11.2. The accelerating effect of stabiliser is clearly evident from these results, as also is the fact that, at levels of about 1 pphm and above, the rate is little dependent upon the amount of stabiliser. Barrett and Thomas conclude from this that a mechanism of the Harkins type is not applicable to this system, since such a mechanism requires that the rate of polymerisation should increase with increasing number of reaction loci, and therefore with increasing amount of surface-active additive (although, of course, the precise way in which the rate varies with amount of additive is not in

TABLE 11.2

Effect of stabiliser level upon apparent first-order rate coefficient for dispersion polymerisation of methyl methacrylate in *n*-dodecane at 80°C (according to Barrett and Thomas[7])

(reaction system:† methyl methacrylate 100, *n*-dodecane 100 − *x*, butyl acetate *x*, stabiliser *x*, azobis*iso*butyronitrile 2)

Stabiliser level (pphm) (x)	*Apparent first-order rate coefficient* sec^{-1}
0	$6{\cdot}3 \times 10^{-3}$
0·5	$7{\cdot}6 \times 10^{-3}$
1·0	$11{\cdot}6 \times 10^{-3}$
2·5	$10{\cdot}8 \times 10^{-3}$
5·0	$10{\cdot}2 \times 10^{-3}$

† There is some doubt as to the precise composition of the various reaction systems used. Thus:

(i) The stabiliser is said to have been used as a 50% solution solution in butyl acetate. The above levels are assumed to refer to dry polymer, and not to solution.

(ii) In consequence of (i), the butyl acetate level apparently varied from system to system, being equal in each case to the stabiliser level.

(iii) In consequence of (ii), the dodecane level presumably varied from system to system, because the initial monomer concentration was apparently maintained at 50% of the total reaction system.

(iv) The initiator level is presumed to have been 2 pphm. It is stated to have been 1%, but whether on the reaction system as a whole or on the monomer is not clear. The former seems more likely.

itself an essential feature of the Harkins theory). A further conclusion drawn by Barrett and Thomas from the lack of dependence of rate upon stabiliser level is that diffusion of monomer into the interior of the particles is probably not a rate-determining step, as long as sufficient stabiliser is present. Granted the truth of this supposition, it is possible to give an explanation for the lack of dependence of the rate coefficient for the fast reaction upon the stabiliser level, provided the assumption is made that sufficient stabiliser is present to ensure the formation of stable particles. If the rate of initiation in the system as a whole is essentially the rate of generation of propagating radicals in the external phase, then the number of propagating centres will clearly be independent of the number of particles in which propagation is occurring. Addition of further stabiliser

TABLE 11.3

Effect of initiator concentration upon apparent first-order rate coefficient for dispersion polymerisation of methyl methacrylate in *n*-dodecane at 80°C (according to Barrett and Thomas[7])

(reaction system:† methyl methacrylate 100, *n*-dodecane 90, butyl acetate 5, stabiliser 5, azobis*iso*butyronitrile—variable)

Initiator level (pphm) ‡	*Apparent first-order rate coefficient* (sec^{-1})
0·10	$1{\cdot}5 \times 10^{-3}$
0·20	$4{\cdot}4 \times 10^{-3}$
0·40	$4{\cdot}9 \times 10^{-3}$
0·60	$6{\cdot}5 \times 10^{-3}$
1·00	$8{\cdot}1 \times 10^{-3}$
2·00	$9{\cdot}8 \times 10^{-3}$

† See footnote to Table 11.2.
‡ In the original paper, the initiator levels are merely given as '%', but it is not clear whether this refers to the reaction system as a whole or to the monomer component. The former seems more likely, and the above levels are pphm on this assumption.

will merely lead to the creation of more particles, each of which contains, on average, fewer radicals, and each of which will therefore grow correspondingly more slowly.

Confirmation of the essential correctness of the interpretation of the reaction kinetics in terms of the classical Trommsdorff 'gel' effect is provided by the results of an analysis of the rate data in the light of a theory, an outline of which is given below. Barrett and Thomas find certain interesting parallels between the results of their analysis and those of Schulz[9] for bulk polymerisation. The conclusion is therefore reached that similar reaction mechanisms are operative in both cases. Such kinetic analysis also leads to an estimate for the partition coefficient of monomer between dispersion medium and polymer particles. The estimate of approximately unity accords reasonably well with the value of 0·8 obtained by direct measurement.

The dependence of apparent first-order rate coefficient upon initiator concentration is shown in Table 11·3. Barrett and Thomas state that 'A square-root relationship fits the data approximately, though if the lowest result is excluded an initiator exponent of 0·4 gives a much closer fit.' Table 11.4 shows the effect of initial monomer concentration upon the apparent first-order rate coefficient. It is observed that changes in initial

TABLE 11.4

Effect of initial monomer concentration upon apparent first-order rate coefficient for dispersion polymerisation of methyl methacrylate in *n*-dodecane at 80°C (according to Barrett and Thomas[7])

(reaction system:† methyl methacrylate *x*, *n*-dodecane 190 − *x*, butyl acetate 5, stabiliser 5, azobis*iso*butyronitrile 1)

Initial monomer level (*x*) ‡	*Apparent first-order rate coefficient* (sec^{-1})
40	$0{\cdot}4 \times 10^{-3}$
60	$2{\cdot}5 \times 10^{-3}$
80	$5{\cdot}4 \times 10^{-3}$
100	$3{\cdot}5 \times 10^{-3}$

† See footnote to Table 11.2.
‡ The various reaction systems are presumed to have had the compositions given by substituting these values of *x* in the generalised composition given above.

monomer content produce changes not only in overall rate, but also in the form of the conversion–time curve. As the initial monomer content is reduced, so the conversion at which the maximum rate is observed is reduced and the initial period of acceleration is progressively shortened.

The addition of solvents for the polymer slows down the polymerisation and markedly distorts the shape of the conversion–time curve. The reduction in rate is entirely in accordance with expectation. In effect, the addition of increasing amount of a polymer solvent or plasticiser effects a gradual transition from a dispersion to a solution polymerisation. The viscosity of the reaction loci is reduced by such additions, and the restrictions upon mutual termination are progressively removed. The conversion–time curves of Fig. 11.8(a) show the effect of gradually replacing the monomer by a polymer solvent, butyl acetate. The retarding effect is far greater than can be accounted for by mere dilution of the monomer. The curves of Fig. 11.8(b) show the effect of gradually replacing the *n*-dodecane diluent with butyl acetate, the monomer concentration remaining constant. These latter curves show clearly the retarding effect of the solvent. That this effect is not peculiar to butyl acetate is indicated by the fact that similar results have been observed with benzene as the solvent.

The effect of reaction temperature upon the apparent first-order rate coefficient is shown in Table 11.5. Barrett and Thomas state that these values correspond to an activation energy of 11·4 kcal $mole^{-1}$, and that

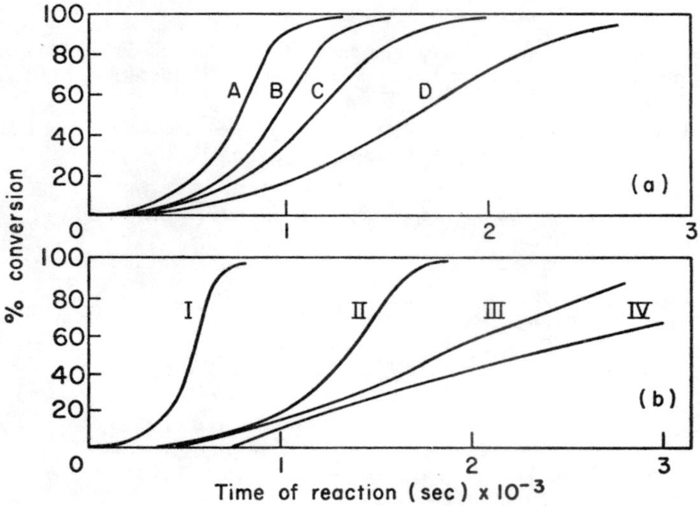

FIG. 11.8. (a) Effect upon conversion–time curve, for dispersion polymerisation of methyl methacrylate in *n*-dodecane at 80°C, of progressively replacing monomer by butyl acetate, according to Barnett and Thomas.[7] Curves A, B, C and D refer respectively to reaction systems containing increasing amounts of butyl acetate. (Reaction systems believed to be as follows: methyl methacrylate 100 − x, butyl acetate x + 5. *n*-dodecane 95, stabiliser 5, azobis*iso*butyronitrile 1, where x was respectively 0, 20, 40 and 60 for systems A, B, C and D.) (b) Effect upon conversion–time curve, for dispersion polymerisation of methyl methacrylate in *n*-dodecane at 80°C, of progressively replacing the *n*-dodecane diluent by butyl acetate, according to Barrett and Thomas.[7] Curves I, II, III and IV refer respectively to reaction systems containing increasing amounts of butyl acetate. (Reaction systems believed to be as follows: methyl methacrylate 100, *n*-dodecane 95 − x, butyl acetate x + 5, stabiliser 5, azobis*iso*butyronitrile 2, where x was respectively 0, 20, 40 and 60 for systems I, II, III and IV.)

this is some 8 kcal mole^{-1} lower than the activation energy estimated from published values for the decomposition of the initiator, the propagation stage of methyl methacrylate polymerisation and the termination stage of methyl methacrylate polymerisation. The explanation given is that the reaction rate increases less sharply with rising temperature because the viscosity of the reaction medium falls with rising temperature, and so termination is less retarded at higher temperatures than at lower temperatures, other factors remaining unchanged. In effect, Barrett and Thomas are postulating that the apparent activation energy for termination is considerably greater than that for a conventional low-conversion reaction, because of the effect of temperature upon the viscosity of the

TABLE 11.5

Effect of reaction temperature upon apparent first-order rate coefficient for dispersion polymerisation of methyl methacrylate in *n*-dodecane at 80°C (according to Barrett and Thomas[7])

(reaction system:† methyl methacrylate 100, *n*-dodecane 100 − *x*, butyl acetate *x*, stabiliser *x*(?),‡ azobis*iso*butyronitrile 2§)

Temperature (°C)	*Apparent first-order rate coefficient* (sec^{-1})
70	$4{\cdot}7 \times 10^{-3}$
80	$8{\cdot}9 \times 10^{-3}$
90	$11{\cdot}8 \times 10^{-3}$

† See footnote to Table 11.2.
‡ The stabiliser level is not given in the original paper. By comparison with the results given in Table 11.2, *x* appears to have been about 0·5.
§ See footnote ‡ to Table 11.3.

medium in which reaction is occurring. In confirmation of this view, they find that the 'deficit' of about 8 kcal $mole^{-1}$ for the reaction as a whole corresponds very closely to half the activation energy for polymer diffusion extrapolated to zero conversion. (The significance of the factor of one-half is that the rate of free-radical polymerisation is usually proportional to the inverse square root of the rate coefficient for termination.)

It is of interest to examine the approximate kinetic analysis which Barrett and Thomas used for the interpretation of some of their results for the fast stage of the reaction. Let X denote the rate of formation of propagating radicals expressed in moles per unit time per unit volume of dispersion as a whole, and X' the same rate expressed in terms of unit volume of dispersed monomer-swollen particles. Then the fact that at any instant the product of X and the dispersion volume must equal that of X' and the volume of particles at that instant (since all propagating radicals generated in the dispersion medium during the fast stage of the reaction are presumed to become quickly incorporated in particles) leads immediately to the conclusion that $X' = X/\phi$, where ϕ is the volume fraction of particles in the dispersion at that instant. If $[M]$ denotes the monomer concentration in the particles, then the application of conventional free-radical polymerisation kinetics gives

$$-\frac{d[M]}{dt} = k_p[M]\sqrt{\left(\frac{X'}{k_t}\right)} = k_p[M]\sqrt{\left(\frac{X}{\phi k_t}\right)}$$

where k_p and k_t are the rate coefficients for propagation and termination, respectively. It should be noted that the appropriate values for k_p and k_t are *those for the particular medium in which the reaction is occurring*. These will certainly not be the conventional values for low-conversion polymerisations, and the discrepancy will be especially large in the case of k_t. $-\mathrm{d}[M]/\mathrm{d}t$ is, of course, the rate of polymerisation expressed in moles of monomer polymerised per unit time per unit volume of particle phase. If this quantity be denoted by P', and P used to indicate the corresponding quantity expressed in terms of the dispersion as a whole, then an argument analogous to that given above for rates of initiation shows that $P' = P/\phi$. Thus it follows that

$$P = \phi P' = k_p[M]\sqrt{\left(\frac{\phi X}{k_t}\right)}$$

Putting $[M] = \alpha c$, where c is the concentration of monomer in the dispersion medium and α is the partition coefficient for monomer between particles and external phase, gives

$$P = \alpha k_p c\sqrt{\left(\frac{\phi X}{k_t}\right)}$$

α is strictly the partition coefficient at the instant to which this equation refers; but, as a good approximation, it may be taken as independent of monomer concentration.

A further development is effected by letting y denote the fractional conversion of monomer at any instant, and c_0 the initial (molar) concentration of monomer in the dispersion medium. Consider an initial unit volume of the medium. It contained c_0 moles of monomer, and all monomer initially present was dissolved in the dispersion medium (there being no other phase present). After polymerisation to the extent y, the initial unit volume of reaction medium has become approximately unit volume of dispersion (ignoring any volume changes accompanying polymerisation or mixing). The total amount of monomer remaining in this unit volume is $c_0(1 - y)$, and the concentration in the particle phase can be written as $[M] = \alpha c_0(1 - y)$. Furthermore, the volume fraction of the particle phase is set equal to $yc_0\overline{V}$, where $\overline{V}$ is the molar volume of a single monomer unit in the polymer, and P is set equal to $c_0\,\mathrm{d}y/\mathrm{d}t$. These substitutions in the above expression for P give

$$c_0\frac{\mathrm{d}y}{\mathrm{d}t} = k_p\alpha c_0(1 - y)\sqrt{\left(\frac{yc_0\overline{V}X}{k_t}\right)}$$

i.e.

$$\frac{\mathrm{d}y}{\mathrm{d}t} = \alpha k_p(1 - y)\sqrt{\left(\frac{yc_0\bar{V}X}{k_t}\right)}$$

It is desirable to examine the substitutions of the preceding paragraph more closely in order to note the nature of the approximations which they assume. The setting of P to equal to $c_0\, \mathrm{d}y/\mathrm{d}t$ seems to involve no approximation at all, other than that the reaction mixture undergoes no overall change in volume as a consequence of polymerisation. For consider an arbitrary volume of dispersion, V. Initially this contained c_0V moles of monomer. The number of moles of monomer polymerised up to an instant of time t after the start of the reaction can be expressed in the form $\int_0^t PV\,\mathrm{d}t$. Thus y, which is the ratio of the number of moles of monomer polymerised to the number present initially, can be expressed in the form

$$y = \frac{\int_0^t PV\,\mathrm{d}t}{c_0V} = \frac{1}{c_0}\int_0^t P\,\mathrm{d}t$$

Thus $\mathrm{d}y/\mathrm{d}t = P/c_0$, where P is of course the rate of polymerisation at the same instant as that to which $\mathrm{d}y/\mathrm{d}t$ refers.

Consider, next, the substitution $\phi = yc_0\bar{V}$. This seems to imply the two-fold assumption that all the polymer formed at any instant is precipitated into the particles, and that the particles do not imbibe monomer and therefore swell in size. (In fact, Barrett and Thomas do not state that they have made these approximations in their discussion of this approximate theory. However, it becomes clear that they have done so during their subsequent presentation of a more exact theory.) For again, considering an arbitrary volume of dispersion V, this initially contained c_0V moles of polymer, and so must contain yc_0V moles of polymerised monomer after time t. The volume of the latter will be $yc_0V\bar{V}$. (Note that, if it is assumed that no change of volume occurs as a result of polymerisation, then $\bar{V}$ is the molar volume of the monomer as well as that of the monomer unit within the polymer.) Thus ϕ, which is then the ratio of volume of polymer to total volume of dispersion, becomes equal to $yc_0V\bar{V}/V$, *i.e.* to $yc_0\bar{V}$.

The third substitution to be considered is $[M] = \alpha c_0(1 - y)$. This seems to be a reasonable approximation if, in addition to the assumption of no volume change during polymerisation, it is also assumed that α is near unity. (Of course, the above derivation of $\phi = yc_0\bar{V}$ assumed, in in effect, that $\alpha = 0$. But the discrepancy likely to be incurred by this inconsistency is probably not serious.) Consider again an arbitrary volume

V. The number of moles of monomer remaining when the fractional extent of reaction is y is obviously $c_0V(1 - y)$. The average concentration of monomer throughout the dispersion is therefore $c_0(1 - y)$. If α is near unity, the concentration throughout the dispersion will be almost uniform, and thus the concentration in the dispersion medium can be taken as $c_0(1 - y)$. This being so, the concentration in the particles is $\alpha c_0(1 - y)$. Of course, an exactly similar argument could be used to show that the concentration of monomer in the dispersion medium is $c_0(1 - y)/\alpha$. Both results are approximations which become progressively better as α approaches unity; when α is exactly unity, the two results coincide.

Further development is possibly by (a) putting $X = 2fk_dc_i$, where k_d is the first-order rate coefficient for the thermal decomposition of the initiator, c_0 is the initiator concentration, and f is the initiation efficiency of the radicals produced by such decomposition; and (b) putting $c_i = (c_i)_0e^{-k_dt}$, in order to allow for the effect of initiator decomposition upon its concentration at any instant. It is unlikely that the latter substitution will make much difference to the kinetics of any practical dispersion polymerisation system, since polymerisation occurs much more rapidly than does initiator decomposition. Making the first substitution, but not the second, and rearranging gives

$$\frac{\mathrm{d}y/\mathrm{d}t}{(1 - y)\sqrt{(yc_0\bar{V}k_dc_i)}} = \alpha k_p\sqrt{\left(\frac{2f}{k_t}\right)}$$

If a value of k_d is assumed for the decomposition of the initiator, then it is possible to calculate the expression on the left-hand side of this equation for all stages of a given reaction. If the above analysis is correct, it follows that $\alpha k_p\sqrt{(2f/k_t)}$ can be found for all stages of the reaction. Barrett and Thomas show that it is possible to draw several inferences from the variation of this quantity with conversion and with reaction conditions. In their paper, Barrett and Thomas also give a more exact analysis in which they allow for the change of volume which accompanies polymerisation, the increase of the volume of the particles due to swelling by the monomer, and also the effect of a corrected partition relationship. The original paper should be consulted for a full discussion of this more exact analysis.

Attention is also drawn to a recent paper by Bueche[10] which describes what is apparently a dispersion polymerisation. In this process, a monomer is polymerised in a liquid, such as heptane, in which the polymer is insoluble. Polymerisation is effected purely thermally by heating under nitrogen without added initiator. A soluble rubber, such as poly*iso*butylene or butyl rubber, is present to ensure the formation of a stable dispersion. Bueche uses the name 'poloid' to denote the product of such a reaction. The diluent he refers to simply as the 'nonsolvent', and the stabilising polymer as the 'shield'.

Although Bueche seems to ascribe stabilisation simply to the presence of unattached soluble polymer in the dispersion medium, Barrett[5] suggests that a graft copolymer may form during the course of the reaction through, for example, the formation of peroxides on the rubber molecules, and that this graft copolymer contributes to the stability of the dispersion. Bueche recognises this possibility, but appears to rule it out for some at least of his systems. He also recognises the possibility of adsorption by entanglement. 'In the poloid system', he says, 'the shield polymer is a homopolymer and is insoluble in the particle phase. Hence the situation is quite different from what it is for an amphipathic stabilizer. It is possible that entanglements of the shield chains with chains in the particles causes them to be bound as a layer at the surface. Or perhaps enough chain grafting occurs during the polymerization to effectively form a block copolymer.' However, it is certainly known that, in early experiments on dispersion polymerisation using natural rubber as the added stabiliser, additional colloid stability was conferred by the graft copolymer formed simultaneously with homopolymer by hydrogen abstraction.

A brief review of the characteristics of polymer dispersions in non-aqueous media has been given recently by Fitch.[11] In this paper, he makes reference to the mechanism of the dispersion polymerisation process. In particular, he gives a summary of his own theory of the mechanism of particles nucleation. This theory is expounded in much greater detail in two papers by Fitch and Tsai.[12,13] Although the experimental investigations in connection with this theory were carried out using non-micellar, aqueous reaction systems, it is claimed that the theory is applicable to particle-initiation reactions in organic media. Further reference to the Fitch theory of particle formation will be found in Section 12.2 of Chapter 12.

REFERENCES

1. E. L. Carr and P. H. Johnson, *Ind. Engng Chem.*, 1949, **41**, 1588.
2. W. H. Carothers, U.S. Patent 2,080,558 (1937).
3. J. W. Vanderhoff, E. B. Bradford, H. L. Tarkowski, J. B. Shaffer and R. M. Wiley, 'Inverse emulsion polymerization', in *Polymerization and Polycondensation*

Processes, Advances in Chemistry Series, American Chemical Society, U.S.A., 1962, No. 34, pp. 32–51.
4. D. H. Napper, *Ind. Engng Chem. Prod. Res. Dev.*, 1970, **9**, 467.
5. K. E. J. Barrett, *Br. Polym. J.*, 1973, **5**, 259.
6. F. A. Waite, paper presented to meeting of the Colloid and Interface Science Group, Faraday Division of the Chemical Society, held at Bristol, June 1974.
7. K. E. J. Barrett and H. R. Thomas, *J. Polym. Sci.*, Pt A-1, 1969, **7**, 2621.
8. K. E. J. Barrett, *J. appl. Polym. Sci.*, 1967, **11**, 1617.
9. G. V. Schulz, *Z. phys. Chem. Frankf. Ausg.*, 1956, **8**, 290.
10. F. Bueche, *J. Colloid and Interface Sci.*, 1972, **41**, 374.
11. R. M. Fitch, *J. Elastoplastics*, 1971, **3**, 146.
12. R. M. Fitch and C. H. Tsai, 'Particle formation in polymer colloids, III: Prediction of the number of particles by a homogeneous nucleation theory', in *Polymer Colloids*, (ed. R. M. Fitch), Plenum Press, New York, 1971, pp. 73–102.
13. R. M. Fitch and C. H. Tsai, 'Homogeneous nucleation of polymer colloids, IV: The role of soluble oligomeric radicals', in *Polymer Colloids*, (ed. R. M. Fitch), Plenum Press, New York, 1971, pp. 103–16.

CHAPTER 12

SOME UNRESOLVED PROBLEMS

12.1 INTRODUCTION

In this final chapter are discussed what appear to the author to be the more important of the unresolved, or only partially resolved, problems concerned with emulsion polymerisation. First and foremost are fundamental questions which relate to the mechanism of the reaction. Does one generalised mechanism apply to all emulsion polymerisation reactions, with perhaps different emphases depending upon the water-solubility and reactivity of the monomer? Or does each group of monomers have to be considered separately? What is the true mechanism of particle nucleation? Where exactly is the principal locus for the propagation reaction? What mechanistic model best fits the facts for any given monomer? How far does the available evidence allow discrimination between one model and another? Thus although extensive evidence has been assembled in earlier chapters in support of the conventional view of the micelles and particles as the principal reaction locus, it will be seen from what follows that the matter cannot yet be said to be fully resolved. Rather, the conventional view is to be regarded as a limiting case which will only pertain if certain conditions are fulfilled, and to which real systems may approximate to a greater or lesser extent.

The second topic with which this chapter will deal concerns problems associated with the micelle-generating substance and the colloid stability of the latex which is produced. The third problem is that of the control of the isomeric microstructure of the polymeric product.

12.2 PROBLEMS RELATING TO REACTION MECHANISM

The reader who is unfamiliar with the current literature on the subject of emulsion polymerisation may be forgiven if he gains the impression from the succeeding paragraph that the discussion of emulsion polymerisation reaction mechanism has gone full circle. The early chapters of this book have been concerned with the consideration of possible principal loci for the propagation step of the reaction, namely, monomer droplets, droplet–aqueous phase interface, true aqueous phase, and the micelles formed from surface-active substances present in the aqueous phase. Experimental evidence has been presented which has led to the conclusion that propagation begins in the micelles and continues within the polymer particles which are produced from the micelles by a gradual transition. The theory of Harkins (Chapter 3) gave the first coherent and comprehensive exposition of this view of the reaction mechanism and of its consequences. The Smith–Ewart theory (Chapter 4) concerned itself with the quantitative development of the micellar mechanism. Many of the subsequent contributions to the literature of emulsion polymerisation have subscribed (implicitly or explicitly) to the main features of the micellar mechanism, if not to all the quantitative conclusions of the Smith–Ewart theory.

The Gardon theory (Chapter 5), which appeared in 1968, assumes the basic Harkins–Smith–Ewart view of the mechanism, and proceeds to an extensive reconsideration of the quantitative aspects. (Perhaps it would be more accurate to say that Gardon has performed extensive re-calculations for a reaction system which conforms to the Harkins–Smith–Ewart model.) This book has also been written mainly from the point of view of the Harkins–Smith–Ewart theory; to have written it otherwise would have entailed giving a distorted view of the main body of literature which exists on the subject. In particular, it will be recalled that the surface-active substances which are deliberately added to emulsion polymerisation systems have been consistently referred to as 'micelle generators', thereby implying a commitment to a particular view of the function of these substances in the system.

It is now appropriate to give consideration to some of those papers which have appeared in recent years which have questioned the validity of what might be fairly described as the 'mainstream' view of the mechanism of emulsion polymerisation. We begin with a paper by Roe and Brass[1] which was published in 1957. This paper is concerned with the rate of growth of particles in the persulphate-initiated emulsion polymerisation of styrene using potassium laurate as the soap. Experiments were carried out to test the assumption of constant rate of increase of particle volume, which is made in the Smith–Ewart theory and of which a discussion is given in Chapter 4, Section 4.2. The effect of both particle size and initiator concentration upon growth rate was investigated.

For large-particle systems, the rate of increase of volume was found to increase with time, after the fashion of an autocatalytic reaction. This is explained by the hypothesis that, when a particle becomes sufficiently large, the probability of the coexistence of two or more radicals within a single particle is no longer negligible. Reference to this effect has already been made in earlier chapters. It can be regarded as a refinement to the Smith–Ewart theory (analogous to that necessitated by the possibility of a finite termination rate) which has to be taken into account when particles are large. The second finding reported by Roe and Brass is that the pattern of growth of the particles depends upon the initiator concentration. This is quite contrary to the prediction of the Smith–Ewart theory, bearing in mind that these results refer to that stage of the reaction which follows the cessation of locus nucleation. Roe and Brass then show that their observations are consistent with the hypothesis that the predominant process by which the radicals enter the particles is by the dissociation of initiator in the immediate vicinity of the particle–aqueous phase interface. Dissociation within the bulk of the aqueous phase, followed by diffusion to the particles, is discarded as being of minor importance only. This hypothesis represents a serious and fundamental departure from the Harkins–Smith–Ewart theory.

The following is a brief summary of the considerations which led Roe and Brass to formulate the above hypothesis. They first obtained an explicit solution to the Smith–Ewart recursion formula which is obtained by equating expressions (4.32) and (4.33) on p. 109 above. The recursion formula itself gives rise to an unlimited series of linear simultaneous equations for which solutions are indeterminate unless some additional constraint is imposed. For 'Case 2' of the Smith–Ewart theory, the additional constraint is, in effect, the postulate that no particle can contain more than one radical. Roe and Brass obtained their solution by making the less exacting assumption that no particle can contain more than six radicals. They then obtained a result for the fractions of the total particle population which, at any one instant, contain respectively 0, 1, 2, 3, 4, 5 and 6 radicals. But this distribution of radical number, taken with the Smith–Ewart hypothesis of uniform generation of radicals throughout the aqueous phase, led to lower rates of growth than were actually observed in the larger particles. The new hypothesis removed the inconsistency.

The mechanistic importance of the interfacial region between particles and aqueous phase has for many years been maintained by a group of Russian workers led by Medvedev. By implication, this group dissents from the micellar mechanism as it has usually been understood. An exposition of the Medvedev view (in English translation) is available in a paper presented in 1957.[2] Specifically, the view is that the surface-active molecules which form the micelles and which are adsorbed on to the particle surface play a part in the initiation reaction as well as providing reaction loci and imparting colloid stability. The following observations are cited in support of the view: the fact that the overall activation energy for the reaction can be lower than that for comparable bulk reactions; the effect of pH of the aqueous phase; the effect of inorganic electrolytes; and the dependence of polymerisation rate upon the nature of the surface-active substance. Medvedev considers that the kinetic features of many emulsion polymerisation reactions support his view. Thus he says: 'The most characteristic and general kinetic feature of emulsion polymerisation is the constancy of the rate which is maintained throughout the whole process, given constant concentrations of monomer and initiator. This property, together with the probability that the initiating reaction takes place at the interface between the adsorption layer and the aqueous phase, gives rise to the assumption that the adsorption layer of soap is the main factor determining the kinetics of the process. By adsorption layer we here understand the surface layers of soap both in the micelles and on the surface of the polymer particles. This function of the adsorption layer lies in its catalytic effect on initiation, and determines the site at which polymerisation takes place. In fact, if the reaction between free radicals and the monomer—which is always present in the polymer particles in considerable concentrations—is rapid, and diffusion in the viscous medium slow, then it can further be concluded that the whole process is concentrated near the interface in the narrow zone of the adsorption layer. The thickness of this zone is governed by the diffusion conditions and remains constant throughout the whole process. The volume of this zone—the effective volume in which polymerisation takes place—will be proportional to the total surface area of the adsorption layers, *i.e.* to the total amount of emulsifier present in the system.'

Medvedev goes on to consider the kinetic consequences of three distinct reaction models. The model which is of particular interest is that which assumes the initiator to be water-soluble, but which envisages the primary initiating radicals as formed by the decomposition of initiator molecules which have been adsorbed into the particle–aqueous phase interface. It will be seen from what follows that this model predicts orders of reaction of unity for the dependence of overall rate upon monomer concentration and upon the concentration of surface-active substance, and an order of reaction of 1/2 for the dependence of rate upon initiator concentration. Conformity of prediction with observation (in certain instances at least) is taken as supporting evidence for the mechanism. But as Dunn[3] rightly observes, 'it is not logical to suppose that because the equations deduced have been verified experimentally, the scheme on the basis of which they were deduced must therefore be correct since it may be possible to deduce the same result from alternative propositions...'. (It should be noted that Dunn gives a table of Medvedev's reaction schemes and conclusions which seems to be partly at variance with those given in Medvedev's 1957 paper. The scheme and conclusions which Medvedev gives for the case where the generation of primary initiating radicals occurs by decomposition of adsorbed initiator correspond to those given by Dunn for initiation by an oil-soluble initiator which decomposes in the adsorption layer. But Dunn is apparently quoting from a later paper by the Medvedev group. It is possible that confusion has arisen through translation difficulties.)

The reaction scheme which Medvedev uses for the case of generation of primary initiating radicals within the adsorption layer is essentially as follows:

$$\mathrm{I} \underset{k_2}{\overset{k_1}{\rightleftharpoons}} \mathrm{I}' \tag{i}$$

$$\mathrm{I}' \xrightarrow{k_d} 2\mathrm{R}\cdot \tag{ii}$$

$$\mathrm{R}\cdot + \mathrm{M} \xrightarrow{k_i} \mathrm{M}_1\cdot \tag{iii}$$

$$\mathrm{M}_n\cdot + \mathrm{M} \xrightarrow{k_p} \mathrm{M}_{n+1}\cdot \tag{iv}$$

$$\mathrm{M}_m\cdot + \mathrm{M}_n\cdot \xrightarrow{k_t} \text{inactive products} \tag{v}$$

In this scheme, I denotes an initiator molecule in the aqueous phase, I′ denotes an adsorbed initiator molecule, R· denotes a primary initiating radical formed by decomposition of an adsorbed initiator molecule, M is a monomer molecule in the interfacial region, and $\mathrm{M}_n\cdot$ a propagating radical which contains *n* monomer units. The *k*'s are the rate coefficients for the various processes. Reaction (i) represents reversible adsorption and desorption of the initiator. This reaction is assumed to be kinetically of the first order in initiator concentration, although it is possible that the adsorption reaction may involve a complex formed with a molecule of the surface-active substance. Medvedev then assumes that the concentration of adsorbed initiator is stationary, and so obtains the equation

$$k_1[\mathrm{I}] = k_2[\mathrm{I}'] + k_d[\mathrm{I}'] \tag{12.1}$$

Likewise, if the total concentration of radicals is assumed to be stationary, then

$$k_t\left(\sum [\mathrm{M}_n\cdot]\right)^2 = 2k_d[\mathrm{I}'] \tag{12.2}$$

The rate of disappearance of monomer per unit volume of reaction system per unit time is then

$$-\frac{\mathrm{d}[\mathrm{M}]}{\mathrm{d}t} = k_p[\mathrm{M}]\left(\sum [\mathrm{M}_n\cdot]\right) \tag{12.3}$$

Substitution from eqns (12.2) and (12.1) readily gives

$$-\frac{\mathrm{d}[\mathrm{M}]}{\mathrm{d}t} = k_p[\mathrm{M}]\sqrt{\left(\frac{2k_dk_1}{k_t(k_2 + k_d)}\right)}[\mathrm{I}]^{1/2} \tag{12.4}$$

The absolute rate of disappearance of monomer will then be the product of $-\mathrm{d}[\mathrm{M}]/\mathrm{d}t$ and the volume of the system in which reaction is occurring. If the latter is proportional to the concentration of surface-active substance, as proposed by Medvedev, then the orders of reaction are as stated above. The precise nature of the kinetic relationship obtained depends upon the model assumed. Thus, if it is assumed that the initiator decomposes in the aqueous phase, and that most of the radicals formed become adsorbed on to the particles, *e.g.* because they first interact with soap molecules, and that the reaction is confined to a surface layer of definite thickness, then one obtains orders of reaction of 1/2 for initiator and soap, but of unity for monomer. But, in any event, it is quite clear that Medvedev regards the emulsion polymerisation reaction as being essentially a micro-bulk polymerisation which takes place in the surface layers of the particles. This point of view is adopted regardless of the details of the assumed mechanism. This carries the further implication that, whatever may be the main causes for the peculiar features of emulsion polymerisation reactions, it is certainly not compartmentalisation of the individual propagation steps.

A certain amount of experimental information is available which appears to be consistent with the particular Medvedev model in which initiation occurs by way of the decomposition of an adsorbed water-soluble initiator. Thus Brodnyan *et al.*,[4] working with methyl methacrylate and with *n*-butyl methacrylate at 30°C, came to the conclusion that the rate of polymerisation varies linearly with the specific surface area of the latex produced, as required by the Medvedev theory. One of their conclusions from this study is that 'The Medvedev mechanism may be more realistic than the Smith–Ewart theory in describing the emulsion polymerization data presented here'. It will also be recalled (see Chapter 10,

Section 10.4) that Elgood, Gulbekian and Kinsler[5] concluded that the surface area of the growing particles is an important variable determining the rate at which vinyl acetate emulsion polymerisation occurs. The state of confusion surrounding this matter is well illustrated by the following comment made by Ryšánek[6] following the presentation of the 1957 Medvedev paper: 'Emulsion polymerisation proceeding according to the rules derived by Smith and Ewart is to be considered as basic. Working with entirely pure chemicals is supposed. Deviations from the Smith–Ewart theory mentioned by Prof. Medvedev... can be explained by other influences than by a catalysed reaction in the adsorption layer of the particles.'

Further consideration of the question of reaction locus has been given by Roe[7] in a paper which appeared in 1968. He reconsiders the various possibilities for chain initiation–monomer droplets, monomer solubilised in micelles, and monomer dissolved in the aqueous phase—and comes to the conclusion that the presence of soap micelles is irrelevant for initiation although the presence of the soap which makes up the micelles is far from irrelevant. The main experimental evidence for this view is summarised in Figs. 12.1 and 12.2 The first of these diagrams shows the effect of soap

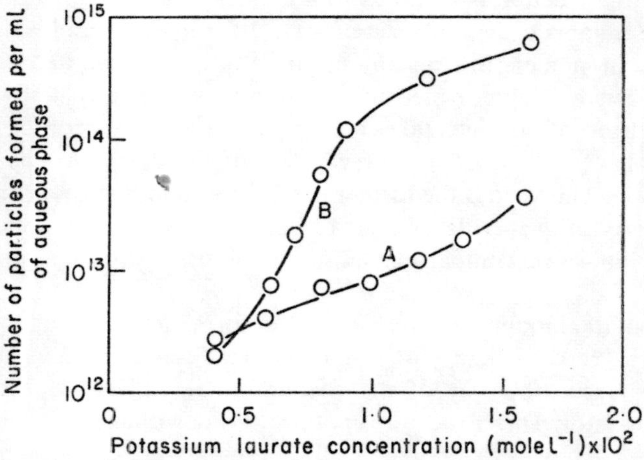

FIG. 12.1. Effect of soap level and electrolyte level upon number of particles formed during persulphate-initiated polymerisation of styrene at 50°C, according to Roe.[7] Curve A refers to a total $[K^+]$ of 0·01 g-ion l^{-1}; curve B to a total $[K^+]$ of 0·1 g-ion l^{-1}. (Reaction system: styrene 20, water *ca.* 250, potassium laurate—variable, $K_2S_2O_8$ 0·3, K_2CO_3–$KHCO_3$ buffer (pH 10) to give requisite total $[K^+]$. See note in caption to Fig. 9.1 regarding significance of total potassium-ion concentration.)

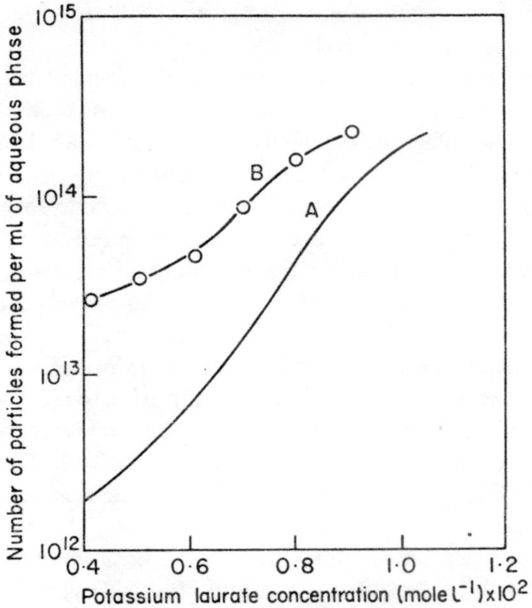

FIG. 12.2. Effect of presence of a more water-soluble monomer (acrylonitrile) upon relationship between soap concentration and number of particles formed during persulphate-initiated emulsion polymerisation, according to Roe.[7] Curve A refers to styrene homopolymerisation; curve B to the copolymerisation of a 9/4 styrene/acrylonitrile mixture. (Reaction system was as given in caption to Fig. 12.1, with replacement of styrene by a comonomer mixture of appropriate composition. Total $[K^+]$ was 0·1 g-ion l^{-1}.)

level upon the number of particles formed during the emulsion polymerisation of styrene at 50°C, initiated by potassium persulphate. The soap was potassium laurate, for which the critical micelle concentration is fairly high, being about 2×10^{-2} mole l^{-1} in the absence of added electrolyte, and about $0{\cdot}8 \times 10^{-2}$ mole l^{-1} at a total potassium-ion concentration of 0·1 mole l^{-1}. Results are given for two levels of total potassium-ion concentration. Three significant findings are illustrated by these results. Firstly, the results for the lower level of potassium-ion concentration show that initiation of particles is possible at soap levels well below that required for micelle formation. Secondly, the results for the higher level of potassium-ion concentration show that no discontinuity in particle formation occurs as the critical soap concentration is exceeded. And thirdly, the effect of increasing the level of potassium ions is to increase the number of particles formed, not only when the extra electrolyte induces micelle formation, but also when the soap level is insufficient for micelles to form even at the higher electrolyte level. All these observations are at variance with the Harkins micellar hypothesis.

Fig. 12.2 shows that the number of particles formed can be increased by adding a more water-soluble monomer, in this case acrylonitrile. Furthermore, the effect is especially marked at soap levels below the critical concentration. Other observations reported by Roe which conflict with the micellar hypothesis are that (a) it is possible to generate equal particle populations in styrene emulsion polymerisation systems containing widely different concentrations of soap molecules, and (b) experiments on mixed soaps show that it is the total soap level, and not the level of micellar soap, which is important in determining the number of particles. Also, experiments on systems containing non-ionogenic surface-active substances in excess of the critical concentration have shown that the addition of small amounts of sodium lauryl sulphate increases the particle population manyfold, even though the number of micelles cannot have been affected because the sodium lauryl sulphate was below its critical concentration.

Roe then proposes a mechanism for particle initiation which appears to be consistent with the above observations. Radicals generated within the aqueous phase are envisaged as reacting with monomer dissolved in the true aqueous phase. Roe then outlines his theory as follows:

> 'Continued development of this sort causes the growing chain to assume the aspect of a polymer particle stabilized against flocculation by adsorbed emulsifier and swollen with absorbed monomer. From this point on, conversion of monomer to polymer within the particle is assumed to occur largely at the expense of swelling monomer, until chain termination occurs through entry of a second radical into the particle from the aqueous phase.
>
> 'Alternatively, at any stage in its history prior to adsorption of an adequate protective layer of emulsifier, the growing radical may be swept up by a pre-existing polymer particle or by a swollen emulsifier micelle, provided micelles are present. In either of the latter cases, the growing polymer radical is in an environment exactly the same as that of a swollen polymer particle developed according to the argument in the preceding paragraph.
>
> 'A particle nucleus or aggregate may be said to have identity as a polymer particle when further growth takes place only by chain propagation and combination within its own structure and when flocculation with other molecules or agglomerates of a similar size is excluded.

The realization and persistence of these conditions depend upon the effectiveness of the protective layer of stabilizer adsorbed at the polymer–water interface. Clearly this will, in turn, depend on the amount of stabilizer adsorbed and on its specific effectiveness. Stability of a particle toward agglomeration may be defined as proportional to the amount of stabilizer adsorbed multiplied by its specific effectiveness.

'It is implicit in the proposed theory that the generation of particle nuclei persists at a rate which depends on concentration of monomer and free radicals but not on other constitutive variables. *Achievement of permanent identity as polymer particles by nuclei depends on the availability of emulsifier. The distinguishing point in the proposed theory is thus that ultimate particle population is determined, with respect to emulsifier, by the total amount of emulsifier available and by its intrinsic effectiveness. The presence and concentration of micelles are of trivial importance.*' (The italics are not in the original.)

Thus, according to the Roe theory, the soap molecules are important not because they form micelles but because they stabilise the oligomeric radicals which form in the aqueous solution, and enable them to form, in turn, small, colloidally-stable, reaction loci. In effect, the Roe theory represents the extension to styrene of ideas which have been developed to explain the emulsion polymerisation behaviour of other more water-soluble monomers such as vinyl acetate.

Roe then addresses himself to the implications of his theory for the Smith–Ewart 'Case 2' calculation. He points out that, despite appearances to the contrary, the hypothesis of micellar initiation is not in fact an essential feature of that calculation. The essential assumption for the Smith–Ewart theory is that the nucleation of new reaction loci takes place until some critical event occurs which causes it to cease. In the original formulation of the Smith–Ewart theory, that event is the disappearance of soap micelles. For the Roe theory, it is the depletion of the soap concentration to a level below which new particles cannot achieve stability through adequate adsorption of soap. Roe comments that 'In the case of a soap having a very low cmc, the soap level at which particle population becomes constant coincides with the cmc, but this is not true in soaps having high cmc values.'

It is unnecessary to reproduce the formal mathematics of the Roe theory. This is identical with that of the Smith–Ewart 'Case 2' calculations. All that is necessary is to consider the Roe interpretation of the critical time t^* of the Smith–Ewart calculation (see Chapter 4, Section 4.2). This time is taken as that for which the concentration of monomeric soap dissolved in the aqueous phase first becomes too low for new particles to be stabilised. This critical concentration is considered to be rather lower than that for micelle formation. Roe's result for the dependence of number of reaction loci upon soap and initiator concentrations is the same as that of Smith and Ewart. A necessary corollary is that it is impossible to discriminate between the two theories by investigating the dependence of particle number upon these concentrations.†

Considerable interest has been shown recently in the mechanism whereby discrete polymer particles form when a monomer dissolved in water polymerises to give rise to a polymer which is insoluble in water. Whereas this subject was formerly considered to be of relevance only to the emulsion polymerisation of the more water-soluble monomers such as vinyl acetate, it is now becoming increasingly realised that it may also be important for the emulsion polymerisation of relatively insoluble monomers such as styrene. The view has been expressed that the matter of particle nucleation during aqueous solution polymerisation is of quite general importance for all emulsion polymerisation reactions.

A theory for the homogeneous nucleation of colloid particles by solution polymerisation has recently been developed by Fitch and his co-workers.[8,12] Reference to the work of this group on the formation of particles during the aqueous solution polymerisation of methyl methacrylate has already been made in Section 10.5 of Chapter 10. The theory endeavours to predict quantitatively the number of particles which form, using a model homogeneous nucleation reaction. It requires a knowledge of the rate of radical generation and the average distance which a primary oligomeric radical can diffuse in solution before it self-nucleates.

The view is taken that, at the outset of the reaction, one has a homogeneous solution of monomer, surface-active substance and initiator. Polymerisation starts and oligomeric chains appear in solution. These grow by free-radical polymerisation until, at some critical chain length, these soluble oligomers become insoluble and form a separate polymer

† From information which has come to hand since this section was written, it seems that the realisation that aqueous-phase solution polymerisation might be the nucleation mechanism for monomers of low water-solubility, as well as for those of relatively high water-solubility, was first made public by Roe and Brass in 1956. Apparently, however, this notion was found to be unacceptable to the scientific community to which it was addressed. Thus White[61] has recently made the interesting comment that 'The Harkins theory of micelle initiation was so widely accepted and so firmly entrenched that in 1956 when Doctors C. P. Roe and P. D. Brass of Uniroyal presented data at the 130th National Meeting of the A.C.S. which refuted it editors and referees refused to accept the paper for publication.'

phase by collapse of the chains upon themselves, thereby forming primary particles. Soluble oligomers subsequently formed will then have a choice of forming more new particles by this process, or of being captured by pre-existing particles. Initially, the rate of appearance of new particles should be equal to the rate of generation of effective radicals. As capture of the oligomeric radicals becomes important, so the rate of particle nucleation will be correspondingly reduced. Fitch concludes that a reasonable estimate of the rate of capture of oligomers by existing particles can be made from diffusion theory, but that the capture process probably occurs by a mechanism which is diffusion-controlled.

Whether or not the details of this particular theory of particle nucleation are correct, there seems little doubt that studies of the aqueous solution polymerisation of sparingly-soluble monomers will play an increasingly prominent part in deepening our understanding of the processes by which the latent latex particles are formed in an emulsion polymerisation system.

It is now necessary to refer briefly to a controversy which concerns the site of the propagation reaction within the swollen particle. The basic correctness of the Harkins–Smith–Ewart theory is taken for granted, but the concern is with the fine details of the reaction locus in the stage of constant polymerisation rate which succeeds that of particle nucleation. A stage of constant rate of polymerisation (*i.e.* one which is apparently of zero order with respect to monomer concentration) is readily explained if reaction is proceeding in a fixed number of reaction loci in which the concentration of monomer is constant and in which a single free radical is present for one-half of the time. But is the monomer concentration in fact constant? A case has been made out in Chapter 4, Section 4.2, for supposing that it might well be. But, as Dunn[3] has pointed out, there has long been evidence in the literature to suggest that in fact the concentration falls with conversion. This evidence is essentially that summarised in Fig. 3.12, which shows the effect of conversion upon the monomer:polymer ratio in the particles of an emulsion polymerisation system. Dunn remarks that although this information has been available for many years, it has generally been ignored. However, Robb[13] has recently found that, for styrene emulsion polymerisation, the particle concentration during the stage of constant polymerisation rate does not remain constant, but increases. Thus it is possible that the constant polymerisation rate arises because of the presence of two opposing tendencies which fortuitously balance one another: increasing particle population and decreasing monomer concentration at the reaction loci.

The fact that the monomer:polymer ratio in the reaction loci falls with conversion during the emulsion polymerisation of styrene has been confirmed recently by Grancio and Williams.[14] Their conclusion is that 'the practice of assuming a constant particle monomer concentration in analytic treatment of emulsion polymerization ought to be followed with increased caution'. They conclude from their kinetic observations that the monomer concentration in the actual polymerisation locus is constant, but that the reaction locus does not include the whole particle. They propose what they describe as a 'core–shell' model[14,15] for the growing particle, thereby implying that the growing particle is heterogeneous (in contrast to the assumption of the Smith–Ewart theory that it is homogeneous). Their suggestion is that 'the growing particle consists of an expanding polymer-rich (monomer-starved) core surrounded by a monomer-rich (polymer-starved) spherical shell. In this core–shell model the outer shell serves as the major locus of polymerization while virtually none occurs in the core because of its monomer-starved condition. A Smith–Ewart on–off mechanism necessarily prevails within the monomer-rich shell. The core is pictured as growing outward somewhat as a ball of string constructed from many single strands. In such a model a changing overall particle monomer concentration is consistent with constant particle rates because reaction takes place in an essentially pure monomer environment.' Supporting evidence for the suggested particle morphology is said to have been obtained from the examination of ultra-thin sections of latex particles by electron microscope.

So much for the statement of the 'core–shell' theory (which, incidentally, bears a certain superficial resemblance to the Medvedev theory outlined earlier, but which conceptually is really quite different). The next stage in the controversy is contained in a paper by Napper.[16] He maintains that the requirements of the 'core–shell' model, as given by Grancio and Williams, conflict with the 'ordinary' laws of diffusion. He affirms that 'if, as assumed by Grancio and Williams, polymerisation occurs primarily in the peripheral zones of the particles, any simple diffusion approach predicts that these outer regions would, if anything, be monomer-deficient in any dynamic situation because of the consumption of monomer therein. The inner regions would thus function as monomer reservoirs, the chemical potential of the monomer decreasing in the outwards direction.' Napper appears to accept the need to postulate a heterogeneous particle, but proposes an alternative model which is the inverse of that proposed by Grancio and Williams. The major locus of polymerization is regarded as being inside the particle, apparently 'in the immediate subsurface regions'.

Williams[17] has then purported to meet Napper's objections in a further paper. He does not consider that his 'core–shell' model violates the laws of diffusion, but states that the model merely implicitly assumes that monomer diffusion is not the rate-controlling process. Additional evidence is cited from a paper by Keusch and Williams[18] to the effect that the 'core–shell' morphology even prevails under conditions of equilibrium saturation. Thus, if this evidence is to be believed, homogeneous swelling of latex particles never occurs; instead a situation of equilibrium encapsulation exists. A further exposition of these ideas is contained in a paper by Keusch, Prince and Williams.[19]

The whole question of the existence or otherwise of monomer-concentration gradients in polymerising latex particles has been reconsidered recently by Gardon.[20] He considers the case of a single-charge emulsion polymerisation system involving a monomer which is a good solvent for its polymer. His conclusion is that there are unlikely to exist, within the growing particles, concentration gradients of sufficient magnitude to affect the kinetics of the reaction. An estimate has been made of the average displacement of monomer due to Brownian motion within the latex particles and in the absence of concentration gradients. This diffusive mean free path corresponds in duration to a time interval which is less than that required for 1% change in conversion, and appears to be much longer in length than the radius of the latex particle. Consequently, loci where monomer concentration is perturbed by conversion to polymer are immediately swamped by unreacted monomer. Gardon further states that direct experimental evidence exists to show that the monomer concentration in latex particles is about the same when non-polymerising latex particles are saturated with monomer as when excess monomer is present during polymerisation. Gardon has also re-examined the conversion-time data which Grancio and Williams maintained compelled them to postulate their 'core–shell' model. He concludes that the data do not in fact compel acceptance of such a model.

As a further example of the uncertainty and confusion which currently surrounds the question of the principal reaction locus in an emulsion polymerisation, brief reference will be made to recent papers by Ugelstad, El-Aasser and Vanderhoff,[21] and by Ugelstad, Hansen and Lange,[22] in which it is shown that, given the appropriate conditions, significant polymerisation can occur within the emulsified monomer droplets. Ugelstad, El-Aasser and Vanderhoff[21] give the results of calculations which appear to show that, if the droplet size of a monomer emulsion can be reduced to a sufficiently low value, then the initiation of polymerisation within those droplets can become sufficiently important for the droplets to be regarded as the principal reaction locus. Experimental results are presented for the persulphate-initiated emulsion polymerisation of styrene at 60°C using, as surface-active substances, (a) sodium lauryl sulphate and (b) a mixture of sodium lauryl sulphate and cetyl alcohol.

It is known that the addition of long-chain fatty alcohols, such as cetyl alcohol, to anionic emulsifiers, such as sodium sulphate, drastically enhances the ability to disperse and stabilise oil-in-water emulsion. Significant differences are reported by Ugelstad, El-Aasser and Vanderhoff for the two emulsion polymerisation systems. Using sodium lauryl sulphate alone, the stability of the styrene emulsion is poor, its average droplet size is large, and the average particle size of the latex produced is very small. These observations can be reasonably explained in terms of either a micelle-initiation mechanism or the formation, precipitation and stabilisation of oligomeric free radicals within the aqueous phase. However using the combination of sodium lauryl sulphate and cetyl alcohol, a stable styrene emulsion is formed in which the droplet size is small. The average size of the particles in the latex produced is approximately that of the monomer emulsion, being much larger than that obtained using sodium lauryl sulphate alone. The inference is drawn—quite reasonably—that the monomer droplets are the principal polymerisation loci. The explanation given is that reduction of the monomer emulsion droplet size makes the droplets more competitive in capturing radicals which are generated in the aqueous phase. In turn, two reasons are suggested for this enhanced competitiveness. In the first place, the surface area of interface between monomer droplets and aqueous phase is increased by reducing the droplet size. But, secondly, an increased interfacial area implies an increased proportion of emulsifier adsorbed at the monomer–aqueous phase interface, and therefore a reduced proportion present in the bulk of the aqueous phase. In consequence, the probability of initiation, growth, precipitation and stabilisation of oligomeric radicals in the aqueous phase is therefore reduced.

Ugelstad, Hansen and Lange[22] have published the results of a more extensive study of the polymerisation of styrene under conditions such that the emulsified monomer droplets are the probable principal loci of reaction. Again, a mixture of an anionic emulsifier (sodium hexadecyl sulphate) and a fatty alcohol (hexadecanol) was used to disperse the styrene. The effect of the fatty alcohol in promoting emulsification apparently depends very much upon the method used for the preparation of the emulsion. The recommended procedure is to heat the anionic emulsifier and fatty alcohol with water prior to adding the styrene. In this way, it is possible to prepare emulsions with droplet sizes in the range 0·5–1 μm, using relatively small amounts of fatty alcohol and only moderate stirring. Fine monomer emulsions prepared by this method are apparently able to adsorb most of the anionic emulsifier in the system. Polymerisations were carried out at 60°C, using potassium persulphate as initiator. Experiments in which the level of fatty alcohol was varied, keeping the level of sodium hexadecyl sulphate constant, showed that the addition of hexadecanol to the reaction system causes a reduction in the rate of polymerisation. Furthermore, the rate decreases markedly as the ratio of fatty alcohol to anionic emulsifier increases from 1 to 2, whereas further increase in the ratio from 2 to 4 is accompanied by only a small further decrease in the rate of polymerisation. The addition of fatty alcohol to the reaction system also leads to a marked increase in the particle size of the latex produced. In particular, an increase in the ratio of fatty alcohol to anionic emulsifier is accompanied by the formation of an increasing number of large particles with diameters in the range 0·5 to 1·5 μm. With increasing amounts of fatty alcohol, the distribution of particle sizes becomes bimodal.

Ugelstad, Hansen and Lange interpret their results in terms of a mechanism whereby polymerisation within the emulsified droplets becomes increasingly more important relative to micellar polymerisation as the droplet size of the monomer emulsion is reduced. The fact that the rate of polymerisation in the presence of hexadecanol decreases with conversion from the start of the reaction is cited as evidence of the monomer droplets as the principal polymerisation locus. For no longer are the droplets behaving as inert reservoirs whose function, as long as they are present, is to maintain a constant monomer concentration in reaction loci external to themselves; they are themselves now the loci, and the concentration of monomer within them falls as the reaction proceeds.

The extent to which studies of emulsion copolymerisations can assist in the elucidation of the mechanism of emulsion polymerisation has been recently reviewed by Fowler.[23] The general impression conveyed by this review is that emulsion copolymerisation is an extremely complex matter in its own right; far from throwing light upon the general mechanism of emulsion polymerisation, it serves rather to confuse and confound matters by superimposing the complexities and uncertainties of copolymerisation upon those of emulsion polymerisation.

At the present time, the only valid reason for studying emulsion copolymerisation is its industrial importance. As an example of the difficulty of interpreting emulsion copolymerisation data in terms of reaction mechanism, the early work of Fordyce *et al.*[24,29] may be cited. Reference has already been made to some of this work (see Chapter 1, Section 1.5), where typical results are given. It will be recalled that this group of investigators compared the relationship between monomer feed and copolymer compositions for copolymers prepared in emulsion with that for the same copolymers prepared at the same temperature by either bulk or solution reaction. An equivalent way of regarding these experiments is as a comparison between the apparent monomer reactivity ratios of a given pair of monomers copolymerising in emulsion, and the monomer reactivity ratios for the same pair of monomers polymerising at the same temperature in bulk or solution. The overall conclusion which emerges from this series of investigations is that comparisons between the feed–copolymer composition relationships (and therefore the monomer reactivity ratios) for emulsion and bulk/solution copolymerisations are inconsistent with the view that the emulsion copolymerisation occurs in the true aqueous phase of the reaction system.

It is interesting to note that, on various occasions, Fordyce and his co-workers draw two quite distinct conclusions from their results. On the one hand, they sometimes regard their observations as indicating that emulsion copolymerisation takes place mainly in what they term the 'oil phase'. They do not, however, specify clearly which phase they mean by this term, although the general tenor of their text is such as to suggest that they mean the emulsified droplets of monomer. On the other hand, on other occasions they place a less definite interpretation upon their results and regard them as indicating that emulsion polymerisation occurs predominantly in a non-aqueous phase. In the present writer's view, this purely negative conclusion is the only one which can legitimately be drawn from their results; indeed, it will be recalled that some of these results were cited in Chapter 1 solely as evidence that the propagation stage of an emulsion polymerisation does not take place within the true aqueous phase of the reaction system. It appears that it is very difficult to decide, on the basis of copolymer studies alone, which of the various non-aqueous phases within the reaction system (*i.e.* monomer droplets, micelles, or monomer-swollen particles) is the principal locus of the propagation reaction. For all are non-polar in character, and at equilibrium the monomers would be expected to partition themselves in roughly the same ratio in each phase. The composition of the copolymer formed cannot therefore be regarded as diagnostic of the phase in which it was formed. This problem of interpretation has been well summarised by Wall, Florin and Delbecq[30] in the following manner: 'From the point of view of monomer reactivity ratios, it would appear from the above results that the oil phase would be the principal reaction locus for the system studied. According to Harkins, however, the reaction starts in the soap micelles, but as soon as some polymer forms, monomer will dissolve in the polymer, and the reaction will continue in the resulting monomer polymer particles. Since the composition of the monomer dissolved in the monomer–polymer particles can reasonably be expected to be nearly the same as that of the oil phase, it is easily seen why the polymerization would appear to take place in the "oil phase" as far as monomer reactivity ratios are concerned.

According to Vanderhoff,[31] Brown[32] has developed a theory of emulsion polymerisation in which the process is treated as one of diffusion of monomer and radicals to the reaction sites, which may be monomer-swollen micelles, monomer-swollen polymer particles, polymer particles with an adsorbed layer of monomer, or monomer emulsion droplets. Vanderhoff states that this theory predicts that the rate of polymerisation should be directly proportional to both the initiator concentration and the concentration of micelle generator.

It is appropriate to conclude this discussion of, as yet, unresolved questions concerning reaction mechanism and locus by drawing attention to a recent paper by Gilbert and Napper[33] in which is questioned—apparently for the first time—the assumption that the populations of particles containing 0, 1, 2,. . .radicals are stationary. It will be recalled that the truth of this assumption has been accepted hitherto at all stages in the consideration of the quantitative aspects of emulsion polymerisation theory. The Smith–Ewart regression formula, which has formed the basis of so many discussions of the particle population distribution, is obtained by equating the expressions (4.32) and (4.33) (p. 109). Alternatively—and it amounts to the same thing—the difference between the two expressions may be set equal to zero. The one expression gives the rate at which particles containing i radicals are formed; the other gives the rate at which they disappear. Thus the difference gives the net rate of increase (or decrease) of particles containing i radicals; equating this to zero implies a stationary (although possibly interchanging) population.

If the assumption of stationary populations is no longer made, then, with the notation of Chapter 4, we must have that

$$\frac{\mathrm{d}n_i}{\mathrm{d}t} = n_{i-1}\frac{\rho}{N} + n_{i+1}k_0 a\frac{i+1}{v} + n_{i+2}k_t\frac{(i+2)(i+1)}{v} - n_i\frac{\rho}{N} - n_i k_0 a\frac{i}{v} - n_i k_t\frac{i(i-1)}{v} \tag{12.5}$$

Making the simplifying substitutions $\sigma = \rho/N$ and $k = ak_0/v$, this equation may be written

$$\frac{\mathrm{d}n_i}{\mathrm{d}t} = (n_{i-1} - n_i)\sigma + \{(i+1)n_{i+1} - in_i\}k + \{(i+2)(i+1)n_{i+2} - i(i-1)n_i\}\frac{k_t}{v} \tag{12.6}$$

On the grounds of mathematical expediency, Gilbert and Napper now make the assumption that the third term of the right-hand side of eqn (12.6) can be neglected. Physically, this seems to imply that termination within the particles is a slow process relative to both the absorption of radicals from the aqueous phase represented by the first term, and the first-order desorption from the particles represented by the second. Making this assumption gives

$$\frac{\mathrm{d}n_i}{\mathrm{d}t} = (n_{i-1} - n_i)\sigma + \{(i+1)n_{i+1} - in_i\}k \tag{12.7}$$

for $i = 1, 2, 3, \ldots$, whilst for $i = 0$,

$$\frac{\mathrm{d}n_0}{\mathrm{d}t} = -n_0\sigma + n_1 k \tag{12.8}$$

Equation (12.7) can be rewritten as

$$\frac{\mathrm{d}n_i}{\mathrm{d}t} = n_{i-1}\sigma - n_i(\sigma + ik) + n_{i+1}(i+1)k \tag{12.9}$$

The totality of these equations can readily be written in matrix form as

$$\frac{\mathrm{d}[n_i]}{\mathrm{d}t} = [\Psi][n_i] \tag{12.10}$$

where

$$[n_i]^{\mathrm{T}} = [n_0\, n_1\, n_2 \ldots] \tag{12.11}$$

and

$$[\Psi] = \begin{bmatrix} -\sigma & k & k & 0 & 0 & \cdots & 0 & 0 & 0 \\ \sigma & -(\sigma+k) & 2k & 0 & 0 & \cdots & 0 & 0 & 0 \\ 0 & \sigma & -(\sigma+2k) & 3k & 0 & \cdots & 0 & 0 & 0 \\ 0 & 0 & \sigma & -(\sigma+3k) & 4k & \cdots & 0 & 0 & 0 \\ \vdots & \vdots & \vdots & \vdots & \vdots & & \vdots & \vdots & \vdots \\ 0 & 0 & 0 & 0 & 0 & \cdots & \sigma & -(\sigma+\overline{r-1}k) & rk \\ 0 & 0 & 0 & 0 & 0 & \cdots & 0 & \sigma & -(\sigma+rk) \end{bmatrix}$$

(12.1

The orders of these matrices are, respectively, $(r + 1) \times 1$ for $[n_i]$ and $(r + 1) \times (r + 1)$ for $[\Psi]$, where r is the maximum number of radicals which a single particle can contain. No restriction need be placed upon r at this stage, although it should be noted that the last equation of the set comprising (12.8) and (12.9) has to be modified to allow for the fact that $n_{r+1} = 0$.

It is possible immediately to write down the solution in matrix form of the differential matrix eqn (12.10), since the elements of the matrix $[\Psi]$ are assumed not to depend upon either time or the n_i. However, Gilbert and Napper prefer to transform eqn (12.10) to

$$\frac{\mathrm{d}[y_i]}{\mathrm{d}t} = [B][y_i] \tag{12.13}$$

where $[B]$ is obtained from $[\Psi]$ by the similarity transformation

$$[B] = [S][\Psi][S]^{-1} \tag{12.14}$$

and in consequence $[y_i]$ is obtained from $[n_i]$ by the transformation

$$[y_i] = [S][n_i] \tag{12.15}$$

$[B]$ is said to be Hermitian, and therefore easier to manipulate than $[\Psi]$. The matrix $[S]$ is taken as the diagonal matrix:

$$[S] = [\delta_{i+1,j+1}\sqrt{(i!)}/\alpha^i] \tag{12.16}$$

where $i, j = 0, 1, 2, \ldots, r$, δ_{ij} is the Kronecker delta, and $\alpha = \sqrt{(\sigma/k)}$. The solution of eqn (12.13) is

$$[y_i] = \mathrm{e}^{[B]t}[y_i{}^0] \tag{12.17}$$

where $[y_i{}^0]$ is an $(r + 1)\times 1$ column matrix whose elements are the initial values of y_i. As it stands, eqn (12.17) gives the elements of the matrix $[y_i]$ in the form of infinite series. These elements can be reduced to finite series of $r + 1$ terms by making use of the eigenvalues, λ_i, and eigenvectors, $[X_i{}^j]$, of the matrix $[B]$. It can be shown that

$$[y_i] = \sum_{j=0}^{r}\left\{\mathrm{e}^{\lambda_j t}[X_i{}^j]\sum_{i=0}^{r} X_i{}^j y_i{}^0\right\} \tag{12.18}$$

where $\sum_{i=0}^{r} X_i{}^j y_i{}^0$ is the inner product of the elements of $[X_i]$ and those of $[y_i{}^0]$.

Gilbert and Napper give curves which show the variation of n_0, n_1 and n_2 with time predicted by their theory, taking plausible values for σ and k. It appears that n_0 decreases monotonically with time to a steady-state value, whilst both n_1 and n_2 increase monotonically to steady-state values. The times required to achieve the steady state appear to be quite long, being of the order of 10^4 seconds, *i.e.* of the order of 3 h. Gilbert and Napper also conclude that it is wrong to make the usual assumption that a linear conversion–time plot implies that the average value of i is 0·5. Applying their theory to the special case of vinyl acetate emulsion polymerisation, they conclude that the efficiency of radical capture by particles is very low: the prediction is that only one radical in $2{\cdot}6 \times 10^3$ of those produced in the aqueous phase enters the particles. Their low efficiency of capture is attributed to the relatively high solubility of the monomer in water.

12.3 PROBLEMS RELATING TO THE MICELLE-GENERATING SUBSTANCE

Much of the discussion of the preceding section has been concerned, *inter alia*, with the general role which the so-called micelle generator plays in an emulsion polymerisation reaction. The concern of this section is not directly with the general role of the micelle generator, but rather with more specific problems which appear to attend the use of certain types of surface-active substance. Of course, it may be that the resolution of some of these specific problems will throw light upon the more fundamental questions which have been discussed in the preceding section. This remains to be seen; all that can be attempted here is a statement of some of those specific problems.

It will be noted that, notwithstanding the controversies summarised in the preceding section, the term 'micelle-generating substance' has been used in the heading of this section. There are several reasons for the retention of this name. In the first place, continuity is thereby maintained with the terminology employed elsewhere in this book. Elsewhere the term has been used because, whatever may be the true function of the surface-active substances which are added to a particular emulsion polymerisation system, it seems that in the majority of cases these substances have been added in the belief that they generate micelles, some of which subsequently become polymerisation loci. A further reason for continuing to refer to these substances as 'micelle generators' is that, in many cases, there is little doubt that the added surface-active substances are present in sufficient amount to form micelles. There is therefore every reason to believe that micelles are present in the majority of emulsion polymerisation systems. As the discussion in the previous section has shown, what has been questioned recently is not the presence of micelles in the system but their relevance for the process of particle nucleation. It is also appropriate to note that it is useful to make a distinction between those surface-active substances which are deliberately added to the reaction system (for whatever ostensible purpose) and those which are formed during the course of the reaction or, for that matter, which are present adventitiously.

This section will be quite brief for various reasons quite apart from the fact that several of the topics which might have been considered under this heading have already been discussed in the previous section. The principal reason for the brevity of the present discussion is the relative scarcity of published documentary evidence for the problems with which it deals. Furthermore, most of the published documentary evidence which does exist has been alluded to earlier in this book.

The first problem to which attention is directed is that of the variability in the conversion–time relationship obtained using certain micelle generators, notably those of the sulphate and sulphonate types. Variation is observed between batches made by the same process, sometimes within batches, and certainly between batches made by different processes. The present writer is of the opinion that this variability has its origin in the extreme difficulty of preparing surface-active substances in a pure form, using the term 'pure' to denote a standard which would meet the criteria normally recognised as appropriate to a reagent of analytical standard.

The expected impurities can be catalogued under several headings. Some will represent homologues of the main component, and are present because, say, the starting material was not sufficiently homogeneous as regards its hydrocarbon moiety. Other impurities take the form of starting material, or an intermediate, which has not been completely eliminated from the product. It would be expected that such impurities might profoundly affect the properties of adsorbed monolayers of the surface-active substance. A third category of impurity consists of isomers of the main component. These may be isomeric either in respect of the hydrophobic moiety or in respect of the placement of a hydrophilic substituent. Thus, for instance, considerable doubt seems to surround the location of the sulphonate group in an alkylarylsulphonate material. Related to these impurities are those in which there is an oversubstitution of hydrophilic group. Thus nominally monosulphonated alkylaryl hydrocarbons inevitably contain some di-sulphonated material as well. Again, a profound effect upon the behaviour of adsorbed monolayers could be expected. Impurities of a quite different class include simple inorganic electrolytes which are formed during the preparation of many anionic surface-active substances.

The reference to the possibility of simple inorganic electrolytes being present in surface-active substances leads immediately to the second type of problem to which reference will be made, namely, the effect of such electrolytes upon the course of emulsion polymerisation reactions and upon the nature of the product produced. A description of some of the effects observed has been given in Chapter 9, Section 9.2. The control which such effects can afford in respect of rate of polymerisation, and of the particle size and viscosity of the latex produced, has been widely exploited in the industrial production of synthetic latices. By contrast, the fundamental processes which underlie these effects seem to be but little understood. In very general terms, the observed variations are probably to be interpreted as being due to the effect of an enhanced ionic strength upon the behaviour of the micelle-generating substances in the reaction system. But any detailed understanding is lacking, and must probably await general agreement concerning the more fundamental aspects of emulsion polymerisation reaction mechanisms.

Probably closely related is the problem of the effect of pH upon the rate of emulsion polymerisation. Attention has been drawn to this effect in Chapter 7, Section 7.6. It is especially marked in the case of carboxylates, where, as the degree of neutralisation of the carboxylic acid is increased progressively, so the polymerisation rate passes through a maximum at the point corresponding to about 100% neutralisation. It is not difficult to understand why the rate should increase as the degree of neutralisation is increased. It is more difficult to understand why over-neutralisation should cause a pronounced and immediate reduction in the reaction rate. It is possible that the excess alkali is functioning as an electrolyte, but this is a matter which would warrant much closer investigation than has hitherto been apparently undertaken.

Elucidation of the problems listed in the preceding paragraphs might go some way towards understanding and perhaps solving a problem which has from time to time plagued the users of micelle generators of the alkylarylsulphonate type. This concerns the tendency for systems containing these substances to suffer pronounced variations both in polymerisation rate and in the colloid stability of the product. The effects can be quite dramatic, the more so in that they can apparently be associated with different samples of material drawn from the same production batch.

The final problem to which brief reference will be made in this section concerns the observation that surface-active substances can sometimes apparently initiate emulsion polymerisation without any added initiator being present. As far as can be seen, the initiation is due to the surface-active substance and not to, say, hydroperoxidised monomer. Hydroperoxidation of the hydrophobic moiety of the surface-active substance seems to be the obvious cause, but, as far as is known, the matter has not received serious attention, and thus the cause cannot be regarded as having been established.

12.4 STEREOSPECIFIC EMULSION POLYMERISATION

In a recent review of outstanding practical problems in emulsion polymerisation, Duck[34] states that 'One of the unsolved problems in the emulsion polymerisation of dienes is the stereoregulation of the polymerisation to obtain stereospecific polymers.' Indeed, a strong case can be argued for the proposition that control of the stereoregularity of the product of an emulsion polymerisation reaction is currently the most important unsolved problem of preparative polymer chemistry, if not of polymer chemistry as a whole. Ideally it would be desirable to have emulsion polymerisation systems available for the polymerisation of vinyl monomers which would enable atactic, isotactic and syndiotactic polymers to be produced as desired. Control over the head-to-tail and head-to-head modes of addition would also be desirable. Taken together, these requirements imply control of both the structural isomerism of the entering unit and its optical isomerism.

Likewise, for diene monomers, it would be desirable to be able to control the structural, optical and geometrical isomerism of the entering units. Control of structural isomerism would imply the ability selectively to enchain units formed by the 1,4, 1,2 and 3,4 modes of addition. It would also imply the ability, in the case of 1,2- and 3,4-polydienes, to select the head-to-tail and head-to-head possibilities. Control of optical isomerism, *i.e.* selection between isotactic, syndiotactic and atactic sequences, would be desirable in the case of 1,2- and 3,4-polydienes. Control of geometrical isomerism would be desirable in the case of 1,4-polydienes, where it would imply the ability to select the ratio of *cis*- to *trans*-1,4 units in the polymer. Yet another aspect of stereoregulation which is common to both vinyl and diene polymerisations is the control of the extent of branching and of crosslinking in the polymer produced.

Many of these problems have been solved for bulk and solution polymerisations by using organo-metallic catalysts of the Ziegler–Natta type. Polymerisations are carried out in non-polar hydrocarbon media from which all traces of polar compounds have been removed. Whilst the details of the reaction mechanism are still disputed, it seems clear that the origin of the stereoregulating effect arises from the presence of highly-polarised metal–carbon bonds in which the degree of dissociation of the bond is relatively small. Control of the orientation of the incoming monomer molecule is undoubtedly enhanced in many instances because the propagating centre is part of an interface, and therefore severe spatial constraints are imposed upon the incoming monomer. It is not fortuitous that many catalyst systems which effect bulk and solution stereospecific polymerisation are insoluble in the reaction medium.

Having stated the problem and the way in which it has been solved in the case of bulk and solution polymerisations, it is easy to understand why apparently insuperable difficulties have been encountered in obtaining similar control in emulsion polymerisation systems. Conventional systems contain large amounts of water. Metal–carbon bonds tend to be hydrolysed by water with great rapidity. Ions which are not hydrolysed tend to be heavily solvated. The electric fields which surround them are in consequence weakened, and ion pairs therefore tend to dissociate. This tendency is enhanced by the high dielectric constant of water. It therefore seems likely that, if stereoregulation in emulsion polymerisation is to be achieved, it will be by way of metal-complex catalysts or initiators which are stable towards water yet reactive towards the monomer. Furthermore, the surface-active substances in the system may have an important role if it turns out that stereospecific emulsion on polymerisation can be persuaded to take place in the interfacial region separating particles from aqueous phase, rather as envisaged in the Medvedev mechanism for a conventional free-radical emulsion polymerisation.

If it is enquired why it is desirable to be able to effect control of stereoregularity in emulsion polymerisations, then the answer is twofold. In the first place, there is the entirely academic reason that it would be desirable to know if and how a possibility which is conceivable to the human mind can actually be achieved. But, in the second place, there is great industrial interest in the possibility. Stereoregular polymers possess physical properties which distinguish them from each other and from the non-stereoregular versions of the same polymer. It is inappropriate in this book to discuss the effect of polymer stereoregularity upon polymer properties; it is sufficient to note that some of the properties conferred by stereospecificity are advantageous in particular applications. There being a demand for stereoregular polymers, it is understandable that, if possible, industry would wish to make use of the conveniences of the emulsion polymerisation process for their production. The situation is admirably summarised in a British Patent granted to the United States Rubber Company in 1964.[35] Referring to processes for the production of *cis*-1,4-polybutadiene using Ziegler–Natta catalysts, such processes are said to 'suffer certain disadvantages in that they require a monomer of exceptional purity, solvent media of similar high purity, and catalysts prepared in the absence of air and/or water. In addition, before and during the reaction step of these methods precautions must be taken to remove and exclude air and water from the equipment used and from the reaction zone. Otherwise the reaction will stop or proceed slowly. Furthermore, the polymer is obtained in a rather dilute solution contaminated by catalyst residues at the end of the reaction, requiring difficult and costly removal of both the solvent and catalyst residues from the product polymer before it is in useable form. *It would be highly advantageous if polybutadiene having a high cis content... could be prepared by conventional emulsion polymerization, a process not having the disadvantages above mentioned.*' (The italics are not in the original.)

Having considered the generalities of the situation, it is now necessary to note two of the cases so far recorded in which control of stereoregularity has proved possible in emulsion polymerisation reactions. The first is a trivial (though industrially-important) example, namely, chloroprene (2-chlorobutadiene). When prepared using conventional free-radical initiators, polychloroprene can contain more than 95% of its monomer units as 1,4 structural isomers. In most of these, the configuration of substituents about the carbon–carbon double bond is *trans* rather than *cis*. This is a feature of polychloroprenes produced by free-radical addition polymerisation, whether in bulk, in solution or in emulsion. It appears that the directive force arises from the presence of the chlorine atom attached to the second carbon atom. The effect is to minimise the formation of 1,2 and 3,4 structural units, rather than to encourage the formation of 1,4 units.

The second example is potentially of more fundamental significance. This concerns the aqueous-phase polymerisation of butadiene in the presence of certain compounds of Group VIII (of the Periodic Table) metals and of suitable surface-active substances. The notable example is that of rhodium(III) chloride in combination with alkylarylsulphonates, where the product appears to be a highly-stereoregular and highly-crystalline *trans*-1,4-polybutadiene. The latex produced by this reaction is

TABLE 12.1

Aqueous-phase polymerisation of butadiene catalysed by various rhodium(III) salts (according to Rinehart *et al.*[36])

(reaction system: butadiene 100, water 200, rhodium salt 1, sodium dodecylbenzenesulphonate—as indicated)

Rhodium salt	*Sodium dodecylbenzene sulphonate level* (pphm)	*Reaction temperature* (°C)	*Average rate of polymerisation* ($g\,h^{-1}$)	*Intrinsic viscosity of polymer*† ($dl\,g^{-1}$)	*% crystallinity in polymer*‡	*% of various microstructures in polymer*§		
						cis-1,4	*trans*-1,4	1,2
$RhCl_3 . 3H_2O$	5	5	0·02	0·5	43	<1	99	0·2
$RhCl_3 . 3H_2O$	5	50	2·4	0·4	37	<1	99	0·3
$RhCl_3 . 3H_2O$	5	80	21	0·1	37	1·2	>98	0·2
$Rh(NO_3)_3 . 2H_2O$	0	50	0·1	0·1	37	<2	>96	<2
$Rh(NO_3)_3 . 2H_2O$	5	50	0·4	0·1	60	<1	98	1
$(NH_4)_3RhCl_6 . 1{\cdot}5H_2O$	5	50	1·3	0·3	36	<1	99	0·2
$Na_3RhCl_6 . 18H_2O$	5	50	1·5	0·3	49	<1	>98	<1

† Determined in tetralin at 135°C.

‡ Determined by X-ray diffraction using the method of Matthews, Peiser and Richards.[38]

§ Determined by infrared spectrophotometry using the method of Hampton;[39] the percentages of the various microstructures were adjusted so as to add up to approximately 100%.

unusual in that its particles are oblate spheroids instead of being spherical in shape. This peculiarity presumably arises because the polymer crystallises as it forms.

Interest in this type of polymerisation was first aroused by the appearance in the literature of two brief reports[36,37] which suggested that certain compounds of the Group VIII metals could function as catalysts for the polymerisation of butadiene in aqueous media, and that, furthermore, the polymer produced could be stereoregular. Thus Rinehart *et al.*[36] reported the preparation of crystalline *trans*-1,4-polybutadiene by the action of certain rhodium(III) salts on butadiene, not only in water but also in other polar media such as ethyl alcohol and dimethyl formamide. Their observations on aqueous systems are summarised in Table 12.1. Several interesting features of the reaction emerge from the work reported in this paper, quite apart from the stereoregularity of the polymer product. Thus the rate of polymerisation increases sharply with rise in polymerisation temperature. The intrinsic viscosity of the polymer (determined in tetralin) decreases as the polymerisation temperature rises, and as the rhodium salt concentration is increased. The presence of a suitable surface-active substance increases the polymerisation rate, although some polymerisation can occur in the absence of such a substance. The polymerisation rate is markedly affected by the nature of the anion or ligand which is associated with the rhodium ion, whereas the anion or ligand appears to have little effect upon the stereoregularity of the product. No induction periods were observed. Neither air nor hydroquinone reduces the polymerisation rate. The filtrate from the reacted system retains its ability to catalyse the polymerisation of further monomer, although the catalytic activity is slightly reduced, due presumably to the loss of rhodium compound. The polymers formed are completely free from gel, being soluble in hot xylene. They are not, however, soluble in cold benzene, presumably because of their high crystallinity.

In a second communication from this group, Rinehart *et al.*[40] reported on the effects of varying the amount of catalyst in a reaction system comprising butadiene, water, sodium dodecylbenzenesulphonate and rhodium(III) nitrate. They also studied a system containing rhodium(III) chloride plus a free-radical initiator. Linear conversion–time curves were obtained. The rate of polymerisation becomes independent of catalyst concentration at levels above about 0·5 pphm. The decreasing catalyst efficiency with increasing level was attributed to an increase in the proportion of the rhodium compound present as inactive species. Always the polymerisation rate is increased by the presence of the surface-active substance. An interesting phenomenon was observed when systems containing a free-radical initiator (azobis*iso*butyronitrile) and varying amounts of rhodium chloride were investigated. At low rhodium chloride concentrations, two competing polymerisation mechanisms appear to be operative. One is of a free-radical nature. An initial fall in polymerisation rate with increasing amount of rhodium catalyst is most plausibly explained as being due to inhibition of the free-radical mechanism. Rinehart and his co-workers have also reported on the rhodium(III) chloride-catalysed isomerisation of polybutadienes. Their objective was to determine whether the high *trans*-1,4 content of the polybutadiene formed during the aqueous-phase polymerisation of butadiene using this catalyst could be accounted for by the isomerisation of other microstructures which might have been formed initially in the polymerisation reaction, notably the *cis*-1,4 microstructure. The isomerisations were carried out by dissolving polybutadiene in a solvent (benzene or xylene), emulsifying aliquots with an aqueous solution containing rhodium(III) chloride and dosium dodecylbenzene-sulphonate, and then heating the emulsion for varying lengths of time under an inert nitrogen atmosphere. In general, it was found that the *trans*-1,4 content increased, and the effect was particularly marked for the high *cis*-1,4 polybutadiene. However, the equilibrium concentration of the *trans*-1,4 isomer after isomerisation was not as high as that present in the product of the rhodium(III) chloride-catalysed polymerisation. Furthermore, it was observed that the rhodium chloride-catalysed isomerisation of *cis*-2-butene was inhibited by the addition of butadiene. Instead of isomerisation of the butene, polymerisation of the butadiene was observed. For these reasons, isomerisation of either monomer or polymer was discounted as the underlying cause of the polymer stereospecificity.

The results reported by Canale *et al.*[37] confirm that butadiene can be polymerised to a *trans*-1,4 polymer by aqueous solutions of rhodium salts. Their work also indicates that much broader steric control of the mode of enchainment of butadiene units may be possible by varying the transition-metal ion. Their observations are summarised in Table 12.2. A typical reaction system using rhodium(III) chloride as catalyst gave 95% conversion to a polymer having an iodine number of 445 (the theoretical value for polybutadiene being 470), and infrared absorption characteristics which suggest that the microstructure is essentially 100% *trans*-1,4. The weight-average molecular weight by light scattering was 26 000. The polymerisation rate was found to be strongly dependent upon the hydrogen ion concentration of the aqueous phase, the maximum rate occurring in

TABLE 12.2

Aqueous-phase polymerisation of butadiene catalysed by various Group VIII metal salts (according to Canale *et al.*[37])

Catalyst	Reaction temperature (°C)	Mole ratio catalyst: butadiene	% of various micro-structures in polymer†		
			cis-1,4	*trans*-1,4	1,2
$RhCl_3$	50	1:2000	—	95·5+	<0·5
$(C_8H_{12}RhCl)_2$‡	50	1:20 000	—	95·5+	<0·5
$PdCl_2$	50	1:20	—	17	83
$(NH_4)_2PdCl_4$	50	1:20	—	2	98
$IrCl_3$	50	1:30	—	99–100	—
$RuCl_3.6\phi_3P$	25	1:200	68	17	15
$RuCl_3.10(n\text{-}C_4H_9)_3P$	50	1:200	15	15	70
$CoSiF_6$	50	1:200	88	8	4

† These were determined by infrared spectrophotometry. The percentages of the various microstructures were adjusted so as to add up to approximately 100%. The degree of unsaturation, as determined by iodine monochloride titration, ranged from 94 to 98%.

‡ Presumed to be a complex of rhodium(III?) chloride and cycloöctadiene.

the pH range from 2 to 4. The results for the effect of pH upon polymerisation rate are shown in Fig. 12.3.

Particular interest has been shown in the alleged possibility of obtaining a polybutadiene containing 88% *cis*-1,4 microstructure using cobalt silicofluoride as a catalyst. Unfortunately, this observation has never been confirmed. Attempts by Morton, Piirma and Das[41] were unsuccessful. It has been suggested that strong infrared absorptions in the region 650–830 cm^{-1} were wrongly assigned to *cis*-1,4 polymer microstructures, whereas in fact they were due to catalyst residues.

Canale and Hewett[42] have also reported that steric control over the enchainment of butadiene units can be achieved by using palladium compounds containing various anions or ligands. Their observations are summarised in Table 12.3. The effect of varying the nature of the surface-active substance in the system was also reported. The results are summarised in Table 12.4. It appears that significant polymerisations can be obtained using palladium salts in the absence of an added surface-active substance. However, in certain instances, the nature of the substance seems to influence the structure of the polymer produced. Canale and Hewett attribute this effect to the ability of the surface-active substance to

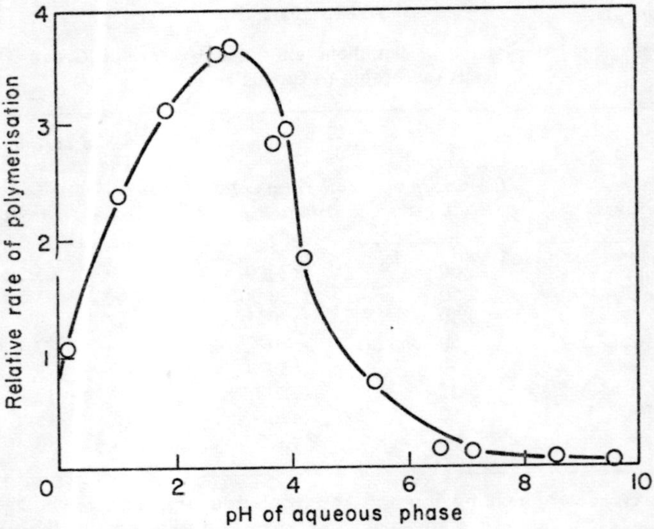

FIG. 12.3. Effect of pH upon rate of stereospecific aqueous-phase polymerisation of butadiene at 50°C catalysed by rhodium(III) chloride, according to Canale *et al.*[37] (Reaction system: butadiene 19, water 50, sodium hexadecylbenzenesulphonate 1·5, $RhCl_3.3H_2O$ 0·46. Mode of varying pH is not specified; presumably it was by addition of HCl.)

TABLE 12.3

Aqueous-phase polymerisation of butadiene at 50°C catalysed by various palladium compounds (according to Canale and Hewett[42])

(reaction system: butadiene 20, water 50–60, sodium hexadecylbenzenesulphonate 1–2, palladium compound 1:20 mole ratio to butadiene)

Palladium compound	*Yield of polymer after 88 h reaction (%)*	*% of various microstructures in polymer*†	
		trans-1,4	1,2
$PdCl_2$	25	17	83
M_2PdCl_4‡	25–31	2	98
K_2PdCl_6	30	2	98
$PdBr_2$	1	9	91
PdI_2	1	85	15
$Pd(CN)_2$	1	78	22
$PdSO_4$	>95	Low-molecular-weight, coloured polymers, for which purification and analysis were not possible	
$Pd(NO_3)_2$	>95		

† Determined by infrared spectrophotometry.
‡ $M^+ = Na^+, K^+$ and NH_4^+.

TABLE 12.4

Effect of nature of added surface-active substance upon aqueous-phase polymerisation of butadiene at 50°C catalysed by various palladium compounds (according to Canale and Hewett[42])

(reaction system: see Table 12.3)

Palladium compound	*Surface-active substance*	*Yield of polymer after* 88 h *reaction* (%)	*% of various microstructures in polymer*†	
			trans-1,4	1,2
$PdCl_2$	None	29	2	98
$PdCl_2$	Sodium hexadecylbenzenesulphonate	25	17	83
$PdCl_2$	*iso*Octylphenyl polyethoxyethanol	18	6	94
$PdCl_2$	Dodecyltrimethylammonium chloride	24	2	98
$(NH_4)_2PdCl_4$	None	30	2	98
$(NH_4)_2PdCl_4$	Sodium hexadecylbenzenesulphonate	34	2	98
$(NH_4)_2PdCl_4$	Dodecyltrimethylammonium chloride	14	2	98

† Determined by infrared spectrophotometry.

act as a ligand for the metal of the catalyst, rather than to its effect upon the physical structure of the monomer emulsion. Canale and Hewett also examined the effect on butadiene polymerisation of catalysts prepared *in situ.* These catalysts were prepared by reaction between $(NH_4)_2PdCl_4$ and each of the following compounds: dimethyl sulphide, dimethyl sulphoxide, tri-*n*-butyl phosphine, triphenyl phosphine, triethyl phosphite and thiocyanate anions. Ligand/palladium ratios of 1, 2 and 4 were used. It was found that, whereas the first ligand had no effect, the second, third and fourth reduced the rate, whilst the fifth and sixth de-activated the catalyst completely. From the work of Canale and Hewett, it appears that, in contrast to the aqueous-phase polymerisation of butadiene catalysed by rhodium compounds, the catalytic effect of palladium compounds is decreased by the presence of a surface-active substance.

Several patents have been granted for aqueous-phase butadiene polymerisations using compounds of the Group VIII metals. Thus Smith and Wilkinson with the United States Rubber Company[43] have a patent for the preparation of *trans*-1,4-polybutadiene using rhodium(III) chloride as the catalyst. A comparison is given of the effects of polymerisation temperature upon the microstructures obtained by free-radical initiation and by rhodium chloride catalysis. The point is made that temperature has much less effect upon microstructure in the latter case. A patent to Shryne and the Shell Oil Company[44] claims that the rate of polymerisation can be substantially increased, without affecting the stereoregularity of the product, by adding formic acid to the system. Canale and the Shell Oil Company[48] have claimed the use of ruthenium salts of mineral acids and various of their complexes as aqueous-phase polymerisation catalysts. Complexes specifically mentioned are those with diene hydrocarbons, trihydrocarbyl phosphines, trihydrocarbyl arsines and trihydrocarbyl stibines. Mixtures of these salts and complexes with hypophosphorous acid are also mentioned, the claim being that the acid increases both the rate and the extent of the reaction. It is preferred that the polymerisation be carried out in the presence of a surface-active substance. Sulphonates and sulphates are specifically mentioned as suitable. The polymerisation is carried out on the acid side of neutrality, preferably between pH 1·5 and 6·0.

In a paper which appeared in 1963, Teyssié[46] reported that certain mono- and poly-dentate ligands able to chelate with rhodium ions have a suppressive effect upon the aqueous-phase polymerisation of butadiene catalysed by rhodium chloride. He observed that, whereas acetylacetone did not influence the polymerisation rate significantly, both pyridine and ethylenediamine tetraäcetic acid were able completely to inhibit the reaction. Teyssié accounted for the difference in behaviour of the various ligands in terms of differences in the electronegativity of the complexing atoms of the ligand. It may be, however, that acetylacetone is incapable of forming a complex with rhodium under the reaction conditions used by Teyssié.

Dauby and Teyssié[47,48] have reported that the addition of 1,3-cyclohexadiene to the butadiene–rhodium(III) chloride system enhances the rate of polymerisation. The microstructure of the polymer is little affected, but the degree of crystallinity and molecular weight are apparently increased. Broad maxima in polymerisation rate and polymer crystallinity were observed when the ratio of cyclohexadiene to rhodium(III) chloride was in the range between 12 and 32, the peak values occurring at a ratio of about 20. Dauby and Teyssié suggest that the increase in polymerisation rate is due to the reducing action of the cyclohexadiene, and to its ability to stabilise the lower oxidation states of rhodium. This explanation is not accepted by Rinehart[49] for two reasons. In the first place, he states that the complex which 1,3-cyclohexadiene forms with rhodium(I) chloride is rather unstable. Secondly, he claims that 1,3-cyclohexadiene promotes polymerisation by stable rhodium(I) complexes.

In a further paper, Dauby, Dawans and Teyssié[50] have reported further on the aqueous-phase homopolymerisation of butadiene catalysed by rhodium(III) chloride, and also upon the copolymerisation of butadiene and piperylene under the same reaction conditions. The overall rate of polymerisation was found to be unaffected by the amount of monomer in the organic phase. This was inferred from the linearity of the conversion–time relationship. Dauby, Dawans and Teyssié state that this 'would indicate a diffusion-controlled polymerization process, which agrees with an interfacial micellar system'. But it is not clear just what is meant by this inference; it is therefore difficult to judge whether or not it is justified. In order to achieve an optimum rate of polymerisation, it was necessary to maintain a mole ratio of 2:1 for the balance between surface-active agent and rhodium chloride. Elemental analysis showed the presence of sulphur equivalent to about one emulsifier molecule per polymer molecule. One rhodium atom per eight polymer molecules was detected, but combined chlorine was not observed. The rate of polymerisation was found to be first-order with respect to the catalyst concentration, at least up to 0·5% on the aqueous phase. The overall activation energy for the rhodium(III) chloride–butadiene system was found to be 16·4 kcal mole^{-1}. A much higher figure of 34·6 kcal mole^{-1} was reported for a system containing 1,3-cyclohexadiene as well, the mole ratio to butadiene being 1:16. This difference in activation energies is attributed to different complexes being involved in the polymerisation reaction. Dauby, Dawans and Teyssié propose a reaction mechanism in which chain initiation occurs by the formation of a metal–carbon link with the elimination of HCl. The rhodium atom then forms successive π-bonds with incoming monomer molecules. The rhodium atom is regarded as having a molecule of the surface-active substance associated with it as a ligand.

The results of a study of the kinetics of the rhodium(III) chloride–butadiene and rhodium(III) nitrate–butadiene systems have been published by Morton, Piirma and Das.[41] Using sodium lauryl sulphate as surface-active substance, and rhodium(III) chloride as catalyst, it is claimed that, as the ratio of the two is varied, a maximum rate occurs at a weight ratio of sodium lauryl sulphate to rhodium chloride in the range 2·0–2·5. No such optimum ratio was observed when rhodium(III) nitrate was used as the catalyst. Inhibitors of free radical polymerisation were found to be ineffective in preventing reaction; rhodium chloride was itself found to act as an inhibitor of free-radical polymerisation. According to these workers, if sufficient surface-active substance is present, the overall rate of polymerisation is first-order with respect to rhodium chloride concentration. Addition of ethanol to the system has different effects depending upon whether the catalyst is rhodium chloride or rhodium nitrate. With rhodium chloride, the rate is reduced to zero at 25% ethanol concentration, whereas with rhodium nitrate there is an initial reduction in rate as ethanol is added, but beyond 25% ethanol concentration there is an increase in rate, although it does not reach its level in the absence of ethanol. Only rhodium nitrate is effective as a catalyst in the presence of other alcohols, such as methanol and butanol. It is ineffective in the presence of higher alcohols such as *n*-hexanol.

A limited investigation was made of the influence of the structure of

TABLE 12.5

Effect of nature of added surface-active substance upon aqueous-phase polymerisation of butadiene at 40°C catalysed by rhodium(III) chloride (according to Morton, Piirma and Das[41])

(reaction system: butadiene 100, water 200, surface-active substance 5, $RhCl_3.3H_2O$ 1; pH 2–4 except where otherwise indicated)

Surface-active substance	% *conversion after* 24 h *reaction*
Sodium lauryl sulphate	23
Sodium decylbenzenesulphonate	15
Sodium oleate (pH 10)	0
Sodium oleate + polyoxyethylated fatty alcohol	0
polyoxyethylated fatty alcohol	0
Sodium *p*-toluenesulphonate	0
Sodium laurylsulphonate	0
n-C_8 to *n*-C_{16} sodium alkylsulphonates	0
n-C_2 to *n*-C_{10} sodium alkyl sulphates	0
n-C_{14} to *n*-C_{18} sodium alkyl sulphates	0
Sodium *n*-decyl sulphate + polyoxyethylated fatty alcohol	0
Sodium branched-alkyl sulphates from	
2,2-dimethyldecan-3-ol	0
dodecan-5-ol	0
dodecan-6-ol	0
hexadecan-5-ol	0
Sodium alkyl phosphates:	
n-C_8, *n*-C_{10}, *n*-C_{16}	0
n-C_{12}	~6
Lauryltrimethylammonium chloride	0
Lauryltriethylammonium chloride	0

TABLE 12.6

Effect of nature of added surface-active substance upon aqueous-phase polymerisation of butadiene at 40°C catalysed by rhodium(III) nitrate (according to Morton, Piirma and Das[41])

(reaction system: butadiene 100, water 200, surface-active substance 5, $Rh(NO_3)_3.2H_2O$ 1·23; ph 2–4)

Surface-active substance	*% conversion after reaction time indicated*
Sodium lauryl sulphate	18 (24 h)
Sodium decylbenzenesulphonate	21 (24 h)
Polyoxyethylated fatty alcohol	3 (24 h)
n-C_6, *n*-C_8, *n*-C_{10}, *n*-C_{14}, *n*-C_{16} sodium alkyl sulphates	trace or nil (36 h)
n-C_6, *n*-C_8, *n*-C_{10}, *n*-C_{14}, *n*-C_{16} sodium alkylsulphonates	nil (48 h)

the surface-active substance on the course of the reaction. It is claimed that the chain length of the hydrophobic part of the molecule is of critical importance, a chain length of twelve carbon atoms being especially effective. Their observations are summarised in Table 12.5 for rhodium chloride and Table 12.6 for rhodium nitrate. It is evident from these results that, of the substances studied, activity is limited to sodium dodecyl sulphate and sodium decylbenzenesulphonate, with slight activity shewn by sodium *n*-dodecyl phosphate. The inactivity of sodium dodecylsulphonate is surprising and is ascribed by Morton, Piirma and Das to an alleged inability of sulphonates to act as ligands for rhodium ions.

In all cases, when polymer was obtained, it was predominantly of *trans*-1,4 microstructure. Polymer molecular weight was observed to be independent of the type and amount of catalyst, of the type and amount of the surface-active agent (insofar as polymer was obtained at all), and of extent of conversion. The molecular weight decreases with increasing polymerisation temperature. Polymer prepared using rhodium nitrate is of lower molecular weight than that prepared using rhodium chloride. Attempted polymerisations of butadiene, isoprene, styrene and piperylene in aqueous media using palladium(II) chloride, osmium(II) chloride, ruthenium(III) chloride and iridium(III) chloride were unsuccessful. Cobalt(II) silicofluoride and rhodium(III) sulphate were ineffective for butadiene, and rhodium(III) chloride was ineffective for isoprene, styrene and piperylene.

As a result of this investigation, Morton, Piirma and Das proposed a mechanism in which a complex is formed between one molecule of rhodium and two molecules of the surface-active substance. The complex is envisaged as being adsorbed at the interface between monomer-swollen micelles and the aqueous phase, the mode of catalysis being regarded as essentially heterogeneous. The concentration of the catalyst at the interface remains constant throughout the reaction, whilst the concentration of monomer is determined by the solubility in or adsorption upon the nascent polymer particle. Termination occurs by detachment of the propagating chain from the complex, leaving the site free for further initiation. Polymer chain length is governed by the relative frequency of termination to propagation steps. In a further attempt to elucidate the mechanism of the reaction, Morton and Das[57] have carried out polymerisation in homogeneous solution in solvents such as tetrahydrofuran, dimethoxyethane and the dimethyl ester of ethylene glycol. Polymerisation occurred much more slowly than in the heterogeneous aqueous systems, and, not surprisingly, the rate of reaction seemed to be independent of the presence of a surface-active substance. Morton and Das suggest that the solvent acts as a ligand for the rhodium, thus forming complexes which are less active as catalysts.

Many of the above features of the rhodium(III) chloride–butadiene aqueous-phase polymerisation have been confirmed by Blackley and Matthan.[52] The variation of the polymerisation rate with temperature was found to be consistent with an activation energy of 16 kcal mole^{-1}, this value being in good agreement with that reported by Dauby, Dawans and Teyssié.[50] The addition of isoprene was found to have a retarding effect in excess of that expected if it were merely acting as an inert diluent. Increase of the chloride ion concentration by the addition of sodium chloride was found to reduce the rate of polymerisation quite markedly. The effect of adding inorganic electrolytes has been investigated in greater detail by Blackley and Mustafa.[53] All the electrolytes studied showed a similar retarding effect to that of sodium chloride. No systematic correlations between behaviour and the nature of either cation or anion were observed, except that di- and tri-valent cations tend to have a more marked effect than do univalent cations. The results are summarised in Fig. 12.4(a) and (b).

Blackley and Mustafa have also investigated the effect of varying the alkyl chain length in both sodium alkyl sulphate and sodium alkylsulphonates. In each case, the range covered was *n*-octyl to *n*-hexadecyl. The respective families of conversion–time curves are shown in Fig. 12.5(a) and (b). In both series, the reaction rate increases progressively with the chain length of the alkyl group. Actually, the efficiency of each molecule

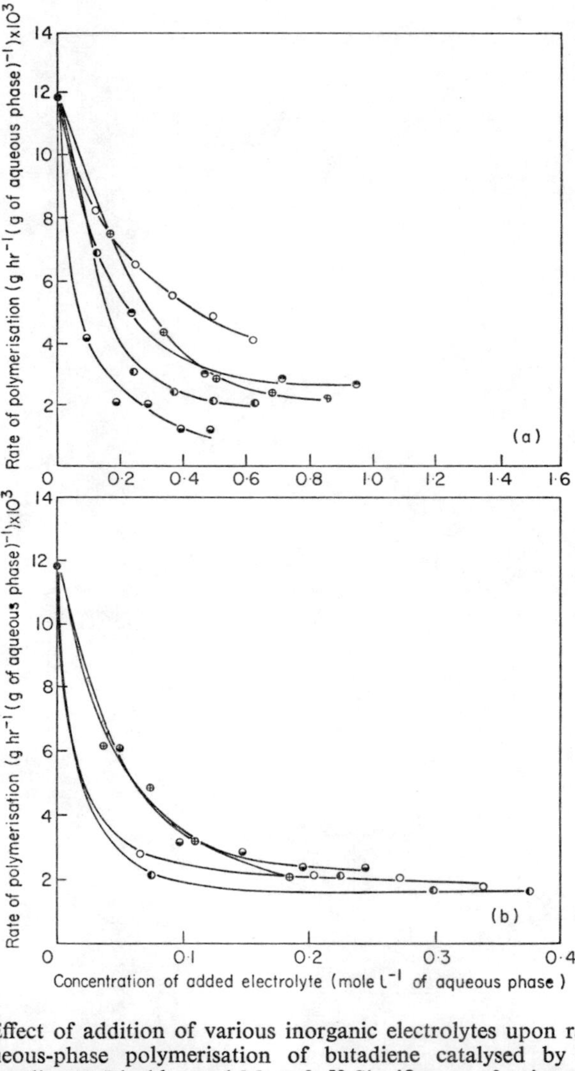

FIG. 12.4. Effect of addition of various inorganic electrolytes upon rate of stereospecific aqueous-phase polymerisation of butadiene catalysed by rhodium(III) chloride, according to Blackley and Mustafa.[53] Significance of points: (a) ⊕ NaCl; ◐ KCl; ◑ NaF; ○ KNO_3; ◒ NaBr; (b) ○ $CaCl_2$; ◒ $MgCl_2$; ◐ $AlCl_3$; ⊕ $FeCl_3$. (Reaction system: butadiene 100, water 200, sodium dodecylbenzenesulphonate 5, $RhCl_3.3H_2O$ 2, electrolyte—variable.)

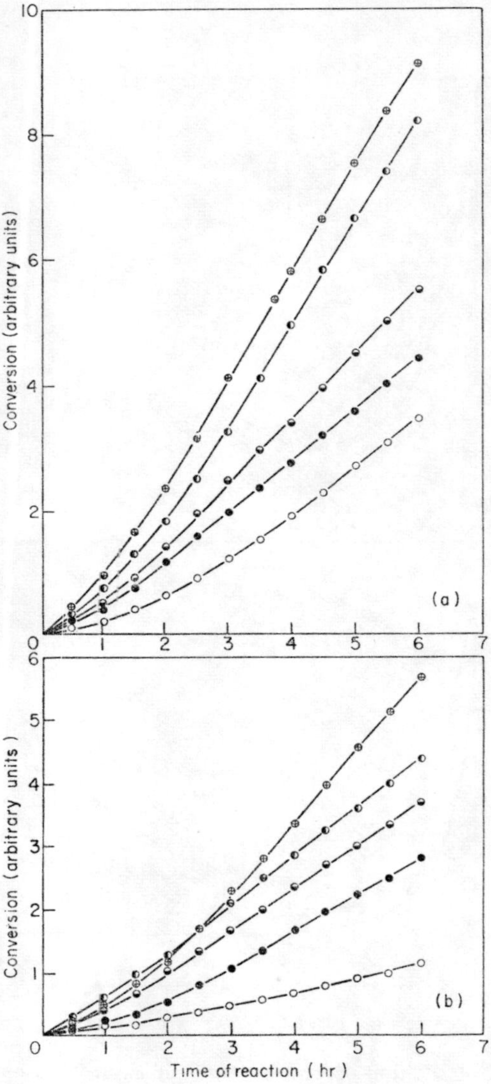

FIG. 12.5. Effect of chain length of soap upon conversion–time relationship for stereospecific aqueous-phase polymerisation of butadiene at 50°C catalysed by rhodium(III) chloride, according to Blackley and Mustafa.[53] The curves in (a) are for a range of *n*-alkyl sulphates; those in (b) are for a similar range of *n*-alkylsulphonates. Significance of points is as follows: ○ octyl; ● decyl; ◒ dodecyl; ◐ tetradecyl; ⊕ hexadecyl. (Reaction system: butadiene 100, water 200, sulphate or sulphonate 5, $RhCl_3.3H_2O$ 2.)

of surface-active substance increases more rapidly with chain length than these curves suggest at first sight. This is because each system contained the same weight of soap (5 pphm), no attempt being made to keep the molar concentrations constant. It therefore follows that the actual number of molecules of soap present in the system fell quite sharply as the alkyl chain length increased. These results are at variance with those of Morton, Piirma and Das,[41] noted above, but are consistent with those reported by Kalal and Mikes.[54] A detailed explanation of the effect of increasing chain length of soap keeping the weight constant would probably have to take into account many factors, among which would be included the following:

(1) the fall in the critical micelle concentration, thereby affecting the concentration of 'monomeric' soap present in the aqueous phase available for complexing with the rhodium;
(2) the increase in aggregation number, and therefore size of the micelles;
(3) the increased solubilising power of the micelles;
(4) the decrease in the total number of soap molecules present in the system; and
(5) the net effect of factors (1), (2) and (3) upon the concentration of micelles in the aqueous phase.

The observation that polymerisation occurs in the presence of sodium alkylsulphonates also conflicts with the finding of Morton, Piirma and Das (see Table 12.5). It is difficult to comment upon possible reasons for these discrepancies. They may be due to differences in the purity of the sulphates and sulphonates used, in particular in respect of the amounts of residual inorganic electrolyte present. Morton, Piirma and Das give few details regarding the preparation of their materials. The sulphates used by Blackley and Mustafa were prepared by esterification of the corresponding alcohol with sulphuric acid in the presence of thionyl chloride, followed by neutralisation with sodium ethoxide and subsequent purification. The sulphonates were prepared by reaction between the corresponding alkyl bromide and sodium sulphite, again followed by purification. The full details of the methods are given in the reference cited.

The results of Blackley and Mustafa also show clearly that, for a given alkyl chain length, the sulphate promotes a higher rate of polymerisation than does the sulphonate. No question of molecular efficiency arises here, because the molecular weights are almost identical. It is possible that the sulphates are more effective ligands for the rhodium ion than are the sulphonates, because of the additional oxygen atom in the former. If the initiating species is a complex between the rhodium ion and the surface-active substance, then it is possible that the concentration of this species is higher when a sulphate is used than when a sulphonate is used.

Several of the characteristics of the rhodium(III) chloride–butadiene system can be explained in a general way by the 'ligand-displacement' mechanism proposed by Blackley and Matthan.[52] This mechanism is summarised in the scheme given below (I). It is assumed that the surface-active substance is a sulphonate (either alkyl or alkylaryl), but the mechanism applies equally to sulphates. The first step is envisaged as the reversible replacement of a chloride ion of the rhodium chloride by a sulphonate molecule. The effect of inorganic electrolytes in retarding the reaction is explained as being due either to the displacing of this equilibrium to the left-hand-side, if the additive is a chloride, or to competition from other anion-displacement reactions if the additive is other than a chloride. The additive would also have the effect of reducing the critical micelle concentration of the surface-active substance, and therefore the concentration of 'monomeric' soap in the aqueous phase. This also would tend to displace the equilibrium to the left-hand side. The propagation step is then seen as a succession of 'ligand-displacement' reactions which take place via six-membered cyclic transition complexes. Each butadiene unit inserts itself across a highly-polarised $Rh\cdots CH_2$ bond, thereby forming a six-membered ring. Subsequent electronic rearrangement causes the butadiene unit to become incorporated in the polymer chain, and the $Rh\cdots CH_2$ bond to be regenerated. Termination might occur as a result of dissociation of the $Rh\cdots CH_2$ bond

Such a mechanism accounts for the retarding effect of chelating substances such as ethylenediamine tetraäcetic acid, and also of that of a high concentration of hydroxyl ions. In both cases, the coordination sites of the rhodium ion are blocked, and the initiation step thereby prevented. The reduction of the rate at low pH may be due to one or both of two causes: (a) suppression of the formation of the initiating complex, through the enhanced concentration of anions which must inevitably be added with the hydrogen ions; and (b) enhancement of the rate of termination. The formation of a 1,4-polybutadiene by this mechanism is readily understandable, since a six-membered cyclic transition state is postulated as a precursor to the incorporation of each butadiene unit. What this mechanism fails to explain satisfactorily is why the butadiene units should be incorporated as *trans* units rather than *cis*. The most likely explanation is that the propagation step takes place in the interface between the particle and the aqueous phase, since this is the region where the reactants are brought together. If so, then it is possible that the surface applies

$$-\overset{\delta+}{\mathrm{Rh}}\mathrm{Cl} + \mathrm{RSO_3^-} \rightleftharpoons -\overset{\delta+}{\mathrm{Rh}}\cdots\overset{\delta-}{\mathrm{O}}\mathrm{SO_2R} + \mathrm{Cl^-}$$

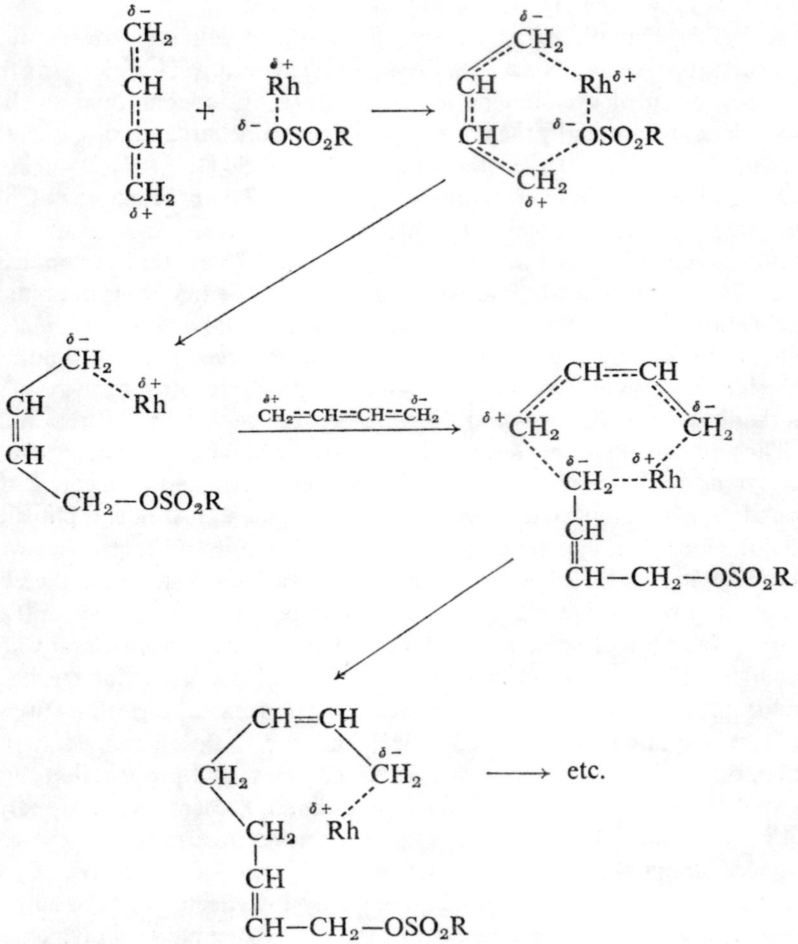

$$-\mathrm{CH_2{-}CH{=}CH{-}\overset{\delta-}{C}H_2\cdots\overset{\delta+}{Rh}} + \mathrm{H^+} \rightleftharpoons -\mathrm{CH_2{-}CH{=}CH{-}CH_3} + -\mathrm{Rh^+}$$

(I)

constraints to the configurations which can be adopted by the atoms of the cyclic transition state as the ring opens. Attention is also drawn to a paper by Duck[55] which seeks to present a unifying hypothesis for the stereoregular polymerisation of conjugated dienes in bulk and solution using a variety of catalysts, and in aqueous phase.

A further example of an aqueous-phase stereoregular polymerisation has been reported by Natta, Dall'Asta and Motroni.[56] This concerns the polymerisation of cyclobutene in the presence of aqueous rhodium(III) chloride using essentially the same conditions as Rinehart and co-workers[40] for butadiene. After 40 h reaction at 50°C, a low yield (10%) of a crystalline polycyclobutylenomer-2 structure was obtained. The product was highly stereoregular, the type being that described as erythro di-isotactic. The previous experience of this group of workers indicated that cyclobutene cannot be polymerised by mechanisms of the simple free-radical, anionic and cationic types, but only by catalysts which provide a coordinated anionic mechanism. They therefore incline to the view that the rhodium chloride catalysis of cyclobutene polymerisation proceeds by way of a mechanism which involves coordination between monomer and rhodium.

The aqueous-phase polymerisation of various vinyl monomers, notably of styrene and methyl methacrylate, by certain transition metal compounds has been reported by Berger and Youngman.[57] Although polymer was obtained in a number of instances, the structure of the product was found to be typical of a free-radical propagation mechanism, thereby contrasting with behaviour in the butadiene polymerisation system. The catalysis was found to be somewhat selective in that a given metal compound is active for some monomers but not for others. A reaction mechanism is proposed in which a coordination complex is formed between catalyst and monomer. A free-radical initiator is then assumed to be generated by a redox reaction within the complex, and propagation then to proceed from this site by a non-coordination mechanism. Copolymerisation studies are said to support such a mechanism in that the copolymer compositions obtained are always typical of those given by a free-radical mechanism, even when the catalyst is effective for the homopolymerisation of only one member of the pair. Attempted copolymerisations with butadiene gave only polybutadiene; the comonomer had no effect upon the microstructure of the polymer, but generally reduced the polymerisation rate.

It is convenient to conclude this review of stereoregulated emulsion polymerisation by reference to two other areas in which work has been carried out. The first of these concerns the persulphate-initiated emulsion polymerisation of isoprene in the presence of both an alkylsulphonate and a sulphonated low-molecular-weight polystyrene as surface-active substances, reported by Chernikhov and Margaritova.[58] The alkylsulphonate component of the micelle generator comprised the sodium salts of the sulphonic acids of paraffin hydrocarbons having chain lengths in the range C_{15} to C_{18}. The other component was the sodium salt of the product obtained by the chlorosulphonic acid sulphonation of a styrene oligomer of average degree of polymerisation 4. The elementary composition of this product is given as 2·9% H, 26·2% C, 13·6% S and 17·7% Br, although it is not at all clear where the bromine originates from. Polymerisations were carried out mainly at 60°C. The claim is made that the ratio of *cis* to *trans* units incorporated in the polyisoprene formed can be controlled to some extent by the balance of the two surface-active substances. The contents of 1,2 and 3,4 structures are always small, but are also somewhat affected by the composition of the emulsifier system used. Rather remarkably, the overall activation energy for the polymerisation seems to depend upon the composition of the emulsifier. Using the alkylsulphonate alone, a value of 7·25 kcal $mole^{-1}$ was observed; but using a 3:1 mixture of alkylsulphonate and sulphonated polystyrene, the much higher value of 13·75 kcal $mole^{-1}$ was obtained. Some information is also given for the polymerisation of dimethylbutadiene, methyl methacrylate and styrene in this system.

The fact that the emulsifier composition seems to affect the microstructure of the polymer produced, together with the observation that the order of reaction is 0·5 with respect to initiator concentration and unity with respect to emulsifier concentration, is regarded as evidence for the essential correctness of the Medvedev theory of the mechanism of emulsion polymerisation, to which reference has been made in Section 12.2. If the view that propagation takes place in the adsorption layer between particle and aqueous phase is correct, then it is certainly plausible that the course of the propagation step should depend to some extent upon the nature of the substances adsorbed in the interface. Certainly, interesting possibilities are raised if it proves possible to obtain effective control of polymer microstructure by choice of surface-active substance.

The second type of system to which reference will be made is that which contains the metal chelates of β-diketones as initiator for aqueous-phase polymerisations. These substances have been known for several years as initiators for the bulk polymerisation of vinyl monomers. Their potentialities as initiators for emulsion polymerisation do not appear to have been explored in detail. A patent granted to the United States Rubber

Company[35] in 1964 claims that the aqueous-phase polymerisation and copolymerisation of butadiene can be effected by using cobalt(II) complexes of this type. Furthermore, whilst the proportion of 1,2 units in the resultant polymer is said to be virtually that of a typical free-radical polybutadiene, the *cis*:*trans* ratio is rather higher and is little affected by change in polymerisation temperature. Specifically claimed as initiators are cobalt(II) acetylacetonate and sodium cobalt(II) acetylacetonate. On the other hand, Blackley and Matthan[52] failed to obtain polymer when using cobalt(II) acetylacetonate with butadiene, but did obtain a high yield of a rubbery, typically free-radical polybutadiene using iron(III) acetylacetonate. Subsequent work has shown, however, that this latter chelate complex is extremely variable in both appearance and catalytic ability, and that therefore reproducibility is hard to attain.

Cobalt(III) acetylacetonate is also effective as an initiator of the aqueous-phase polymerisation of butadiene. This chelate seems to be much less variable in character than is the iron analogue, and its behaviour as a polymerisation initiator is reasonably reproducible. The polymer produced seems to be of typical free-radical character, but is very highly crosslinked. An extensive study of the characteristics of this system has been carried out by Blackley and Burgar.[59] The reaction rate was found to be rather sensitive to the presence of oxygen. If reproducible rates are to be obtained, it is necessary to bring the reaction system together under high vacuum, having first purged all the vessels and reactants with nitrogen. It is worth noting that, under these conditions, no polymerisation was observed using cobalt(III) acetylacetonate as catalyst, nor, for that matter, when the iron(III) complex was used.

Regarding the mechanism of initiation, Blackley[60] has proposed an 'internal redox' reaction as the process whereby the free-radical activity is generated. A possible reaction scheme (II) is given below. It agrees in a general way with mechanisms which have been proposed to account for the initiating ability of these substances in bulk polymerisations; but, of course, in an emulsion system many other complications are possible because of the polar nature of the surrounding aqueous phase. Briefly, the basic assumption of the 'internal redox' mechanism is that the metal ion, which is originally in one of its higher valency states, is reduced to a lower valency state by a drift of electrons from the ligand. In the case of acetylacetonates, such a drift is facilitated by the presence of a 'quasi-aromatic' system of conjugated unsaturation. A free-radical site is formed by rearrangement, from which propagation takes place. Two possibilities are indicated in the above scheme: in one, the metal atom remains attached to the propagating

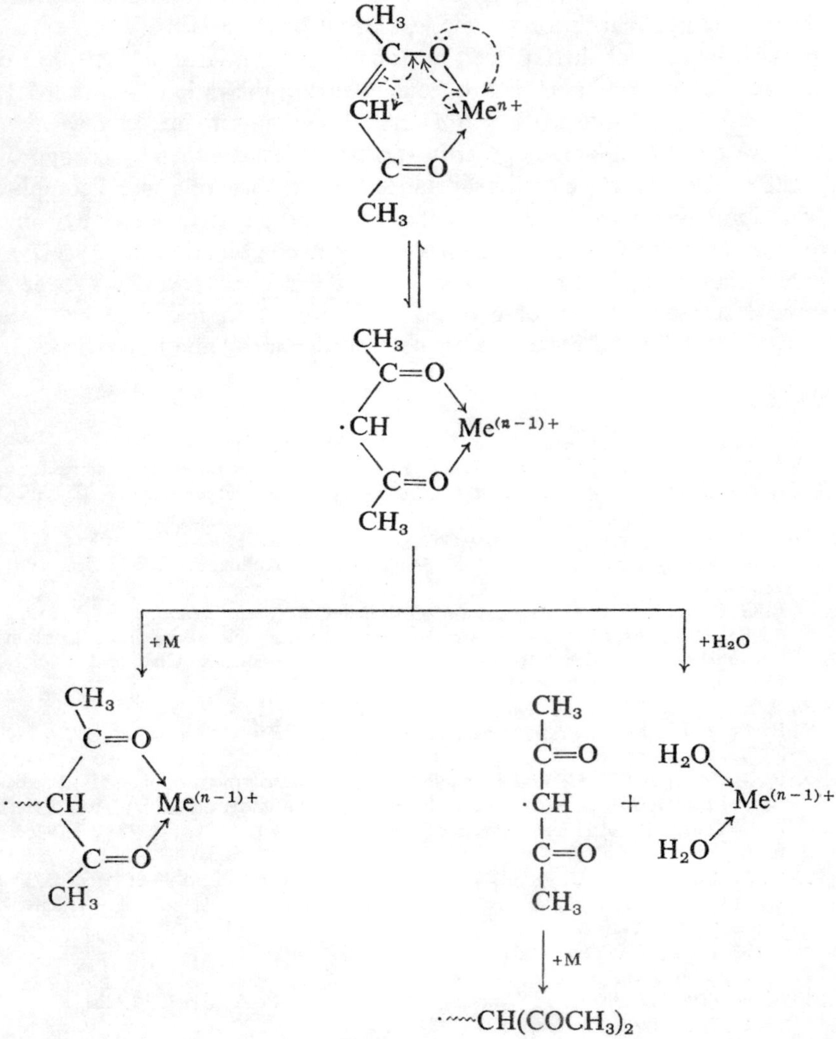

Me = metal capable of existing in more than one valency state
M = monomer molecule

(II)

species, whereas in the other it becomes detached through interaction with the water of the aqueous phase. The extent to which conventional theories of emulsion polymerisation are applicable to this system is in doubt because, in the case of cobalt(III) acetylacetonate, the initiator is partitioned between the hydrocarbon and aqueous phases in the ratio of about 6·5:1.

What of the future in these matters? It remains to be seen whether effective control of stereoregularity can be achieved either by an appropriate choice of surface-active substance or by the use of a metal complex as an initiator which retains control over the propagating centre by some type of coordination mechanism. Possibly a combination of these two approaches will be effective. There is little doubt, however, that if a general stereoregulation of emulsion polymerisation reactions becomes possible, it will be of great interest both academically and industrially.

REFERENCES

1. C. P. Roe and P. D. Brass, *J. Polym. Sci.*, 1957, **24**, 401.
2. S. S. Medvedev, *Proc. Int. Symp. Macromolec. Chem.*, *Prague*, Pergamon, London/Czechoslovak Academy of Sciences, and Czechoslovak Chemical Society, 1957, pp. 174–90.
3. A. S. Dunn, *Chem. Ind.*, 1971, 1406.
4. J. G. Brodnyan, J. A. Cala, T. Konen and E. L. Kelley, *J. Colloid Sci.*, 1963, **18**, 73.
5. B. G. Elgood, E. V. Gulbekian and D. Kinsler, *Polymer Letters*, 1964, **2**, 257.
6. A. Ryšánek, *Proc. Int. Symp. Macromolec, Chem. Prague*, Pergamon, London/ Czechoslovak Academy of Sciences, and Czechoslovak Chemical Society, 1957, p. 187.
7. C. P. Roe, *Ind. Engng Chem.*, 1968, **60**(9), 20.
8. R. M. Fitch, M. B. Prenosil and K. J. Sprick, *J. Polym. Sci.*, Pt C, 1969, No. **27**, 95.
9. R. M. Fitch and C. H. Tsai 'Particle formation in Polymer colloids, III: Prediction of the number of particles by a homogeneous nucleation theory', in *Polymer Colloids* (ed. R. M. Fitch), Plenum Press, New York, 1971, pp. 73–102.
10. R. M. Fitch and C. H. Tsai, *Polymer Letters*, 1970, **8**, 703.
11. R. M. Fitch and C. H. Tsai, 'Homogeneous nucleation of polymer colloids, IV: The role of soluble oligomeric radicals', in *Polymer Colloids* (ed. R. M. Fitch), Plenum Press, New York, 1971, pp. 103–16.
12. R. M. Fitch, *Br. Polym. J.*, 1973, **5**, 467.
13. I. D. Robb, *J. Polym. Sci.*, Pt A-1, 1969, **7**, 417.
14. M. R. Grancio and D. J. Williams, *J. Polym. Sci.*, Pt A-1, 1970, 8, 2617.
15. D. J. Williams, *J. Elastoplastics*, 1971, **3**, 187.
16. D. H. Napper, *J. Polym. Sci.*, Pt A-1, 1971, **9**, 2089.
17. D. J. Williams, *J. Polym. Sci.*, Polymer Chemistry Edition, 1973, **11**, 301.
18. P. Keusch and D. J. Williams, *J. Polym. Sci.*, Polymer Chemistry Edition, 1973, **11**, 143.
19. P. Keusch, J. Prince and D. J. Williams, *J. Macromol. Sci.-Chem.*, 1973, **A7**, 623.
20. J. L. Gardon, *J. Polym. Sci.*, Polymer Chemistry Edition, 1973, **11**, 241.
21. J. Ugelstad, M. S. El-Aasser and J. W. Vanderhoff, *Polymer Letters*, 1973, **11**, 503.
22. J. Ugelstad, F. K. Hansen and S. Lange, *Makromolek. Chem.*, 1974, **175**, 507.

23. W. F. Fowler, 'Elucidation of emulsion polymerization mechanism based upon copolymer studies', in *Vinyl Polymerization* (ed. G. E. Ham), Dekker, New York, 1969, Vol. 1, Pt II, pp. 139–74.
24. R. G. Fordyce and E. C. Chapin, *J. Am. chem. Soc.*, 1947, **69**, 581.
25. R. G. Fordyce and G. E. Ham, *J. Am. chem. Soc.*, 1947, **69**, 695.
26. R. G. Fordyce, *J. Am. chem. Soc.*, 1947, **69**, 1903.
27. E. C. Chapin, G. E. Ham and R. G. Fordyce, *J. Am. chem. Soc.*, 1948, **70**, 538.
28. R. G. Fordyce and G. E. Ham, *J. Polym. Sci.*, 1948, **3**, 891.
29. R. G. Fordyce and G. E. Ham, *J. Am. chem. Soc.*, 1951, **73**, 1186.
30. F. T. Wall, R. E. Florin and C. J. Delbecq, *J. Am. chem. Soc.*, 1950, **72**, 4769.
31. J. W. Vanderhoff, 'The mechanism of emulsion polymerization', in *Vinyl Polymerization* (ed. G. E. Ham), Dekker, New York, 1969, Vol. 1, Pt II, pp. 1–138.
32. G. L. Brown, private communication to J. W. Vanderhoff, 1964; see ref. 27, p. **25**.
33. R. G. Gilbert and D. H. Napper, *J. chem. Soc., Faraday Trans.* I, 1974, **70**, 391.
34. E. W. Duck, *Br. Polym. J.*, 1970, **2**, 60.
35. United States Rubber Company, B. Pat. 970,764 (1964).
36. R. E. Rinehart, H. P. Smith, H. S. Witt and H. Romeyn, *J. Am. chem. Soc.*, 1961, **83**, 4864.
37. A. J. Canale, W. A. Hewett, T. M. Shryne and E. A. Youngman, *Chem. & Ind.*, 1962, 1054.
38. J. L. Matthews, H. S. Pieser and R. B. Richards, *Acta crystallogr.*, 1949, **2**, 85.
39. R. R. Hampton, *Anal. Chem.*, 1949, **21**, 923.
40. R. E. Rinehart, H. P. Smith, H. S. Witt and H. Romeyn, *J. Am. chem. Soc.*, 1962, **84**, 4145.
41. M. Morton, I. Piirma and B. Das, *Rubb. Plast. Age*, 1965, **46**, 404.
42. A. J. Canale and W. A. Hewett, *Polymer Letters*, 1964, **2**, 1041.
43. H. P. Smith and G. Wilkinson/United States Rubber Company, U.S. Pat. 3,025,286 (1962).
44. T. M. Shryne/Shell Oil Company, U.S. Pat. 3,168,507 (1965).
45. A. J. Canale/Shell Oil Company, U.S. Pat. 3,168,508 (1965).
46. P. Teyssié, *Compt. Rendu*, 1963, 2846.
47. R. Dauby and P. Teyssié/Institut Français du Pétrole, des Carburants et Lubrifiants, French Pat. 1,402,816 (1965).
48. P. Teyssié and R. Dauby, *Polymer Letters*, 1964, **2**, 413.
49. R. E. Rinehart, *Polymer Preprints*, 1966, **7**, 556.
50. R. Dauby, F. Dawans and P. Teyssié, *J. Polym. Sci.*, Pt C, 1967, No. 16, 1989.
51. M. Morton and B. Das, *J. Polym. Sci.*, Pt C, 1969, No. 27, 1.
52. D. C. Blackley and R. K. Matthan, *Br. Polym. J.*, 1970, **2**, 25.
53. D. C. Blackley and I. Mustafa, unpublished results; I. Mustafa, Ph.D. Thesis, C.N.A.A., 1972.
54. J. Kalal and F. Mikes, *Collection Czechoslovak Chem. Commun.*, 1970, **35**, 788.
55. E. W. Duck, *Rubb. Plast. Weekly*, 18 Aug. 1962, p. 3.
56. G. Natta, G. Dall'Asta and G. Motroni, *Polymer Letters*, 1964, **2**, 349.
57. R. S. Berger and E. A. Youngman, *J. Polym. Sci.*, Pt A, 1964, **2**, 357.
58. A. Ya Chernikhov and M. F. Margaritova *Polymer Science, U.S.S.R.*, 1964, **6**(2), 261.
59. D. C. Blackley and W. F. H. Burgar, unpublished results; W. F. H. Burgar, Ph.D. Thesis, C.N.A.A., 1973.
60. D. C. Blackley, *Proceedings of International Rubber Symposium*, 1969, p. 15.
61. W. W. White, 'Latex polymerisation: theory and practice', paper presented to the Polymer Group, Chicago Section of the American Chemical Society, 22 March 1974.

Index